Proteins

Structure, Dynamics and Design

Proteins

Structure, Dynamics and Design

Edited by

V. Renugopalakrishnan
Harvard Medical School and The Children's Hospital
Boston, Massachusetts 02115, USA

Paul R. Carey
Division of Biological Sciences
National Research Council of Canada
Ottawa, Ontario, Canada K1A OR6

Ian C.P. Smith
Division of Biological Sciences
National Research Council of Canada
Ottawa, Ontario, Canada K1A OR6

Shaw G. Huang
Department of Chemistry
Harvard University
Cambridge, Massachusetts 02138, USA

and

Andrew C. Storer
Biotechnology Research Institute
National Research Council of Canada
Montreal, Quebec, Canada H4P 2R2

ESCOM ■ Leiden ■ 1991

CIP-Data Koninklijke Bibliotheek, Den Haag

Proteins

Proteins: Structure, Dynamics and Design / ed. by V. Renugopalakrishnan, Paul R. Carey, Ian C.P. Smith, Shaw G. Huang and Andrew C. Storer - Leiden : ESCOM. - Ill. With index, ref. Subject heading: Proteins.

ISBN-13:978-94-010-9065-0   e-ISBN-13:978-94-010-9063-6 DOI: 10.1007/978-94-010-9063-6

Published by:

ESCOM Science Publishers B.V. P.O. Box 214 2300 AE Leiden The Netherlands

Preface

The major goal of 'Expanding Frontiers in Polypeptide and Protein Structural Research' has been to bring the various avenues for the exploration of protein structures to a single forum. The idea of organizing the symposium was conceived by one of the editors, V. Renugopalakrishnan, during the 9th International Biophysics Congress satellite symposium at Kibbutz Nof-Ginosar, Israel in 1987. It was originally supposed to dwell on 2D NMR and molecular dynamics of polypeptides and proteins. During the earlier part of the last decade, these two approaches began to emerge as powerful tools to probe protein structures at the atomic level in solution. The developments in molecular biology ushered in the capability to design polypeptides and proteins for specific application in science and technology. The emergence of 2D NMR and molecular dynamics was greatly facilitated by contemporary developments in molecular biology and protein engineering. Therefore an international symposium devoted exclusively to 2D NMR and molecular dynamics studies of proteins was felt necessary to bring two major approaches in a single forum.

In addition to emphasis on 2D NMR and molecular dynamics simulation, the scope of the symposium included optical spectroscopy, protein design, and new horizons in protein structure. The symposium consisted of five plenary sessions devoted to NMR and optical spectroscopy as probes for protein structure, protein dynamics, computational methods in protein design, and new horizons in protein structure. In addition, five workshops in related areas, viz., NMR: spectra to structure, free energy calculations, protein folding, parameters in molecular mechanics calculations, and advances in computer technology and its impact on protein structural studies, were held to complement the plenary lectures.

The symposium was organized as a satellite meeting to the International Biophysics Congress. It was held in Whistler, British Columbia, Canada between 23 and 27 July, 1990. The Organizing Committee consisted of V. Renugopalakrishnan (Co-Chairman), Harvard Medical School, Shaw G. Huang, Department of Chemistry, Harvard University, Paul R. Carey (Co-Chairman) and Ian C.P. Smith, Division of Biological Sciences, National Research Council of Canada, Ottawa and Andrew C. Storer, Biotechnology Research Institute, National Research Council of Canada, Montreal. The Organizing Committee was supported by an International Advisory Committee and the members of this committee are listed on a separate page.

The present volume was compiled from the plenary lectures, workshops, short invited talks, poster sessions and some chapters which were solicited from distinguished scientists. The format for this volume was modified so as to divide it into four broad sections: NMR as a probe for protein structure, Optical spectroscopy as a probe for protein structure, Protein dynamics: Theory and experiment, and Protein design. The sections on NMR, Optical spectroscopy and Protein design are dedicated to the memories of V.F. Bystrov, R.C. Lord and C. Levinthal, respectively.

The symposium was sponsored by numerous Government and Private agencies. Among the Government agencies we would like to mention the National Research Council of Canada; the National Institute of Diabetes and Digestive and Kidney Diseases, National Institutes of Health, Bethesda, Maryland; Office of Naval Research, Department of the Navy; Army Research Office, Department of the Army; and the Department of Energy. We would like to express our grateful thanks to Professor J. Jaz, United Nations Educational, Scientific and Cultural Organization, Paris, France and Professor Bernard Pullman, University of Paris, Paris, France for their invaluable help. Numerous private companies and organizations have helped us generously and we have acknowledged them separately.

The burden of secretarial work in Boston was carried out by Everett A. Carlson who acted as the symposium coordinator. We are especially thankful to him and Ms. Muriel Chase for coordinating the symposium activities. Mr. Laurier Forget of the National Research Council of Canada and his staff carried out all work related to the actual organization of the symposium in Canada and we would like to thank them for their excellent contribution to the successful organization of the symposium.

We would like to express our sincere thanks to the authors and the publisher, Dr. Elizabeth Schram, for contributing their time and effort in enabling us to publish this volume in a timely fashion after the symposium.

Venkatesan Renugopalakrishnan
Boston, USA

Paul R. Carey
Ottawa, Canada

Ian C.P. Smith
Ottawa, Canada

Shaw G. Huang
Cambridge, USA

Andrew C. Storer
Montreal, Canada

Expanding Frontiers in Polypeptide and Protein Structural Research
Whistler, British Columbia, Canada
July 23-27, 1990

Organizing Committee

V. Renugopalakrishnan
Harvard Medical School
and The Children's Hospital
Boston, MA 02115, USA

Andrew C. Storer
National Research Council Canada
Montreal, Quebec, Canada H4P 2R2

Ian C.P. Smith
National Research Council Canada
Ottawa, Ontario, Canada K1A OR6

Paul R. Carey
National Research Council Canada
Ottawa, Ontario, Canada K1A OR6

Shaw G. Huang
Harvard University
Cambridge, MA 02138, USA

International Advisory Committee

E. Benedetti, Naples, Italy
V.F. Bystrov, Moscow, U.S.S.R.
M.J. Glimcher, Boston, MA, USA
N. Gō, Kyoto, Japan
V.J. Hruby, Tucson, AZ, USA
R.S. Rapaka, Rockville, MD, USA
K. Wüthrich, Zürich, Switzerland

B. Brooks, Bethesda, MD, USA
L.M. Gierasch, Dallas, TX, USA
M. Goodman, La Jolla, CA, USA
G. Govil, Bombay, India
B. Pullman, Paris, France
R.H. Sarma, Albany, NY, USA
K.T. Yasunobu, Honolulu, HI, USA

Abbreviations

Abbreviations used in this volume are defined below:

ACP   acyl carrier protein
Aib   α-amino isobutyric acid
AMBER   Assisted Model Building with Energy Refinement
BCP   blood complement protein porcine C5a(des-Arg)
CHARMM   Chemistry at Harvard Molecular Mechanics
Che-Y protein   a 14 kDa cytoplasmic protein
CMV   cucumber mosaic virus
DPDPE   (D-Pen^2,D-Pen^5) enkephalin
FT-IR   Fourier-transform infrared spectroscopy
GRF   growth hormone releasing factor
GROMOS   Groningen Molecular Simulation
IL-1   interleukin-1
IL-2   interleukin-2
MHC   major histocompatibility complex
MO   molecular orbital
NPY   neuropeptide Y
REDOR ^{13}C-^{15}N-rotational echo double resonance

Contents

Contents

Section II: Optical spectroscopy as a probe for protein structure

Section III: Protein dynamics: Theory and experiment

Section IV: Polypeptide and protein design: Protein engineering

Contents

Section I

NMR as a probe for protein structure

Dedicated to the memory of

V.F. Bystrov

This section is dedicated to the memory of Professor Vladimir F. Bystrov, Shemyakin Institute of Bioorganic Chemistry, U.S.S.R. Academy of Sciences, Moscow, U.S.S.R. for significant contributions to NMR studies of biological macromolecules.

NMR structures of proteins: Improved precision through stereospecific resonance assignments

Kurt Wüthrich

Institut für Molekularbiologie und Biophysik,
Eidgenössische Technische Hochschule-Hönggerberg, CH-8093 Zürich, Switzerland

Introduction

During the last decade a method for the determination of the complete three-dimensional structure of proteins was developed, which uses nuclear magnetic resonance (NMR) spectroscopy for the data collection, and distance geometry, or possibly other mathematical techniques, for the structural interpretation of the NMR data [1-7]. At present, NMR is the only approach, besides diffraction techniques with single crystals [8], that is available for protein structure determination at atomic resolution. Its introduction is of fundamental interest since, in addition to generating new structures of proteins that have not been crystallized for diffraction studies, NMR can provide data that are in many ways complementary to those obtained from X-ray crystallography. It thus promises to widen our view of protein molecules with regard to a better grasp of the relations between structure and function, and to contribute to a sound basis for protein design and protein engineering.

Survey of the NMR Method for Protein Structure Determination

The scheme in Fig. 1 presents an outline of a protein structure determination by NMR spectroscopy [2]. For the data collection the protein is dissolved in an aqueous solvent. The protein concentration must be of the order of 1-5 mM (or possibly higher). The sample volume is 0.5 ml, so that 5-25 mg of a protein with molecular weight 10,000 is needed for the preparation of an NMR sample. The pH, ionic strength and temperature can be adjusted so as to mimic the physiological milieu of the protein.

The foundations of the method are: (i) *Measurements of proton-proton distances* by nuclear Overhauser enhancement (NOE) experiments suitably executed to minimize adverse effects of spin diffusion [9]. This can be achieved by determination of NOE buildup curves in one-dimensional [10] or multi-dimensional [11] experiments. Such NOE measurements can provide information on proton-proton distances in the range from about 2.0 to 5.0 Å. In a qualitative interpretation the observation of each ^1H-^1H NOE manifests an upper bound on a ^1H-^1H distance. The information contained in a small region of a two-dimensional NOE (NOESY) spectrum (Fig. 2a) can then be represented by the scheme in Fig. 3a. (ii) *Sequential resonance assignment* as an efficient technique for obtaining sequence-specific ^1H NMR assignments. By the fact that polypeptide chains in proteins generally contain multiple units of each amino acid type, NMR spectral assignments are non-trivial and no generally applicable assignment procedure was available until 1982, when the sequential assignment strategy was introduced [12–15]. The crucial importance of resonance assignments as a basis for protein structure determination [1] is illustrated with Figs. 2 and 3: The small spectral region from a NOESY spectrum shown in Fig. 2a contains 30 cross peaks. In the absence of sequence-specific resonance assignments each of these peaks merely indicates the presence of two nearby hydrogen atoms, which may be located anywhere in the polypeptide chain (Fig. 3a). When resonance assignments have been obtained, each cross peak identifies an upper limit on the distance between two distinct locations along the polypeptide chain (Fig. 3b), which is the information needed for the determination of the three-dimensional protein structure. Overall, in its impact on the NMR structure determination method the sequential assignment strategy can be compared to the use of isomorphous heavy atom derivatives for solving the phase problem in protein crystallography [8,16]. (iii) *Methods for the structural interpretation of the NMR data.* The maximum possible number of conformational constraints must be collected to prepare the input for the calculation of a complete three-dimensional protein structure. Since only a small percentage of all cross peaks in the NOESY spectra are needed for the sequential assignment procedure, the individual chemical shifts obtained as a result of the resonance assignments can subsequently be used to further assign all, or nearly all remaining NOESY cross peaks (Fig. 2b). The set of NOE distance constraints thus obtained can be supplemented by measurements of scalar spin-spin coupling constants, and by the identification of hydrogen bonds in regular secondary structures identified in the course of the sequential resonance assignment process [2]. Since these NMR

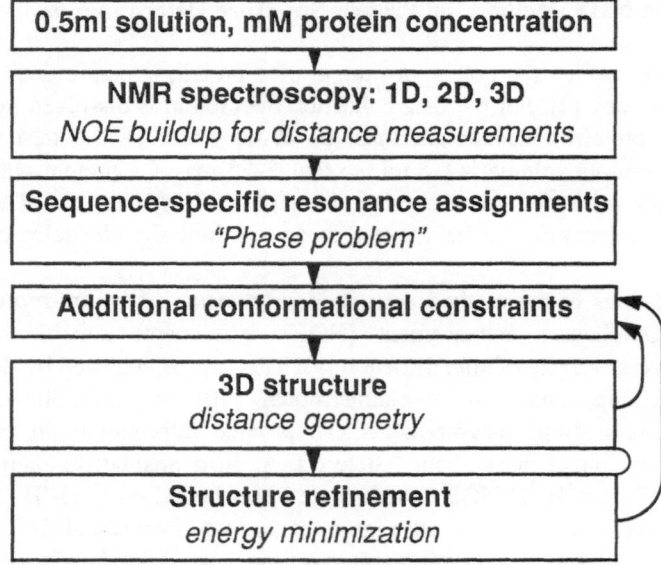

Fig. 1. *Diagram outlining the course of a protein structure determination by NMR (see text for further explanations).*

data are of an entirely different kind than the data obtained by X-ray diffraction [8], new techniques had to be developed for their structural interpretation. Initial structure

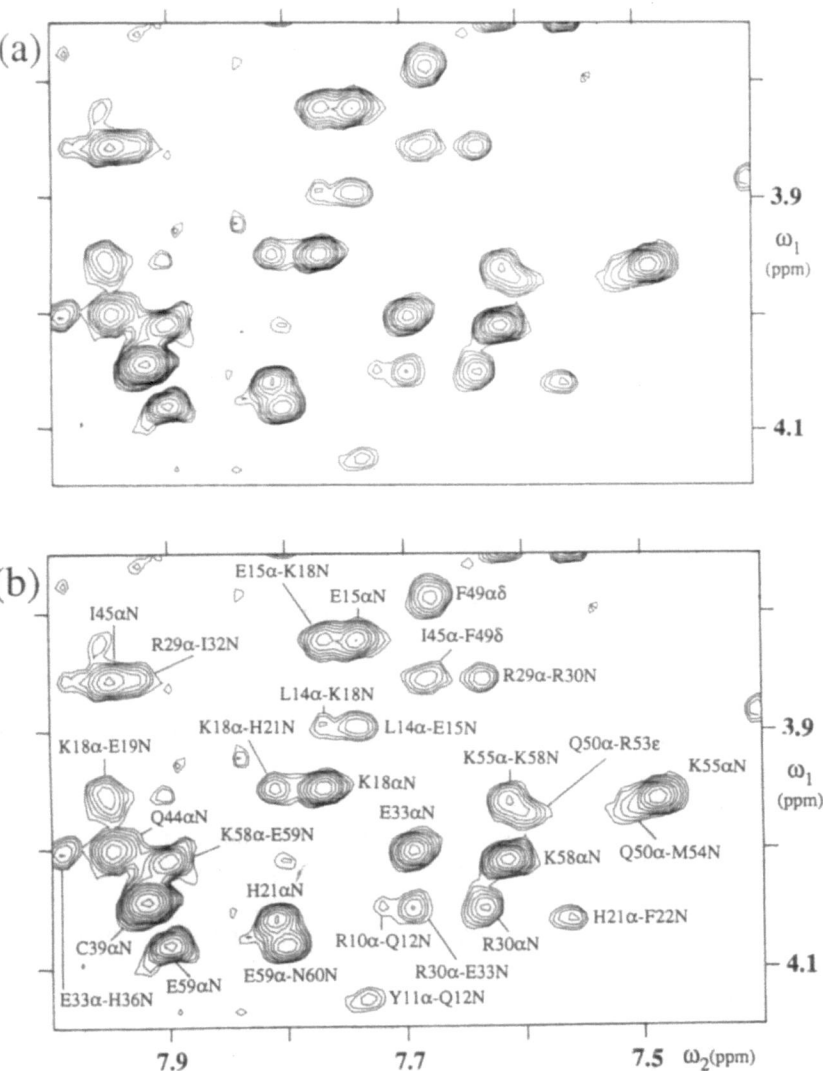

Fig. 2. (a) Contour plot of a small region from a ¹H soft-NOESY spectrum of the Antennapedia homeodomain from Drosophila. (b) Same spectral region with sequence-specific resonance assignments. For interresidual NOEs the cross peaks are identified by the one-letter amino acid symbols of the two residues, their sequence positions and the proton types, and for intraresidual NOEs by the identification of one residue and two proton types.

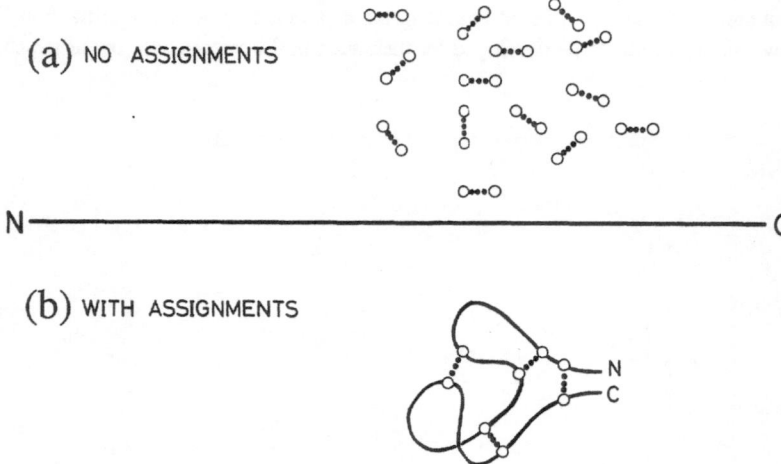

Fig. 3. *Scheme illustrating the information content of ^1H-^1H NOEs in a polypeptide chain (represented by the horizontal line in the center) with and without sequence-specific resonance assignments. Open circles represent hydrogen atoms of the polypeptide, and dotted lines represent short ^1H-^1H distances manifested by the NOEs (see text). (Reproduced from ref. 2).*

calculations from NMR data used metric matrix distance geometry [17–21]. Subsequently a variable target function algorithm was implemented for the same purpose [22], as well as restrained molecular dynamics calculations [23]. In present practice most structure determination protocols include either a structural interpretation of the NMR data with a variable target function algorithm (e.g., [24]; Fig. 4), which can be supplemented by molecular mechanics energy minimization (e.g., [25]), or an initial analysis with metric matrix distance geometry followed by molecular dynamics calculations (e.g., [26]).

As is indicated by the arrows in the lower right of Fig. 1, a structure determination usually goes through several cycles of data collection and structure calculations. One reason for this is that 2D and even 3D ^1H NMR spectra of proteins often contain groups of two or several cross peaks with identical chemical shifts along one of the frequency axes. Although this initially prevents unambiguous assignments for these cross peaks, most of the ambiguities can usually be resolved by reference to preliminary structures calculated without using these NOEs in the input. Multiple cycles of data collection and structure calculations include also checks of the protein structure against the experimental NOESY spectra [2,3,6,27].

Stereospecific Assignments for the Methyl Groups of Valine and Leucine by Biosynthetically Directed Fractional ^{13}C Labeling

With test calculations [28,29] as well as from experience with structure calculations from experimental NMR data (e.g., [24]) it was clearly established that the quality of an NMR structure of a protein could be significantly improved if instead of using the pseudoatom concept [30], individual assignments were obtained for the diastereotopic

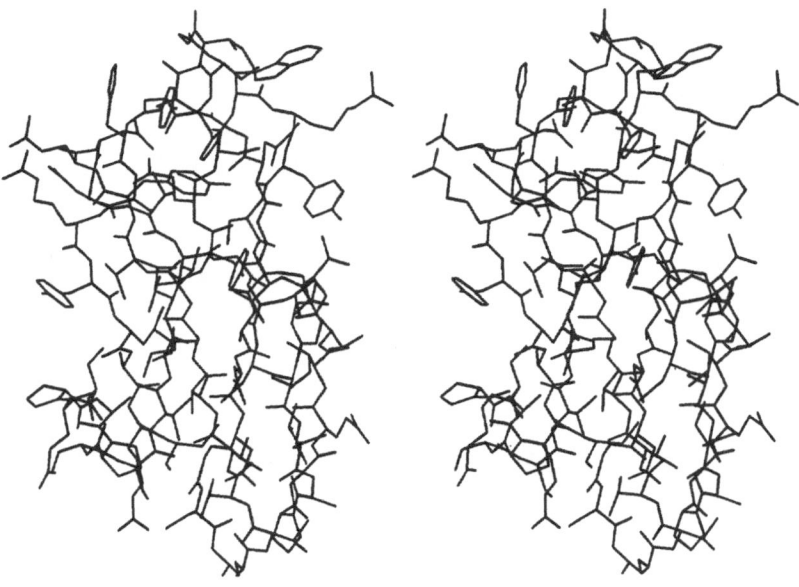

Fig. 4. Stereo view of the three-dimensional structure of the protein Tendamistat determined from NMR measurements in aqueous solution [24]. All bonds connecting heavy atoms are drawn for the residues 5–73. The complete molecule contains 74 amino acid residues and has a molecular weight of 9000.

substituents of prochiral centers. To a certain extent (for example, 50% of all β-methylene groups and all or nearly all isopropyl groups of Val and Leu), such stereospecific assignments can be obtained in the course of a structure calculation from NMR data (e.g., [24,26,28]). However, there is evidence that the convergence of a structure calculation can be improved if stereospecific assignments are available at the outset of the calculation (unpublished). This section surveys a biosynthetic ^{13}C labeling technique [31,32] that enables complete stereospecific assignments of the methyl groups of Val and Leu at an early stage of a protein structure determination, so that these will be available during the collection of NOE distance constraints to prepare the input for the structure calculations.

The method used is based on the fact that in the biosynthesis of Val and Leu with glucose as the carbon source, the isopropyl group is made up of a two-carbon fragment from one pyruvate unit, while the second methyl group is transferred from another pyruvate unit. This methyl migration has been shown to be stereoselective, and the migrating methyl group becomes *pro-S* in both valine and leucine; i.e., it is γ^2-CH$_3$, or δ^2CH$_3$, respectively. Direct proof for this stereoselective biosynthetic pathway was obtained for *Escherichia coli* [33, 34].

Biosynthetic fractionally ^{13}C-labeled proteins can be obtained from microorganisms grown on minimal media containing a mixture of roughly 10% [^{13}C$_6$]-glucose and 90% unlabeled glucose as the sole carbon source. The carbon positions in such preparations are uniformly ^{13}C labeled to an extent of about 10%. Disregarding the natural ^{13}C abundance of 1.1% in the unlabeled glucose, the probability that two adjacent carbon positions are labeled in the same molecule is then 1%, unless the two carbon atoms originate from the same carbon source molecule, whence this probability becomes 10%. From the aforementioned knowledge on the biosynthesis of Val and Leu it follows that these two distinct situations prevail for the isopropyl group in these amino

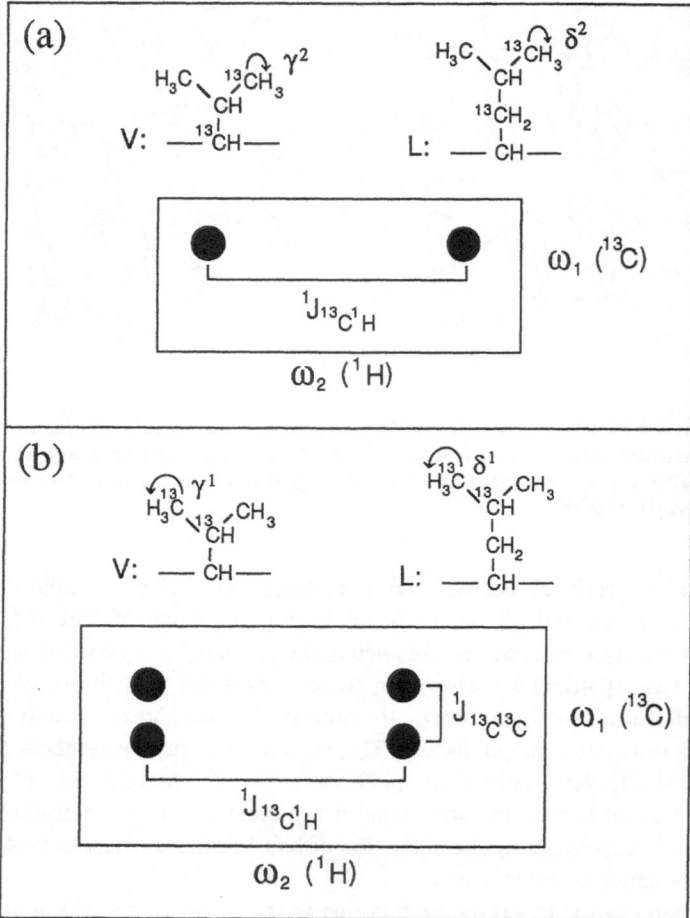

Fig. 5. *Schematic representation of the dominant multiplet fine structures expected for the methyl ^{13}C-^{1}H cross peaks of Val and Leu in [^{13}C,^{1}H]-COSY spectra recorded with a protein preparation fractionally labeled with ^{13}C. The arrows indicate the coherence transfer that is relevant in the present context. (a) For γ $^{2}CH_3$ of Val or δ $^{2}CH_3$ of Leu the multiplet consists of two components along ω_2 separated by the $^{1}J_{13_C 1_H}$ coupling constant. No splitting is observed along ω_1. This pattern is the same as for methyl groups in a protein with natural abundance of ^{13}C. (b) In addition to the $^{1}J_{13_C 1_H}$ splitting along ω_2 γ $^{1}CH_3$ of Val and δ $^{1}CH_3$ of Leu have a splitting of $^{1}J_{13_C 13_C}$ along ω_1. In an alternative experiment, which has the advantage of improved sensitivity, the fine structure along ω_2 can be removed by ^{13}C decoupling. (Reproduced from ref. 32).*

acids. The *pro-R* methyl group (γ^1, or δ^1, respectively) and the adjacent > CH– group originate from the same pyruvate molecule and are, in the absence of isotope scrambling, labeled with ^{13}C in the same molecules (Fig. 5b). On the other hand, the *pro-S* methyl group and the adjacent carbon atom originate from two different pyruvate molecules. Therefore, if the *pro-S* methyl group is enriched with ^{13}C, there is a probability of only 1% that the adjoining > CH– group in the same molecule is also labeled (Fig. 5a).

The stereospecific distinction between the pairs of diastereotopic isopropyl methyl groups in a fractionally ^{13}C-labeled biosynthetic protein is most clearly evidenced in ^{1}H-decoupled ^{13}C NMR spectra, where the ^{13}C resonance of the *pro-R* methyl group is a doublet with a splitting of about 33 Hz due to the one-bond ^{13}C-^{13}C coupling with the neighboring ^{13}C spin, while the ^{13}C NMR signal of the *pro-S* methyl group is a singlet. These characteristic features can be observed, for example, in two-dimensional ^{1}H-detected heteronuclear correlation experiments ([^{13}C,^{1}H]-COSY). A pulse sequence originally devised by Bodenhausen and Ruben [35] ensures both ^{1}H-^{1}H and ^{1}H-^{13}C decoupling in the ^{13}C dimension, so that the aforementioned differences between the *pro-R* and *pro-S* methyl groups can readily be observed along the ω_1 frequency axis (Fig. 5).

The method of biosynthetic fractional ^{13}C labeling for obtaining stereospecific assignments for Val and Leu methyls can generally be employed with recombinant proteins. Provided that a reasonably efficient overexpression system is available, the expense for [^{13}C$_6$]-glucose is of the order of 1000 dollars. The method has been used with complete success, i.e., 100% assignments of the Val and Leu methyl groups, for 434 repressor [32] and for other proteins (unpublished).

Acknowledgements

The author's research is supported by the Schweizerischer Nationalfonds (project Nr. 31.25174.88). I thank Mr. R. Marani for the careful processing of the manuscript.

References

1. Wüthrich, K., Wider, G., Wagner, G. and Braun, W. (1982) *J. Mol. Biol.* **155**, 311.
2. Wüthrich, K. (1986) *NMR of Proteins and Nucleic Acids*, Wiley, New York.
3. Wüthrich, K. (1989) *Science* **243**, 45.
4. Kaptein, R., Boelens, R., Scheek, R.M. and van Gunsteren, W.F. (1988) *Biochemistry* **27**, 5389.
5. Bax, A. (1989) *Annu. Rev. Biochem.* **58**, 223.
6. Clore, G.M. and Gronenborn, A.M. (1989) *Crit. Rev. Biochem. Mol. Biol.* **24**, 479.
7. Oppenheimer, N.J. and James, T.L., Eds. (1989) *Methods in Enzymology* **177**, p. 125.
8. Blundell, T.L. and Johnson, L.N. (1976) *Protein Crystallography*, Academic Press, New York.
9. Solomon, I. (1955) *Phys. Rev.* **99**, 559.
10. Gordon, S.L. and Wüthrich, K. (1978) *J. Am. Chem. Soc.* **100**, 7094.
11. Anil Kumar, Wagner, G., Ernst, R.R. and Wüthrich, K. (1981) *J. Am. Chem. Soc.* **103**, 3654.
12. Dubs, A., Wagner, G. and Wüthrich, K. (1979) *Biochim. Biophys. Acta* **577**, 177.

13. Billeter, M., Braun, W. and Wüthrich, K. (1982) *J. Mol. Biol.* **155**, 321.
14. Wagner, G. and Wüthrich, K. (1982) *J. Mol. Biol.* **155**, 347.
15. Wider, G., Lee, K.H. and Wüthrich, K. (1982) *J. Mol. Biol.* **155**, 367.
16. Green, D.W., Ingram, V.M. and Perutz, M.F. (1954) *Proc. Roy. Soc.* **A225**, 287.
17. Braun, W., Bösch, C., Brown, L.R., Gō, N. and Wüthrich, K. (1981) *Biochim. Biophys. Acta* **667**, 377.
18. Braun, W., Wider, G., Lee, K.H. and Wüthrich, K. (1983) *J. Mol. Biol.* **169**, 921.
19. Havel, T.F. and Wüthrich, K. (1984) *Bull. Math. Biol.* **46**, 673.
20. Havel, T.F. and Wüthrich, K. (1985) *J. Mol. Biol.* **182**, 281.
21. Williamson, M.P., Havel, T.F. and Wüthrich, K. (1985) *J. Mol. Biol.* **182**, 295.
22. Braun, W. and Gō, N. (1985) *J. Mol. Biol.* **186**, 611.
23. Brünger, A.T., Clore, G.M., Gronenborn, A.M. and Karplus, M. (1986) *Proc. Natl. Acad. Sci. USA* **83**, 3801.
24. Kline, A.D., Braun, W. and Wüthrich, K. (1988) *J. Mol. Biol.* **204**, 675.
25. Billeter, M., Schaumann, T., Braun, W. and Wüthrich, K. (1990) *Biopolymers* **29**, 695.
26. Kraulis, P.J., Clore, G.M., Nilges, M., Jones, T.A., Pettersson, G., Knowles, J. and Gronenborn, A.M. (1989) *Biochemistry* **28**, 7241.
27. Braun, W. (1987) *Quart. Rev. Biophys.* **19**, 115.
28. Güntert, P., Braun, W., Billeter, M. and Wüthrich, K. (1989) *J. Am. Chem. Soc.* **111**, 3997.
29. Driscoll, P., Gronenborn, A.M. and Clore, G.M. (1989) *FEBS Lett.* **243**, 223.
30. Wüthrich, K., Billeter, M. and Braun, W. (1983) *J. Mol. Biol.* **169**, 949.
31. Senn, H., Werner, B., Messerle, B., Weber, C., Traber, R. and Wüthrich, K. (1989) *FEBS Lett.* **249**, 113.
32. Neri, D., Szyperski, Th., Otting, G., Senn, H. and Wüthrich, K. (1989) *Biochemistry* **28**, 7510.
33. Sylvester, S.R. and Stevens, C.M. (1979) *Biochemistry* **18**, 4529.
34. Hill, R.K., Yan, S. and Arfins, S.M. (1973) *J. Am. Chem. Soc.* **95**, 7857.
35. Bodenhausen, G. and Ruben, D. (1980) *Chem. Phys. Lett.* **69**, 185.

Molecular dynamics of peptides and proteins investigated by NMR

**R.R. Ernst, M. Blackledge, S. Boentges, J. Briand, R. Brüschweiler,
M. Ernst, C. Griesinger*, Z.L. Mádi, T. Schulte-Herbrüggen
and O.W. Sørensen**
*Laboratorium für Physikalische Chemie, Eidgenössische Technische Hochschule,
CH-8092 Zürich, Switzerland*

Abstract

Possible methods for studying the intramolecular dynamics of peptides and proteins are discussed and applied to the cyclic decapeptide antamanide. A backbone motional mode is analyzed, the proline ring-puckering motion is studied, and the influence of intramolecular motion on cross-relaxation measurements is discussed. Possibilities for the reduction of the multiplet complexity in 2D spectra for accurate J-coupling measurements, useful in dynamics studies, are described.

Introduction

In the course of the past ten years, nuclear magnetic resonance (NMR) has become a powerful and accepted tool for the determination of biomolecular structure in solution [1,2]. For medium size molecules, NMR structure determinations have become almost routine, and for a sizable number of proteins and nucleic acid fragments, structural models have been derived.

While molecular structure in solution is well and easily characterized by NMR methods, it is considerably more difficult to determine satisfactory motional models. This is true, taking into account all presently available techniques. On the other hand, dynamic information would be highly desirable for the following reasons:

(i) Any biological function necessarily involves motion, such as molecular association, conformational changes, breaking or formation of bonds, etc.

(ii) Most biologically active molecules require for their activity flexibility or internal motional degrees of freedom, e.g. for enabling access of a substrate to the active site, or to adapt the shape of the binding site to the shape of the ligand for maximum interaction.

(iii) 'Molecular structure' strictly refers to a snapshot picture of a possibly mobile molecule. In fact, however, most structure determinations deliver a motionally averaged structure that is not intelligible unless at least certain aspects of mobility are characterized.

(iv) NMR structure determinations are based on distance constraints that are derived

---

*Present address: Institut für Organische Chemie, Universität Frankfurt, Niederurseler Hang, D-6000 Frankfurt 50, Germany.

from cross-relaxation rates with an r_{kl}^{-6} dependence on the internuclear distance r_{kl}. In the presence of motion, averaged distances $<r_{kl}^{-6}>$ or $<r_{kl}^{-3}>^2$ are measured, depending on the rate of the motional process. A structure determination based on these averaged values does not always lead merely to an averaged structure, but the distance constraints may, in more fortunate situations, be contradictory and not interpretable in terms of a static model, or, in less fortunate cases, be misleading, causing conclusions in terms of incorrect static structures.

Studies of molecular dynamics are therefore important for their own sake as well as an indispensable complement of structural investigations.

The mentioned difficulties of deriving structural models from experimental data are due to a mismatch between available and required data:

(i) While the static structure of a molecule with N atoms is fully defined as a point in a (3N-6)-dimensional space of internal coordinates, continuous dynamics cannot be described completely by any finite set of parameters. For an N-atom cluster, it is for example possible to define 3N-6 independent auto-correlation functions that measure the decay of intramolecular order in time. Each of these functions requires in the simplest case two parameters for its description. For more accurate models of correlation functions, the number of parameters can be increased without limits. But auto-correlation functions of internuclear distances or angles do not provide a full description of motion, cross-correlation functions of two or more molecular parameters are also needed. Obviously, the number of independent motional parameters is huge.

(ii) On the other hand, experimental sources of dynamic information are rather scarce. Neutron scattering [3] and fluorescence depolarization [4] allow one to determine directly correlation functions that provide immediate information on the rate of molecular dynamics. However, the motional processes cannot easily be localized within the molecule unless specific scattering centers are introduced at strategic points in the molecule. X-ray diffraction [5] provides highly localized motional information through the Debye-Waller factors which, however, are insensitive to the rate of the processes. Nuclear magnetic resonance allows the detection of local dynamics through the observation of relaxation and cross-relaxation rates that are caused by motionally modulated internuclear dipolar interactions. Each nuclear pair delivers, for one kind of relaxation measurement, just one single number that is itself determined by two or three sample values of the power spectral density function $J_{kl}(\omega)$ of the random process. By means of magnetic field-dependent measurements it is in principle possible to trace out the entire power spectral density function $J_{kl}(\omega)$. However, in practice, the accessible range of magnetic field strengths B_0 or Larmor frequencies $\omega_0 = -\gamma B_0$ is limited and hardly exceeds a factor 10 when high-resolution spectral information is required. On the other hand, motional rates in molecules easily cover a range of more than 10 orders of magnitude.

In summary, while structural studies (disregarding the influence of motion) are comparably straightforward and simple, dynamical studies can be quite challenging and difficult. Although dynamic information is of great importance, motional models based on experimental data remain always incomplete due to the lack of sufficiently informative measurements. In many situations, it is necessary to supply missing information by numerical model calculations, rendering the dynamic models dependent on the inherent assumptions.

In the following, we briefly describe three NMR studies of motion in the cyclic decapeptide antamanide that cover three different ranges of dynamic rates. Antamanide, (-Val^1-Pro^2-Pro^3-Ala^4-Phe^5-Phe^6-Pro^7-Pro^8-Phe^9-Phe^{10}-), is known to occur in chloroform and in other solvents in only one dynamically interconverting conformation [6-10]. The NMR spectrum contains at all accessible temperatures only a single set of resonance lines. The following motions can be distinguished:

(i) Conformational dynamics of the peptide ring, if it occurs, is expected to be relatively slow.

(ii) Side-chain motion is likely to occur on a more rapid time scale, e.g. phenyl ring rotation and proline ring puckering.

(iii) Local vibrational dynamics will have very short correlation times τ_c.

Nuclear magnetic resonance provides access to motional processes on a wide time scale:

Very slow processes: $1s < \tau_c$ — Real-time observation after an initial perturbation.

Slow processes: $1ms < \tau_c < 10 s$ — Line shape effects, exchange broadening and exchange narrowing.

Medium rate processes: $1 \mu s < \tau_c < 10 ms$ — Rotating frame relaxation time $T_{1\rho}$ measurements with variable rf field strength.

Rapid processes: $30 ps < \tau_c < 1 \mu s$ — Laboratory frame relaxation time T_1, rotating frame relaxation time $T_{1\rho}$, and cross-relaxation rate Γ_{kl} measurements.

Extremely rapid processes: $\tau_c < 100 ps$ — Observation of averaged geometric parameters, governed by molecular order parameters, that enter the relaxation rates.

Conformational Peptide-ring Dynamics in Antamanide

There has been early evidence for conformational dynamics in antamanide from ultrasonic absorption measurements of antamanide dissolved in dioxane where an otherwise unexplained absorption near 1 MHz was found [10]. Obviously, no firm assignment of the motional mode could be made based on this global measurement. Patel [6] and Kessler et al. [7,9] have obtained evidence for a conformational equilibrium from chemical shifts and from the incompatibility of NMR distance constraints, respectively. Kessler et al. [9] concluded the presence of two or four rapidly exchanging conformations which differ most significantly in the dihedral HNCH angles ϕ_5 and ϕ_{10}. Two of the four suggested conformations correspond to the

X-ray structure [11] and to a structure obtained by a restrained molecular dynamics simulation, respectively.

We decided to further investigate the dynamics of this process by rotating frame relaxation $T_{1\rho}$ measurements [12,13]. It is known that $T_{1\rho}$ is sensitive to dynamics with intermediate rates as it samples the power spectral density $J(\omega_1)$ at the frequency $\omega_1 = -\gamma B_1$ where B_1 is the strength of the applied rotating rf field [14]:

$$1/T_{1\rho} = 1/T_{1\rho}^{dipolar} + P_1 P_2 \frac{\Delta\omega^2 \tau_{ex}}{1 + \omega_1^2 \tau_{ex}^2} \tag{1}$$

$T_{1\rho}$ refers to one particular nucleus for which the Larmor frequency changes by $\Delta\omega$ during the conformational transition between the two states with probabilities p_1 and p_2 and exchange lifetime τ_{ex}. At several temperatures, $T_{1\rho}$ was measured as a function of the rf field strength ω_1. It was found that only the $T_{1\rho}$ of the NH protons of Val[1] and Phe[6] are measurably dependent on ω_1, as can be seen in Fig. 1. These two protons seem to undergo a large chemical shift change. This could be due to breaking and formation of hydrogen bonds. From the temperature dependence of the exchange rate deduced from $T_{1\rho}$, an activation energy of 25 kJ/mol with an exchange time constant of 27 μs at 320 K was determined [13].

The exclusive behavior of the NH protons of Val[1] and Phe[6] is difficult, although not impossible to reconcile with the structural proposal by Kessler et al. [9] where NH

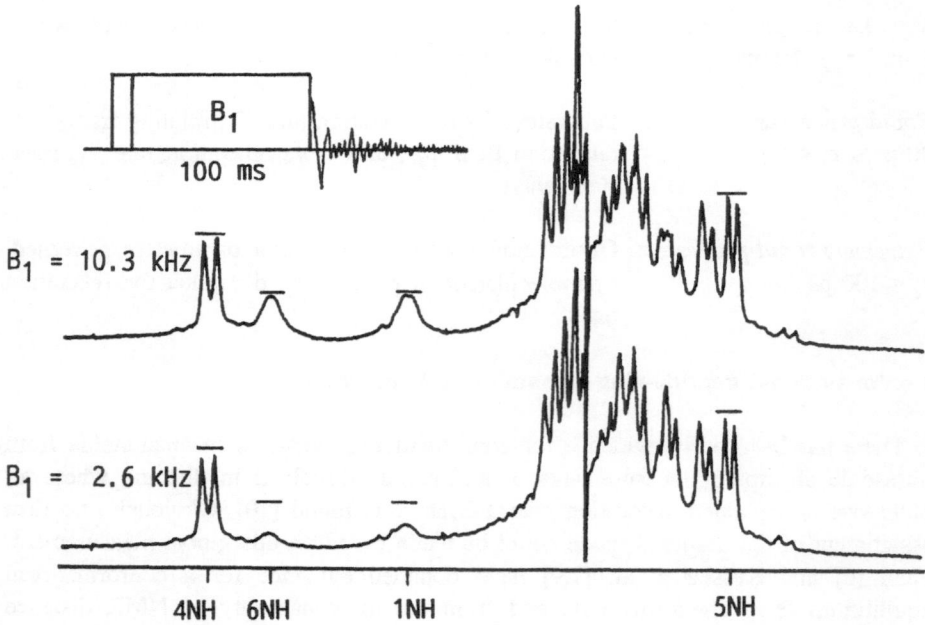

Fig. 1. NH resonance region in antamanide spectra (chloroform solution at 303 K and 400 MHz) recorded after spin-locking for 100 ms with a variable rf field strength γ B_1/2π = 2.6 kHz and 10.3 kHz. The NH resonances of Val[1], Ala[4], Phe[5], and Phe[6] are indicated.

hydrogen-bond breaking occurs for Val[1], Ala[4] Phe[6] and Phe[9], while hydrogen-bond rearrangement takes place for Phe[5] and Phe[10] (see Fig. 2b). One would have to assume negligible chemical shift changes for all NH protons except for those of Val[1] and Phe[6]. To obtain clarity on this point, we performed a careful structural study of antamanide based on NOE distance constraints and vicinal J-coupling constants, that is described in detail elsewhere [13].

A novel approach was applied to determine feasible structures involved in a conformational exchange process [15]. The philosophy is based on the fact that an observed, conformationally averaged internuclear distance r_{kl}^{NOE} predominantly reflects the conformation(s) j with short distance $r_{kl}^{(j)} \leq r_{kl}^{NOE}$ while for other conformations p, $r_{kl}^{(p)} > r_{kl}^{NOE}$. We may use the measured value r_{kl}^{NOE} as an *upper limit constraint* for $r_{kl}^{(j)}$ in order to find the conformation j. This implies that for other conformations p this constraint will possibly be violated.

The procedure starts with an initial structure that is selected by an educated guess influenced either by the X-ray structure, a molecular dynamics simulation run, or some other source of information. The list of NOE constraints is randomly ordered. The first NOE from the list is introduced in the form of an upper limit constraint represented by a semiparabolic energy term in the classical molecular potential. The energy is then minimized by a variation of the structural parameters. When the internal energy (without NOE constraints) remains above a preset threshold, the constraint is dropped from the list and the next constraint introduced. When the energy drops below the threshold, the constraint is left in the list and in the potential and the next constraint introduced. This procedure is continued to the end of the list of constraints. This produces an energetically feasible structure that fulfills a subset of the constraints and violates the rest.

In this manner, 1176 conformational structures have been generated for antamanide. In order to fulfill all NOE and J-coupling constraints, the structures have been combined in pairs with a population ratio that minimizes the deviations from the experimental constraints. A large number of pairs is obtained that can be ordered according to their rms violation of the constraints. The conformational pair with minimum deviation is sketched in Fig. 2 together with the proposal by Kessler et al. [9]. This figure shows that the proposed structure fits the hydrogen-bonding network expected from the rotating frame relaxation study with the NH protons of Val[1] and Phe[6] being involved in the hydrogen-bond dynamics.

Several reservations have to be put on a possible claim this model being an accurate and complete description of antamanide in solution. First of all, other pairs of the set have also a real chance to represent the antamanide dynamics and, secondly, it cannot be excluded that more than two structures are involved in the dynamics. At most, a distribution of possible structures, that fit the available data, with corresponding probabilities can be given. Generally, the proof of the uniqueness of a dynamic model is a serious problem in such studies. The entire set of feasible structural pairs is described in ref. 13.

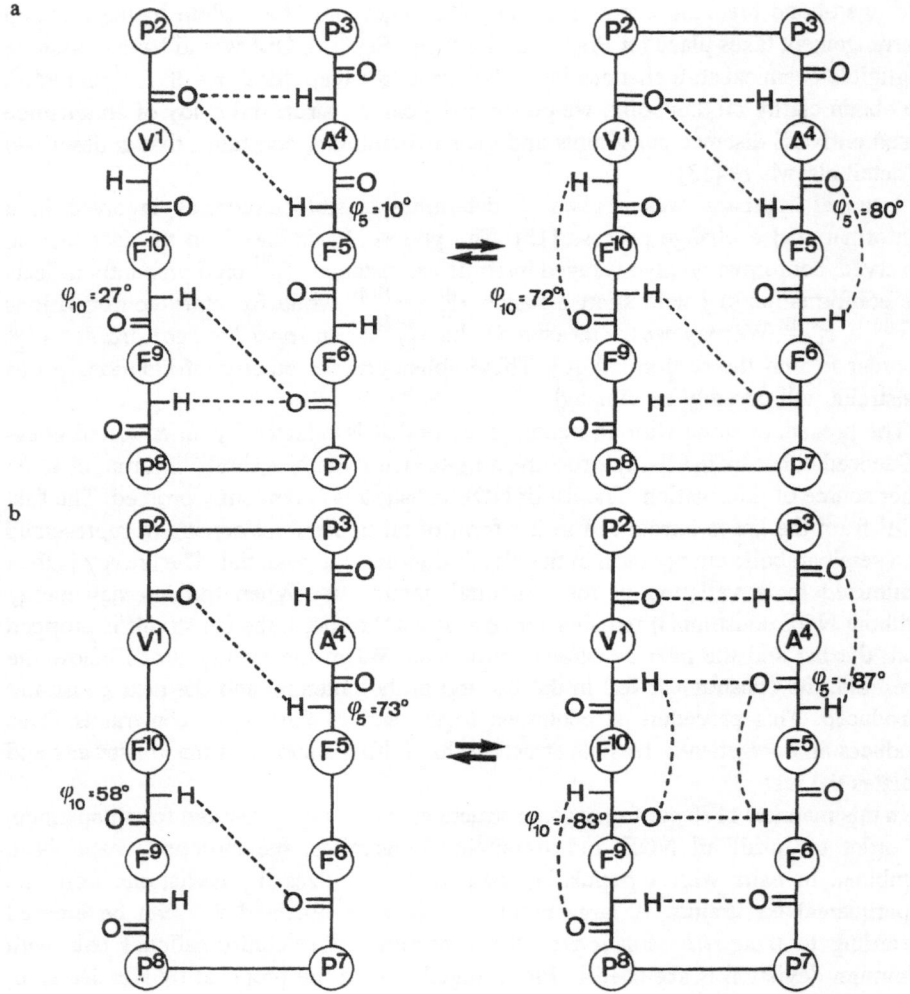

Fig. 2. *Hydrogen-bonding network of antamanide conformations. (a) Conformational exchange pair with minimum rms violation of the NOE and J-coupling constraints. (b) Conformational exchange pair proposed by Kessler et al. [9].*

Proline Ring Dynamics in Antamanide

It is known that there is no dynamic *cis-trans* isomerism of the proline amide bonds in antamanide. While the amide bonds Pro[2]-Pro[3] and Pro[7]-Pro[8] are *cis*-, the bonds Val[1]-Pro[2] and Phe[6]-Pro[7] are *trans*-configurated [9]. Nevertheless there can be conformational dynamics of the proline side-chain rings. It could even be that the backbone dynamics, discussed in the previous section, induces a conformational change of the proline rings. Proline ring dynamics could thus be correlated with the backbone dynamics.

Vicinal proton-proton couplings are predestinate for a conformational analysis of proline rings. Each ring contains ten couplings with angular information according to the Karplus relation [16] or variations of it [17]. Unfortunately, the proline spectra are generally complicated by a narrow chemical shift range leading to strong coupling effects. One-dimensional spectra of antamanide can hardly be analyzed and even 2D COSY spectra contain overlapping cross peaks and strong coupling features that defy a straightforward analysis by one of the multiplet analysis computer procedures using a recursive contraction algorithm [17] or symmetry arguments [19–21].

We decided [22] to base the analysis on a least-squares fitting procedure of 2D multiplets [23]. Maximum simplicity of the multiplet patterns is of advantage for a computer analysis as the error surface on which the search has to be performed becomes simpler. For this reason, E.COSY spectra are of advantage, even in the presence of fairly strong coupling. For a weakly coupled N-spin system, a COSY cross peak contains 2^{2N-2} multiplet components while for an E.COSY cross peak only 2^N multiplet components remain [24–26]. In the case of strong coupling, the ratio is somewhat less favorable for E.COSY. The least-squares analysis of an E.COSY spectrum of antamanide was performed stepwise [22]. At first, fairly well separated cross peaks of Pro[3] were localized and fitted by a computer-simulated E.COSY pattern using a modified Levenberg-Marquardt search algorithm [23]. The optimum computer multiplet pattern was then subtracted from the experimental spectrum, and in the remainder the cross peaks of Pro[2] were selected for the next fitting step. In this manner, all four proline spin systems were fully analyzed for a total of 82 two- to five-bond coupling constants [22] of which the 40 vicinal coupling constants were used for the conformational analysis. The assignment of the resonances to *cis-* and *trans-*protons (with respect to the position of the α proton) was left open as the chemical shifts do not permit an unequivocal choice.

Based on a modified Karplus relation [17], it was possible to determine dihedral angles and, finally, again with a least-squares procedure, an optimum ring structure for each of the four prolines with minimum rms angular deviation to the experimental values. As there are eight possible assignments of the β, γ and δ protons, eight most feasible structures were obtained for each of the four prolines. For Pro[3] and for Pro[8], indeed unique structures with a satisfactory fit were found with rms errors of 0.69 Hz and 0.57 Hz, respectively, for the fitted J-coupling constants. On the other hand, for Pro[2] and Pro[7], the rms error could not be reduced below 2.2 Hz. This led to the conclusion that two (or more) conformations must be involved and the fitting procedure has been extended to take into account two rapidly interchanging structures. While for Pro[2] and Pro[7], the fits improved significantly to rms errors of 0.44 Hz and 0.41 Hz, respectively, the errors for Pro[3] and Pro[8] did not change much (to 0.61 Hz and 0.53 Hz, respectively).

The conclusion to be drawn from this analysis is that Pro[3] and Pro[8] appear to be conformationally stable whereas Pro[2] and Pro[7] show conformational dynamics with at least two conformations involved. Both prolines undergo envelope-type motions where the flap of the envelope (C_γ) undergoes maximum amplitude excursions as Fig. 3 shows.

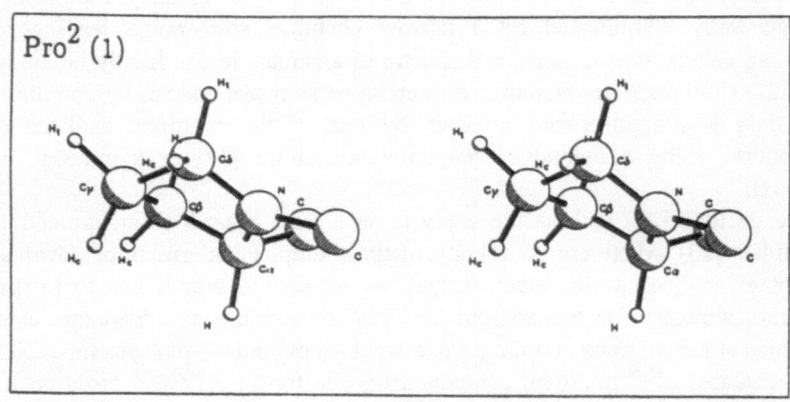

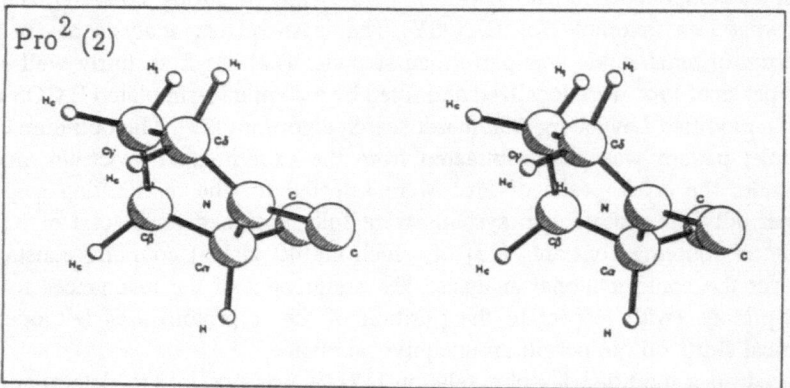

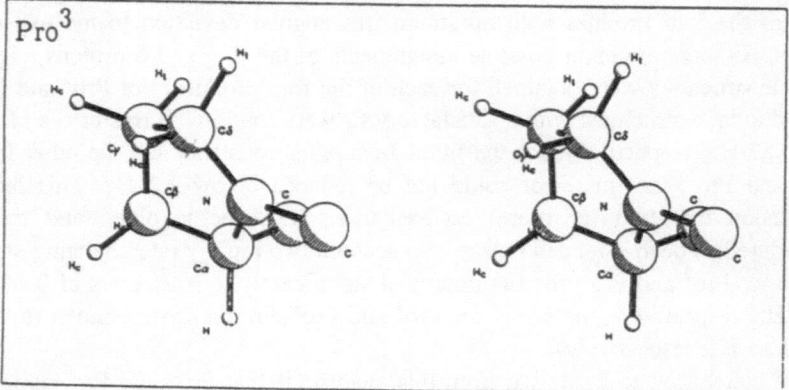

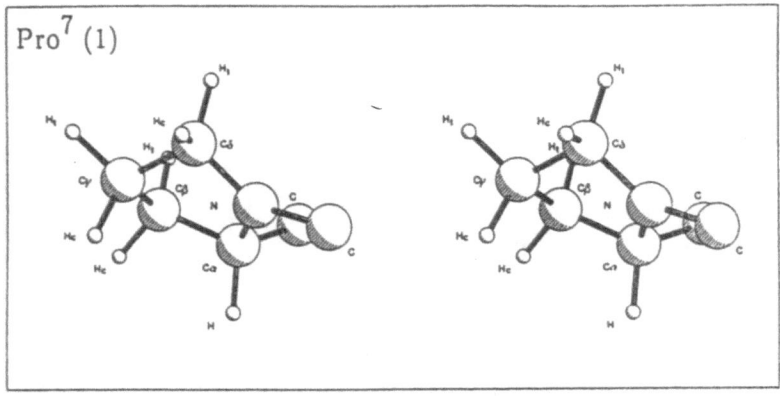

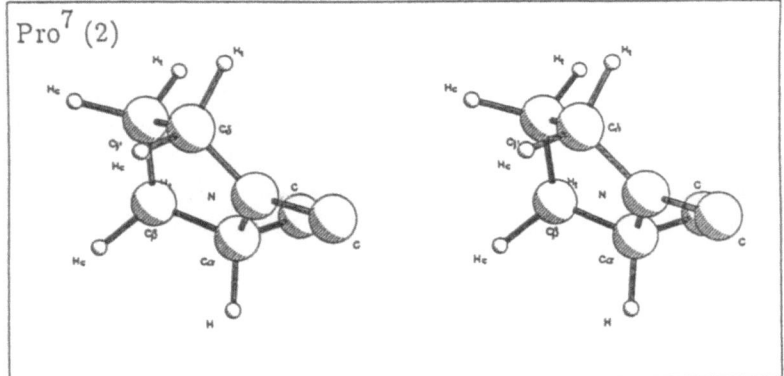

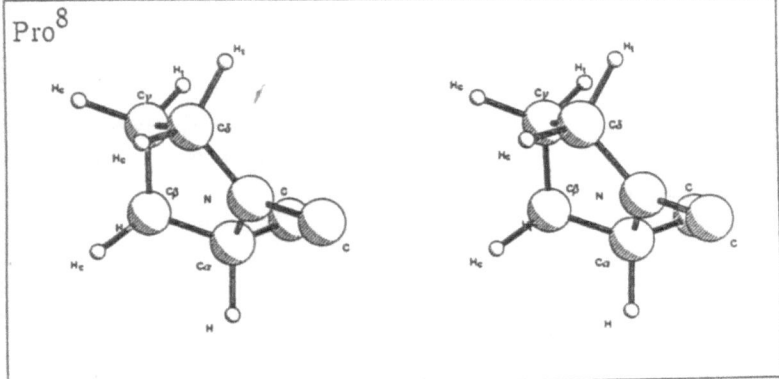

Fig. 3. *Stereographic representation of the proline ring conformations in antamanide deduced from J-coupling measurements. For Pro[2] and Pro[7], there are pairs of rapidly exchanging conformations whereas Pro[3] and Pro[8] are conformationally stable (from ref. 22).*

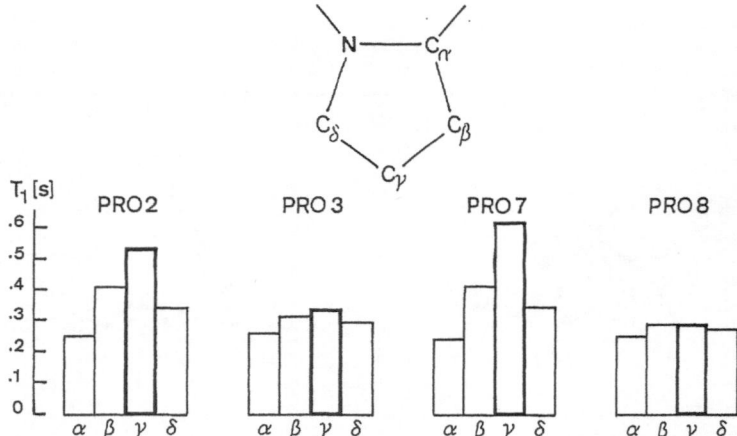

Fig. 4. ^{13}C relaxation times of proline-ring carbons in antamanide (at 280 K and 400 MHz proton resonance frequency). The increased relaxation times of C_γ are due to the proposed envelope motion of the rings of Pro^2 and Pro^7.

These conclusions are confirmed by the measurement of the ^{13}C relaxation times, illustrated in Fig. 4. While the relaxation times for Pro^3 and Pro^8 vary little around the ring, C_γ in Pro^2 and Pro^7 has the largest T_1 value, indicating maximum averaging of the intramolecular internuclear dipolar interaction. This averaging already indicates that the conformational motion must be significantly faster than the overall molecular tumbling with a correlation time of $\tau_c \simeq 300$ ps. A detailed analysis of the relaxation time leads to correlation times for the intramolecular proline dynamics of $\tau_{ex} \simeq 30$–40 ps for Pro^2 and Pro^7 [22]. Obviously, there is no correlation between the peptide-ring and the proline-ring dynamics.

Rapid Dynamics in Antamanide

Here the influence of rapid dynamics on NOE and ROE cross-relaxation rates is discussed. Dynamical processes that are slow compared to the overall molecular tumbling rate $1/\tau_c$ lead to motionally averaged cross-relaxation rates, each momentary conformation $\mathbf{X}$ present in the ensemble with probability $p(\mathbf{X})$ having its own cross-relaxation rate constant $\Gamma_{kl}(\mathbf{X})$ between spins I_k and I_l. This leads to the averaged observed rate constant

$$\overline{\Gamma}_{kl} = \int p(\mathbf{X}) \, \Gamma_{kl}(\mathbf{X}) \, d\mathbf{X} \tag{2}$$

On the other hand, very rapid processes with rates exceeding the overall molecular tumbling rate cause primarily partially averaged internuclear dipolar interactions that can lead to an enhancement or a reduction of the relaxation rates. The correlation function $C_{kl}(t)$ of the dipolar interaction between spins I_k and I_l

$$C_{kl}(t) = \left\langle \frac{Y_{20}(\theta_{kl}^{lab}(t_0 + t)) \cdot Y_{20}^*(\theta_{kl}^{lab}(t_0))}{r_{kl}^3(t_0 + t) \cdot r_{kl}^3(t_0)} \right\rangle \tag{3}$$

can be written, following ref. 27, as a product, separating the influences of overall and intramolecular motion and becomes in the case of isotropic rotational tumbling

$$C_{kl}(t) = C^{tumbling}(t) \cdot C_{kl}^{internal}(t) \qquad (4)$$

The intramolecular correlation function $C_{kl}^{internal}(t)$ is often an exponentially decaying function with time constant τ_{kl} that approaches a constant value for $t \to \infty$. This implies that some of the intramolecular structure or order is preserved. This order can be described by an order parameter S_{kl}^2 and the correlation function may be approximated by

$$C_{kl}^{internal}(t) \approx < \frac{1}{r_{kl}^6} > \{S_{kl}^2 + (1 - S_{kl}^2)\, e^{-t/\tau_{kl}}\} \qquad (5)$$

For very rapid processes with $\tau_{kl} \ll \tau_c$, the second term in this expression is relaxation-inactive, and the relaxation rate is merely determined by the product $<r_{kl}^{-6}> S_{kl}^2$.

It is physically plausible that there must be opposing effects caused by the angular and radial motion of the internuclear vector r_{kl} on the cross-relaxation rate $\bar{\Gamma}_{kl}$ [28,29]. While any variation of the angular term, represented by the spherical harmonic $Y_{20}(\theta_{kl}^{lab}(t))$ in Eq. (3), necessarily leads to a reduction of the cross-relaxation rate, the variation of r_{kl} about its equilibrium value $<r_{kl}>$ can lead to an enhancement of the cross-relaxation rate as the short distances will have a dominating influence. This suggests to attempt a separation of angular and radial motions in the discussion of the cross-relaxation rate $\bar{\Gamma}_{kl}$ [28,29].

The order parameter S_{kl}^2 shall tentatively be written as a product of radial order $S_{r,kl}^2$ and angular order $S_{\Omega,kl}^2$

$$S_{kl}^2 \approx S_{r,kl}^2 \cdot S_{\Omega,kl}^2 \qquad (6)$$

with $\quad S_{r,kl}^2 = <r_{kl}^{-3}>^2 / <r_{kl}^{-6}> \qquad (7)$

and $\quad S_{\Omega,kl}^2 = \frac{4\pi}{5} \sum_{m=-2}^{2} |\, <Y_{2m}(\theta_{kl}^{mol}, \phi_{kl}^{mol})> \,|^2 \qquad (8)$

where $\theta_{kl}^{mol}(t)$ and $\phi_{kl}^{mol}(t)$ are the time-dependent angles that orient the internuclear vector within the intramolecular frame. An experimental test of this approximation is rather difficult as it is not easily possible to selectively vary radial or angular order. For this reason, we have performed a molecular dynamics simulation for antamanide and computed correlation functions and cross-relaxation rates. These computations are discussed in detail in ref. 29.

From the trajectory of an 800 ps molecular dynamics simulation using CHARMM [30], the total, radial, and angular intramolecular order parameters have been computed for all proton pairs with average separation below 4.5Å and therefore measurable cross-relaxation rates. Figure 5 shows the product $S_r^2 \cdot S_\Omega^2$ plotted against the 'correct' order parameter S^2. It is seen that the product approximation is valid especially for internuclear vectors with $S^2 > 0.5$. Noticeable deviations occur particularly for interactions with phenyl-group protons. It is expected that the assumption of independent radial and angular motion is the better fulfilled the more motional processes affect the interaction. When there is only one dominant process, some noticeable correlation between radial and angular motion can be caused.

A plot of the relative cross-relaxation rates for a flexible compared to a rigid molecule, measured by $\gamma_{NOE,kl} = NOE(flex)_{kl} / NOE(rigid)_{kl}$, versus the total order parameter S^2 shows very little correlation (Fig. 6). As mentioned before, reduced intramolecular order can increase or decrease the cross-relaxation rate. On the other hand, when plotting versus the ratio S_Ω^2 / S_r^2, a clear correlation is obtained with increased cross-relaxation rate for increasing angular and decreasing radial order as shown in Fig. 7. Qualitatively the same behavior is also shown by cross-relaxation on the rotating frame [29].

Rapid intramolecular motion is a complicating factor for the interpretation of cross-relaxation data derived from NOESY or ROESY spectra. A proper interpretation requires knowledge about possible motional processes. On the other hand, it is in principle possible to deduce information on molecular motion by a careful and combined analysis of NOESY and ROESY intensities as is discussed in ref. 29.

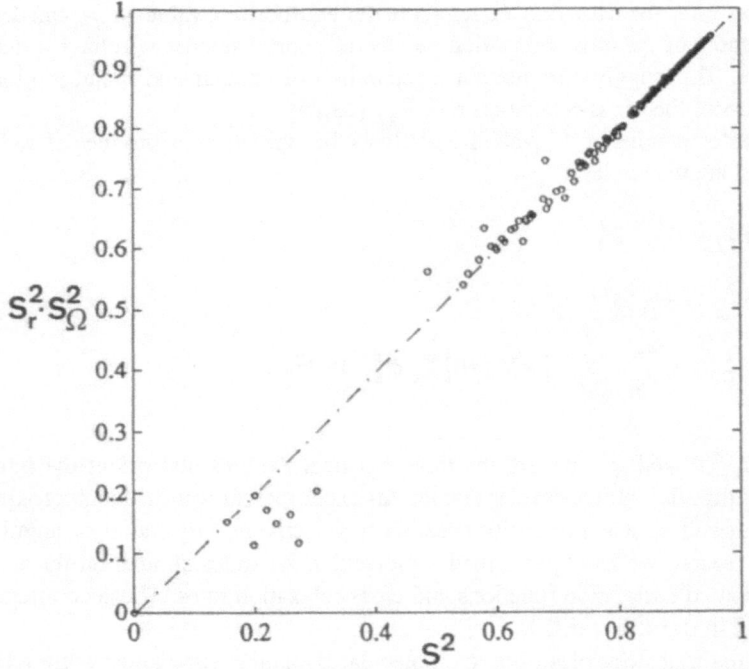

Fig. 5. Test of product approximation of order parameter S^2. The product of radial order parameter S_r^2 and angular order parameter S_Ω^2 is plotted against S^2 for all short-distance proton pairs in antamanide $(r_{kl} < 4.5\text{Å})$ (see also ref. 29).

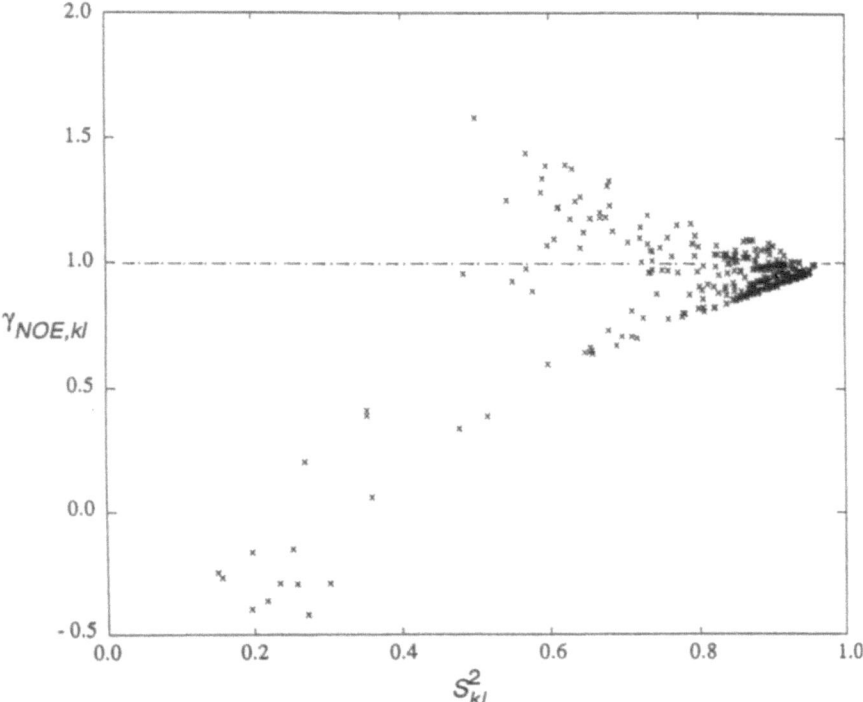

Fig. 6. *Plot of the relative NOE rates* $\gamma_{NOE,kl}$ = $NOE(flex)_{kl}/NOE(rigid)_{kl}$ *in antamanide against the total order parameter* S^2. *There is no correlation of the two quantities (see also ref. 29).*

2D Multiplet Structures of Maximum Simplicity

The discussion on dynamics of proline rings has demonstrated a practical need for procedures that produce multiplets of maximum simplicity in 2D (or higher dimensional) spectra for easing their manual or computer analysis to measure accurate J-coupling constants, for example to study dynamic conformational averaging as in the case of the proline rings. Obviously the full information content must be retained. The step from COSY to E.COSY multiplets is already a significant one, reducing the number of multiplet components from 2^{2N-2} to 2^N [24–26]. Here the question arises how far further simplifications can proceed [31,32].

This question can be answered without considering at first the required pulse sequences [32]. A (kl) multiplet of maximum simplicity can be defined as one in which each incoming coherence of spin I_k is transformed into a single coherence of spin I_l (apart from its transformation into single coherences of other spins). This means that in a 2D spectrum with ω_1 in the vertical direction, each row in a multiplet must contain a single multiplet component. In comparison, an E.COSY multiplet contains in each row two components in antiphase. This implies that, in principle, a further simplification by a factor 2 is feasible, leading to 2D multiplets with 2^{N-1} components.

23

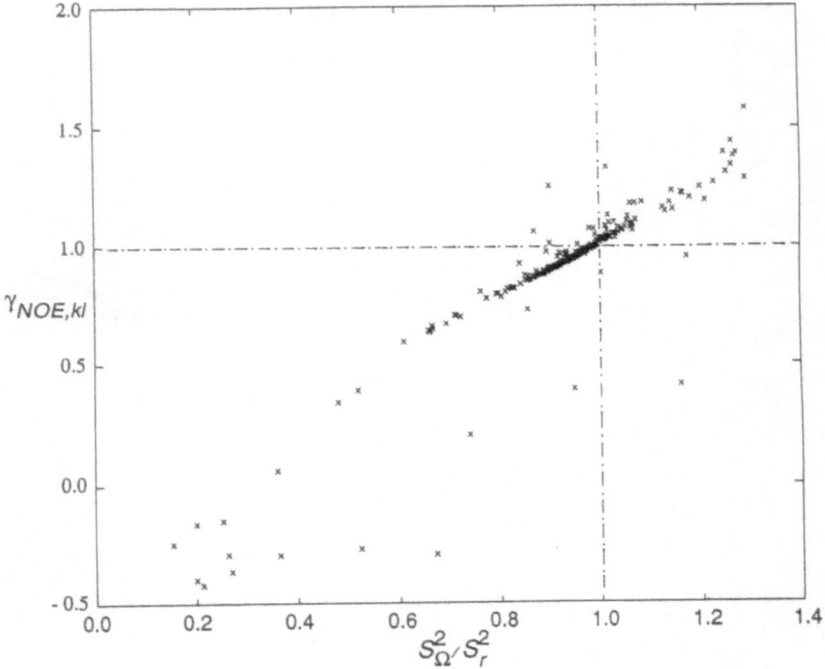

Fig. 7. Plot of the relative NOE rates $\gamma_{NOE,kl}$ in antamanide against the ratio of angular to radial order parameters. In contrast to Fig. 6, there is a clear correlation of the two quantities (see also ref. 29).

The mixing pulse sequence that performs the additional simplification must reduce the basic antiphase square of COSY and E.COSY spectra to a diagonal two-peak pattern. This can be discussed for the two active spins I_k and I_l alone in the absence of passive spins. The simplification requires the separation of regressive and progressive connectivities. This condition is satisfied if the mixing process conforms to the symmetry operator

$$\Pi = 2I_{kz}I_{lz} \tag{9}$$

This operator classifies the four states $\alpha\alpha$, $\alpha\beta$, $\beta\alpha$, and $\beta\beta$ into two classes according to

$$\Pi \mid \psi> = \pi^{(\psi)} \mid \psi> \tag{10}$$

with the quantum numbers $\pi^{(\psi)} = \frac{1}{2}$ for $\psi = \alpha\alpha$ and $\beta\beta$ and $\pi^{(\psi)} = -\frac{1}{2}$ for $\psi = \alpha\beta$ and $\beta\alpha$. For a mixing operator R that preserves the Π quantum numbers, regressive transitions for the two-spin system can occur only in the anti-echo spectrum while progressive transitions are confined to the echo spectrum as the following examples show.

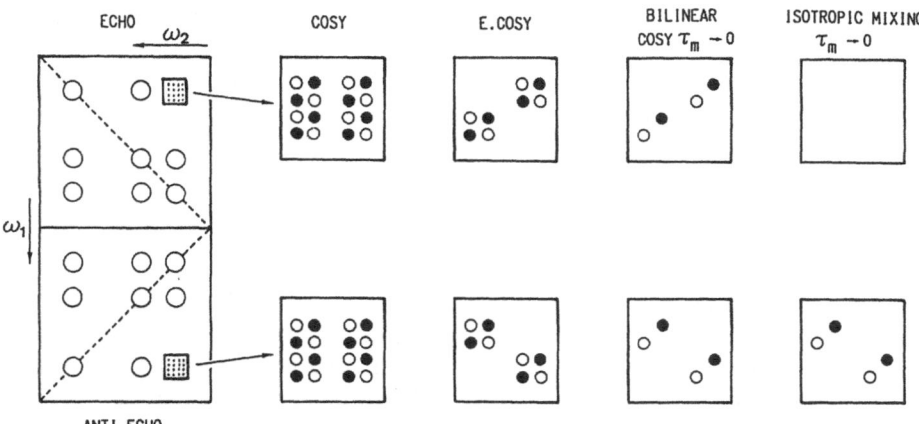

Fig. 8. Multiplet patterns in various correlation experiments for a 3-spin system with weak coupling among all spins. The multiplet patterns in echo and anti-echo parts are shown for COSY, E.COSY (or small flip angle COSY), bilinear COSY for short mixing time $\tau_m << 2\pi J_{kl}$, and for isotropic mixing (TOCSY) for $\tau_m << 2\pi J_{kl}$.

Note that for an allowed coherence transfer $|\psi_2> <\psi_1| \xrightarrow{\text{R}} |\psi_2'> <\psi_1'|$ the conversion of origin *and* destination functions under R must be permitted: $\psi_1 \xrightarrow{\text{R}} \psi_1'$, and $\psi_2 \xrightarrow{\text{R}} \psi_2'$. This requires that ψ_1 and ψ_1' as well as ψ_2 and ψ_2' must possess equal π quantum numbers. This applies to the following transfers:

$$|\alpha\beta> <\alpha\alpha| \xrightarrow{\text{R}} |\beta\alpha> <\alpha\alpha|$$

regressive, anti-echo connectivity

$$|\beta\beta> <\beta\alpha| \xrightarrow{\text{R}} |\beta\beta> <\alpha\beta|$$

regressive, anti-echo connectivity

$$|\alpha\alpha> <\alpha\beta| \xrightarrow{\text{R}} |\beta\beta> <\alpha\beta|$$

progressive, echo connectivity

25

$$|\beta\alpha> <\beta\beta| \xrightarrow{\;R\;} |\beta\alpha> <\alpha\alpha|$$

progressive, echo connectivity

For a regressive connectivity, two of the three involved levels have the same magnetic quantum number M, while for a progressive connectivity, all three levels possess different magnetic quantum numbers. For an anti-echo connectivity, the arrows point in the same ΔM direction, whereas for an echo connectivity, they point in opposite ΔM directions. The four transfers shown are the only allowed ones that transfer coherence from I_k to I_l. Both echo and anti-echo cross-peak multiplets contain just two components as desired. The relations described above are less easily appreciated when operator representations of coherence are used, such as $|\alpha\beta> <\alpha\alpha| = I_k^\alpha I_l^-$ and $|\alpha\beta> <\beta\beta| = I_k^+ I_l^\beta$.

The propagator R of a pulse sequence that preserves the quantum number π, and therefore commutes with the symmetry operator Π, must have the general form

$$R = \exp\{-i(\alpha_{xx}I_{kx}I_{lx}+\alpha_{xy}I_{kx}I_{ly}+\alpha_{yx}I_{ky}I_{lx}+\alpha_{yy}I_{ky}I_{ly}+\alpha_{zz}I_{kz}I_{lz}+\alpha_{zl}I_{kz}+\alpha_{lz}I_{lz})\} \tag{11}$$

Except for the last two, normally absent terms, R has the form of a bilinear rotation. Two simple sequences are known that lead to such a propagator: (i) the bilinear rotation pulse sequence $(\pi/2)_x-\tau_m/2-\pi_x-\tau_m/2-(\pi/2)_x$ with the propagator

$$R = \exp\{-i\,2\pi\,J_{kl}\,\tau_m\,I_{ky}\,I_{ly}\} \tag{12}$$

and (ii) a spin locking sequence with zero effective rf field leading to the isotropic propagator

$$R = \exp\{-i\,2\pi\,J_{kl}\,\tau_m\,\mathbf{I}_k\,\mathbf{I}_l\} \tag{13}$$

Both sequences cause the desired simplified basic pattern for a two-spin system with only two multiplet components.

Finally, it is necessary to take into account also the passive spins so that an arbitrary spin system can be treated as a superposition of two-spin systems. In order to lead to an E.COSY-type sequential duplication of the basic pattern, the effect of the propagator R on the passive spins should be negligible. This requires either an E.COSY-type linear combination of different experiments or, more simply, a small bilinear rotation angle, i.e. a short mixing time τ_m, leading to small-angle bilinear COSY or to small-angle TOCSY experiments [31,32]. Both cause 2D multiplet structures with just one component in each row. Small-angle TOCSY causes another simplification by removing echo-type peaks as it preserves the order of coherence [31,32].

In addition to the advantage of multiplet structures of maximum simplicity, easing their analysis, these experiments have the inherent disadvantage to produce mixed-phase peak shapes that cannot be corrected to pure absorption by a mere phase adjustment. This is inherent to all experiments that lead to echo and anti-echo peaks of unequal intensity [2,33].

Conclusions

This essay has been focused on the possibilities to study intramolecular dynamics by NMR. The selected example, antamanide, may not exhibit dynamic features of obvious biological relevance; it rather served as an example of a small molecule having defined conformations exchanging on several time scales.

Other subjects that were intended to be presented could not be discussed due to lack of space. This concerns in particular computer techniques for the display, manipulation and analysis of three-dimensional spectra, exemplified by heteronuclear 3D spectra of ^{15}N-labelled ribonuclease A [34], improved heteronuclear cross-polarization experiments in liquid phase as an alternative to pulsed heteronuclear coherence transfer [35], and improved spin-locking sequences [36].

Acknowledgements

These projects have been supported by the Swiss National Science Foundation. The molecular dynamics simulations have been done in collaboration with Dr. B. Roux and Prof. M. Karplus. One of the antamanide samples used in the experimental work has been kindly supplied by Prof. H. Kessler. The generous allowance to use AMX 500 and AMX 600 spectrometers at Spectrospin AG in Fällanden is gratefully acknowledged. The manuscript has been processed by Mrs. I. Müller.

References

1. Wüthrich, K. (1986) *NMR of Proteins and Nucleic Acids*, Wiley-Interscience, New York.
2. Ernst, R.R., Bodenhausen, G. and Wokaun, A. (1987) *Principles of NMR in One and Two Dimensions*, Clarendon Press, Oxford.
3. Cusack, S. (1989) *Chemica Scripta* **29A**, 103.
4. Beechem, J.M. and Brand, L. (1985) *Annu. Rev. Biochem.* **54**, 43.
5. Petsko, G.A. and Ringe, D. (1984) *Annu. Rev. Biophys. Bioeng.* **13**, 331; Ringe, D and Petsko, G.A. (1985) *Progr. Biophys. Mol. Biol.* **47**, 197.
6. Patel, D.J. (1973) *Biochemistry* **12**, 667.
7. Kessler, H., Griesinger, C., Lautz, J., Müller, A., Van Gunsteren, W.F. and Berendsen, H.J.C. (1988) *J. Am. Chem. Soc.* **110**, 3399.
8. Kessler, H., Müller, A. and Pook, K.-H. (1989) *Liebig Ann.*, 903.
9. Kessler, H., Bats, J.W., Lautz, J. and Müller, A. (1989) *Liebigs Ann.*, 913.
10. Burgermeister, W., Wieland, T. and Winkler, R. (1974) *Eur. J. Biochem.* **44**, 311.
11. Karle, I.L., Wieland, T., Schermer, D. and Ottenheym, H.J.C. (1979) *Proc. Natl. Acad. Sci. USA* **76**, 1532.
12. Blackledge, M.J., Brüschweiler, R., Griesinger, C. and Ernst, R.R. (1990) 10th Int. Biophysics Congress, Vancouver.

13. Blackledge, M.J., Brüschweiler, R., Griesinger, C. and Ernst, R.R., submitted for publication.
14. Deverell, C., Morgan, R.E. and Strange, J.M. (1970) *Mol. Phys.* **18**, 553.
15. Brüschweiler, R., Blackledge, M.J. and Ernst, R.R., submitted for publication.
16. Karplus, M. (1959) *J. Chem. Phys.* **30**, 11.
17. Haasnoot, C.A.G., De Leeuw, F.A.A.M., De Leeuw, H.P.M. and Altona, C. (1981) *Biopolymers* **20**, 1211.
18. Meier, B.U. and Ernst, R.R. (1988) *J. Magn. Reson.* **79**, 540.
19. Hoch, J.S.C., Hengyi, S., Kjaer, M., Ludvigsen, S. and Poulsen, F.M. (1987) *Carlsberg Res. Commun.* **52**, 111.
20. Glaser, S. and Kalbitzer, H.R. (1987) *J. Magn. Reson.* **74**, 450; Pfändler, P. and Bodenhausen, G. (1988) *J. Magn. Reson.* **79**, 99.
21. Boentges, S., Meier, B.U., Griesinger, C. and Ernst, R.R. (1989) *J. Magn. Reson.* **85**, 337.
22. Mádi, Z.L., Griesinger, C. and Ernst, R.R. (1990) *J. Am. Chem. Soc.* **112**, 2908.
23. Mádi, Z.L. and Ernst, R.R. (1988) *J. Magn. Reson.* **79**, 513.
24. Griesinger, C., Sørensen, O.W. and Ernst, R.R. (1985) *J. Am. Chem. Soc.* **107**, 6394.
25. Griesinger, C., Sørensen, O.W. and Ernst, R.R. (1986) *J. Chem. Phys.* **85**, 6837.
26. Griesinger, C., Sørensen, O.W. and Ernst, R.R. (1987) *J. Magn. Reson.* **75**, 474.
27. Lipari, G. and Szabo, A. (1982) *J. Am. Chem. Soc.* **104**, 4546.
28. Brüschweiler, R., Roux, B., Blackledge, M., Griesinger, C., Karplus, M. and Ernst, R.R. (1990) 10th Biophys. Congress, Vancouver.
29. Brüschweiler, R., Roux, B., Griesinger, C., Karplus, M. and Ernst, R.R., submitted for publication.
30. Brooks, B.R., Bruccoleri, R.E., Olafson, B.D., States, D.J., Swaminathan, S. and Karplus, M. (1983) *J. Comput. Chem.* **4**, 187.
31. Schulte-Herbrüggen, T., Sørensen, O.W. and Ernst, R.R. (1990) 10th European Experimental NMR Conference, Veldhoven, The Netherlands.
32. Schulte-Herbrüggen, T., Mádi, Z.L., Sørensen, O.W. and Ernst, R.R., *Mol. Phys.* (in press).
33. Bodenhausen, G., Kogler, H. and Ernst, R.R. (1984) *J. Magn. Reson.* **58**, 370.
34. Boentges, S., Griesinger, C. and Ernst, R.R., submitted for publication.
35. Ernst, M., Griesinger, C. and Ernst, R.R., submitted for publication.
36. Briand, J. and Ernst, R.R., in preparation.

Stable-isotope-assisted multinuclear NMR investigations of proteins*

John L. Markley

National Magnetic Resonance Facility at Madison, Biochemistry Department, College of Agricultural and Life Sciences, University of Wisconsin-Madison, 420 Henry Mall, Madison, WI 53706, U.S.A.

Abstract

We have explored various strategies for labeling proteins with ^{2}H, ^{13}C, and ^{15}N for NMR analysis. Uniform ^{15}N and/or ^{13}C labeling is efficient for initial assignments since many connectivities can be established with a single sample. The wealth of data can be used to answer questions concerning a protein's covalent structure and cofactors as well as its structure and internal mobility. Two- and three-dimensional versions of the ^{1}H-^{15}N correlation – ^{1}H-^{1}H NOE experiment are particularly effective for sequential assignment purposes and for determining α-helical structure. Two- and three-dimensional versions of the ^{1}H-^{13}C correlation – ^{1}H-^{1}H NOE experiment are useful for determining β-sheet. Uniform double labeling with ^{13}C and ^{15}N establishes coupling pathways through the peptide bond which offer great promise for efficient sequential assignments of larger proteins. Incorporation of single amino acid types labeled with ^{13}C or ^{15}N can simplify crowded spectral regions or help confirm assignments. Atom specific single and double labeling yield still higher levels of discrimination. In addition to facilitating assignments, stable isotope labeling can open windows on functionally important regions of the protein. Measurements of heteronuclear coupling constants provide dihedral angle constraints, and isotope-assisted proton-proton NOEs provide distance constraints. Selective labeling of species involved in molecular interactions establishes the means of discriminating inter- and intra-molecular interactions. Studies of kinetic processes, such as hydrogen exchange, internal rotations, conformational transitions, and protein folding can be facilitated by labeling with ^{13}C or ^{15}N. Illustrations of these approaches are drawn from our 2D and 3D NMR studies of staphylococcal nuclease (17 kDa), cytochrome c_{553} (10 kDa), ferredoxin (10 kDa), and flavodoxin (19 kDa).

Principles of stable-isotope-assisted NMR

Multinuclear NMR spectroscopy of larger proteins

The classical approach to NMR structural studies of proteins involves (1) the determination of sequence-specific backbone assignments, (2) the extension of these assignments to side chains, (3) the analysis of cross relaxation patterns and measurement of coupling constants to provide input constraints, and (4) iterated refinement of the input constraints along with structural models consistent with them [1]. Although this method has been highly successful for small proteins (up to 10 kDa), the approach becomes increasingly difficult with larger proteins because of peak overlap in spectral regions used for determining sequential connectivities and distance constraints. With smaller proteins, side chains are commonly assigned by

*Supported by grants from the NIH (RR02301 and GM35976, USDA (85-CRCR-1-1589), and NEDO.

use of ^{1}H{^{1}H}RELAY, HOHAHA, or NOESY data. These are less effective with larger proteins because of overlap or inefficient coherence transfer. Severe overlap in 2D ^{1}H-^{1}H NOE spectra of larger proteins render them largely uninterpretable. Proton homonuclear 3D NMR methods do not offer clear solutions to these problems; they do provide additional information, but at the cost of increased spectral complexity.

Over the past few years, an array of techniques has emerged that take advantage of coupling among ^{1}H, ^{13}C, and ^{15}N nuclei [2,3]. These methods promise to extend solution NMR structural analysis into the 20 kDa range. Unlike the proton-only methods, multinuclear methods give increasingly simplified 3D and 4D spectra. Thus overlap problems can be solved by increasing the dimensionality and by making use of the chemical shift dispersions of the different nuclei (^{1}H, ^{13}C, and ^{15}N). As described below, dual labeling with ^{13}C and ^{15}N allows for new sequential assignment strategies that make use of one-bond couplings along the peptide backbone. The multinuclear approach also can assist in the resolution of coupling constants and ^{1}H-^{1}H NOEs needed for structural analysis.

Progress in multidimensional, multinuclear NMR methods has been driven by advances in large-scale production of proteins and methods of introducing stable isotopes [4,5] along with rapid development of powerful new NMR techniques that exploit the labeling patterns [6–8]. The success of these methods has fulfilled the promise shown by early attempts to label proteins with ^{2}H [9], ^{13}C [10], ^{15}N [11], and dual ^{13}C and ^{15}N [12] so that specific spectral information could be extracted.

The advantages of detecting heteronuclear correlations were readily apparent in the first of these protein spectra [2,13]. Implementation of ^{1}H (indirect, inverse) detection schemes [14,15] made such experiments feasible at lower protein concentrations and paved the way for more sophisticated pulse sequences that would combine heteronuclear correlations with NOE [16] or HOHAHA [17] steps. Inverse detection made it possible to observe ^{1}H-^{13}C [18] or ^{1}H-^{15}N [19] correlations in proteins at natural isotopic abundance. However, this approach proved feasible only at very high protein concentrations. Uniform ^{15}N and/or ^{13}C labeling have become the methods of choice for 3D and 4D NMR investigations of larger proteins.

Labeling strategies

Because of the low natural abundance of ^{13}C (1.1%) and ^{15}N (0.37%), proteins must be labeled with these isotopes for most efficient data collection. The optimal labeling pattern depends on the kind of information desired. For the most general concerted assignment approach, it is desirable to have > 98% ^{13}C and ^{15}N doubly labeled protein. Such proteins can be expensive to prepare unless they are efficiently overexpressed in a system that utilizes inexpensive precursors such as [UL ^{13}C]glucose and [^{15}N]ammonium salt. Lower levels of uniform ^{13}C enrichment have proven useful for the determination of two- and three-bond ^{1}H-^{13}C coupling constants by 2D [20] or 3D [21] methods or for the tracing of carbon skeletons by the ^{13}C{^{13}C}DQC (INADEQUATE) experiment [22,23]. In most cases it is less expensive to label proteins with ^{15}N than with ^{13}C. Uniform labeling with ^{15}N alone (usually at the highest level obtainable) provides the means for simplifying the (^{1}H^N,

$^1H^{\alpha}$) fingerprint region by means of ^1H-^{15}N coupling. Backbone sequential assignments can be deduced from 2D [24] or 3D [25] versions of $^1H\{^{15}N\}$SBC-NOE experiments. Uniform ^{15}N labeling is useful for deducing helical regions. Sometimes the amino acid spin system of an amino acid can be determined from $^1H\{^{15}N\}$SBC-NOE or $^1H\{^{15}N\}$SBC-TOCSY data. In general, however, full side-chain assignments in larger proteins are most easily accomplished by uniform ^{13}C labeling [17,22-24,26-31]. Uniform ^{13}C labeling facilitates the analysis of β-sheets and turns [24].

What is the molecular weight limit for stable-isotope assisted studies of proteins? If 1D ^{13}C detection is employed, the size limit is remarkably high. For example, selective double labeling of immunoglobulins with one amino acid type labeled with ^{13}C and one amino acid type labeled with ^{15}N has shown that $^{13}C_i^1$-$^{15}N_{i+1}$ coupling across the peptide bond can be resolved with a protein as large as 150 kDa [32]. This general approach [12] is tedious, but may be the only viable alternative for the largest proteins amenable to study by NMR. With 2D or 3D methods that rely on ^1H detection, the proton T_2 relaxation rate appears to be the limiting factor [8]. A possible way around this is to combine deuteration with labeling with ^{15}N [16] and/or ^{13}C.

Selective labeling can be put to a variety of uses. Stereospecific labeling of amino acids with ^2H [5] or ^{13}C [33] provides the most reliable methods for stereospecific assignment of resonances from prochiral atoms in proteins. Selective labeling is a convenient way to identify types of amino acid residues, particularly in crowded spectral regions. The sequential assignment problem can be simplified considerably if a number of resonances can be identified with particular amino acid types on the basis of selective labeling.

In deciding whether to use uniform or selective labeling, one needs to consider whether the information desired would be obscured or less readily extracted from a uniformly labeled sample. In general, uniform labeling provides the largest amount of information per labeled protein analogue and per NMR experiment. However, the particular information (e.g., coupling constants, relaxation rates, distances) desired may be less readily detected with a uniformly labeled sample than with a selectively labeled sample.

Methods for introducing labels

Our first experience in preparing proteins labeled uniformly with ^{13}C and ^{15}N was with proteins produced by *Anabaena*, a photosynthetic bacterium. ^{13}C was introduced by growing the cells on $^{13}CO_2$ as the sole carbon source [34]. We have produced proteins and amino acids from *Anabaena* at 15%, 26%, and 45% ^{13}C. The *Anabaena* proteins were labeled uniformly with ^{15}N by feeding the organism [98-98% ^{15}N]nitrate ion. Double labeling was accomplished by feeding both [98-98% ^{15}N]nitrate and $^{13}CO_2$ [35].

After taking out the proteins of interest (plastocyanin, ferredoxin, flavodoxin, cytochrome c_{553}) the remaining protein was hydrolyzed to amino acids [36]. This amino acid mixture can be used in growing non-photosynthetic cells such as *E. coli* [36]. We have separated individual amino acids from the mixture by scaling up the displacement chromatography procedure of LeMaster and Richards [37]. Although tedious, this is an excellent way to produce single, uniformly labeled amino acids.

One always prefers to have the highest level of ^{15}N enrichment obtainable so as to increase the sensitivity of data collection. With ^{13}C, the question of level is more complicated since ^{13}C-^{13}C coupling can complicate spectra and broaden resonances. Enrichment to ^{13}C levels of 15–26% proved optimal for most 2D experiments [2,3]. It also has been a good uniform enrichment level for the accurate determination of ^{13}C chemical shifts and coupling constants or for ^{13}C detection of ^{13}C-^{15}N connectivities [38,39]. 3D data acquisitions, particularly those involving triple resonance (^{1}H, ^{13}C, ^{15}N) correlations [40] and ^{13}C-^{13}C coherence transfer [41,42], are best carried out at the highest available ^{13}C enrichment. Carbon is used as a coarse dimension in such experiments so the line broadening caused by couplings is less of a problem.

Recent progress in protein engineering has opened up several options for labeling proteins. Efficient overproduction of proteins has been achieved with insect cells, fungi, plant cells, yeast, and bacteria. Because they can be grown on a variety of simple organic molecules, bacteria such as *E. coli* offer an ideal system for stable isotope labeling. We are now producing several proteins in *E. coli* at yields of 20–80 mg purified protein per liter culture: staphylococcal nuclease [43], flavodoxin [R.J. Reedstrom, R. Chylla, N. Straus and J.L. Markley, unpublished], and ferredoxin [Y.-K. Chae, B.-H. Oh and J.L. Markley, unpublished]. The genetics and metabolic pathways of *E. coli* are well understood, and one can often construct auxotropic strains for selective labeling [5,44].

For uniform ^{13}C labeling, we grow the *E. coli* on [98% UL ^{13}C]glucose as the sole carbon source. For ^{15}N labeling we use [^{15}N]ammonium salts in place of other nitrogen sources. For selective labeling, we use amino acids isolated from *Anabaena* as described above or amino acids labeled by chemical synthesis.

Recent Stable-Isotope-Assisted NMR Studies from our Laboratory

Staphylococcal nuclease

Our initial approach was to use the classical proton methods to obtain full sequence-specific assignments for the nuclease H124L·pdTp·Ca^{2+} ternary complex (Nuclease H124L is recombinant staphylococcal nuclease produced in *E. coli*; its sequence is identical to the nuclease isolated from the V8 strain of *Staphylococcus aureus*; pdTp is thymidine-3′,5′-bisphosphate, a competitive enzyme inhibitor). When it became clear that the proton methods would not yield more than a few assignments, we used the random fractional deuteration method [5] to improve the quality of the fingerprint region. With the added resolution afforded by suppressing passive couplings and spin diffusion effects, a large number of sequential walks could be worked out. Data from uniform and amino-acid-selective ^{13}C and ^{15}N labeling were needed to complete the assignments of the ternary complex [24,30,31,43].

Our recent investigations of unligated nuclease H124L have made extensive use of heteronuclear 2D and 3D spectroscopy of protein samples labeled with ^{13}C and ^{15}N. These results have enabled us to detect conformational changes that occur on complex formation (J. Wang, A.P. Hinck, S.N. Loh and J.L. Markley, unpublished; J.F. Wang, E. Mooberry, W.F. Walkenhorst and J.L. Markley, unpublished). The prochiral methyl groups of valines and leucines have been assigned by incorporating block ^{13}C-labeled amino acids into nuclease H124L (A.P. Hinck, M. Kainosho and J.L. Markley, unpublished).

We are employing amino-acid-specific and atom-specific labeling to investigate conformational transitions in staphylococcal nuclease. Labeled [^{13}C$^{\delta 1}$]tryptophan was incorporated into the single Trp residue. We found that the folding/unfolding transition could be followed by ^{13}C exchange spectroscopy, either by observing the ^{13}C directly or by indirect detection of attached ^{1}H [45]. In order to study the configuration at the Lys^{116}-Pro^{117} peptide bond, [99% ^{13}C$^{\alpha}$]L-lysine and [99% ^{13}C$^{\alpha}$]proline have been incorporated into nuclease H124L. The double isotope edited 2D NOE spectrum reveals the ^{1}H$^{\alpha}{}_i$...^{1}H$^{\alpha}{}_{i+1}$ NOE expected for a *cis* peptide bond (A.P. Hinck and J.L. Markley, unpublished). The next step will be to incorporate the same amino acids into mutant proteins, such as nuclease H124L+G79S, for which this peptide bond is expected to be *trans* [46].

As a selective label, [^{13}C$^{\alpha}$]glycine was introduced into nuclease H124L. Metabolic reactions transferred some of the label to serine. Direct ^{13}C relaxation measurements were carried out at three field strengths, on unligated nuclease and on the ternary complex. The glycine and serine relaxation data were analyzed by the 'model-free' approach of Lipari and Szabo [48]. Although the results showed sequence- and ligation-state-dependent correlation times and order parameters, the study revealed difficulties in accurate relaxation measurements in proteins of this size. These may be overcome by increasing the efficiency of ^{13}C labeling and by utilizing ^{1}H-detection of the ^{13}C relaxation.

Cytochrome c_{553}

We used the multinuclear assignment strategy to derive complete ^{1}H, ^{13}C, and ^{15}N resonance assignments for the prosthetic heme group of the ferrocytochrome c_{553} from the photocyanobacterium *Anabaena 7120* [49]. Previous assignments made on the basis of ^{1}H-^{1}H NOEs had led to ^{1}H NMR assignments of the hemes in several cytochromes c, but few ^{13}C and ^{15}N NMR assignments have been published. The key to understanding the underlying causes of functional differences among cytochromes c, e.g., the wide range of redox potentials (50 to 500 mV), is a detailed knowledge of the electronic and chemical properties of the heme and how they are influenced by the protein environment. A complete and reliable NMR analysis, however, depends on resolution and assignment of signals from all the relevant atoms.

The cytochrome$_{553}$ sample was isolated from *Anabaena* 7120 labeled uniformly with 26% ^{13}C or 98% ^{15}N as described above. The assignment strategy was based on multinuclear 2D NMR experiments that correlate pairs of nuclei that are directly or distantly scalar coupled: ^{13}C{^{13}C} double quantum coherence spectroscopy (DQC), ^{1}H{^{13}C} single bond correlation spectroscopy (SBC), ^{1}H{^{13}C} multiple-bond correlation spectroscopy (SBC), and ^{1}H{^{15}N}MBC. The carbon spin system was delineated first by using the ^{13}C{^{13}C}DQC experiment which correlates coupled carbon pairs [23]. The proton resonances were then assigned on the basis of one-bond

^{1}H-^{13}C coupling and multiple-bond ^{1}H-^{13}C coupling with the ^{1}H{^{13}C}SBC and ^{1}H{^{13}C}MBC experiments, respectively [27]. Finally, ^{1}H{^{15}N}MBC was used to assign the heme nitrogens based on the three-bond coupling between ^{15}N and the heme meso protons.

Several of the assigned resonances appeared at unusual chemical shifts. Carbons 3, 4a, 12 and S-meso and protons 3a, 4a, 4b and S-meso were well downfield from the analogous carbon and proton resonances in other parts of the heme. This indicates a relatively lower electron density in the region of the heme around pyrrole ring II. The nitrogen resonances of pyrrole rings II and IV both appeared near 186 ppm, whereas the nitrogens of pyrrole rings I and III resonate at significantly higher field (184.2 and 179.0 ppm, respectively). Cytochrome c_{553} from *A. flos-aquae*, a closely related cyanobacterium, is known to have an R configuration about the ligand Met sulfur atom in which the Met CH$_3$ group is oriented pointing toward the pyrrole I N and the Met non-bonded lone pair is pointing toward the pyrrole IV N [50]. It is questionable whether this asymmetry alone can account for the different ^{15}N chemical shifts. Steric and/or electronic interactions with other protein groups, including the axial histidine ligand, probably contribute as well. It is widely believed that physiological electron transport occurs at the solvent-exposed heme edge which includes pyrrole rings II and IV. Our finding that pyrrole ring II is generally the most electron-poor region of the heme suggests that this part of the heme may be important in the function of the cytochrome c_{553}.

Ferredoxin

Our multinuclear studies of oxidized *Anabaena* 7120 vegetative ferredoxin have yielded a low-resolution solution structure of this paramagnetic protein (B.-H. Oh, A.M. Krezel, Y.-K. Chae and J.L. Markley, unpublished). A fundamental problem in the analysis is that the paramagnetic center provides an efficient relaxation mechanism for nuclei in its vicinity; this causes signal broadening and also hyperfine (contact or pseudo contact) shifts. As a result, signals from nuclei near paramagnetic centers do not show up in ordinary 2D NMR spectra. Signals from protons (and other nuclei) that are 7 Å or more distant from the iron-sulfur cluster can be resolved and assigned, and we recently reported comprehensive assignments from the diamagnetic ^{1}H, ^{13}C, and ^{15}N resonances of *Anabaena* 7120 vegetative ferredoxin [28,29].

Many of the hyperfine ^{1}H, ^{13}C, and ^{15}N signals have been resolved [51] although only a few have been assigned. Recently, NOE connectivities from hyperfine resonances have been observed in 1D [52] and 2D (L. Skjeldal, B.-H. Oh, W.M. Westler and J.L. Markley, unpublished) spectra of 2Fe-2S* ferredoxins. These NOEs are mainly intraresidual and, although promising for the assignment of hyperfine signals, do not yield much information about the conformation of the iron-sulfur cluster.

E.COSY data and 2D NOE data were analyzed to determine stereospecific assignments of prochiral geminal β-proton resonances. Distance constraints were generated from analysis of NOESY cross peak intensities and from measurements of coupling constants. These data were used, along with information about the covalent chemistry of the iron-sulfur cluster, to determine the solution structure of the ferredoxin by a distance geometry approach (DSPACE, Hare Research). In this structure, we observed the same global fold and secondary structure found in the crystal structure of the same protein [53]. Differences were found however between the solution structure of *Anabaena* 7120 ferredoxin and the X-ray structures reported for *Spirulina platensis* [54] and *Aphanoceca sacrum* [55] ferredoxins. These differences are mainly in the region of residues 68–75 which are disordered in these structures but helical in the solution and crystal [53] structures of *Anabaena* 7120 ferredoxin.

Flavodoxin

Our initial studies of *Anabaena* 7120 flavodoxin focused on the FMN group at its active site. The concerted, multinuclear approach was employed in assigning NMR signals from the apoprotein and FMN moieties of the holoprotein [27]. Sequence-specific resonance assignments were obtained for the consensus binding site of the isoalloxazine ring of the cofactor and from those of the 33 residues that form the five-stranded β-sheet core of the protein. The pattern of interstrand NOEs indicated that the relative orientation of the strands is similar to that observed in the X-ray structure of the 70% homologous flavodoxin from *Anacystis nidulans* (3). Complete assignments of flavin ^{13}C chemical shifts were obtained by comparing the 1D ^{13}C spectrum with the 2D ^{13}C{^{13}C} double-quantum correlation spectrum.

The flavin ^{1}H resonances, which had been assigned by extending the ^{13}C assignments to their attached protons, provided a window on the cofactor binding site. Normal and ^{13}C-assisted NOE spectroscopy led to identification of 25 protein-flavin NOEs. Additional distance constraints came from NOEs between 29 pairs of protons on residues in the flavin binding site. These distances were used in modeling the flavin binding site by restrained molecular dynamics (DISCOVER software package, Biosym) (Fig. 1) [20].

The relative orientations of the flavin, tryptophan, and tyrosine rings in the *Anabaena* 7120 flavodoxin structure are similar to those in the *Anacystis nidulans* flavodoxin X-ray structure [56]. The indole ring of W57 is more coplanar with the flavin ring in *Anacystis nidulans* flavodoxin than in *Anabaena* 7120 flavodoxin; in the latter, the outer edge of the benzene ring points toward the flavin methyl groups. The Y94 ring overlaps the flavin ring slightly less in *Anabaena* 7120 than in *Anacystis nidulans* flavodoxin, but is still positioned to form good ring-stacking interactions. Both carbonyl oxygens of the flavin ring appear to be hydrogen bonded, and H^{N3} forms a strong hydrogen bond to the main-chain carbonyl oxygen of N97. N^5 may be weakly hydrogen bonded to the backbone amide proton of I59, but no hydrogen bonds were observed to N^1. The $^{13}C^2$ and $^{13}C^4$ chemical shifts of *Anabaena* 7120 flavodoxin fall between those observed in model compounds in polar and apolar environments [57,58]. This suggests that the carbonyl oxygens of these two carbons may participate in hydrogen bonds. This hypothesis is supported by the experimental one-bond $^{13}C^4$-$^{13}C^{4a}$ (74.9 Hz) and $^{13}C^{4a}$-$^{13}C^{10a}$ (56.9 Hz) coupling constants, which resemble more closely those of FMN in a polar solvent than TARF in an apolar solvent [58]. The magnitudes of the ^{15}N chemical shifts [59], suggest that this edge of the flavin ring is polarized by interactions with residues in the flavin binding site. Hydrogen

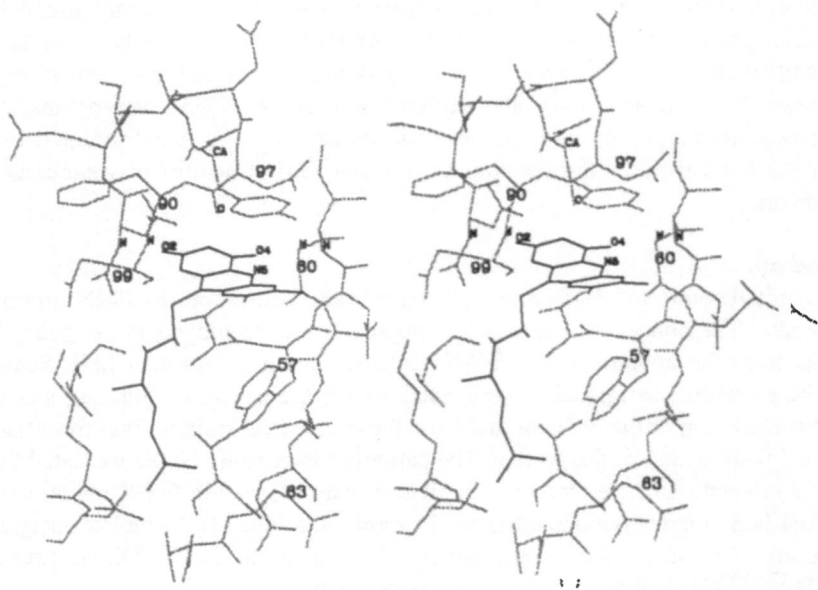

Fig. 1. Stereoscopic view of the FMN binding site of Anabaena *7120 flavodoxin (19 kDa) as determined from solution NMR data [20]. In modeling the binding site, we started with the refined X-ray crystallography coordinates of* Anacystis nidulans *flavodoxin (Dr. M.L. Ludwig, personal communication). First, side chains were replaced and one residue was inserted in order to convert the sequence to that of* Anabaena *7120 flavodoxin. Then, the energy was minimized (500 steepest descent iterations followed by 5000 conjugate gradient iterations). The resulting 'structure' was then subjected to restrained energy minimization in which the NMR distance constraints were represented by square-well harmonic functions with force constants of 25.15 kcal/(mol Å). Calculations were limited to residues within 12 Å of $C^{2'}$ of FMN. Only three of the NMR-constrained distances had violations greater than 0.1 Å. None of these three was larger than 0.3 Å.*

bonding to the FMN carbonyl groups has a systematic effect on the chemical shifts of the other carbon atoms of the isoalloxazine ring [57,58]. The chemical shifts (corrected for ring-current effects) of the non-carbonyl flavin ring carbons, as well as that of N^{10} [59], resemble those observed in *Azotobacter vinelandii* flavodoxin (which lacks an aromatic group corresponding to W57 of *Anabaena* 7120 flavodoxin) more closely than those of other flavodoxins studied [57,58]. Our results suggest that the redox potential of the flavin may be influenced more by polarization of the isoalloxazine ring than by interactions with neighboring aromatic residues.

We have recently applied the series of new 3D experiments developed by Bax and co-workers [8] to a sample of flavodoxin, 99% labeled with ^{13}C and ^{15}N in order to complete the *Anabaena* 7120 flavodoxin assignments (A.S. Edison, R. Chylla, R. Reedstrom, W.M. Westler, E.S. Mooberry and J.L. Markley, unpublished). This new isotope-assisted technology should facilitate future experiments in which we plan to

Table 1   *Labeling strategies for stable-isotope-assisted multinuclear NMR spectroscopy*

Labeling pattern	Isotope	Percent (optimal)	Application [reference]
protein, uniform			
	^{2}H	>98	total deuteration [5]
		50–75	random fractional deuteration [5]
	^{13}C	>98	^{1}H-^{13}C correlations and NOE editing [8]; ^{13}C-^{13}C TOCSY correlations [41,42]
		25	^{1}H-^{13}C coupling and NOE editing [24]; ^{13}C-^{13}C DQ correlations [22,23]
	^{15}N	>98	^{1}H-^{15}N correlations and NOE editing
	^{13}C+^{15}N	>98/>98	^{1}H-^{13}C-^{15}N correlations and NOE editing [40]
		25/>98	^{13}C-^{15}N coupling and ^{13}C-^{15}N correlations [38,39]
amino acid, uniform			
	^{2}H	>98	selective deuteration [5,9]
	^{13}C	>98	spectral simplification [22]
		25	spectral simplification [24,30,31,36]
	^{15}N	>98	spectral simplification [24]
amino acid, non-uniform			
single atom	^{13}C	>98	spectral simplification; 2D exchange spectroscopy [45]
multiple atom (block)	$(^{13}$C$)_n$	>98/>98	J_{CC} coupling
	^{13}C/^{15}N	>98/>98	J_{CN} coupling
chiral	^{2}H	variable	prochiral assignments [5]
	^{13}C	variable	prochiral assignments [33]
nearest neighbor analysis			
$[^{13}$C′]αα$_1$ + $[^{15}$N]aa$_2$	^{13}C/^{15}N	>98/>98	sequential assignment ^{13}C′$_i$-^{15}N$_{i+1}$ coupling
$[^{13}$C$^α]$aa$_1$ + $[^{13}$C$^α]$aa$_2$	^{13}C/^{13}C	>98/>98	analysis of peptide bond configuration (*cis/trans*) ^{13}C^α-^{1}H$_i$-^{13}C^α-^{1}H$_{i+1}$ double isotope edited NOE

use site-directed mutagenesis to test hypotheses concerning the roles of individual residues in controlling the redox properties of the flavin and in interacting with its redox partner, the flavodoxin oxidoreductase.

Conclusion

Protein NMR spectroscopy is undergoing a rapid growth spurt as the result of applying new multidimensional, multinuclear NMR techniques to stable isotope-labeled proteins. Fine tuning of the methodology will certainly follow. What are optimal data collection strategies for structural studies? What can be done to make the analysis of multidimensional data sets more efficient? Beyond that, have we now exhausted the range of multinuclear correlations that can be profitably mined? Is the scope of protein NMR well defined by existing experiments? If the history of protein NMR is any guide, we should expect more surprises ahead.

Acknowledgements

I thank my co-workers, whose names are given in the text and references, who collaborated in the studies reviewed here.

References

1. Wüthrich, K. (1986) *NMR of Proteins and Nucleic Acids*, Wiley, New York.
2. Markley, J.L., Westler, W.M., Chan, T.-M., Kojiro, C.L. and Ulrich, E.L. (1984) *Fed. Proc.* **43**, 2648.
3. Stockman, B.J. and Markley, J.L. (1990) *Advances in Biophysical Chemistry*, Vol. 1, (Bush, C.A., Ed.) JAI Press Inc., Greenwich, Connecticut, p. 1.
4. McIntosh, L.P. and Dahlquist, F.W. (1990) *Quart. Rev. Biophys.* **23**, 1.
5. LeMaster, D.R. (1989) *Annu. Rev. Biophys. Chem.* **19**, 243.
6. Markley, J.L. (1989) *Methods Enzymol.* **176**, 12.
7. Fesik, S.W. and Zuiderweg, E.R.P. (1990) *Quart. Rev. Biophys.* **23**, 97.
8. Kay, L.E., Ikura, M. and Bax, A. (1991) *J. Magn. Reson.* **91**, 84; and references therein.
9. Markley, J.L., Putter, I. and Jardetzky, O. (1968) *Science* **161**, 1249.
10. Browne, D.T., Kenyon, G.L., Packer, E.O., Sternlicht, H. and Wilson, D.M. (1973) *J. Am. Chem. Soc.* **95**, 1316.
11. Bachovchin, W.W. and Roberts, J.D. (1978) *J. Am. Chem. Soc.* **100**, 8041.
12. Kainosho, M. and Tsuji, T. (1982) *Biochemistry* **24**, 6273.
13. Chan, T.-M. and Markley, J.L. (1982) *J. Am. Chem. Soc.* **104**, 4010.
14. Bax. A., Griffey, R.H. and Hawkins, B.L. (1983) *J. Magn. Reson.* **55**, 301.
15. Griffey, R.H. and Redfield, A.G. (1987) *Quart. Rev. Biophys.* **19**, 51.
16. Shon, K. and Opella, S.J. (1989) *J. Magn. Reson.* **82**, 193.
17. Oh, B.-H., Westler, W.M. and Markley, J.L. (1989) *J. Am. Chem. Soc.* **111**, 3083.
18. Wagner, G. and Brüwiler, D. (1986) *Biochemistry* **25**, 5839.
19. Ortiz-Polo, G., Krishnamoorthi, R., Markley, J.L., Live, D.H., Davis, D.G. and Cowburn, D. (1986) *J. Magn. Res.* **68**, 303.
20. Stockman, B.J., Krezel, A.M., Markley, J.L., Leonhardt, K.G. and Straus, N.A. (1990) *Biochemistry* **29**, 9600.
21. Edison, A.S., Westler, W.M. and Markley, J.L. (1991) *J. Magn. Reson.* **91**, in press.
22. Westler, W.M., Kainosho, M., Nagao, H., Tomonaga, N. and Markley, J.L. (1988) *J. Am. Chem. Soc.* **110**, 4093.
23. Oh, B.-H., Westler, W.M., Darba, P. and Markley, J.L. (1988) *Science* **240**, 908.
24. Wang, J., Hinck, A.P., Loh, S.N. and Markley, J.L. (1990) *Biochemistry* **29**, 102.
25. Zuiderweg, E.R.P. and Fesik, S.W. (1989) *Biochemistry* **28**, 2387.
26. Stockman, B.J., Westler, W.M., Darba, P. and Markley, J.L. (1988) *J. Am. Chem. Soc.* **110**, 4096.
27. Stockman, B.J., Reily, M.D., Westler, W.M., Ulrich, E.L. and Markley, J.L. (1989) *Biochemistry* **28**, 230.
28. Oh, B.-H. and Markley, J.L. (1990) *Biochemistry* **29**, 3993.
29. Oh, B.-H., Mooberry, E.S. and Markley, J.L. (1990) *Biochemistry* **29**, 4004.
30. Wang, J., Hinck, A.P., Loh, S.N. and Markley, J.L. (1990) *Biochemistry* **29**, 4242.
31. Hinck, A.P., Loh, S.N., Wang, J. and Markley, J.L. (1990) *J. Am. Chem. Soc.* **112**, 9031.
32. Kato, K., Matsunaga, C., Nishimura, Y., Waelchi, M., Kainosho, M. and Arata, Y. (1989) *J. Biochem. (Tokyo)* **105**, 867.
33. Neri, D., Szyperski, T., Otting, G., Senn, H. and Wüthrich, K. (1989) *Biochemistry* **28**, 7510.
34. Kojiro, C.L. (1985) Ph.D. thesis, Purdue University.
35. Oh, B.-H. (1989) Ph.D. thesis, University of Wisconsin-Madison.

36. Grissom, C.B. and Markley, J.L. (1989) *Biochemistry* **28**, 2116.
37. LeMaster, D.M. and Richards, F.M. (1982) *Anal. Biochem.* **122**, 238.
38. Westler, W.M., Stockman, B.J., Hosoya, Y., Miyake, Y. and Kainosho, M. (1988) *J. Am. Chem. Soc.* **110**, 6265.
39. Mooberry, E.S., Oh, B.-H. and Markley, J.L. (1989) *J. Magn. Reson.* **85**, 147.
40. Kay, L.E., Clore, G.M., Bax, A. and Gronenborn, A.M. (1990) *Science* **249**, 411.
41. Fesik, S.W., Eaton, H.L., Olejniczak, E.T., Zuiderweg, E.R.P., McIntosh, L.P. and Dahlquist, F.W. (1990) *J. Am. Chem. Soc.* **112**, 886.
42. Eaton, H.L., Fesik, S.W., Glaser, S.J. and Drobny, G. (1990) *J. Magn. Reson.* **90**, 452.
43. Wang, J., LeMaster, D.M. and Markley, J.L. (1990) *Biochemistry* **29**, 88.
44. Muchmore, D.C., McIntosh, L.P., Russel, C.B., Anderson, D.E. and Dahlquist, F.W. (1989) *Methods Enzymol.* **177**, 44.
45. Alexandrescu, A.T., Loh, S.N. and Markley, J.L. (1990) *J. Magn. Reson.* **87**, 523.
46. Alexandrescu, A.T., Hinck, A.P. and Markley, J.L. (1990) *Biochemistry* **29**, 4516.
47. McCain, D.C., Ulrich, E.L. and Markley, J.L. (1988) *J. Magn. Reson.* **80**, 296.
48. Lipari, G. and Szabo, A. (1982) *J. Am. Chem. Soc.* **104**, 4559.
49. Zehfus, M.H., Reily, M.D., Ulrich, E.L., Westler, W.M. and Markley, J.L. (1990) *Arch. Biochem. Biophys.* **276**, 369.
50. Ulrich, E.L., Krogmann, D.W. and Markley, J.L. (1982) *J. Biol. Chem.* **257**, 9356.
51. Oh, B.-H. and Markley, J.L. (1990) *Biochemistry* **29**, 4012.
52. Dugad, L.B., LaMar, G.N., Banci, L. and Bertini, I. (1990) *Biochemistry* **29**, 2263.
53. Rypneiwski, W.R., Breiter, D.R., Wesenberg, G., Oh, B.-H., Markley, J.L., Rayment, I. and Holden, H.M. (1991) *Biochemistry*, in press.
54. Tsukihara, T., Fukuyama, K., Nakamura, M., Katsube, Y., Tanaka, N., Kakudo, M., Wada, K., Hase, T. and Matsubara, H. (1981) *J. Biochem. (Tokyo)* **90**, 1763.
55. Tsukihara, T., Fukuyama, K., Mizushima, M., Harioka, T., Kusunoki, M., Katsube, Y., Hase, T. and Matsubara, H. (1990) *J. Mol. Biol.* **216**, 399.
56. Smith, W.W., Pattridge, K.A., Ludwig, M.L., Petsko, G.A., Tsernoglou, D., Tanaka, M. and Yasunobu, K.T. (1983) *J. Mol. Biol.* **165**, 737.
57. Moonen, C.T.W., Vervoort, J. and Müller, F. (1984) *Biochemistry* **23**, 4859.
58. Vervoort, J., Müller, F., Mayhew, S.G., van den Berg, W.A.M., Moonen, C.T.W. and Bacher, A. (1986) *Biochemistry* **25**, 6789.
59. Stockman, B.J., Westler, W.M., Mooberry, E.S. and Markley, J.L. (1988) *Biochemistry* **27**, 136.

NMR studies of two related phosphotransfer proteins

Michael G. Wittekind and Rachel E. Klevit

Department of Biochemistry, University of Washington, Seattle, WA 98195, U.S.A.

Introduction

HPr is a small, soluble protein that functions as one of the phosphocarrier proteins in the phosphoenolpyruvate : sugar phosphotransfer system (PTS). This system is responsible for the simultaneous uptake and phosphorylation of numerous sugars in both Gram-positive and Gram-negative bacteria (for a recent review, see ref. 1). HPr serves as a phosphoryl acceptor from Enzyme I (E_I) and as a phosphoryl donor to a factor III^{sugar} or enzyme II^{sugar}. During the phosphotransfer reaction, HPr exists as a phosphoprotein intermediate, phosphorylated on the imidazole N^σ of a conserved histidine, His[15].

We have used high-resolution two-dimensional NMR spectroscopy to study the solution structures for HPrs from both Gram-negative (*E. coli*) and Gram-positive (*B. subtilis*) bacteria. *E. coli* HPr (*ec*HPr) and *B. subtitis* HPr (*bs*HPr) have 29 amino acid residues in common (34% sequence identity) and another 18 residues that share structural/functional side chains (55% sequence similarity). The longest continuous stretch of similarity is from residue 13 through 20 and includes His[15], the site of phosphorylation (see Fig. 1). Although we expected that these homologous proteins

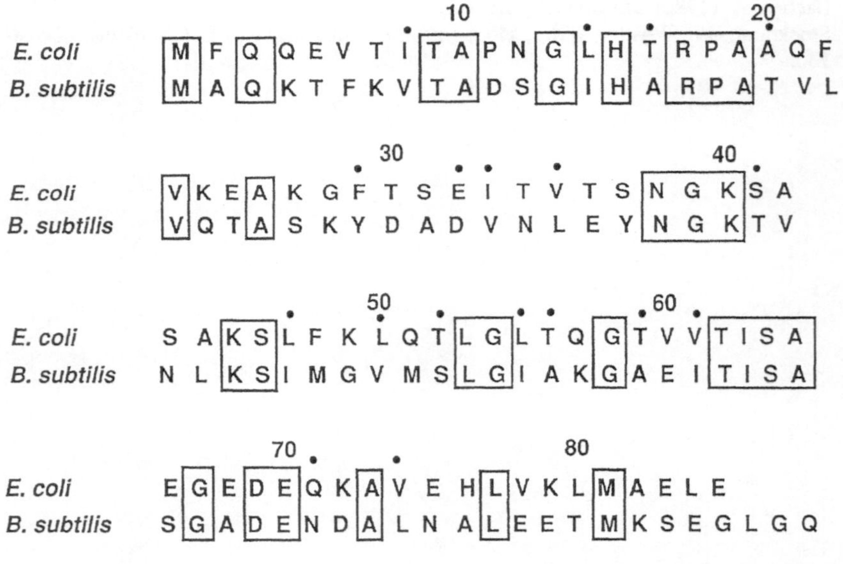

Fig. 1. Primary structure of HPr from E. coli *and* B. subtilis. *The boxes enclose identical residues and the dots identify similar residues.*

will have similar tertiary structures, there are reported differences in the properties of Gram-positive and Gram-negative HPrs that make a detailed comparison of their structures of interest. For example, *bs*HPr can be phosphorylated by an ATP-dependent kinase on Ser[46] [2]. This regulatory phosphorylation has an inhibitory effect on HPr function and has been proposed to give rise to a hierarchy of utilization of the PTS sugars [3]. However, although residue 46 is also a serine in *ec*HPr, this additional regulatory phosphorylation does not occur in the Gram-negative protein. In fact, it cannot even be phosphorylated in vitro using the kinase purified from *B. subtilis* [2].

Results and Discussion

Our first 2D NMR studies were on the *E. coli* protein [4-6]. On the basis of the available homonuclear ^{1}H experiments (COSY, RELAY, and NOESY), complete sequential assignments were obtained for amide, $C^\alpha H$, and $C^\beta H$ resonances as well as for many side-chain protons. The patterns of sequential connectivities were used to identify regions of regular secondary structure, as shown in Fig. 2. *ec*HPr has four β-strands, named A-D from the N-terminus, and three α-helices (A-C). About 66% of the 85 residues of *ec*HPr exist in these regular secondary structural elements, with the remaining 34% residues, including the active site His[15], being in turns or irregular or extended structures.

NOEs observed among protons between two β-strands were used to define the topology of the β-sheet. Some 20 inter-strand main-chain NOEs were observed and these were consistent with only one configuration: a 4-stranded antiparallel sheet with a topology of A-D-B-C. Because of the β-α-β supersecondary structure of the protein (see Fig. 2), this sheet topology places the α-helices in an approximately parallel orientation relative to the sheet and anti-parallel to each other. NOEs observed among side-chain protons on one side of the β-sheet (the hydrophobic side) and the helices defined the direction of the crossovers for the helices over the β-sheet: in each case, the α-helix forms a right-handed crossover.

Shortly after the model for the solution structure of *ec*HPr was published, the structure determined by X-ray diffraction was reported [7]. Although the placement of most of the secondary structure along the primary structure agrees with the NMR results, the overall topology of the tertiary fold is quite different. In particular, the X-ray structure has two 2-stranded antiparallel β-sheets; one consisting of strands A and D and the other consisting of strands B and C. There appears to be no interaction between these two β-sheets. In addition, the orientations of the strands in the sheet are different. Whereas in the NMR structure, strands A and B are parallel to each other, they are anti-parallel in the X-ray structure; the same holds for strands C and D. This difference has two important ramifications to the overall fold. First, the α-helices are not parallel to the sheet, but rather more perpendicular to it. Second, while the 4-stranded sheet is very amphipathic in the NMR structure, with the hydrophobic side covered by the α-helices and the hydrophilic side exposed to solvent, the flipped orientation of the two sheets in the X-ray structure causes the hydrophobic face of the B-C sheet to be exposed to solvent.

41

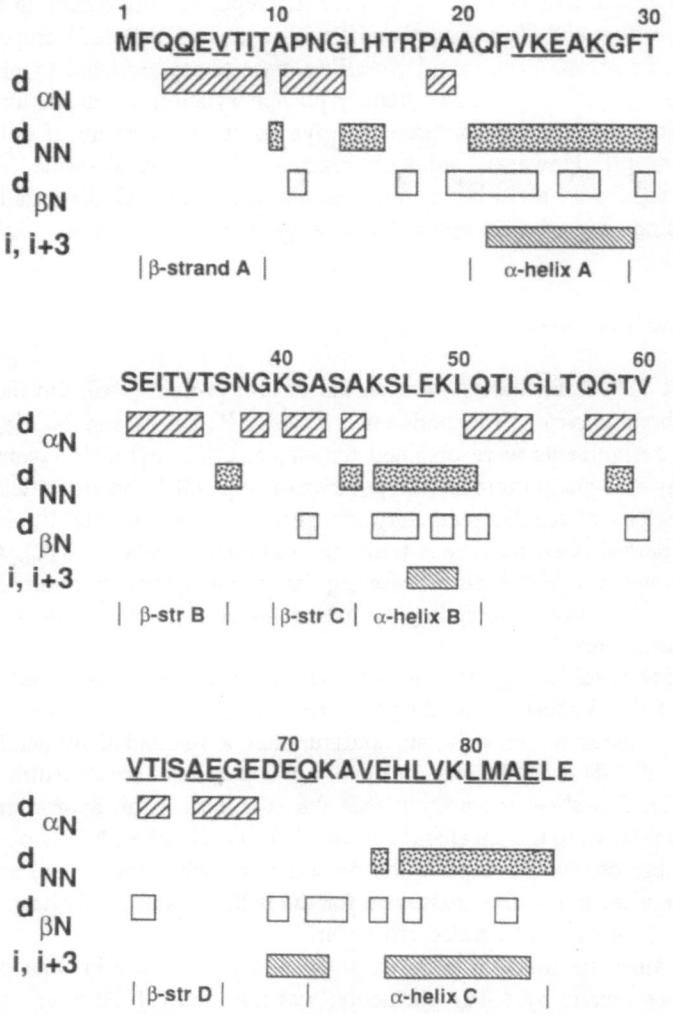

Fig. 2a. Sequential connectivity patterns for E. coli *HPr. Short- and medium-range NOEs observed are summarized. The underlined residues have persistent NH resonances.*

Recently, we have completed the assignments of the spectrum of *bs*HPr [8]. At the level of sequence similarity between *bs*HPr and *ec*HPr, little spectral overlap was expected, and the *bs*HPr spectrum had to be assigned independently from that of *ec*HPr. Once the assignments were completed, however, several patterns emerged. In the fingerprint regions (containing the cross peaks that correlate each NH chemical shift with its corresponding $C^\alpha H$ chemical shifts) the pattern and identities of the 'outliers' are similar in the two spectra. In each case there are two amide resonances that are significantly shifted downfield from the rest. The most downfield amide proton is from residue 30 in both spectra (even though they possess different side chains in the two proteins) and the second most downfield amide proton is that of Gly[67]. A

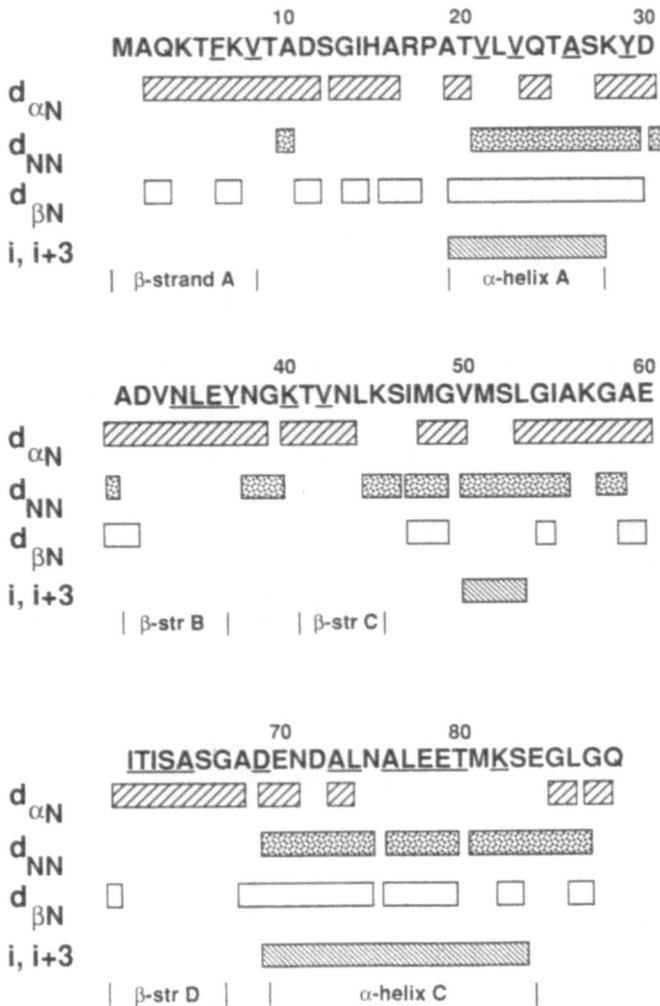

Fig. 2b.   *Sequential connectivity patterns for* B. subtilis *HPr. Short- and medium-range NOEs observed are summarized. The underlined residues have persistent NH resonances.*

similarity also exists for the $C^{\alpha}H$ resonances that are shifted downfield of the solvent signal (~4.8 ppm). Although there is still no one-to-one correspondence of cross peaks in this region for the two spectra, the residue positions with downfield $C^{\alpha}H$'s are related in the two proteins. As has been noted by many workers, these residues are all found in regions of β-strand structure.

As shown in Fig. 2, the pattern of sequential connectivities observed for *bs*HPr is quite similar to that of *ec*HPr. In particular the long stretches of $d_{\alpha N}$ connectivities exhibited by the four β-strands in *ec*HPr are placed identically in *bs*HPr. The d_{NN}, $d_{\beta N}$, and medium-range connectivities indicative of α-helix A are observed. A short segment of helical connectivities was observed between residues 47-52 in *ec*HPr, with a single medium-range interaction between Phe^{48} and Gln^{51}, leading us to suggest the existence of a short α-helix B. d_{NN} connectivities extend from residue 47 to 56 in *bs*HPr, with a single medium-range connectivity between Val^{50} and Leu^{53}. To date, some of the structures that have been calculated for *bs*HPr using a distance geometry algorithm have a short α-helix in this region. Additional information such as coupling constants will help to better define the secondary structure for this region of the molecules. Finally, α-helical connectivities were observed for residues 70–84 in *ec*HPr (helix C) and in *bs*HPr. In fact, d_{NN} connectivities extend all the way to residue Gly^{88} in *bs*HPr (the penultimate residue), but the last $i \rightarrow i+3$ connectivity observed is between Thr^{80}-Ser^{83}. Thus, as expected on the basis of homology, the major structural elements of the two related proteins are indeed similar. Interestingly, the region where the connectivity patterns are the most dissimilar is around the active site, His^{15}, where d_{NN} connectivities indicative of a turn structure were observed for *ec*HPr while $d_{\alpha N}$ connectivities indicative of an extended structure were observed for *bs*HPr. Due to the irregular nature of the backbone structure in this region, a more thorough description must await the structures determined by distance geometry.

In conclusion, the folding topologies of the two HPrs as determined from a qualitative analysis of the 2D NMR spectra are similar to each other and consist of a four-stranded antiparallel β-sheet with two (or three?) α-helices lying on one side of the sheet. This topology is significantly different from that obtained from X-ray diffraction studies of the *ec*HPr. It is worth noting that the conditions under which the crystals were grown were quite different from those used for the NMR studies: 68% sat. lithium sulfate, 0.1 M citric acid, 0.2 M sodium phosphate, pH 3.7 for the crystallization and 50 mM potassium phosphate, pH 6.5 for the NMR. Thus, it is an intriguing possibility that the conditions have dramatically altered the structure. However, speculation regarding this possibility must await further refinement of both structures.

Acknowledgements

This work was supported by NIH NIDDK R01 DK35187 and by an American Heart Association Established Investigatorship (R.E.K.) and a Damon Runyon-Walter Winchell Cancer Research Fund Fellowship DRG-1007 (M.G.W.)

References

1. Meadow, N.D., Fox, D.K. and Roseman, S. (1990) *Annu. Rev. Biochem.* **59**, 497.
2. Reizer, J., Novotny, M.J., Hengstenberg, W. and Saier, M.H. (1984) *J. Bacteriol.* **160**, 333.
3. Deutscher, J., Kessler, V., Alpert, C.A. and Hengstenberg, W. (1984) *Biochemistry* **23**, 4455.
4. Klevit, R.E., Drobny, G. and Waygood, E.B. (1986) *Biochemistry* **25**, 7760.
5. Klevit, R.E. and Drobny, G. (1986) *Biochemistry* **25**, 7770.
6. Klevit, R.E. and Waygood, E.B. (1986) *Biochemistry* **25**, 7774.
7. El-Kabbani, O.A.L., Waygood, E.B. and Delbaere, L.T.J. (1987) *J. Biol. Chem.* **262**, 12926.
8. Wittekind, M., Reizer, J. and Klevit, R.E. (1990) *Biochemistry* **29**, 7191.

Motional properties of acyl carrier protein: Effects on NMR structural data

James H. Prestegard* and Yangmee Kim

Chemistry Department, Yale University, New Haven, CT 06511, U.S.A.

Introduction

Recently advances in NMR technology and computational methods have allowed the determination of three-dimensional structures of small proteins in solution [1–3]. A number of examples of successful application now exist including ones showing excellent agreement of X-ray and NMR structures [4,5]. Most of these determinations assume a rigid model for the protein in order to convert measured intensities from two-dimensional NOESY experiments to interproton distance constraints, and hence a three-dimensional structure. Yet, we know that proteins are dynamic structures that must execute motions over a wide range of timescales in order to function. The explanation for the success of a rigid model in many applications, in part, lies with the insensitivity of proton-proton NOEs to the very rapid, small-amplitude, motions executed by most groups in a protein [6]. However, there are cases where slower motions of one or more groups can lead to difficulties in determination of high-quality solution structures.

We believe such a case arises in the structure determination of acyl carrier proteins (ACP) [7] from both *Escherichia coli* and spinach. These proteins serve to carry acyl chains during synthesis by the fatty acid synthetase system. As such, they must interact with a variety of enzymatic sites. This, along with the fact that the proteins have no interpeptide disulfide bonds and are highly negatively charged, make them prime candidates for dynamic interconversion of conformational forms.

The effects of motion on NOESY data for ACP from *E. coli* are subtle. Motions which exist are rapid enough to produce just a single set of averaged NMR resonances. The only evidence for conformational averaging remains some apparent incompatibilities in distance constraints derived on the basis of a rigid model. Given the tenuous nature of these data, we had turned to ACP from another species, ACP I from spinach, in search of more direct confirmation of dynamic averaging in these functionally and structurally related proteins [8,9]. The protein from spinach, indeed, shows evidence of multiple forms in slow exchange. The reduced protein, which has no fatty acid attached to its phosphopantetheine prosthetic group, shows a second set of minor resonances in the aromatic region of the spectrum, which were demonstrated via NOESY and ROESY spectra to exchange with the major resonances on a timescale of 1 s. The acylated protein, which has a palmitic acid chain attached, shows duplicate sets of resonances in the aromatic region with a nearly 50 : 50

---

* To whom correspondence should be addressed.

intensity ratio. The relationship of these equally populated sets of resonances to the major and minor resonances seen in the reduced protein from spinach, and the relationship of these resolved resonances to the averaging suggested to occur in the *E. coli* protein is, however, not well-established. Regardless of the relationship of resonances seen, these cases illustrate the variety of motional interconversions that can occur in protein systems. Evidence delineating the various cases is presented below.

Results and Discussion

First let us review the existing ROESY data on reduced spinach ACP (ACP-SH). Figure 1 shows an expansion of the aromatic region of a 500 MHz 2D ROESY experiment on a 7 mM ACP-SH sample at pH 5.9, 303 K. A section of the corresponding one-dimensional experiment is shown at the bottom. For the ROESY experiment, a spin-lock field of approximately 0.5 gauss was applied 1000 Hz upfield of peak D for 160 ms. Under these conditions coherent transfers are minimal. NOE peaks appear as negative peaks in the contour map and exchange peaks appear as positive peaks in the map. Negative and positive peaks are plotted to the upper left and lower right of the diagonal respectively. Cross peaks in the lower-right part of the figure clearly show that peak D, assigned to the *para* proton of phenylalanine 31, exchanges with a partially resolved minor resonance (about 15%) at 7.26 ppm. Peak C2, assigned to the *meta* proton of Phe[31], and with a peak under B2 at 7.22 ppm, and peak C1, assigned to the *ortho* protons of Phe[31], exchanges with peak B1 at 7.31 ppm. Evidence for exchange of peaks assigned to Phe[52] is less clear because of interference from t_1 noise and noise near the diagonal, but the evidence for exchange of at least one of the phenylalanines between two distinct environments is unequivocal.

There are some unusual aspects to the demonstrated exchange. The shift in peak positions (as much as 0.6 ppm), is large for a small conformation change. Also, the clustering of *ortho, meta* and *para* resonances for the minor conformer near 7.3 ppm is reminiscent of an unstructured peptide. We, therefore, believe the minor resonances represent a small portion of denatured protein which exists under the conditions of pH and temperatures chosen. This is supported by the fact that the percentage of the minor form decreases as divalent metals are added to stabilize the highly negatively charged protein, and by the fact that raising the temperature increases the intensity of the minor resonances (from ~ 15 to ~ 30% with a 10° rise in temperature).

Since spinach ACP is believed to be less thermodynamically stable than *E. coli* ACP, it is not clear that the relatively slow exchange between structured and unstructured forms in spinach ACP is directly related to the more rapid exchange postulated to exist in *E. coli* ACP. It is also unlikely that these minor resonances relate to the well-resolved, nearly equally populated, pairs of resonances seen for palmitoyl-ACP from spinach [8].

The existing data on ACP from spinach, acylated with palmitic acid, are reviewed in Fig. 2. The figure shows a series of one-dimensional spectra showing the aromatic

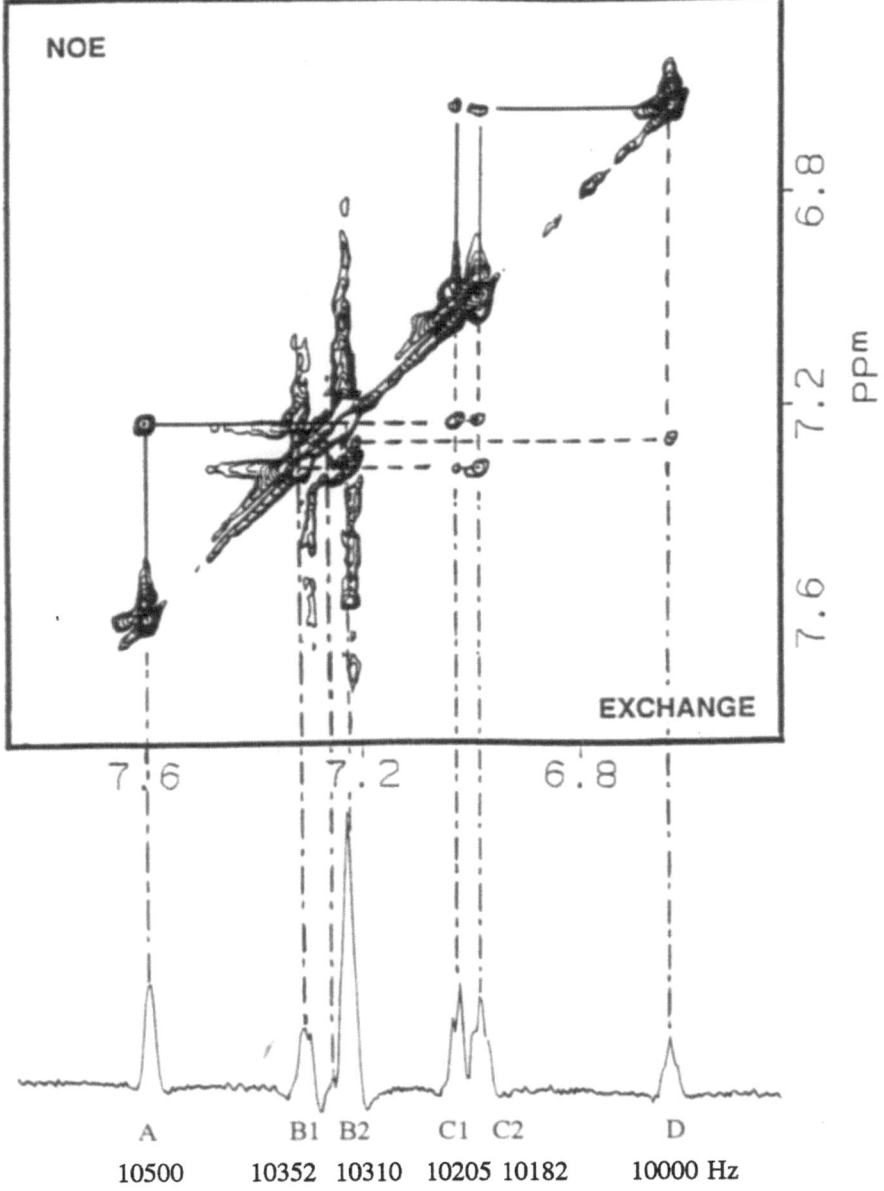

Fig. 1. 500 MHz ROESY spectrum of the aromatic region of spinach ACP in D_2O, pH 5.9, 303 K. 680 t_1 points were acquired with 96 scans per point. Reproduced from ref. 9.

region of palmitoyl ACP under conditions which differ in temperature, pH, and divalent ion content. Both peaks D and A are now split into pairs of lines with approximately equal intensity. These pairs are separated by less than two tenths of a ppm and cannot be shifted in relative intensity by more than a few percent with the addition of divalent ions, alteration in temperature, or variation of pH. The pairs

47

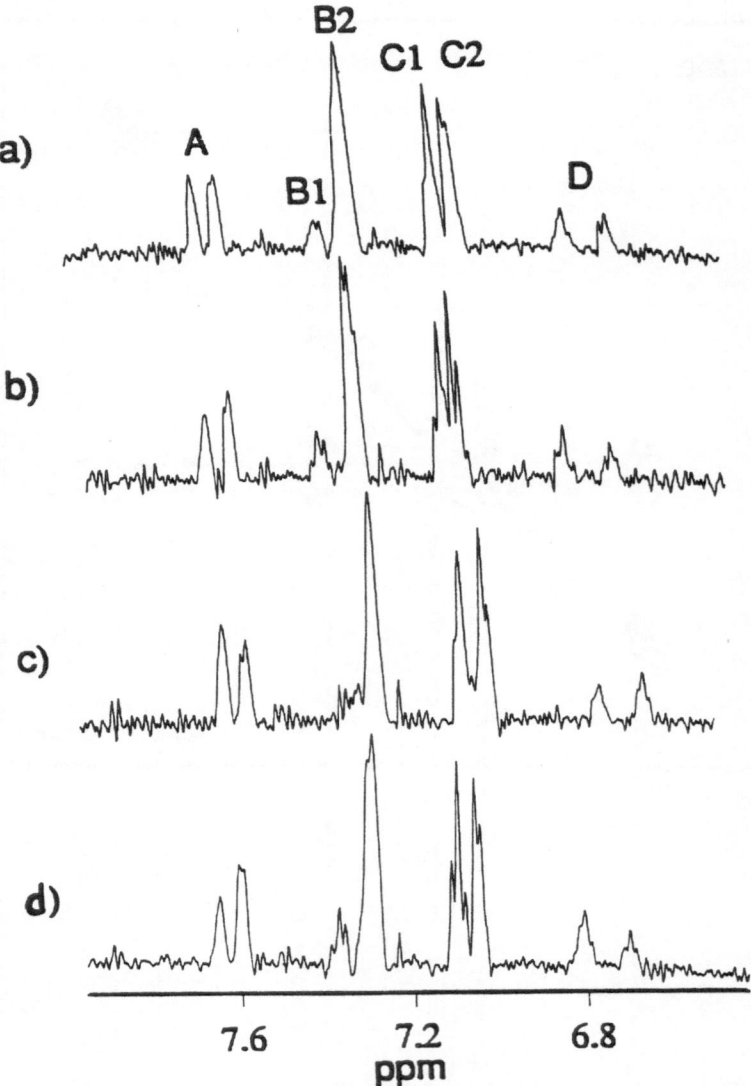

Fig. 2.   500 MHz spectrum of the aromatic region of 1 mM palmitoyl-spinach ACP in D_2O;
a) pH 6.7, 303 K; b) pH 6.7, 313 K; c) pH 6.7, 303 K, 2 mM Mg^{2+}; d) pH 7.8, 303 K.
Reproduced from ref. 8.

of resonances in acylated spinach ACP are therefore more likely to correspond to discrete structural forms than structured and unstructured species.

Does the above observation mean that there is only a single structurally stable form of spinach ACP-SH under conditions where the minor peaks in the aromatic region of the spectrum are absent? Not necessarily; it is possible that exchange could be rapid enough to lead to averaged chemical shifts as postulated for the *E. coli* species, or it is possible that differences in chemical shifts for the phenylalanines in the two conformers are significant in palmitoyl ACP but insignificant in ACP-SH.

We believe the latter to be the case.

There are actually two pieces of evidence for the existence of a slow conformational equilibrium in spinach ACP-SH, even in cases where the minor resonances apparent in the aromatic region of the spectrum are suppressed. Both occur in a very crowded region of the spectrum and are only apparent when multipulse NMR methods are used for spectral simplification. The first piece of evidence is presented in Fig. 3. This is a one-dimensional heteronuclear double-quantum-filtered experiment in which protons coupled to phosphorus are selectively detected [10]. Mg^{2+} has been added to the sample to fully suppress the minor resonances in the aromatic region. ACP has one phosphorus site, a phosphodiester which connects Ser^{38} to the prosthetic group responsible for carrying an acyl chain. There are two pairs of methylene protons significant phosphorus couplings should occur to: the Ser^{38} β-methylenes and the γ-methylenes on the pantoic acid moiety of the prosphetic group. An ABX multiplet is clearly visible for Ser^{38} at 4.18 ppm, but two sets of ABX multiplets are observed for the pantoic acid methylenes, one at 3.76 ppm and one at 3.43 ppm. The samples are homogeneous by HPLC and are fully reduced. Therefore, the most likely explanation for the duplicate sets of peaks is that two

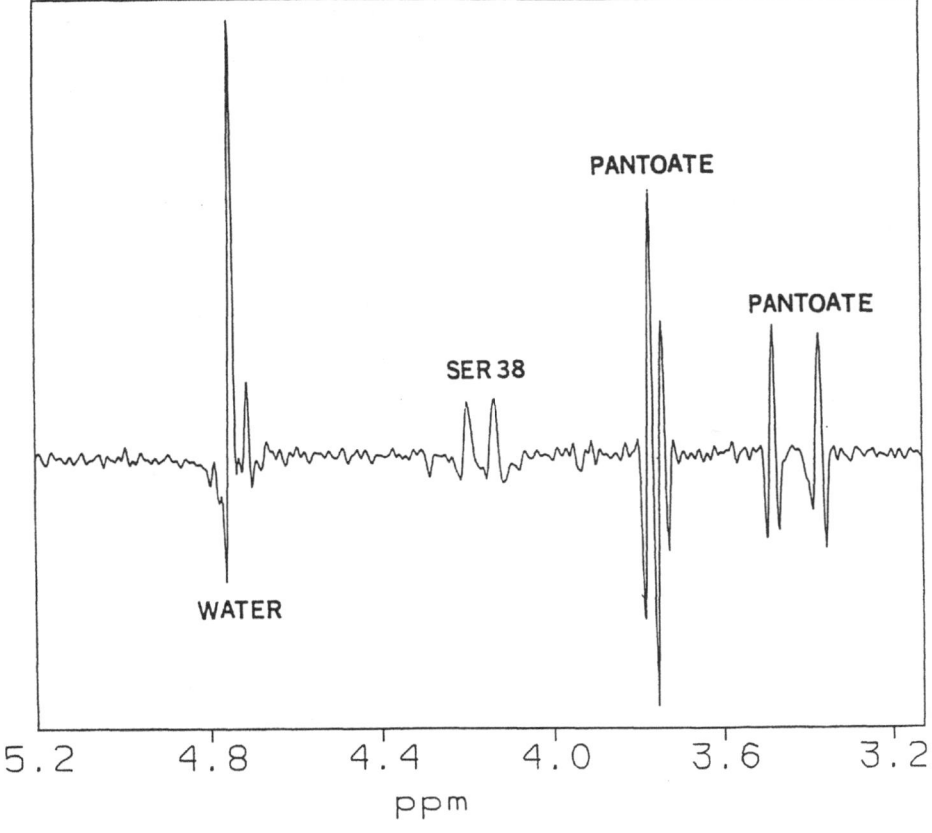

Fig. 3. ^{31}P-^{1}H double-quantum-filtered 500 MHz ^{1}H spectrum of spinach ACP (7 mM) in D_2O, pH 6.0, 303 K, 14 mM Mg^{2+}.

conformational states are indeed present and are reflected in the chemical shifts of prosthetic group protons, but not in chemical shifts of the phenylalanine resonances in the aromatic region of the spectrum.

A second piece of evidence for two states occurs in RELAY spectra of spinach ACP-SH [11]. Figure 4 shows a segment of a region containing NH-α and NH-β connectivities for alanine. Alanine is fairly unique among amino acids in that a far upfield methyl resonance (β) can be connected to an amide proton (NH), by just two through-bond transfers of magnetization. The pairs of resonances connected by vertical lines can therefore be assigned with some confidence to alanines. Moreover, the α-β connectivities can be confirmed in COSY spectra reducing the possibility of accidental degeneracy of shifts on the horizontal axis. Ten such alanine connectivities are labeled in the figure with the clear possibility of establishing one or more additional connectivities. There are only nine alanines in the spinach ACP sequence suggesting that some sites must give rise to more than one set of resonances. It is possible that these extra peaks arise from alanines close to the prosthetic group attachment site, Ala[36] or Ala[27], for example. Localizing a conformational equilibrium to the site of the prosthetic-group attachment in spinach ACP-SH may provide a rationale for the effects seen over a broader piece of the protein when it is acylated. The acyl chain is believed to lie in a hydrophobic pocket between two helices. In the *E. coli* structure, both phenylalanines lie in this pocket, and thus both may experience changes in chemical shifts as the acyl chain alters its binding motif in the two conformational states.

Thus, proteins such as ACP show motional properties that range from ones sufficiently rapid and random to be well-approximated by a rigid protein model, to ones that are slow enough to show discrete resonances for conformational forms. Some of these motions may include structured to unstructured transitions, but others may include subtle equilibria between conformational states of functional significance. Sometimes these conformational equilibria are readily apparent because of well-resolved discrete resonances for each conformer. But, interconversion rates can also be raised enough to produce a single set of average resonances.

Allowing for the possibility of averaged NOEs when multiple states are present is important for proteins such as ACP. A simple example can be used to illustrate the important effects that motion can have on NOE interpretation and structure determination of proteins. Suppose a proton, C, can exist in two equally probable sites located between a pair of protons, A and B, which are fixed by molecular geometry to be 9 Å apart. One site is 3 Å from A and 6 Å from B. The other site is 3 Å from B and 6 Å from A. If proton C moves from one site to the other on a timescale in the range of 10^{-7} to 10^{-3} s, only a single set of resonances will be observed, and the NOEs between A and C and B and C will be equal to approximately one half the intensity of a NOE for a rigid structure with a 3 Å interproton distance. If we use the usual $1/r^6$ distance dependence of a rigid model to interpret our observed NOEs, we will conclude that C is 3.3 Å from A *and* 3.3 Å from B. Since A and B are 9 Å apart, these are geometrically incompatible constraints, but we won't know this as we begin our structure determination.

Methods for structure determination which superimpose harmonic or squarewell

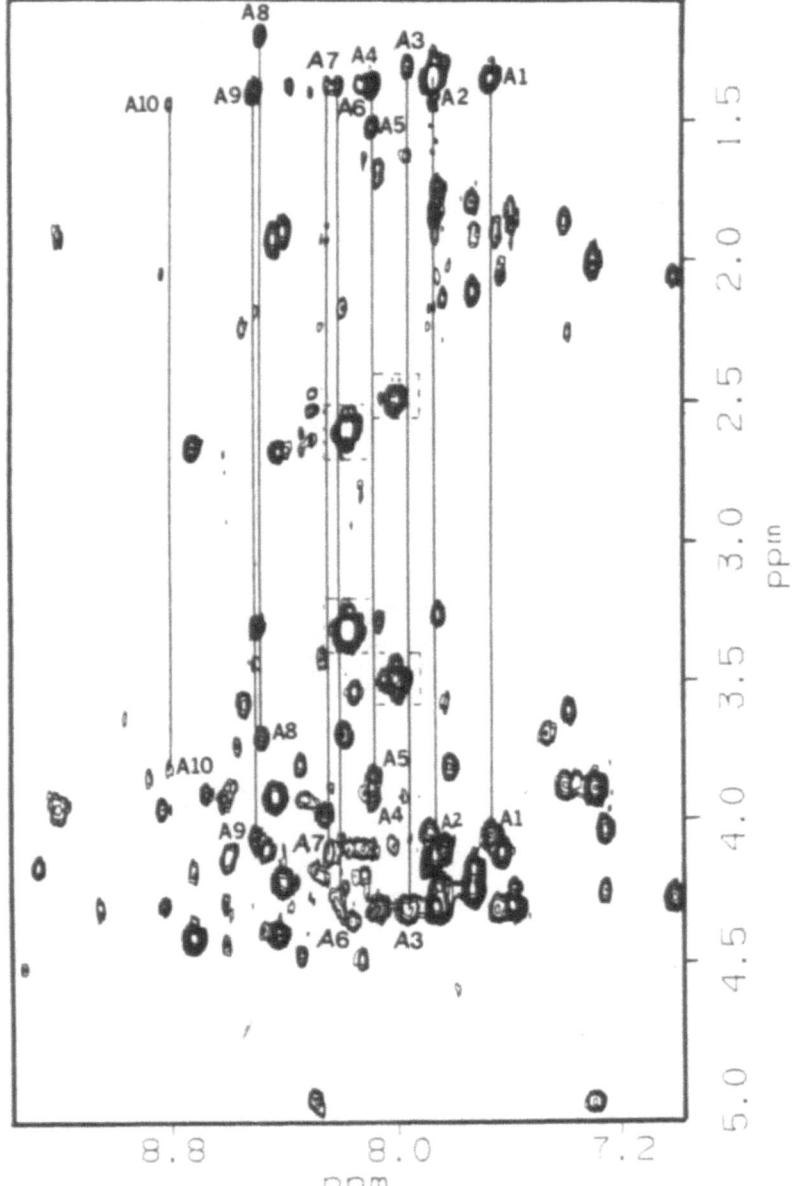

Fig. 4.   490 MHz RELAY spectrum of spinach ACP (7 mM) in H_2O, pH 6.0, 303 K, 14 mM Mg^{2+}, 350 t_1 points were acquired over 25 hours using a mixing time of 18 ms.

representations of constraints will tend to produce a single set of structures with compromise A-C and B-C distances of 4.5 Å. Only if structures produced in this way are carefully examined for unrealistic molecular energies, will the problem be apparent. An alternative distance constraint representation which uses a highly anharmonic function provides a means of getting divergent families of structures that

51

more directly suggest the presence of motional averaging. The structures produced in this way may also serve as good starting points for more elaborate structure determination protocols [7].

Acknowledgement

We are grateful to Dr. J. Ohlrogge for supplying the spinach ACP and to the NIH for its financial support through GM32243.

References

1. Wüthrich, K. (1989) *Acc. Chem. Res.* **22**, 36.
2. Markley, J.L. (1989) *Methods Enzymol.* **176**, 112.
3. Clore, G.M. and Gronenborn, A.M. (1989) *Crit. Rev. Biochem. Mol. Biol* **24**, 479.
4. Kline, A.D., Braun, W. and Wüthrich, K. (1988) *J. Mol. Biol.* **204**, 675.
5. Wagner, G., Braun, W., Havel, T.F., Schaumann, T., Go, W. and Wüthrich, K. (1987) *J. Mol. Biol.* **196**, 611.
6. LeMaster, D.M., Kay, L.E., Brünger, A.T. and Prestegard, J.H. (1988) *FEBS. Lett.* **236**, 71.
7. Kim, Yangmee and Prestegard, J.H. (1989) *Biochemistry* **28**, 8792.
8. Kim, Yangmee, Ohlrogge, J.B. and Prestegard, J.H. (1990) *Biochem. Pharm.* **40**, 7.
9. Kim, Yangmee and Prestegard, J.H. (1990) *J. Am. Chem. Soc.* **112**, 3707.
10. Griffey, R.H. and Redfield, A.G. (1987) *Quart. Rev. Biophys.* **19**, 51.
11. Bax, A. and Drobny, G. (1985) *J. Magn. Reson.* **61**, 306.

Analysis of backbone dynamics of interleukin-1β

G. Marius Clore and Angela M. Gronenborn

Laboratory of Chemical Physics, National Institute of Diabetes and Digestive and Kidney Diseases, National Institutes of Health, Bethesda, MD 20892, U.S.A.

[13]C and ^{15}N nuclear magnetic relaxation data provide a wealth of information on the nature of internal motions of macromolecules in solution. In general, the fast internal motions can be described by two model-independent quantities: a generalized order parameter S which provides a measure of the magnitude of the motion, and an effective correlation time τ_e [1,2]. This simple formalism has proved remarkably successful in accounting for relaxation data on small molecules and simple polymers, as well as for fragmentary data obtained from one-dimensional NMR measurements on peptides and proteins [3–5]. With the advent of new two-dimensional techniques for measuring heteronuclear relaxation [6–8] it has now become possible to obtain a detailed and comprehensive picture of these fast motions in proteins. We therefore undertook an in-depth analysis [8] of the backbone dynamics of IL-1β using inverse detected ^{1}H-^{15}N 2D NMR methods. [15]N T_1, T_2 and NOE data were obtained for 90% of the backbone amide groups (128 out of a total of 144). In the course of this study we noticed that the [15]N T_1, T_2 and NOE data for the backbone amide groups of certain residues cannot be accounted for by the simple two parameter model-free approach and require the introduction of two distinct internal motions, one with a time scale less than 100 ps, the other with an effective correlation time in the range 1–3 ns.

We were able to show that all measurable residues exhibit very fast motions on a time scale < 20–50 ps. In addition, 32 residues display a second motion of significant amplitude on a time scale of 0.5–4 ns which is less than an order of magnitude smaller than the overall rotational correlation time (8.3 ns), while another 42 residues are characterized by an additional motion on the 30 ns to 10 ms time scale which leads to [15]N T_2 exchange line broadening.

The T_1 and T_2 relaxation times and the NOE enhancement of an amide ^{15}N spin relaxed by dipolar coupling to a directly bonded proton and by chemical shift anisotropy are given by Abragham [9]:

$$1/T_1 = d^2\{J(\omega_H - \omega_N) + 3J(\omega_N) + 6(J(\omega_H + \omega_N)\} + c^2J(\omega_N) \tag{1}$$

$$1/T_2 = 0.5d^2\{4J(0) + J(\omega_H - \omega_N) + 3J(\omega_N) + 6J(\omega_H) + 6J(\omega_H + \omega_N)\}$$
$$1/T_2 = + c^2\{3J(\omega_N) + 4J(0)\}/6 \tag{2}$$

$$NOE = 1 + \{T_1(\gamma_H/\gamma_N)d^2[6J(\omega_H + \omega_N) - J(\omega_H - \omega_N)]\} \tag{3}$$

where $d^2 = 0.1\gamma_H^2\gamma_N^2h^2<1/r_{HN}^3>^2$ and $c^2 = (2/15)\gamma_N^2H_0^2(\sigma_\parallel - \sigma_\perp)$. γ_H and γ_N are the gyromagnetic ratios of ^{1}H and ^{15}N, respectively, r_{HN} is the NH bond length (1.02 Å from neutron diffraction), H_0 is the magnetic field strength, $\sigma_\parallel$ and $\sigma_\perp$ are the parallel and perpendicular components of the axially symmetric ^{15}N chemical shift tensor which differ by -160 ppm, and $J(\omega_i)$ is the spectral density function.

In the model-free formalism of Lipari and Szabo [1,2], the spectral density function for a molecule undergoing isotropic tumbling is given by

$$J(\omega_i) = S^2\tau_R/(1 + \omega_i^2\tau_R^2) + (1 - S^2)\tau/(1 + \omega_i^2\tau^2) \tag{4}$$

which corresponds to an internal correlation function of $C_I(t) = S^2 + (1 - S^2)e^{-t/\tau_e}$, where S is the generalized order parameter, τ_R the overall isotropic rotational correlation time of the molecule, and $\tau = \tau_R\tau_e/(\tau_R + \tau_e)$ where τ_e is a single effective correlation time describing the internal motions (Note that from the X-ray structure of Il-1β [10], the three principal components of the inertia tensor are calculated to be in the ratio of $1.00 : 0.77 : 0.93$, indicating that IL-1β is a globular, almost spherical, protein which should behave isotropically in solution).

For 53 of the 153 residues of IL-1β, τ_e is sufficiently small ($\ll 100$ ps) that the T_1 and T_2 data can be accounted for by the simplified spectral density function $J(\omega_i) = S^2\tau_R/(1 + \omega_i^2\tau_R^2)$ and the data for another 42 residues which had larger than average 15N T_1/T_2 ratios with concomitantly larger linewidths could be fitted using the simplified spectral density function in conjunction with an additional chemical exchange term. For those residues, however, where the relaxation data cannot be fitted with this simplified spectral density function, we find that Eq. (4) can account for the T_1 and T_2 data at several spectrometer frequencies, but fails to account for the NOE data. This behaviour was found for the remaining 32 residues. In particular, the calculated values for the NOE are either too small or negative, whereas the observed ones are positive. To illustrate this point, the data for several amino acids are listed in Table 1 and compared to the best fit calculated values using Eq. (4).

The solution to this discrepancy between theory and experiment is obtained in a simple model-free manner by expanding the internal correlation function to a two exponential decay given by $C_I(t) = S_f^2S_s^2 + (1 - S_f^2)e^{-t/\tau_f} + S_f^2(1 - S_s^2)e^{-t/\tau_s}$ [8,11]. This corresponds to a spectral density function of the form

$$J(\omega_i) = S_f^2S_s^2\tau_R/(1 + \omega_i^2\tau_R^2) + S_f^2(1 - S_s^2)\tau/(1 + \omega_i^2\tau^2) \tag{5}$$

where S_f^2 is the amplitude of the very fast motion whose effective correlation time τ_f can be neglected (i.e. it is $\ll 100$ ps), and S_s^2 is the amplitude of the slower motion with an effective correlation time $\tau_s = \tau_R\tau/(\tau_R - \tau)$ which is still faster than the overall correlation time τ_R. The best fit values of S_f^2, S_s^2 and τ_s together with the calculated values of the T_1, T_2 and NOE data are given in Table 1. It will be noted that the value of the overall order parameter, $S_f^2S_s^2$, is the same as that of the generalized order parameter S^2 obtained using the spectral density function [4].

The effect of using the spectral density function [5] compared to [4] is easily understood by noting that the calculated values of τ_e (0.2–0.3 ns) are an order of magnitude smaller than those of τ_s (1–3 ns). For $\tau_R \sim 8$ ns and $\omega_N = 2\pi \times 60.8$ MHz, the NOE reaches a minimum value at an internal correlation time of ~ 0.25 ns. Thus the

Table 1 Comparison of calculated and observed ^{15}N relaxation data for some backbone amide groups of Il-1β

Residue	$T_1(600)$ (ms)	$T_2(600)$ (ms)	NOE(600)	S^2	τ_e (ps)	S^2	S_f^2	S_s^2	τ_s (ns)
S17									
expt	653 ± 38	107 ± 8.5	0.78						
calc Eq. 4	643 ± 23	107 ± 5	0.28 ± 0.03	0.73 ± 0.04	574 ± 18				
calc Eq. 5	659 ± 7	107 ± 1	0.76 ± 0.02			0.72 ± 0.01	0.86 ± 0.01	0.84 ± 0.01	2.6 ± 0.8
G22									
expt	784 ± 7	210 ± 11	0.32						
calc Eq. 4	784 ± 4	210 ± 6	-1.2 ± 0.04	0.33 ± 0.014	310 ± 8				
calc Eq. 5	782 ± 2	210 ± 2	0.33 ± 0.02			0.31 ± 0.005	0.68 ± 0.006	0.46 ± 0.007	1.2 ± 0.04
L26									
expt	745 ± 40	106 ± 2.4	0.79						
calc Eq. 4	745 ± 19	105 ± 1.1	0.42 ± 0.1	0.77 ± 0.01	107 ± 43				
calc Eq. 5	747 ± 7	106 ± 0.5	0.78 ± 0.02			0.76 ± 0.005	0.81 ± 0.008	0.93 ± 0.009	2.0 ± 1.0
D35									
expt	739 ± 6	128 ± 1.6	0.68						
calc Eq. 4	739 ± 3	128 ± 0.7	-0.24 ± 0.01	0.61 ± 0.004	250 ± 0.7				
calc Eq. 5	736 ± 2	128 ± 0.5	0.66 ± 0.02			0.59 ± 0.004	0.76 ± 0.008	0.78 ± 0.005	2.0 ± 0.3
D54									
expt	899 ± 35	142 ± 3.3	0.58						
calc Eq. 4	892 ± 16	142 ± 1.5	-0.008 ± 0.07	0.56 ± 0.007	104 ± 11				
calc Eq. 5	896 ± 9	142 ± 0.7	0.58 ± 0.02			0.55 ± 0.004	0.68 ± 0.006	0.81 ± 0.009	0.95 ± 0.08

The errors in the experimental NOEs are ± 0.1. The errors for the calculated values of the various parameters are obtained from analysis of the variance-covariance matrix obtained from the least-squares Powell optimization procedure. In all the fits, a value of 8.3 ns is used for τ_R which was obtained by a best fit of the T_1/T_2 ratios of all residues *simultaneously* (excluding 16 which had T_1/T_2 ratios outside ± 1 SD from the mean) using Eqs. 1 and 2. In the fits obtained with Eq. 4, S^2 and τ_e were varied; and in the fits obtained with Eq. 5, S_f^2, S_s^2 and τ_s were varied (and S^2 is given by the product $S_f^2 S_s^2$). In the best fits using Eq. 4, only the T_1 and T_2 data were used. In the best fits using Eq. 5 the T_1, T_2 and NOE data were used.

shift in internal correlation time to larger values which accompanies the introduction of the two internal motion formulation results in larger values of the NOE, whilst leaving the values of T_1 and T_2 unaffected.

These results clearly indicate that the ^{15}N relaxation data are sufficient to separate out two internal motions with time scales differing by more than one order of magnitude. Hence the correlation function for the internal motions of these residues can no longer be approximated to a single exponential of the Lipari and Szabo treatment, but rather requires two distinct exponentials. The slower motion, of the order of 1–3 ns, is slightly slower than the overall rotational correlation time (8–9 ns) for Il-1β, and can have a significant magnitude. The very fast motion, which must have a time scale of «100 ps, reflects the fast random thermal motions that are manifested in molecular dynamics calculations.

Acknowledgements

This work was supported by the Intramural AIDS Targeted Antiviral Program of the Office of the Director of the National Institutes of Health. We thank Attila Szabo for enlightening discussions.

References

1. Lipari, G. and Szabo, A. (1982) *J. Am. Chem. Soc.* **104**, 4546.
2. Lipari, G. and Szabo, A. (1982) *J. Am. Chem. Soc.* **104**, 4559.
3. McCain, D.C. and Markley, J.L. (1986) *J. Am. Chem. Soc.* **108**, 4259.
4. Weaver, A.J., Kemple, M.D. and Predergast, F.G. (1988) *Biophys. J.* **54**, 1.
5. Dellwo, M.J. and Wand, A.J. (1989) *J. Am. Chem. Soc.* **111**, 4571.
6. Kay, L.E., Torchia, D.A. and Bax, A. (1989) *Biochemistry* **28**, 8972.
7. Nirmala, N.R. and Wagner, G. (1989) *J. Magn. Reson.* **82**, 659.
8. Clore, G.M., Driscoll, P.C., Wingfield, P.T. and Gronenborn, A.M. (1990) *Biochemistry* **29**, 7387.
9. Abragham, A. (1961) *The Principles of Nuclear Magnetism*, Clarendon Press, Oxford.
10. Finzel, B.C., Clancy, L.L., Holland, D.R., Muchmore, S.W., Watenpaugh, K.D. and Einspahr, H.M. (1989) *J. Mol. Biol.* **209**, 779.
11. Clore, G.M., Szabo, A., Bax, A., Kay, L.E., Driscoll, P.C. and Gronenborn, A.M. (1990) *J. Am. Chem. Soc.* **112**, 4989.

Comparative study of solution conformation of bradykinin and its analogues

Sudha Srivastava, Ratna S. Phadke and Girjesh Govil
Tata Institute of Fundamental Research, Bombay 400 005, India

Introduction

Bradykinin Arg-Pro-Pro-Gly-Phe-Ser-Pro-Phe-Arg (BK) belongs to a group of plasma hormones called kinins. The hormone exhibits several diverse pharmacological activities [1]. It induces dilation of blood vessels and thus reduction of blood pressure [2] and causes contraction of smooth muscles of bronchia, intestines and uterus [3]. The analogues, which differ from the native BK by one amino acid, e.g. [Lys^1]-BK and [Tyr^8]-BK are totally inactive. Since the conformation of hormones is closely linked to its receptor recognition, we have studied conformational aspects of native BK and its analogues. We have used 2D NMR techniques to investigate the solution conformation.

Results and Discussion

Resonance assignments

Complete resonance assignment of three peptides, native BK, [Lys^1]-BK and [Tyr^8]-BK has been achieved using 2D COSY [4] and NOESY [5] techniques. The fingerprint region of the COSY spectrum of BK shows five out of six NH-CαH correlations (Fig. 1). Arg^1 NH, the N-terminal residue, exchanges with traces of D$_2$O present in the solvent. However, in the case of [Lys^1]-BK and [Tyr^8]-BK, the N-terminal NH shows strong NH-CαH correlations. Spin systems Gly, Phe, Ser and Tyr are assigned by their chemical shift positions and coupling correlations (Fig. 1). The CαH protons of three Pro could be traced to their CδH protons. Arg^9 could be assigned starting from the CϵNH-CδH correlation and tracing back to the CαH. Similar correlations can be observed for Arg^1 at higher temperature. For [Lys^1]-BK, the CϵH of Lys provides the starting point and leads to the correlation of CαH. For all three peptides, sequential assignments [6] have been made which confirm the NH and CαH positions. The chemical shift positions are listed in Table 1.

Conformation

Clues to the conformational ordering of the three peptides have been gathered from different NMR parameters. Differences in chemical shift positions from those reported for random coil [7] are large in the case of BK, less in [Lys^1]-BK and negligible for [Tyr^8]-BK. In case of BK, some of the resonances (Arg^1 CαH, Arg^9 NH, Arg^1 CβH, Pro^7 CβH) shows a very large difference ranging from 0.5 to 1 ppm from the shifts corresponding to random coils.

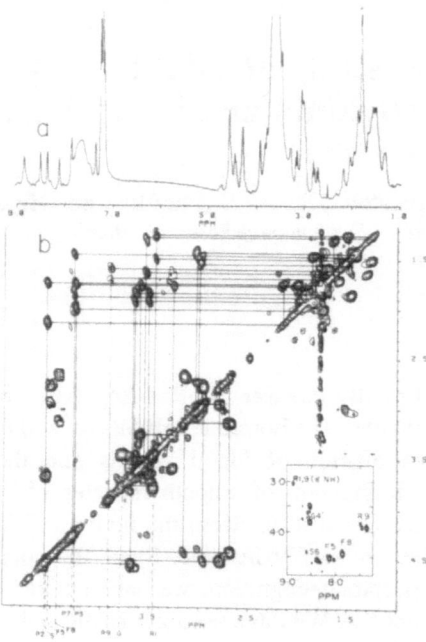

Fig. 1. (a) 500 MHz 1D-^1H spectrum of bradykinin dissolved in DMSO-d_6 (b) Expanded region of COSY spectrum of the peptide using pulse sequence D1-90-t_1-Δ-90-Δ-t_2 [15] indicating spin system correlations. Inset figure indicates fingerprint region. One letter symbols for the amino acid residues along with their numbers in sequence indicate the NH-CαH correlation. Some of the weak intensity peaks are due to the presence of a small amount of a different conformer presumably arising from cis-trans isomerisation in proline residues.

The temperature coefficients (Δδ/ΔT) of the backbone NH protons [8] are indicated in Table 1. For BK, values < 0.002 (indicative of solvent-shielded protons) are observed for Gly[4], Phe[5] and Arg[9] residues. For [Lys[1]]-BK, only Arg[9] NH falls in this range while Phe[5] shows a value of 0.0021 ppm/K. For [Tyr[8]]-BK, all NH protons show values of higher magnitude. This indicates the presence of more than one hydrogen bond in native BK, at least one in [Lys[1]]-BK and none in [Tyr[8]]-BK. $^3J_{NH-CαH}$ coupling constants in conjunction with Bystrov curves [9] are useful though in a limited way as they do not uniquely define dihedral angle φ. However, a comparative study of these along with other NMR parameters can provide clues for the estimation of the φ values. Sequential NOEs (NH$_{i+1}$-CαH$_i$) are observed for the three peptides as shown in Fig. 2 for BK. In case of BK, all NH$_i$-NH$_{i+1}$ NOEs are observed. In addition, NOEs from Pro[3] CδH to Phe[5] NH, Pro[7] CδH to Arg[9] NH and Arg[1] CαH to Pro[2] CδH are observed. These evidences indicate a bend of type II at both ends of BK giving rise to a highly rigid conformation in the molecule [10]. In the case of [Lys[1]]-BK also, both the sequential NOEs (NH$_{i+1}$-CαH$_i$ and NH$_i$-NH$_{i+1}$) are observed. At the N-terminal end, although there is a high probability of a β-bend due to the sequence Pro-Pro-Gly-Phe, yet we do not observe a low Δδ/δT value for Gly[4] and Phe[5] (0.0021 ppm/K could be interpreted as partially shielded proton)

58

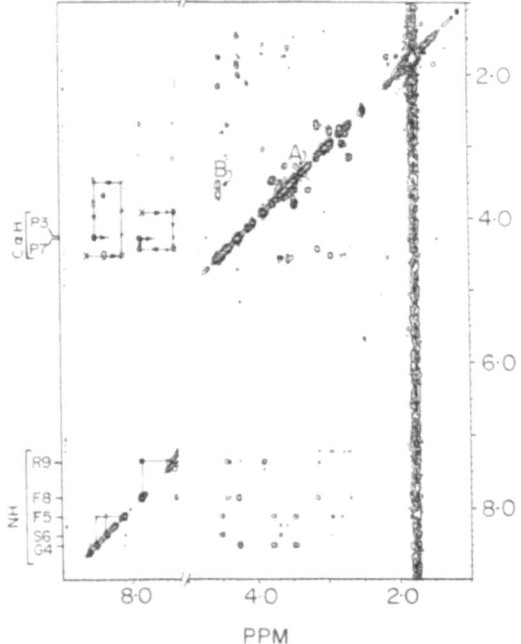

Fig. 2. *Relevant portion of phase-sensitive NOESY spectrum of bradykinin at 500 ms mixing time. The sequential NOEs-NH_{i+1}-$C\alpha H_i$ in the upper left region (X indicates the starting point) and NH_i-NH_{i+1} in the lower left region are indicated. Other cross peaks are (A) between Arg[1] $C\alpha H$ and Pro[2] $C\delta H$ and (B) between Pro[2] $C\alpha H$ and Pro[3] $C\delta H$.*

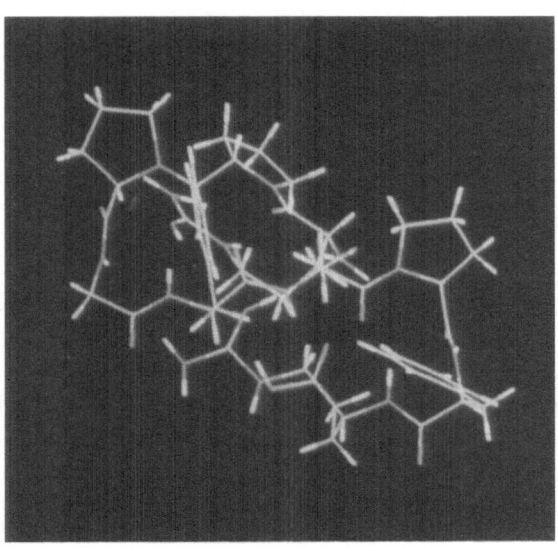

Fig. 3. *Computer-aided molecular graphics model of bradykinin.*

Table 1 ^1H chemical shifts (ppm), temp. coeff. $(\Delta\delta/\Delta T)$ and $^3J_{NH-C\alpha H}$ values in DMSO. I: BK, II: $[Lys^1]$-BK and III: $[Tyr^8]$-BK (TMS used as external reference)

Residues		NH	CαH	CβH	CγH	CδH	CεNH	Δδ/ΔT 10^{-3}ppmK^{-1}	J (Hz)
Arg¹ Lys¹	I	–	3.43	1.26, 1.50	1.27, 1.64	3.57	–	–	–
	II	8.25	3.57	1.89	1.28	1.47	–	–	8.5
	III	8.12	4.18	1.70, 1.57	1.47	3.05	7.66	–	–
Pro²	I	–	4.58	1.75, 2.15	1.76	3.29, 3.62	–	–	–
	II	–	4.57	2.11, 1.74	1.60	3.55	–	–	–
	III	–	4.59	2.22, 1.78	1.88, 1.75	3.72	–	–	–
Pro³	I	–	4.27	1.78, 2.01	1.93	3.54, 3.68	–	–	–
	II	–	4.28	1.98, 1.86	1.77	3.36	–	–	–
	III	–	4.28	2.00, 1.79	1.88, 1.63	3.62	–	–	–
Gly⁴	I	8.53	3.50, 3.80	–	–	–	–	1.7	–
	II	8.60	3.73, 3.51	–	–	–	–	–	–
	III	7.96	3.62	–	–	–	–	3.4	6.1
Phe⁵	I	8.10	4.52	2.78, 2.95	–	–	–	0.3	7.3
	II	8.16	4.46	2.83, 3.00	–	–	–	2.1	6.1
	III	8.36	4.59	2.91, 2.69	–	–	–	6.4	7.3
Ser⁶	I	8.33	4.55	3.68	–	–	–	5.8	5.5
	II	8.43	4.60	3.68	–	–	–	3.3	7.3
	III	7.96	4.59	3.62	–	–	–	4.4	8.5
Pro⁷	I	–	4.28	1.46, 1.89	1.84	3.54, 3.66	–	–	–
	II	–	4.17	2.13, 1.81	1.77	3.26	–	–	–
	III	–	4.32	1.88, 1.60	1.68, 1.46	3.52	–	–	–
Phe⁸ Tyr⁸	I	7.85	4.43	2.68, 3.15	–	–	–	3.9	8.6
	II	7.79	4.43	2.69, 3.17	–	–	–	2.2	7.3
	III	7.66	4.45	2.91, 2.54	–	–	–	2.9	8.5
Arg⁹	I	7.34	3.92	1.60, 1.68	1.45, 1.56	3.02, 3.04	8.86	1.6	6.1
	II	7.30	3.94	1.66	1.38	3.04	–	1.3	7.3
	III	8.12	4.18	1.70, 1.57	1.53	3.10	7.55	4.4	6.1

particularly when compared to the corresponding values in BK. This indicates that although there is an overall rigidity in the molecule, the replacement of Arg imparts a significant amount of flexibility towards the N-terminal end. On the other hand, $[Tyr^8]$-BK exhibits neither the above mentioned NOEs specific for a bend nor the low $\Delta\delta/\Delta T$ values for the backbone NH. Interestingly, none of the sequential NOEs (NH_i-NH_{i+1}) could be observed, which indicates a complete loss of conformational rigidity in the molecule.

Based on these findings, a model of BK has been built where backbone amide protons are involved in 1–4 type H-bonding leading to β-turns at both the termini (Fig. 3). At the N-terminal end, the Phe^5 NH proton forms an H-bond with Pro^2 CO giving rise to a β-turn with predicted torsional angles $\phi3 = -60°$; $\psi3 = 120°$; $\phi4 = 90°$ and $\psi4 = 0°$. At the C-terminal end, the Ser^6 CO is H-bonded with the Arg^9 NH proton giving rise to another β-turn with similar torsion angles. Glycine NH is solvent inaccessible being the central residue in the β-bend and thus exhibits a lower $\Delta\delta/\Delta T$ value. The involvement of Ser^6 CO in the H-bonding agrees with the reported ^{13}C NMR experiment [11]. The results on BK are also in agreement with the Chou Fasman calculations [12] which show a high probability for the occurrence of a β-turn for the sequences Pro-Pro-Gly-Phe and Ser-Pro-Phe-Arg. Replacement of Arg^1 by Lys imparts a certain flexibility to the molecule possibly due to different chain length which in the former case is accessible for electrostatic interaction with C-terminal carboxyl [13]. On the other hand, the loss in conformational rigidity of $[Tyr^8]$-BK may be due to the interaction of Tyr-OH with the neighbouring residue, destabilizing the H-bonds which results in loss of activity [14].

References

1. Mason, D.T. and Melmon, K.L. (1966) *J. Clin. Invest.* **45**, 1685.
2. Denys, L., Bothner-By, A.A., Fisher, G.H. and Ryan, J.W. (1982) *Biochemistry* **21**, 6531.
3. Melmon, K.L. and Morrelli, H.F. (1978) *Clinical Pharmacology – Basic Principles in Therapeutics*, 2nd Ed., MacMillan Pub. Co. Inc., New York, p. 664.
4. Aue, W., Barthold, E. and Ernst, R.R. (1976) *J. Chem. Phys.* **64**, 2229.
5. Jeener, J., Meier, B.H., Bachmann, P. and Ernst, R.R. (1979) *J. Chem. Phys.* **71**, 4546.
6. Wagner, G., Kumar, A. and Wüthrich, K. (1981) *Eur. J. Biochem.* **114**, 375.
7. Bundi, A. and Wüthrich, K. (1979) *Biopolymers* **18**, 285.
8. Kople, K.D., Ohnishi, M. and Go, A. (1969) *J. Am. Chem. Soc.* **91**, 4264.
9. Bystrov, V.F. (1976) in *Progress in Nuclear Magnetic Resonance Spectroscopy*, Vol. 10, (Emsley, J.W., Feeney, J. and Sutclifle, L.H., Eds.) Pergamon Press, Oxford, 41.
10. Mirmira, S.R., Durani, S., Srivastava, S. and Phadke, R.S. (1990) *Magn. Reson. Chem.* **28**, (in press).
11. London, R.E., Stewart, J.M., Cann, J.M. and Matwiyoff, N.A. (1978) *Biochemistry* **17**, 2270.
12. Chou, P.Y. and Fasman, G.D. (1979) *Biophys. J.* **26**, 367.
13. Lintner, K. and Fermandjian, S. (1979) *Biochem. Biophys. Res. Commun.* **91**, 803.
14. Mashford, M.L. and Roberts, M.L. (1971) *Biochem. Pharmacol.* **20**, 969.
15. Kumar, A., Hosur, R.V. and Chandrasekhar, K. (1984) *J. Magn. Reson.* **60**, 143.

Determination of locally accurate solution protein structures and unambiguous stereo-specific ^{1}H NMR assignments

S.A. Sherman and Michael E. Johnson*

Department of Medicinal Chemistry and Pharmacognosy, University of Illinois at Chicago, Box 6998 (M/C 781), Chicago, IL 60680, U.S.A.

Introduction

The determination of stereo-specific ^{1}H NMR assignments and locally accurate protein conformation are very important both for further NOE identification, and for building and refining protein spatial structure.

The possibility of determining protein local conformations from NOE data before building their spatial structure is a fundamental difference between the structural interpretations of NMR and X-ray data for proteins, in that information on the interproton distances obtained from NOE spectra can be correlated with the local conformation of amino acid residues. The sequential interproton distances, $d_{\alpha N}(\psi)$, $d_{NN}(\phi, \psi)$ and $d_{\beta N}(\psi, \chi_1)$ between $N_{i+1}H$ and the protons C^{α}_iH, N_iH and C^{β}_iH, respectively, and the intraresidual interproton distances, $d_{N\alpha}(\phi)$, $d_{N\beta}(\phi, \chi_1)$ and $d_{\alpha\beta}(\chi_1)$ distances between N_iH and C^{α}_iH, N_iH and C^{β}_iH, C^{α}_iH and C^{β}_iH, respectively are functions of the internal rotation angles ϕ, ψ and χ_1 (see Fig. 1). Thus, in principle, it is possible to go directly from distances to angular coordinates without using an intermediate Cartesian coordinate space.

Sherman et al. [1] previously proposed a method for determining the most probable (ϕ, ψ) values of the protein backbone by the set of sequential d connectivities (i.e., the data on availability or absence of the sequential NOE cross peaks related to nearest-neighbor residue $d_{\alpha N}$, d_{NN} and $d_{\beta N}$ connectivities; see [2] and Fig. 1). On the basis of this method, it was shown to be possible to obtain a well-determined local protein backbone conformation before building the full spatial structure.

NMR experiments are no longer limited to recording qualitatively the spatial proximity of separate pairs of protons in a molecule. More recent developments of special methods for interpreting spectra and the expansion of the technical possibilities of NMR spectroscopy have made possible semiquantitative measurements of the interproton distances from NOE data [2].

In the present work we analyze the possibility of using this data to refine the values of ϕ- and ψ-angles, to determine χ_1-conformations, and to implement unambiguous stereo-specific ^{1}H NMR assignments of β-protons. An approach which is intended to determine the most probable values of the angles ϕ, ψ and χ_1 of L-amino acid residues is described.

---

* To whom correspondence should be addressed.

Results and Discussion

The approach used here is similar to [1] and its subsequent development. The method [1] is based on the concept of structural analysis of the protein backbone conformation using the set of sequential d connectivities. These connectivities are used to assign proton signals to particular amino acid residues of the polypeptide chain [2].

In [1] it was proposed to divide the (ϕ, ψ) conformational map into regions (A, B, T, Q, P, R, S, K, R^*, S^* and P^*) such that conformations of the L-amino acid backbone corresponding to the same region had the same set of sequential d connectivities. For example, region A is characterized by availability of $d_{\alpha N}$ and $d_{\beta N}$ connectivities and absence of d_{NN} connectivity; in region B there is only $d_{\alpha N}$ connectivity ($d_{\beta N}$ and d_{NN} are absent) and so on. This decomposition allows the assignment of each amino acid residue of a protein molecule to appropriate regions of (ϕ, ψ) conformational space according to the sequential d connectivities corresponding to the residue.

For every region with the same set of sequential d connectivities, it is possible to specify the point (ϕ, ψ) which is its best representative. The (ϕ, ψ) expectation values calculated from the distribution functions can be used as the best representative coordinates of a region. Probability distribution densities of the conformational states of amino acid residues in (ϕ, ψ) space were determined in [1] from X-ray crystal structures in the Protein Data Bank [3], and used as the distribution functions. The interdependence between sequential d connectivities, weighted means (<ϕ>, <ψ>), and standard deviations (σ_ϕ and σ_ψ) for amino acid residues situated in different regions of conformational space (ϕ, ψ) were determined on this basis.

Subsequent development of the method described in [1] implies joint consideration of the intraresidual and sequential interproton distance dependencies on the dihedral angles ϕ, ψ and χ_1. The new method is based on the combined use of the correlation dependencies of the semiquantitatively estimated NOE cross-peak intensities between protons in sterically allowed conformational space (ϕ, ψ, χ_1), and on statistical processing of protein X-ray data [3].

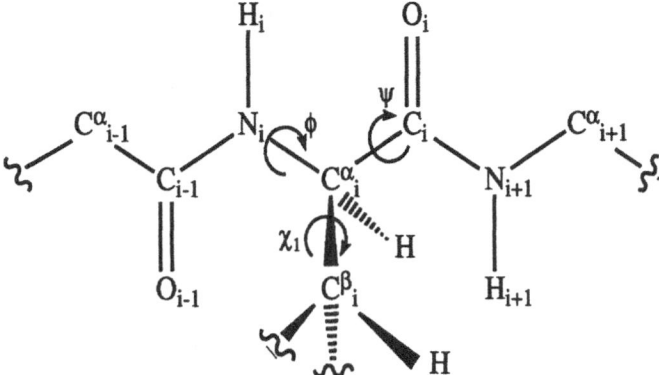

Fig. 1. Fragment of polypeptide chain.

Table 1 *The relationships between values of the χ_1-angle and the corresponding positions of atoms for L-amino acids with non-branched and branched β-carbons*

Value of χ_1-angle (deg.)	Names of atoms (non-branched β-carbons)			Names of atoms (Val, branched β-carbon)			Names of atoms (Ile/Thr, branched β-carbons)		
	$C^{\gamma 1}$	$H^{\beta 2}$	$H^{\beta 3}$	$C^{\gamma 1}$	$C^{\gamma 2}$	H^β	$C^{\gamma 1}/O^{\gamma 1}$	$C^{\gamma 2}$	H^β
60	g^+	t	g^-	g^+	t	g^-	g^+	g^-	t
180	t	g^-	g^+	t	g^-	g^+	t	g^-	g^+
-60	g^-	g^+	t	g^-	g^+	t	g^-	t	g^+

In this work, the atom coordinates are computed with an algorithm [4] assuming standard geometry for amino acid residues. The steric restrictions for atom-atom distances [5] are considered. It is assumed that for amino acid residues locked in a unique conformation within the protein structure, the distance intervals 2.0–2.5, 2.0–3.0, 2.0–4.0 Å between protons can be translated into strong (s), medium (m) and weak (w) NOE intensities [2], and that the corresponding NOE cross peak is absent (a) when the analyzed distance is more than 4.0 Å. These calculations also utilize the observation [6] that the χ_1 conformers have a trimodal distribution.

For L-amino acids in every region (A, B, T, etc., see [1]) of (ϕ, ψ) space the NOE cross-peak intensities between protons were estimated semiquantitatively. The calculations were performed separately for amino acids that are branched and non-branched at the β-carbon. Val, Ile and Thr contain branched β-carbons and a β-methine proton where the other L-amino acids contain two β-methylene protons. Figure 2 and Table 1 show the relationships between values of the χ_1 angle and the corresponding positions of atoms for L-amino acids with non-branched and branched β-carbons.

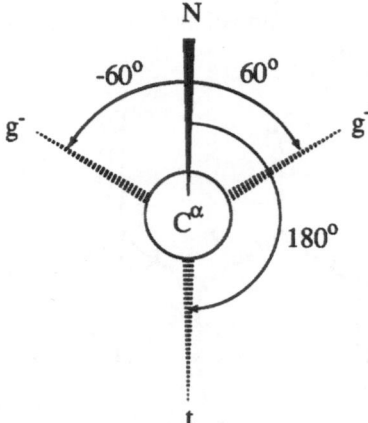

Fig. 2. *Newman projection for definition of the relationships between values of the χ_1-angle and the corresponding positions of atoms in L-amino acids. The view is along the C^α-C^β bond, with the C^α-position in the front.*

Semiquantitative estimates of intraresidual and sequential cross-peak intensities for amino acids with non-branched β-carbons are listed in Table 2. The notation, –, is used in the tables to show when a region is sterically forbidden. Table 2 can also be used for amino acids with branched β-carbons, taking into account only that the β2 (in the case of Val) or β3 (in the cases of Ile and Thr) protons are heavy atoms, and the corresponding records must be deleted from the table.

The data in Table 2 show that a unique set of NOE cross-peak intensities (semiquantitatively estimated from interproton distances) corresponds to every conformation in (φ, ψ, χ_1) space. The cross-peak intensity between intraresidual protons NH and $C^\alpha H$ gives very useful information about the residue backbone conformation: the strong intensity of this cross-peak shows that the backbone conformation of the residue corresponds to the right half of the (φ, ψ) conformational map, while a medium cross-peak intensity suggests a conformation in the left half of the map. All regions of the map except three pairs B and S^*, T and R^*,

Table 2 *Semiquantitative estimate of cross-peak intensities for L-amino acid residues with non-branched β-carbons*

Name of d connectivity	Region of (φ, ψ) space										
	A	B	T	K	Q	P	R	S	R^*	S^*	P^*
Intensities independent of the χ_1-angle											
$d_{\alpha N}$	s/m	s	s/m	m	w	w	w	w	s/m	s	w
d_{NN}	a	w/a	s/m	s/m	s/m	w	s/m	w	m	w	w
$d_{N\alpha}$	m	m	m	m	m	m	m	m	s	s	s
Intensities when the χ_1-angle is locked in the g^+ position											
$d_{\alpha\beta3}$	s	s	s	s	s	s	–	–	–	–	–
$d_{\alpha\beta2}$	s	s	s	s	s	s	–	–	–	–	–
$d_{\beta3N}$	w/a	a	a	w/a	w/a	w/a	–	–	–	–	–
$d_{\beta2N}$	s/m	w	a	m/w	w/a	w/a	–	–	–	–	–
$d_{N\beta3}$	s/m/w	s/m/w	m/w	s/m/w	s/m/w	s/m	–	–	–	–	–
$d_{N\beta2}$	w/a	w/a	w/a	w/a	w/a	w/a	–	–	–	–	–
Intensities when the χ_1-angle is locked in the t position											
$d_{\alpha\beta3}$	–	m	–	m	m	m	m	–	m	m	–
$d_{\alpha\beta2}$	–	s	–	s	s	s	s	–	s	s	–
$d_{\beta3N}$	–	w/a	–	m/w	w/a	w/a	s/m	–	w/a	w/a	–
$d_{\beta2N}$	–	a	–	w/a	w/a	w/a	w/a	–	w/a	w/a	–
$d_{N\beta3}$	–	s/m	–	s/m	s/m	s/m	s/m	–	w	w	–
$d_{N\beta2}$	–	s/m/w	–	s/m/w	s/m/w	s/m/w	s/m/w	–	m	m	–
Intensities when the χ_1-angle is locked in the g^- position											
$d_{\alpha\beta3}$	s	s	s	s	s	–	s	s	s	s	s
$d_{\alpha\beta2}$	m	m	m	m	m	–	m	m	m	m	m
$d_{\beta3N}$	s/m	w/a	w/a	a	w/a	–	w/a	m/w	w/a	w/a	w/a
$d_{\beta2N}$	w/a	a	w/a	m/w	w	–	s/m	s	w/a	w/a	w/a
$d_{N\beta3}$	w/a	w/a	w/a	w/a	w/a	–	w/a	w/a	w/a	w/a	w/a
$d_{N\beta2}$	s/m	s/m	s/m	s/m	s/m	–	s/m	s	w	w	w

and P and P* have different sets of d connectivities [1]. The pairs mentioned above have the same set of d connectivities but are situated in different halves of the map. In this way, using the cross-peak intensity between intraresidual protons NH and $C^{\alpha}H$ in addition to sequential d connectivities is sufficient to provide unambiguous determination of region. The intraresidual cross-peak intensities between the $C^{\alpha}H$ proton and the $C^{\beta 2}H$ and $C^{\beta 3}H$ protons are very useful for determining whether or not the χ_1-angle is locked in the g^+ position: in that case, both of these peaks are strong. The other pair of intraresidual cross peaks between the NH proton and the $C^{\beta 2}H$ and $C^{\beta 3}H$ protons provides complementary information which is useful for determining the t and g^- conformers. Thus Table 2 can be used to refine the backbone conformation, and to determine the type of χ_1-conformer and, consequently, to establish a single prochiral configuration of the β-protons. In practice, due to overlap with other NOE cross peaks, to partial saturation by the water irradiation etc., part of the NOE information can be missed. However, since Table 2 contains information about the relationships between intraresidual and sequential cross-peak intensities, a restoration of some missing cross peaks can be done using the available ones.

The use of the correlation dependencies of the semiquantitatively determined NOE cross-peak intensities between protons, in conjunction with statistical processing of protein X-ray data [3], allows us to determine the most probable values (weighted means) of φ and ψ angles and their standard deviations. The method used in this data calculation is similar to [1]. The results obtained are shown in Table 3.

Table 3 *Weighted means (<φ>, <ψ>) and standard deviations (σ_{ϕ}, σ_{ψ}) of the angles φ and ψ for L-amino acid residues*

	Region of (φ, ψ) space										
	A	B	T	K	Q	P	R	S	R*	S*	P*
	χ_1 is locked in the g^+ position[a]										
<φ>	−155	−155	−130	−130	−65	−55	−	−	−	−	−
σ_{ϕ}	17	18	15	15	18	15	−	−	−	−	−
<ψ>	160	135	30	−10	−5	0	−	−	−	−	−
σ_{ψ}	19	16	18	10	19	12	−	−	−	−	−
	χ_1 is locked in the t position[a]										
<φ>	−	−100	−	−75	−85	−60	−65	−	65	70	−
σ_{ϕ}	−	32	−	26	26	18	29	−	14	12	−
<ψ>	−	130	−	−15	−10	−5	−30	−	70	70	−
σ_{ψ}	−	24	−	10	16	15	17	−	12	12	−
	χ_1 is locked in the g^- position[a]										
<φ>	−105	−100	−85	−75	−85	−	−70	−60	65	65	65
σ_{ϕ}	25	25	21	22	23	−	25	12	14	14	14
<ψ>	160	135	20	−10	−5	−	−35	−55	55	65	45
σ_{ψ}	19	28	22	10	20	−	18	15	19	10	10

[a] For the mean values, $<\chi_1>$, and their standard deviations, σ, it is possible to use the values determined in [6]: g^+: $<\chi_1> = 60°$, σ = 25°; t: $<\chi_1> = -170°$, σ = 24°; g^-: $<\chi_1> = -70°$, σ = 21°.

An analysis of the data in Tables 2 and 3 shows that, taking into account the possibilities of modern NMR, and the uncertainties in experimental data, the semiquantitatively determined intraresidual and sequential NOESY cross-peak intensities are sufficient to determine values of the ϕ-, ψ- and χ_1-angles for residues locked in a unique conformation with approximately the same accuracy as in their determination from intermediate resolution X-ray data; a single prochiral configuration of β-protons is established automatically as a sub-product of χ_1 determination.

The suggested method can be realized as follows: First, using method [1], assign each amino acid residue of the protein to appropriate regions of conformational space (ϕ, ψ) according to the sequential d connectivities corresponding to the residue. Second, refine the region by the cross-peak intensity between intraresidual protons NH and $C^\alpha H$ in the cases when method [1] gives more than one possibility. Third, determine the χ_1-conformer from Table 2 on the basis of the available set of intraresidual and sequential cross-peak intensities. Fourth, knowing the region and the χ_1-conformer, determine the most probable ϕ-, ψ- and χ_1-values and their standard deviations from Table 3.

The ϕ-, ψ- and χ_1-values obtained via this procedure may then be applied as an approximation for segmental building of the spatial protein structure [7]. Establishment of a single prochiral configuration of β-protons and the restoration of missing NOE data (for example, due to overlap with other cross peaks) on the basis of available NMR data are additional important applications of this method.

References

1. Sherman, S.A., Andrianov, A.M. and Akhrem, A.A. (1987) *J. Biomol. Struct. Dyn.* **4**, 869.
2. Wüthrich, K. (1986) *NMR of Proteins and Nucleic Acids*, Wiley, New York, p. 292.
3. Bernstein, F.C., Koetzle, T.F., Williams, G.J.B., Meyers, E.F., Brice, M.D., Rodgers, J.R., Kennard, O., Mouchi, T.S. and Tasumi, M. (1977) *Eur. J. Biochem.* **80**, 319.
4. Galactionov, S.G., Sherman, S.A., Kirnarsky, L.I. and Nikiforovich, G.V. (1970) *Dokl. Akad. Nauk BSSR* **14**, 236.
5. Ramachandran, G.N. and Sasisekharan, V. (1968) *Adv. Prot. Chem.* **23**, 238.
6. Janin, J., Wodak, S., Levitt, M. and Maigret, B. (1978) *J. Mol. Biol.* **125**, 357.
7. Sherman, S.A., Andrianov, A.M. and Akhrem, A.A. (1988) *J. Biomol. Struct. Dyn.* **5**, 785.

Assessment of cheY binding regions using 2D NMR and paramagnetic ligands

L. Kar[a], P.Z. de Croos[a], S.J. Roman[b], P. Matsumura[b] and M.E. Johnson[a],*

[a]*Department of Medicinal Chemistry and Pharmacognosy (M/C 781) and*
[b]*Department of Microbiology and Immunology,*
University of Illinois at Chicago, Box 6998, Chicago, IL 60680, U.S.A.

Introduction

CheY is a 14 kDa cytoplasmic protein that is activated by the transfer of a phosphoryl moiety to a carboxylate side chain (Asp[57]) from phospho-CheA during signal transduction in bacterial chemotaxis [1,3,6]. It has been established that metal ions (possibly magnesium ions) are required for the auto-phosphorylation of CheA, the transfer of phosphate from phospho-CheA to CheY and the auto-dephosphorylation of phospho-CheY [3]. We report here 1D and 2D NMR studies which show that CheY exhibits two conformations corresponding to the metal-free and metal-bound states, and that CheY will bind phosphates only in the metal-bound conformation.

Results and Discussion

Native CheY, purified from *E. coli* [5] may be obtained in either the metal-bound (CheY1) or the metal-free (CheY2) form, with distinguishably different proton NMR spectra (Fig. 1). Addition of a divalent metal ion, like magnesium, in equimolar ratio with the protein, converts CheY2 to the metal-bound form, with an 1D NMR spectrum essentially identical to that of CheY1 (Fig. 1). Similar results are obtained with equimolar amounts of calcium, strontium, zinc and manganese, showing that the conformational change does not require magnesium ions specifically. While CheY2 may always be obtained from CheY1 by the removal of bound metal ions with EDTA, the native form cannot be regained completely by the addition of a divalent metal ion. Differences are noticeable in their binding affinity for phosphates (Fig. 3), and even in the proton NMR spectra of these two different metal-bound forms (Figs. 1b and c, particularly around 1.9 to 2.0 ppm). We shall use CheY1' to refer to the form obtained on adding divalent metal ions back to CheY2.

Comparison of double-quantum-filtered COSY (DQF-COSY) spectra in D_2O shows that Trp[58] (aromatic region shown in Fig. 2), Tyr[106] and Thr[87] (not shown) are particularly affected by the conformational change at the metal-binding site (see Table 1). The nine aromatic residues in CheY consist of six Phe, two Tyr and one Trp. HOHAHA and NOESY in D_2O and in 90% water have been used to assign Trp[58] and the two tyrosines (Tyr[51] is distinguished from Tyr[106] by its NOE to Phe[53]).

---

* To whom correspondence should be addressed.

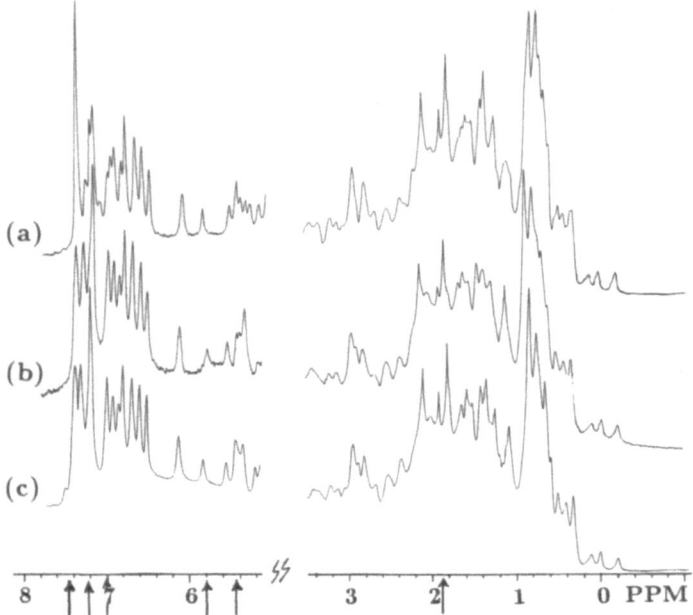

Fig. 1.   500 MHz 1D proton-NMR spectra at 20 °C, illustrating the conformational difference between the metal-free and metal-bound forms of CheY. Spectra are from (a) 0.8 mM metal-free protein, CheY2; (b) 0.8 mM CheY2 + 0.8 mM MgCl₂; (c) 0.8 mM CheY1. All spectra are from D₂O-exchanged protein, at pD 6.4. The regions of most significant spectral difference are pointed out with arrows on the chemical-shift axis.

The assignment of Trp[58] is further confirmed by the differences in NOESY spectra in 90% water for CheY1 and CheY2 (spectra not shown). Changes in chemical shifts of the N_1H of Trp[58], and its associated NOE cross peaks to C_2H and C_7H (Table 1) clearly indicate a conformational change involving Trp[58].

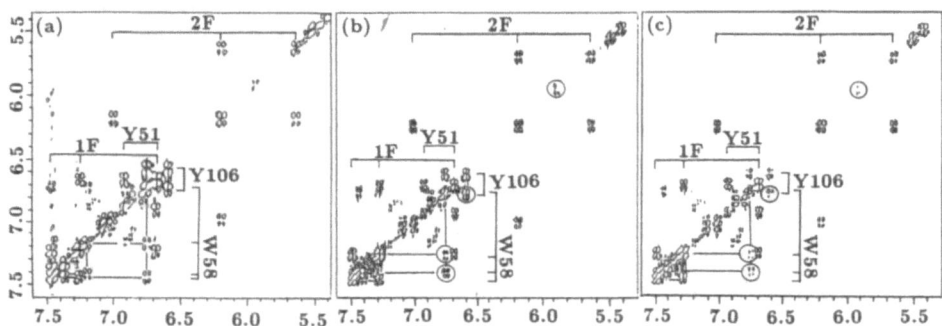

Fig. 2.   Aromatic region of DQF-COSY spectra at 30 °C, from (a) metal-free CheY2; (b) metal-bound CheY1 and (c) CheY1 + SL-ATP (20 : 1). Spectra are from 0.8 mM protein in D₂O at pD 6.4. Assignments are shown for five of the nine aromatic residues. Comparison of (a) and (b) shows that the chemical shifts of the aromatic protons of Trp[58] are affected to a large extent by the conformational change. Cross peaks of Trp[58] and Tyr[106] that are suppressed in CheY1 by SL-ATP (and manganese) are encircled in (b) and (c).

Table 1   *Chemical shifts (at 30°C, referenced to DSS) of assigned resonances of three residues that show significant changes on metal binding*

Assignment	CheY1 (Metal-bound)	CheY2 (Metal-free)
C_α-H	5.46	5.49
$C_{\beta 1}$-H	4.54	–
$C_{\beta 2}$-H	4.27	–
N_1-H	10.32	10.22
C_2-H	7.28	7.32
C_4-H	7.37	7.49
C_5-H	6.75	6.75
C_6-H	7.27	7.19
C_7-H	7.49	7.43
C_α-H	5.87	5.93
$C_{\beta 1}$-H	2.37	2.34
$C_{\beta 2}$-H	2.73	2.77
C_2/C_6-H	6.75	6.75
C_3/C_5-H	6.58	6.58
T^{87} C_α-H	5.10	5.17
C_β-H	4.10	4.03
C_γ-H	0.94	0.94

^{31}P NMR has been used to study the phosphate-binding properties of the three forms of CheY mentioned above. Significant broadening of the three phosphate resonances of ATP is observed on the addition of CheY1 to a solution of ATP (Fig. 3a), indicating binding of ATP to CheY1. While no line broadening is observed with CheY2 (Fig. 3d), CheY1' broadens the phosphate resonances (Fig. 3b), demonstrating that binding divalent metal ions, and the resulting conformational change, restores phosphate binding affinity to the protein.

Similar binding experiments with structural analogs of ATP, other nucleoside phosphates and inorganic phosphates show that the metal-bound forms of CheY will bind any ionized phosphate group, but not a tri-substituted phosphate group, like trimethyl phosphate, indicating that charge interactions are important in the phosphate binding (data not shown).

From the variation of ^{31}P NMR line broadening with substrate concentration [2], a dissociation constant (K_D) of approximately 1 mM is obtained for phosphate binding, assuming a single binding site on CheY1. A qualitative comparison of phosphate line broadening observed with CheY1 and CheY1' shows that CheY1' has a lower affinity for phosphates. An estimate of K_D for phosphate binding was not possible for CheY1' since phosphates added in excess of the protein concentration resulted in the loss of the bound metal, accompanied by the conformational change from CheY1' back to CheY2. Titration of fluorescence quenching with added magnesium ions gave a K_D of approximately 0.6 mM for magnesium binding to CheY2 (at pH 7, 22°C), with a binding stoichiometry of 1 : 1. The affinity of CheY1 for magnesium is expected to be much higher, since the metal remains bound even at a 10 : 1 ratio of phosphate : CheY1.

Addition of low concentrations of the paramagnetic metal ion manganese to the metal-free form of the protein also changes its conformation to the metal-bound form and suppresses several resonances corresponding to residues close to the metal-binding site, including those of Trp[58]. Similarly, paramagnetic relaxation enhance-

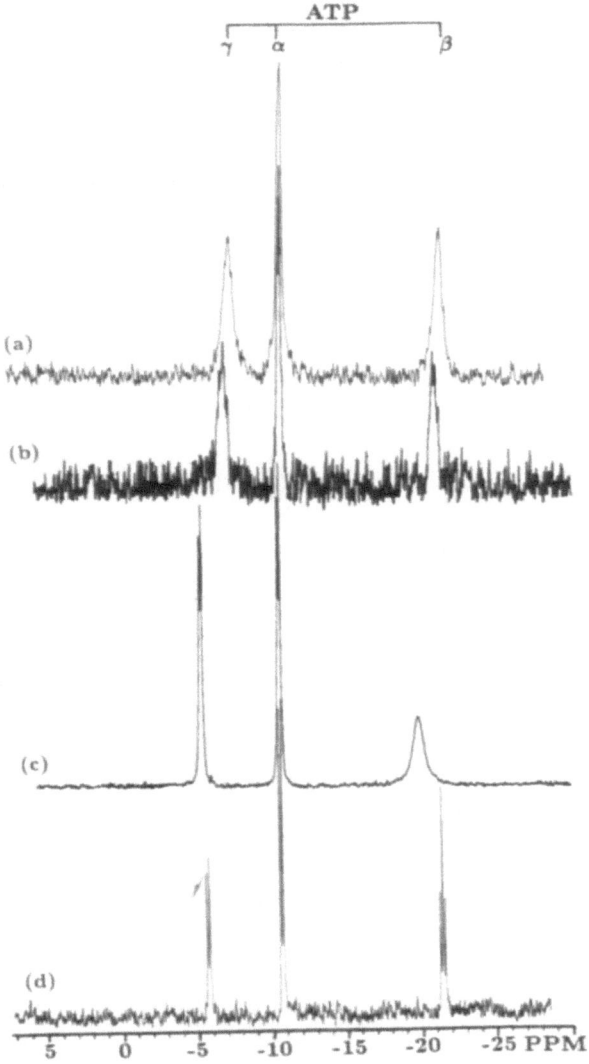

Fig. 3. ^{31}P *NMR spectra at 24 °C, showing line broadening effects of the three forms of CheY on the phosphate lines of ATP. Spectra are from: (a) 0.8 mM CheY1 + 4.0 mM ATP; (b) 0.8 mM CheY2 + 0.8 mM magnesium chloride + 0.8 mM ATP; (c) 10 mM magnesium chloride + 10 mM ATP; (d) 0.8 mM CheY2 + 3.5 mM ATP. Spectrum (d) is essentially identical to a spectrum of ATP in the absence of any protein. All samples are in D_2O at pD 7.0 and contain approximately 0.2 mM EDTA + 0.01% azide. Note the difference in line broadening patterns between (b) and the MgATP control spectrum (c). The small differences in resonance positions are due to small variations in pH between samples.*

ment due to the binding of nitroxide spin-labeled ATP (SL-ATP) is used to identify residues close to the phosphate binding site. Resonances suppressed by SL-ATP are the same as those suppressed by manganese (Fig. 4), indicating that the phosphate and metal-binding sites are in close proximity to each other. This is further evident in parallel DQF-COSY experiments, where all the cross peaks suppressed by SL-ATP (Fig. 2c) are also observed to be suppressed by manganese (spectrum not shown).

A systematic study of effects of mutations at aspartyl residues 12, 13 and 57 points to Asp^{57} as the normal site of phosphorylation in CheY [1]. The cross peaks in the DQF-COSY spectra that are affected most by the metal binding have been assigned to Trp^{58}. Trp^{58} implicates the phosphorylation site at Asp^{57} to be involved in the conformational change. Binding of manganese clearly suppresses the aromatic resonances of Trp^{58}. In addition, cross peaks that are suppressed by manganese in DQF-COSY experiments are also observed to be suppressed by SL-ATP. This is direct evidence that the metal and the phosphate bind in the same region near Asp^{57}, close to the normal site of phosphorylation in CheY.

In summary, 1D and 2D NMR, in conjunction with paramagnetic metal ion and spin label enhanced relaxation, has been used to investigate the observed binding of

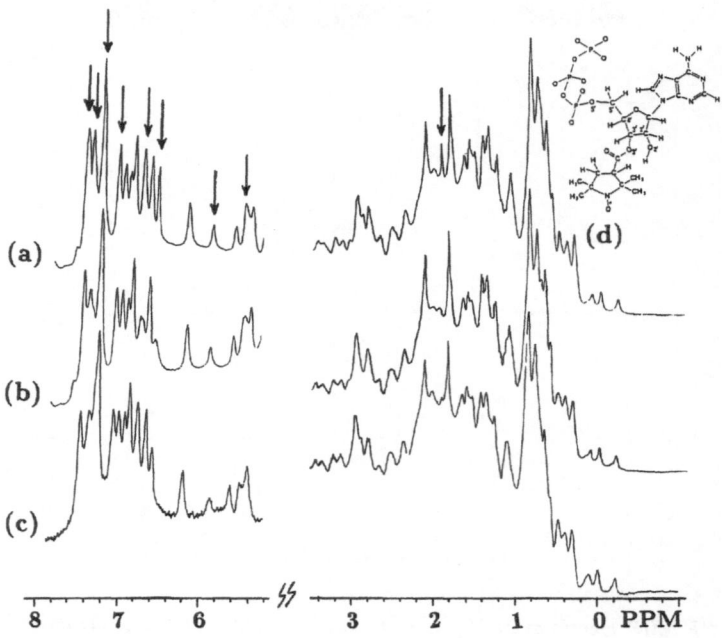

Fig. 4. Proton NMR spectra showing that the same proton resonances of CheY in (a) are suppressed when: (b) SL-ATP binds to CheY1 and (c) manganese binds to CheY2. Spectra are from: (a) 0.8 mM CheY1; (b) 0.8 mM CheY1 + 0.2 mM SL-ATP; and (c) 0.5 mM CheY2 + 0.025 mM manganese chloride. All spectra are in D$_2$O at pD 6.4. Resonances that are suppressed significantly are marked with arrows in (a). The molecular structure of SL-ATP is shown in (d).

divalent metal cations and phosphates to CheY. It is shown that both the metal and the phosphate bind in close proximity to the phosphorylation site at Asp[57], and that the bound metal ion is essential for the interaction of phosphates with CheY. The bound metal may be removed (from CheY1') by excess phosphate and appears to regulate the interaction of phosphates with CheY through a conformational change at the phosphate binding site.

Acknowledgements

This work was supported in part by National Institute of Health research Grant AI18985 to P.M. The GE GNΩ500 NMR spectrometer was purchased in part with a BSRG shared instrumentation grant (RR03295) from the NIH.

References

1. Bourret, R.B., Hess, J.F. and Simon, M.I. (1990) *Proc. Natl. Acad. Sci. USA* **87**, 41.
2. Dwek, R.A. (1973) in *Nuclear Magnetic Resonance (N.M.R.) in Biochemistry*, Clarendon Press, Oxford, pp. 137–138.
3. Hess, J.F., Bourret, R.B., Matsumura, P. and Simon, M.I. (1988) *Cold Spring Harbor Symposia on Quant. Biol.* **LIII**, 41–48.
4. Hess, J.F., Oosawa, K., Matsumura, P. and Simon, M. (1987) *Proc. Natl. Acad. Sci. USA* **84**, 7609.
5. Matsumura, P., Rydel, J.J., Linzmeier, R. and Vacante, D. (1984) *J. Bacteriol.* **160**, 36.
6. Sanders, D.A., Gillece-Castro, B.L., Stock, A.M., Burlingame, A.L. and Koshland, Jr., D.E. (1989) *J. Biol. Chem.* **264**, 21770.

The determination of carbon-to-nitrogen distances in melanostatin using $^{13}C,^{15}N$-REDOR NMR spectroscopy

Joel R. Garbow and Charles A. McWherter

Monsanto Corporate Research, 700 Chesterfield Village Parkway,
St. Louis, MO 63198, U.S.A.

Introduction

The importance of molecular conformation as a determinant of biological specificity is now a central tenet of molecular biology. X-ray crystallography, computational chemistry and high-resolution NMR spectroscopy have contributed mightily to our understanding of the structural basis of molecular recognition. However, direct determination of the conformation of a ligand when it is bound to a macromolecular receptor remains a difficult task because of the requirement for diffracting single crystals [1] (X-ray), or limits on the molecular size of the receptor or its solubility [2] (solution NMR). The recent development by Schaefer and co-workers [3–5] of $^{13}C,^{15}N$-rotational echo double resonance (REDOR) NMR spectroscopy for determining carbon-to-nitrogen distances in solids would seem to be directly applicable to the problem of determining bound ligand conformation without limitations imposed by molecular ordering, molecular weight, or solubility.

We have embarked on a study to define the methods and requirements for constructing structural models from REDOR data. Our model system is the tripeptide neurohormone melanostatin (Pro-Leu-Gly-NH$_2$) originally identified as being responsible for inhibiting the release of melanin-stimulating hormone [6], and more recently for modulating dopamine receptors in the central nervous system [7]. A series of melanostatins, each selectively enriched with both a ^{13}C and ^{15}N label, were prepared and studied by ^{13}C-detected REDOR. ^{13}C-^{15}N dipolar coupling constants were determined and the distances from the carbon-to-nitrogen labeling sites was calculated. A comparison is made of ^{13}C-detected REDOR distances with those calculated from the type II bend structure reported in the X-ray study [8].

Results and Discussion

$^{13}C,^{15}N$-labeled melanostatins

A series of four doubly-labeled melanostatins were prepared using commercial samples of labeled amino acids and the Merrifield solid-phase synthesis method (Table 1). Each peptide was characterized by FAB mass spectrometry and both ^1H and ^{13}C NMR; the details of the synthesis and characterization will be described elsewhere. Each of the labeled melanostatins was diluted with a 9-fold excess of unlabeled melanostatin solid, dissolved in 50% methanol with gentle warming, and cooled to give a supersaturated solution from which the peptide crystallized. The peptide crystals were dried over P_2O_5 in vacuo and were transferred to the sample rotor for solid-state NMR measurements.

Solid-state NMR measurements

^{13}C NMR spectra were collected on a home-built spectrometer (^1H frequency 127 MHz) equipped with a triply-tuned {$^1H, ^{13}C, ^{15}N$} probe. Samples were spun at the magic angle at a rate of 3 kHz in a double-bearing rotor system. REDOR spectra were collected following 2 ms ^1H-^{13}C cross-polarization preparation periods, with matched, 50 kHz rf fields. High-power proton dipolar decoupling ($H_1(H)$ = 75–110 kHz) was used throughout.

Carbon-to-nitrogen distances from REDOR data

REDOR NMR allows accurate measurement of through-space distances between pairs of selectively enriched ^{13}C and ^{15}N sites. The physical quantity which is determined in a REDOR experiment is the heteronuclear dipolar coupling constant, D_{CN}. D_{CN} is itself directly proportional to $(r_{CN})^{-3}$, where r_{CN} is the internuclear distance between ^{13}C and ^{15}N sites. The basic pulse sequence for ^{13}C-detected

^{13}C-Detected REDOR

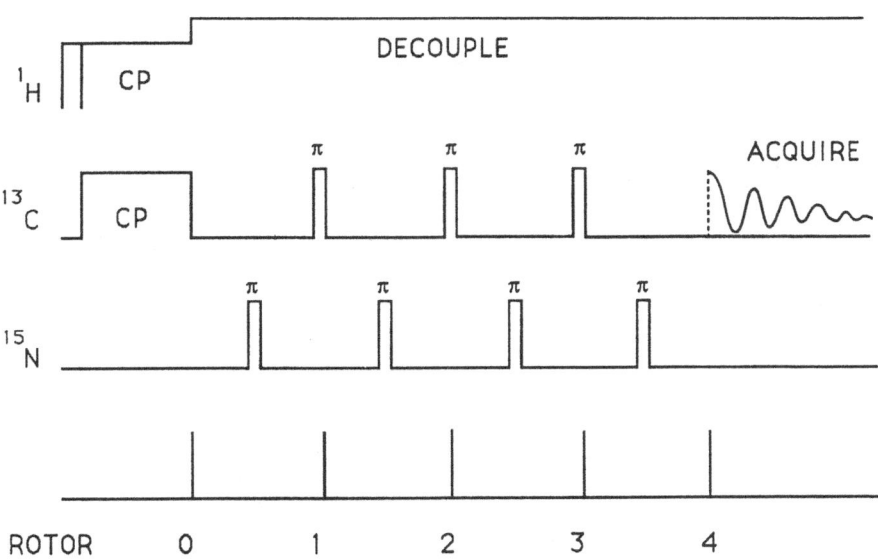

Fig. 1.   Pulse sequence for ^{13}C-observe REDOR NMR. This sequence differs from the original REDOR pulse sequence (ref. 3) in that π-pulses are equally divided between the ^{13}C and ^{15}N rf channels. On alternate scans of the REDOR experiment, the ^{15}N π-pulses are either applied or omitted. This figure illustrates the REDOR pulse sequence with 4 rotor periods of ^{13}C-^{15}N dipolar-coupling evolution (N_C = 4). N_C can be increased (in increments of 2) by adding rotor periods and pairs of ^{13}C and ^{15}N π-pulses between the end of the cross-polarization (CP) preparation and the start of data acquisition.

Table 1 *A comparison of $^{13}C,^{15}N$-REDOR distances with X-ray distances for selectively enriched $^{13}C,^{15}N$-melanostatins*

^{13}C Site	^{15}N Site	N_C	$\Delta S/S_0$	D_{CN} (Hz)	r_{CN} (Å)		
					REDOR	X-ray	($\phi = \psi = 180°$)
[1-^{13}C]Leu	[^{15}N]Leu	16	0.794	202	2.42	2.43	2.43
[1-^{13}C]Gly	[^{15}N]Leu	30	0.070	26	4.80	5.02	6.20
[2-^{13}C]Gly	[^{15}N]Leu	30	0.096	31	4.55	4.71	4.86
[1-^{13}C]Pro	[α-^{15}N]Gly	30	0.283	55	3.74	3.71	4.85

REDOR, which is performed as a difference experiment, is shown in Fig. 1. On alternate scans of the experiment, the π-pulses on the nitrogen channel are either applied or omitted. Signals from alternate scans are accumulated and Fourier-transformed separately. When normalized by the intensity of the signal due to the isotopically-enriched site alone (S_0), the difference between signal with and without π-pulses (ΔS) depends upon D_{CN} and 2 experimental parameters: (1) the spinning rate, ν_R and (2) the rotor periods of ^{13}C-^{15}N dipolar coupling evolution, N_C. The relationship between ($\Delta S/S_0$) and the dimensionless parameter $\lambda = (D_{CN}N_C)/\nu_R$ can be written in closed form [4], but because the expression involved must be integrated over all orientations in the powder sample, a computer is required to evaluate it.

^{13}C-REDOR NMR of $^{13}C,^{15}N$-labeled melanostatins

Spectra representative of the labeled melanostatins are shown in Fig. 2 for a diluted sample of [1-^{13}C]Pro-Leu-[α-^{15}N]Gly-NH$_2$. The bottom panel is the echo spectrum of the full sample and the middle panel is the ^{13}C-enriched carbonyl-carbon signal (S_0), obtained by subtracting the spectrum of the natural-abundance tripeptide (data not shown). The enriched carbonyl-carbon signal appears as an asymmetric doublet because dipolar coupling to its covalently bonded (quadrupolar) ^{14}N neighbor is incompletely removed by magic-angle spinning. The top panel is the REDOR difference signal (ΔS). Using this data, λ, D_{CN} and r_{CN} are readily calculated [4].

Carbon-to-nitrogen distances in $^{13}C,^{15}N$-labeled melanostatins

The results of our REDOR study of crystalline samples of labeled melanostatin are shown in Table 1. The carbon-to-nitrogen distances calculated from REDOR are compared with the distances between the labeled sites calculated from two different melanostatin structures: 1) the published X-ray structure [8] and 2) the structure when backbone dihedrals are placed in a fully extended conformation. The Pro-[1-^{13}C,^{15}N]Leu-Gly-NH$_2$ sample has the enriched labels fixed at a distance of 2.43 Å by the covalent geometry of the leucyl residue, i.e., this distance does not vary with peptide conformation. The REDOR value of 2.42 Å compares extremely well and serves to certify our methodology.

The remaining three peptides of Table 1 contain label pairs whose internuclear distances are quite sensitive to the peptide conformation, as can be seen by comparing the REDOR and X-ray values with those calculated for the peptide when placed in a fully extended conformation. The distances we have measured with

REDOR range from 2.5 to 5 Å. Although we expect the accuracy of our measurements to be greater at shorter distances, we see good agreement between the REDOR and X-ray results, even at longer distances. In the crystal structure, a 10-atom cycle is closed by a hydrogen bond from the *trans* carboxamide proton to the carbonyl oxygen of the prolyl residue, giving rise to a compact structure. A compact structure is also evident in the REDOR data, as seen by the consistently shorter (up to 20%) distances compared to the extended structure.

In addition to labeled $^{13}C,^{15}N$ pairs, ΔS also receives contributions from natural-abundance ^{13}C spins which are close to the labeled ^{15}N sites and, to a lesser extent, from natural-abundance ^{15}N spins adjacent to the labeled ^{13}C sites. For long distances, having small REDOR signals, these natural-abundance signals are comparable in size to those due to the labeled $^{13}C,^{15}N$ pair. Although we have tried to correct for natural-abundance signals in calculating the REDOR distances reported in Table 1, we believe that we have underestimated their magnitudes. Such underestimates lead to REDOR distances which are too short. Experiments are in progress to refine the current data, including experiments performed on ^{15}N-only enriched peptides. These should improve our estimation of natural-abundance effects,

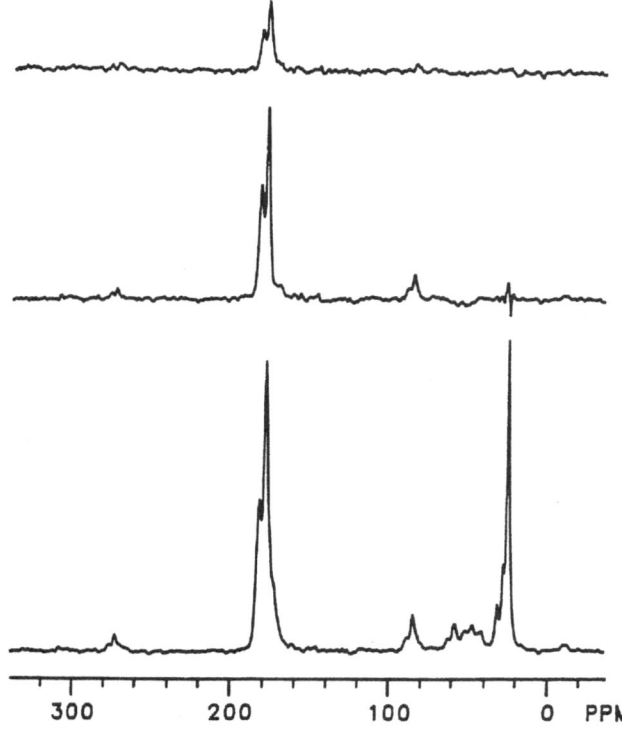

Fig. 2. *REDOR ^{13}C NMR spectra of [1-^{13}C]Pro-Leu-[α-^{15}N]Leu-Gly-NH$_2$ diluted with natural-abundance peptide (1:9): (bottom) Echo spectrum of full sample; (middle) Echo spectrum following subtraction of the spectrum of natural-abundance peptide (data not shown). Signal in this spectrum is due to the ^{13}C-enriched carbonyl carbon only; (top) REDOR difference. Spectra were collected using the pulse sequence of Fig. 1 with ν_R = 3 kHz; N_C = 30.*

leading to even better agreement between REDOR and X-ray results.

The measured distances in Table 1 can be used to progressively constrain the backbone dihedrals: The distance between the labels determined for the Pro-[^{15}N]Leu-[2-^{13}C]Gly-NH$_2$ depends only on the intervening dihedral ψ_{Leu}, that of [1-^{13}C]Pro-Leu-[α-^{15}N]Gly-NH$_2$ on ϕ_{Leu} and ψ_{Leu}, and r_{CN} of Pro-[^{15}N]Leu-[1-^{13}C]Gly-NH$_2$ on ψ_{Leu} and ϕ_{Gly}. A detailed analysis of these constraints, and a generalization to the determination of peptide backbone conformation is in progress.

Melanostatin has proven to be a very useful molecule in working through the technical issues of determining peptide structure from REDOR data. In addition to completing structural work on melanostatin, work is currently in progress on application of REDOR to the conformation of ligands bound to macromolecules.

Acknowledgements

The authors thank Drs. Jacob Schaefer and Terry Gullion for many helpful discussions about REDOR, James Doom and Eric Kolodziej for mass spectral analysis of labeled peptides, and Meheryar Rivetna for useful suggestions on improving the recovery of labeled compounds.

References

1. Blundell, T.L. and Johnson, L.N. (1976) in *Protein Crystallography*, Academic Press, New York, NY.
2. Fesik, S.W. (1989) in *Computer-Aided Drug Design* (Perun, T.J. and Propst, C.L., Eds.), Ch. 5, Marcel Dekker, New York, NY.
3. Gullion, T. and Schaefer, J. (1989) *J. Magn. Reson.* **81**, 196.
4. Gullion, T. and Schaefer, J. (1989) *Adv. Magn. Reson.* **13**, 57.
5. Marshall, G.R., Beusen, D.D., Kociolek, K., Redlinski, A.S., Lepalawy, M.T., Pan, Y. and Schaefer, J. (1990) *J. Am. Chem. Soc.* **112**, 963.
6. Celis, M.E., Taleisnik, S. and Walter, R. (1971) *Proc. Natl. Acad. Sci. USA* **68**, 1428.
7. Mishra, R.K., Chiu, S., Chiu, P. and Mishra, C.P. (1983) *Methods Find. Exp. Clin. Pharmacol.* **5**, 203.
8. Reed, L.L. and Johnson, P.L. (1973) *J. Am. Chem. Soc.* **95**, 7523.

Solution conformation of neuropeptide Y: 2D NMR and molecular dynamics studies

A. Balasubramaniam[a], S.-G. Huang[b], S. Sheriff[a],
M. Prabhakaran[c] and V. Renugopalakrishnan[c]

[a]*Division of G.I. Hormones, Dept. of Surgery, University of Cincinnati Medical Center, Cincinnati, OH 45267, U.S.A.*
[b]*Department of Chemistry, Harvard University, Cambridge, MA 02138, U.S.A.*
[c]*Laboratory for the Study of Skeletal Disorders and Rehabilitation, Dept. of Orthopaedic Surgery, Harvard Medical School, Children's Hospital, Boston, MA 02115, U.S.A.*

Introduction

Neuropeptide Y(NPY), a 36-residue peptide amide isolated from porcine brain [1], is present in higher concentrations in mammalian brain than any other peptides isolated so far. NPY together with the structural homologues, peptide YY [2] and pancreatic polypeptide (PP) [3], constitute the pancreatic polypeptide family of hormones. Avian PP has been crystalized in the presence of zinc and its structure elucidated using X-ray crystallography. X-ray studies of avian PP [4] have shown that it has a compact globular structure composed of a polyproline-like structure at the amino terminal and a C-terminal α-helix comprising residues 14–31. The X-ray crystal structure of avian PP has been used extensively in recent modeling studies of NPY [5]. It is therefore not surprising that all these investigations show that NPY, like avian PP, has a globular structure consisting of a polyproline-like structure and an α-helical domain at the N- and C-terminals, respectively. Our investigations of the solution structure of NPY by CD and Raman spectroscopy [6], augmented by Chou-Fasman analysis, however, showed that NPY has two distinct secondary structural domains composed of an amino-terminal region with β-turns and an α-helical C-terminal region. In order to further probe the secondary structure of NPY, we synthesized NPY, NPY(1–23) and NPY(17–36) and investigated their structures by 2D NMR spectroscopy. We also investigated their binding to rat cerebral cortex.

Materials and Methods

NMR studies were performed on a Bruker AM 500 FT-NMR spectrometer at Harvard University and on a Bruker AM 600 FT-NMR spectrometer at the University of Alabama at Birmingham. Approximately 4 mg. of NPY was dissolved in about 0.4 ml of 100% perdeuterated dimethyl sulfoxide. Two-dimensional COSY, NOESY, and TOCSY experiments were performed using standard pulse sequences [7]. Both magnitude mode and phase-sensitive mode using the time-proportional-phase-increment technique were used for COSY and TOCSY experiments. However, only phase-sensitive mode NOESY experiments were performed in order to provide adequate spectral resolution for better assignment of the cross peaks. A range of mixing times were used for NOESY experiments.

The initial structure of NPY was built manually, using computer graphics choosing the dihedral angles to satisfy local NOE contacts except for Pro residues where $(\phi,\psi) = -70°, 30°$ were assumed.

The structure was then minimized by both steepest descent and conjugate gradient methods. A dynamical simulation at 1000°C was carried out on this structure to arrive at starting molecular models. We have employed constrained minimization followed by dynamics incorporating non-ambiguous constraints and then ambiguous constraints in the later stages of dynamics simulations. The constraint length (R_0) was kept at 3.5 Angstroms. The constraint potential is of the form:

$$E(R) = k_1(R-R_0)^2$$

where k_1 = 1000 kJ/mol; $R > R_0$. The molecular package program Gromos (Groningen Molecular Simulation) was used in the simulation studies [8]. The minimization and dynamics were done alternatively on the system. The constraints also were applied and released alternatively. The conformation of amino acids, packing considerations, hydrogen bond arrangements, retention of NOE constraints and conformational energetics of the structures were monitored during this process.

Results and Discussion

To assign the proton resonances in NPY we have basically utilized COSY and 2D HOHAHA (TOCSY) experiments to identify the proton resonances within the same residue. The ambiguities of assignments due to the existence of like residues in NPY like Tyr were eliminated by a sequential assignment strategy [7] which examines the CH(i) to NH(i+1) and NH(i) to NH(i+1) connectivities in the 2D NOE experiments.

The existence of a large number of NH-NH cross peaks arising from residues at the C-terminal in the 2D NOE spectrum of NPY are suggestive of a largely α-helical structure in this region. 2D NOE studies also demonstrate the occurrence of short proton-proton contacts at the N-terminal which is indicative of the presence of β-turns. Complete details of the 2D NMR studies will be published elsewhere (Huang et al.). Consistent with this observation, 2D NMR studies of NPY (1–23) and NPY (17–36) provided further evidence for the presence of β-turns and α-helical structures, respectively.

The simulated final structure of NPY is shown in stereo in Fig. 1. The final structure consists of a series of β-turns at the N-terminal and a distorted linear α-helical structure at the C-terminus.

NPY and NPY (17–36) inhibited ^{125}I-NPY binding to the rat cerebral cortex in a dose-dependent manner with half maximal inhibitory potencies (IC_{50}) of 0.50 ± 0.05 and 1.00 ± 0.01 nM, respectively. NPY (1–23) did not inhibit ^{125}I-NPY binding even at 1 μM.

Conclusions

In DMSO, NPY has two distinct structural domains, an amino terminal region with

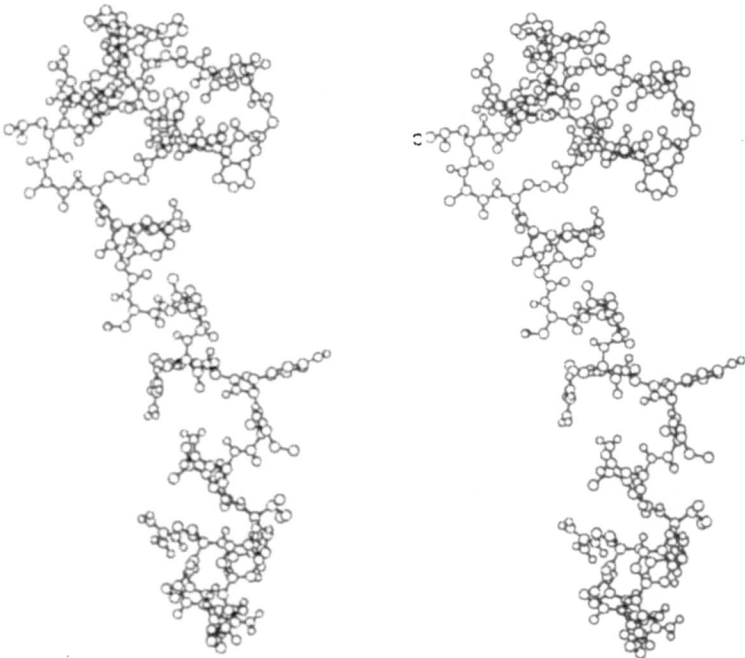

Fig. 1. Secondary structure of NPY from molecular dynamics simulation satisfying NOE constraints.

β-turns and a carboxyl-terminal region with α-helical structures [6]. The C-terminal α-helical region appears to play a crucial role in binding to the cerebral cortex.

Acknowledgements

A.B. is supported by a grant from NIH (GM-38601). M.P. is supported by a grant from the American Heart Association of Greater Chicago.

References

1.   Tatemoto, K. (1982) *Proc. Natl. Acad. Sci. USA* **79**, 5485.
2.   Tatemoto, K. (1982) *Proc. Natl. Acad. Sci. USA* **79**, 2514.
3.   Kimmel, J.R., Hayden, L.J. and Pollock, H.G. (1975) *J. Biol. Chem.* **250**, 9369.
4.   Glover, D., Barlow, D.J., Pitts, J.E., Wood, S.P., Tickle, I.J., Blundell, T.L., Tatemoto, K., Kimmel, J.R., Wollmer, A., Strassburger, W. and Zhang, Y.-S. (1985) *Eur. J. Biochem.* **142**, 379.
5.   Allen, J., Novotny, J., Martin, J. and Heinrich, G. (1987) *Proc. Natl. Acad. Sci. USA* **84**, 2532.
6.   Balasubramaniam, A., Renugopalakrishnan, V., Rigel, D.F., Nussbaum, M.S., Rapaka, R.S., Dobbs, J.C., Carreira, L.A. and Fischer, J.E. (1989) *Biochim. Biophys. Acta* **997**, 176.
7.   Wüthrich, K. (1986) *NMR of Proteins and Nucleic Acids*, Wiley, New York, NY.
8.   Van Gunsteren, W.F., Berendsen, H.J.C., Hermans, J., Hol, W.G. and Postma, J.P.M. (1983) *Proc. Natl. Acad. Sci. USA* **80**, 4315.

Proline residues in bacteriorhodopsin: Conformation and temperature dependence

Charles M. Deber*, Guang-Yi Xu** and Barbara J. Sorrell

Research Institute, Hospital for Sick Children, Toronto M5G 1X8, Canada
Department of Biochemistry, University of Toronto, Toronto M5S 1A8, Canada

Introduction

Proline has been described as a classical breaker of α-helical structure in proteins [1]. Its backbone dihedral angles are limited by the cyclic nature of its pyrrolidine ring; as well, there is no amide proton on a Pro-linked peptide bond to participate in helix stabilization through H-bonding; and, as a secondary amide, Pro is a relatively polar residue [2]. Yet, Pro residues are widely observed in transmembrane (TM) segments of many integral membrane proteins which function as receptor subunits or transporters [3]. Particularly because such segments are expected to contain high contents of α-helices, Pro residues should create thermodynamically less stable structures where they occur within TM segments.

These circumstances have led us to suggest that, in addition to the purely structural attributes which Pro residues confer on peptide chains, further functionality may exist for intramembranous X-Pro peptide bonds [3,4]. Two properties characteristic of X-Pro bonds may be considered in this context. Firstly, while peptide bonds linking amino acids in proteins are overwhelmingly *trans*, those involving proline (i.e., the X-Pro bond where X = any amino acid) occur in both *trans* and *cis* conformations, interconvertable by rotation of 180° about the C-N bond; specific *cis* X-Pro bond sites have been observed in crystal structures of several proteins [e.g., 5], while kinetics of protein re-folding have suggested a step attributable to *cis/trans* interconversion at X-Pro sites [6]. Thus, *trans-cis* interconversion of an X-Pro bond within, or adjacent to, a cellular membrane, could alter the alignment of a segment, and thereby provide the conformational basis for a dynamic, reversible mechanism for signal transduction or transport channel regulation [3].

However, *trans/cis* X-Pro isomerization, particularly if it were to occur in mid-bilayer, could in some instances be either too drastic a change structurally, and/or kinetically inefficient for channel regulation. In such cases, functionality for intramembranous Pro residues may arise from a second property of X-Pro bonds, i.e., the fact that the tertiary amide character of this bond confers increased basicity to the attached carbonyl oxygen atom. Thus, conformations of Pro-containing cyclic peptides are frequently characterized by structures in which the X-Pro bonds remain

---

* To whom correspondence should be addressed.
** Visiting Scientist on leave from the Shanghai Institute of Pharmaceutical Industry.

trans, while the carbonyl group bound to the imino-nitrogen atom of the Pro residue serves as an H-bond acceptor. This situation not only promotes involvement of the X-Pro carbonyl group in the H-bond which closes 10-membered β-turn and 7-membered γ-turn structures [7], but in addition, implies an innate local flexibility of Pro-containing transmembrane segments, and accordingly the possible interconversion between α-helix and structures with an alternate local arrangement of hydrogen bonds [4].

To examine these phenomena directly in a membrane transport protein, we have been studying bacteriorhodopsin (bR) - a well-characterized integral membrane protein of the purple membrane of *H. halobium* which uses light energy to transport protons [8] - by ^{13}C nuclear magnetic resonance spectroscopy (^{13}C NMR). Bacteriorhodopsin has 11 Pro residues, five of which reside within the membrane domain, with three of them (highly conserved [9]) (Pro^{50}, Pro^{91}, Pro^{186}) reported to be deeply membrane-buried [10]. Conformational changes during the bR photocycle have implicated Pro residues [11], while site-directed mutagenesis of Pro^{186} to a leucine residue produced a mutant which displayed a drastically altered chromophore and diminished pumping ability [12].

Results and Discussion

^{13}C NMR provides a method for absolute identification of the isomeric status of X-Pro peptide bonds; peptide bond isomers are normally in slow exchange at ambient temperature, and the diagnostic *cis* and *trans* ^{13}C-chemical shifts of the C_γ carbon resonances of the Pro pyrrolidine side chain are typically separated by ca. 2 ppm, as observed in a wide variety of naturally-occurring and synthetic Pro-containing linear and cyclic peptides [e.g., 13-15]. We have now obtained ^{13}C-labelled bR by replacing unlabelled proline in the culture medium by $^{13}C_\gamma$-proline, as described in the legend to Fig. 1. The 125 MHz spectrum of $^{13}C_\gamma$-Pro-bR (Fig. 1A) at 24° was recorded in the mixed-solvent system $CHCl_3 : CD_3OD$ (1 : 1) + 0.1 M $LiClO_4$, in which bR mobility is sufficient for observation of high-resolution 1D spectra while the bR monomer retains its native secondary structure and lipid environment [16]. The ^{13}C-enriched Pro C_γ residues give rise to three groups of resonances between 25.2-26.9 ppm. This Pro C_γ chemical-shift range indicates that in resting state bR, all 11 X-Pro peptide bonds in the protein are in the *trans* conformation [17]. *Cis* X-Pro bonds, if present, would be expected to produce a Pro C_γ peak at ca. 23.0 to 23.9 ppm, a region which is devoid of signals [13-15,18].

The observed spectral division of the 11 Pro residues of bR into three resonance groups arises not only from local shielding/deshielding effects of the side chains in the residues immediately adjacent to each proline (i.e., there are 11 different X-Pro-Y peptide triads), but also from the influence of the overall secondary/tertiary structure of bR, the latter effect known to give rise to extensive chemical-shift dispersion [19]. In this context, it became of interest to explore the susceptibility of bR Pro-containing segments to temperature-induced conformational changes. Although *cis-trans* isomerization of X-Pro peptide bonds is not known specifically to be temperature-induced, other perturbation(s) in the local environment of these residues sensitive to structural interconversions, as discussed above, may be manifested at temperatures near the physiological range for native bR. Figure 1B displays the changes which occur *within* the Pro C_γ *trans* region when the temperature is increased. The relative areas of the three Pro-C_γ peak groupings are altered: the ratios of the areas of these peaks, calculated both from NMR-integrated intensities and independently by cut-and-weigh procedures, have changed from 3 : 2 : 6 at 24° to 3 : 1 : 7 at 45° (downfield to upfield). Visual inspection of the spectrum, aided by difference spectra (not shown), indicates that the change apparently involves a single bR proline residue, which moves upfield from 26.3 ppm to 25.7 ppm.

The endogenous bR lipid resonances are also affected by the increase in temperature. Comparison of Figs. 1A and 1B shows coalescing of the lipid peaks near 23 ppm at 45°. This observation reflects the increased motion of the two terminal methyl lipid phytanyl groups [20], resulting in averaging of their local environments.

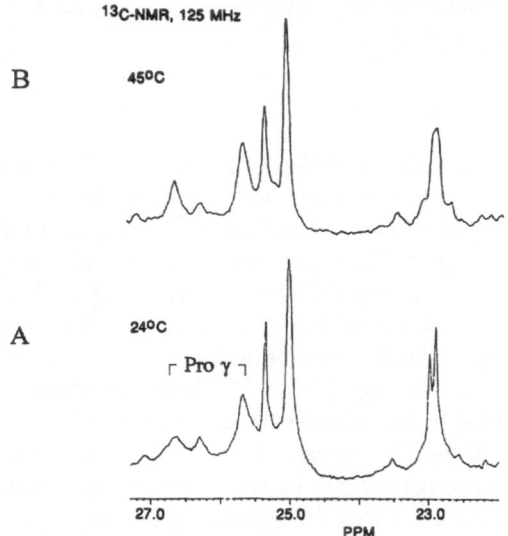

Fig. 1. ^{13}C-NMR spectra (125 MHz) in the region 22–27 ppm of $13C_\gamma$-Pro-bacteriorhodopsin (bR) (21 mg protein/mL). (A) 24°; (B) 45°.

H. halobium *strain ET1001 (a kind gift from Dr. W. Stoeckenius, University of California, San Francisco, CA) was grown and purified as described [23] except that the sucrose gradient was replaced by two sonication steps [24] in the final purification.* ^{13}C-labelled-bR was obtained by replacing unlabelled proline in the culture medium with an equivalent amount of D,L-$^{13}C_\gamma$-proline [25], a kind gift from Dr. D. Torchia, N.I.H., Bethesda, MD. ^{13}C enrichment was >95%. Solvent system: $CHCl_3$: CD_3OD (1 : 1) + 0.1 M $LiClO_4$ [16] (see text). Spectra were obtained on a Bruker 500 MHz spectrometer with continuous proton broadband decoupling in ca. 24,000 transients. Pulse width: 90°. A delay of 4–5 sec. was inserted after each pulse to ensure that Pro ring carbons (expected to display T_1's in the order of 0.1 sec. [26]) are fully relaxed. bR endogenous purple membrane lipid natural abundance peaks near 23 and 25 ppm in (A) and (B) are as reported by Degani et al. [20]. The total intensity of enriched Pro C_γ resonances, corresponding to the 11 Pro residues in bR, occurs in the region expected for trans X-Pro peptide bonds. See text for a further discussion.

Loosening of secondary structure and accordingly increased exposure of some residues to the negatively-charged surrounding environment of the bR monomer (arising from the presence of purple membrane anionic lipids) could generally induce shifts in corresponding ^{13}C resonances. However, since the observed chemical-shift movement arises from a single Pro residue, one interpretation of the shift is that it may be associated with a local re-ordering of bR conformation in the vicinity of the Pro residue involved. Chemical-shift movement within the *trans* Pro C_γ (and Pro C_β) spectral regions has been correlated with γ-turn structure [21]. Formation of the seven-membered H-bonded ring produces eclipsing of the Pro $C_\beta H_2$ by the Pro C=O moiety, and upfield shifts of the Pro C_β proton and carbon resonances; this effect concomitantly produces downfield movement of adjacent Pro C_γ resonances. By analogy, the upfield movement of a bR Pro C_γ resonance with increasing temperature would then reflect the conformational alteration from a pre-existing γ-turn (at 24°) to another, possibly less-ordered local structure. Synthetic ^{13}C-enriched membrane-active peptides containing the Leu-Pro-Phe triad display chemical-shift changes upon membrane interaction which have been similarly interpreted in terms of γ-turn formation [4].

Thus, whether the conformationally-mobile Pro residue moves 'in-and-out' of the membrane, or actually remains membrane-buried (e.g., Pro^{186}), the experiments in Fig. 1 represent a spectral manifestation of the inherent structural flexibility of a Pro-containing segment in bR, and conceivably a dynamic role for such a segment in channel regulatory events. Adaptation of this methodology to the study of bR intermediates (e.g., M_{412} [22]) could thus examine a possible functional role of either *trans/cis* X-Pro isomerization, and/or other photo-induced structural consequences of the presence of intramembranous Pro residues in bacteriorhodopsin.

Acknowledgements

This work was supported, in part, by grants to C.M.D. from the Medical Research Council of Canada, and from the Natural Sciences and Engineering Research Council of Canada. B.J.S. held an M.R.C. Studentship. G.-Y. X. is a Visiting Scientist of the Hospital for Sick Children Foundation. NMR spectra were recorded at the Toronto Biomedical NMR Center, which is supported, in part, by a Medical Research Council Maintenance grant.

References

1. Chou, P.Y. and Fasman, G. (1978) *Adv. Enzymol.* **47**, 45.
2. Rose, G.D., Geselowitz, A.R., Lesser, G., Lee, R. and Zehfus, M.H. (1985) *Science* **229**, 9834.
3. Brandl, C.J. and Deber, C.M. (1986) *Proc. Natl. Acad. Sci. USA* **83**, 917.
4. Deber, C.M., Glibowicka, M. and Woolley, G.A. (1990) *Biopolymers* **29**, 149.
5. Schmid, F.X., Grafl, R., Wrba, A. and Beintema, J.J. (1986) *Proc. Natl. Acad. Sci. USA* **83**, 872.
6. Brandts, J.F., Halvorson, H.R. and Brennan, M. (1975) *Biochemistry* **14**, 4953.
7. Smith, J.A. and Pease, L.G. (1980) *CRC Crit. Rev. Biochem.* **8**, 315.

8. Khorana, H.G. (1988) *J. Biol. Chem.* **263**, 7439.
9. Oesterhelt, D. and Tittor, J. (1989) *Trends Biochem. Sci.* **14**, 57.
10. Katre, N.V., Finer-Moore, J., Stroud, R.M. and Hayward, S. (1984) *Biophys. J.* **46**, 195.
11. Rothschild, K.J., He, Y.-W., Gray, D., Roepe, P.D., Pelletier, S.L., Brown, R.S. and Herzfeld, J. (1989) *Proc. Natl. Acad. Sci. USA* **86**, 9832.
12. Hackett, N.R., Stern, L.J., Chao, B.H., Kronis, K.A. and Khorana, H.G. (1987) *J. Biol. Chem.* **262**, 9277.
13. Meraldi, J.P., Schwyzer, R., Tun-kyi, A. and Wüthrich, K. (1982) *Helv. Chim. Acta* **55**, 1962.
14. Madison, V., Atreyi, M., Deber, C.M. and Blout, E.R. (1974) *J. Am. Chem. Soc.* **96**, 6725.
15. Fraser, P.E. and Deber, C.M. (1985) *Biochemistry* **24**, 4593.
16. Arseniev, A.S., Kuryatov, A.B., Tsetlin, V.I., Bystrov, V.F., Ivanov, V.T. and Ovchinnikov, Yu.A. (1987) *FEBS Lett.* **213**, 283.
17. Deber, C.M., Sorrell, B.J. and Xu, G.-Y. (1990) *Biochem. Biophys. Res. Comm.* **172**, 862.
18. Deber, C.M., Fossel, E.T. and Blout, E.R. (1974) *J. Am. Chem. Soc.* **96**, 4015.
19. Torchia, D.A., Sparks, S.W. and Bax, A. (1988) *Biochemistry* **27**, 5135.
20. Degani, H., Danon, A. and Caplan, S.R. (1980) *Biochemistry* **19**, 1626.
21. Deber, C.M., Madison, V. and Blout, E.R. (1976) *Accts. Chem. Res.* **9**, 106.
22. Smith, S.O., Courtin, J., van den Berg, E., Winkel, C., Lugtenburg, J., Herzfeld, J. and Griffin, R.G. (1989) *Biochemistry* **28**, 237.
23. Oesterhelt, D. and Stoeckenius, W. (1974) *Methods Enzymol.* **31**, 667.
24. Braiman, M. and Mathies, R. (1980) *Biochemistry* **19**, 5421.
25. Young, P.E. and Torchia, D.A. (1983) in *Proceedings of the 8th American Peptide Symposium* (Hruby, V.J. and Rich, D.H., Eds.), Pierce Chem. Co., Rockford, IL, pp. 155–158.
26. Torchia, D.A. and Lyerla, J.R. Jr. (1974) *Biopolymers* **13**, 97.

A magnetic resonance study of solution conformation of Substance P and its N-terminal fragment

D.R. Shukla and Sudha Mahajan

Molecular Biophysics Laboratory, School of Life Sciences, Jawaharlal Nehru University, New Delhi 110 067, India

Introduction

Substance P is an important neuromodulator. The amino acid sequence of Substance P and its N-terminal fragment is as follows:

Arg-Pro-Lys-Pro-Gln-Gln-Phe-Phe-Gly-Leu-Met-NH_2 → Substance P
R - P - K - P - Q - Q - F - F - G - L - M - NH_2
Arg-Pro-Lys-Pro → SP_{1-4}
R - P - K - P

Furthermore, it is known to be involved in hypotension, vasodilation and contraction of various smooth muscles [1]. Structure-activity studies have led to a speculation that undecapeptide is a precursor of more potent shorter sequences which may be released by enzymes acting at its N-terminal end [2]. The shortening of undecapeptide at its N-terminal end by three residues may produce important changes in conformation. It is conceivable that removal of two positive charges Arg^1 and Lys^3 certainly change its physico-chemical properties. The higher lipophilicity of fragment 5–11 favours its diffusion through cell membrane and accounts for higher pharmacological activity [3]. Enzymes cleaving between amino acids 4 and 5 have been investigated in vitro and evidence has been found for their in vivo existence. SP_{1-4} is one of the fragments resulting from this post-proline enzymatic cleavage [4]. SP_{1-4} was found to be 30–50 times less immunoreactive than other fragments [5].

Previous studies on conformation of analogs of Substance P have suggested various possible conformations. Particularly *cis/trans* isomerization is an important phenomenon which occurs in Substance P fragments [6] due to the proline moiety at the N-terminal end. Theoretical studies conducted on SP_{1-4} have proposed a stretched conformation for the molecule [7].

The objective of the present study is to obtain the preferential conformation of Substance P and its N-terminal fragment SP_{1-4} in lipid environment. Here we report the solution conformation of Substance P at various temperature and pH conditions.

Materials and Methods

Substance P and its N-terminal fragment SP_{1-4} were obtained from Sigma Chemicals (U.S.A). The solvents used were D_2O and DMSO-d_6 (99.96% deuterium content).

NMR Experiments

The NMR samples were prepared by dissolving peptides in appropriate solvent. The samples were lyophilized in D_2O or DMSO-d_6 prior to final preparation. The sample concentrations were 10 mM and 25 mM for Substance P and SP_{1-4} respectively. pH of the samples was maintained in the range of 6.4–6.6 in final solution. For higher pH NaOD/D_2O were used. The pH is reported uncorrected for isotopic effects.

All the NMR spectra were recorded on a Bruker AM 500 MHz FT-NMR spectrometer equipped with an Aspect 3000 data system. In every solvent the assignments were achieved with the aid of 2D phase-sensitive COSY experiments. The pulse sequence used in this case was $90°-t_1-90° -$ FID (t_2); 256 increments in t_1 and 1024 points in t_2 were recorded.

Results and Discussion

Distribution of chemical shifts in 500 MHz ^1H NMR spectra of Substance P and SP_{1-4} in D_2O and DMSO-d_6 is shown in Table 1, Table 2.1 and 2.2 respectively. Table 3.1 and Table 3.2 display significant changes in chemical shifts of various protons at different temperature and pH. The assignments for different amino acid residues were made on the basis of 2D phase-sensitive COSY connectivities and one-dimensional ^1H NMR spectra. Although SP_{1-4} is a small molecule, its complete assignment is not possible on the basis of 2D COSY only due to the presence of proline moieties. For complete assignment earlier reports on Substance P and its analogs were consulted [8–11].

In case of SP_{1-4} in D_2O, HY-HS connectivities were clearly detectable (Figs. 3 and 4). This provided us a clue to assign Arg[1] (AH, BH) and Lys[3] (AH4, BH) unambiguously. The SH-EH connectivity in the Lys[3] spin system is also apparent. But due to overcrowding and diagonal intervention in the 1–2 ppm region AH and BH

Table 1   *Distribution of chemical shifts <δ> ppm in ^1H NMR spectra of SP in D_2O at 500 MHz*

Residue	AH	BH	YH	SH	ECH
Arg	4.2	1.94	1.70	3.22	
Pro	4.40	2.30	2.00	3.65	
Lys	4.54	1.8, 1.7	–	1.70	3.20
Pro	4.43	2.35	2.03	3.85	
Gln	4.65	1.85	2.30	–	
Gln	4.45	1.89	2.30	–	
Phe	4.66	2.99	–	–	
Phe	4.66	3.17	–	–	
Gly	–	–	–	–	
Leu	4.36	1.66	1.65	0.93	
Met-NH$_2$	4.49	2.07	2.69	–	

Table 2.1   *Distribution of chemical shifts <δ> ppm in ¹H NMR spectra of SP₁₋₄ in D₂O*

Residue	AH	BH	YH	SH	EH
Arg	4.40	1.98	1.73	3.25	–
Pro	4.56	2.33	2.03	3.82	–
Lys	4.50	1.86	1.54	1.70	3.01
Pro	4.39	2.32	2.02	3.75	–

Table 2.2   *Distribution of chemical shifts <δ> ppm in ¹H NMR spectra of SP₁₋₄ in DMSO-d₆ at 500 MHz*

Residue	NH	AH	BH	YH	SH	EH
Arg	8.19	4.35	1.70	1.63	3.06	–
Pro		4.36	2.07	1.62	3.56	–
Lys	8.21	4.37	1.48	1.35	1.54	2.73
Pro		4.19	2.12	1.80	3.36	

Table 3.1   *Distribution of chemical shifts <δ> ppm in ¹H NMR (500 MHz) spectra of SP in D₂O at 300 K*

Protons (pH)	AR	AP₁	AP₂	AL	BR	BP₁	1	BF	SP₁	SP₂	EK	Aromatic protons
7.0	4.20	4.40	4.43	4.36	1.94	2.3	0	2.99	3.65	3.85	3.02	7.22,7.32
11.0	3.12	3.70	3.80	3.36	0.86	1.4	0	2.16	2.52	2.88	2.50	7.23,7.27
<δ>	1.08	0.70	0.63	1.00	1.08	0.9	0	0.83	1.13	0.97	0.52	0.01,0.05

Table 3.2   *Distribution of chemical shifts <δ> ppm at 500 MHz of various protons of SP in D₂O at different temperatures*

Protons T (K)	AR	AP₁	AP₂	AL	BR	BP₁	1	BF	SP₁	SP₂	EK	Aromatic protons
300	3.12	3.70	3.80	3.36	0.86	1.8	6	1.40	2.16	2.52	2.88	7.25,7.27
310	3.28	3.88	3.98	3.64	1.04	1.6	0	2.04	2.72	3.06	2.30	7.50,7.44
320	3.38	3.96	4.08	3.74	1.24	1.7	2	2.22	2.82	2.90	2.44	7.54,7.64
330	3.48	4.02	4.20	3.38	1.27	1.8	2	2.36	3.12	3.26	2.50	7.64,7.72

A = alpha; B = beta; S = delta; E = epsilon.

were clearly not distinguishable. The YH-SH connectivities in case of Pro² and Pro⁴ are clearly assigned. Based on this observation assignments were completed for AH and BH for two prolines. Pro² and Pro⁴ are also responsible for double resonance lines in Fig. 3 [12]. Pro² and Pro⁴ could be distinguished according to results published by Szollosy and co-workers [13,14].

Doubling of resonance lines is an indication of the existence of various isomers of the oligopeptides containing proline moiety [14]. Earlier studies have already suggested this isomerism due to the presence of a proline moiety at the N-terminal end of the fragment. Our results also show this phenomenon significantly.

In DMSO-d₆ (Figs. 1 and 6) assignments were completed for SP₁₋₄ on the basis of the same argument. NH-AH connectivities could be traced for Arg¹ and Lys³. SH-NH connectivities are only detectable in Arg¹ because under the given conditions Lys³

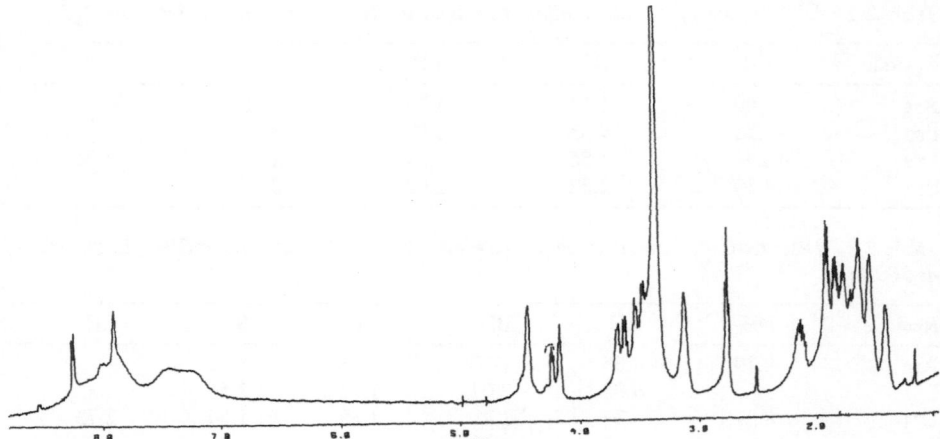

Fig. 1. ^1H NMR spectrum of SP_{1-4} in DMSO-d_6

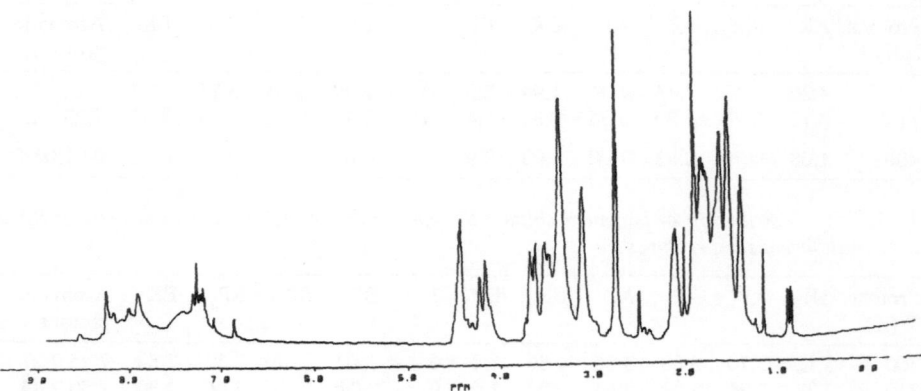

Fig. 2. ^1H NMR spectrum of SP in DMSO-d_6

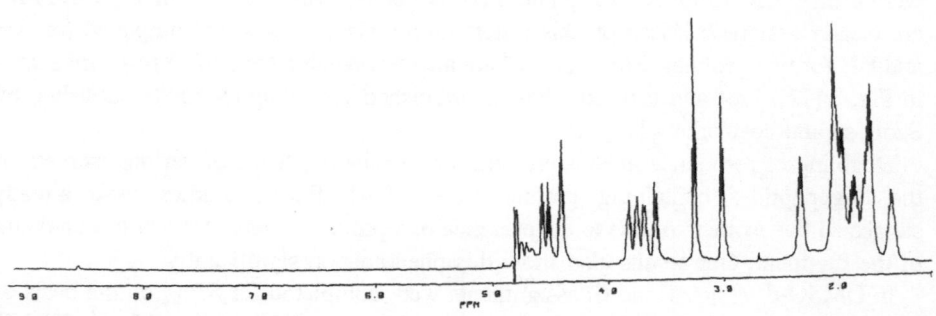

Fig. 3. ^1H NMR spectrum of SP_{1-4} in D_2O.

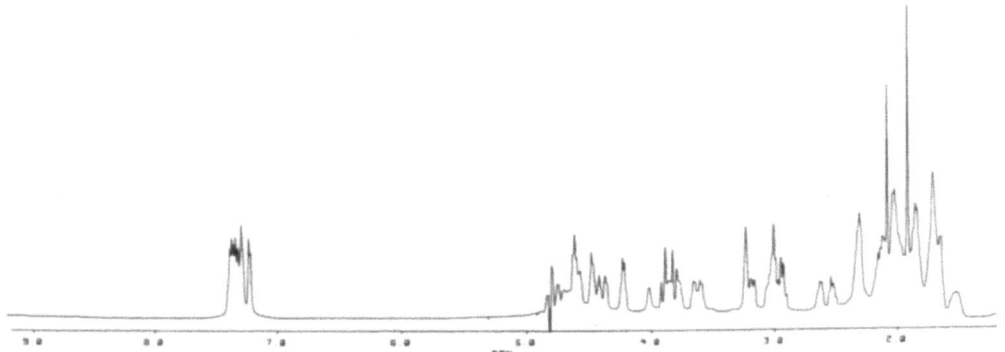

Fig. 4. ^{1}*H NMR spectrum of SP in D$_{2}$O.*

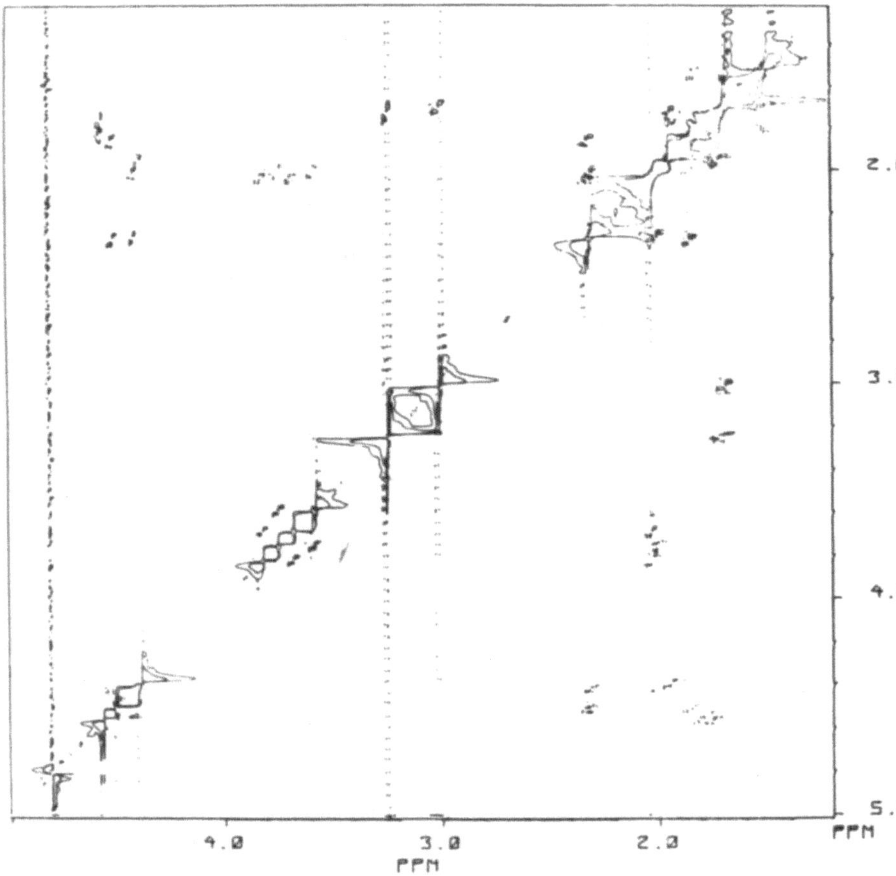

Fig. 5. *2D Phase-sensitive COSY of SP$_{1-4}$ in D$_{2}$O.*

91

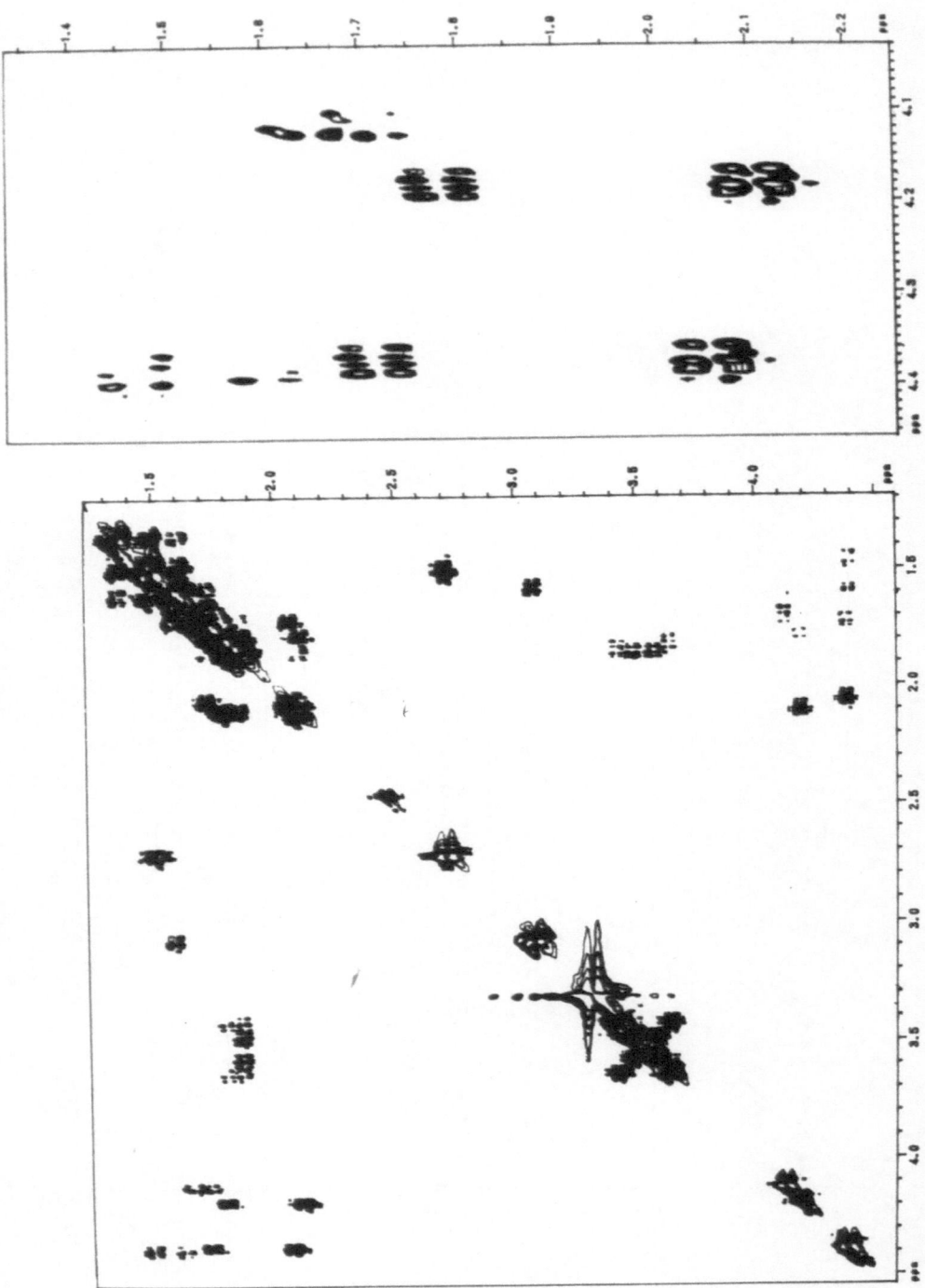

Fig. 6.   2D Phase-sensitive COSY of SP$_{1-4}$ in DMSO-d$_6$.

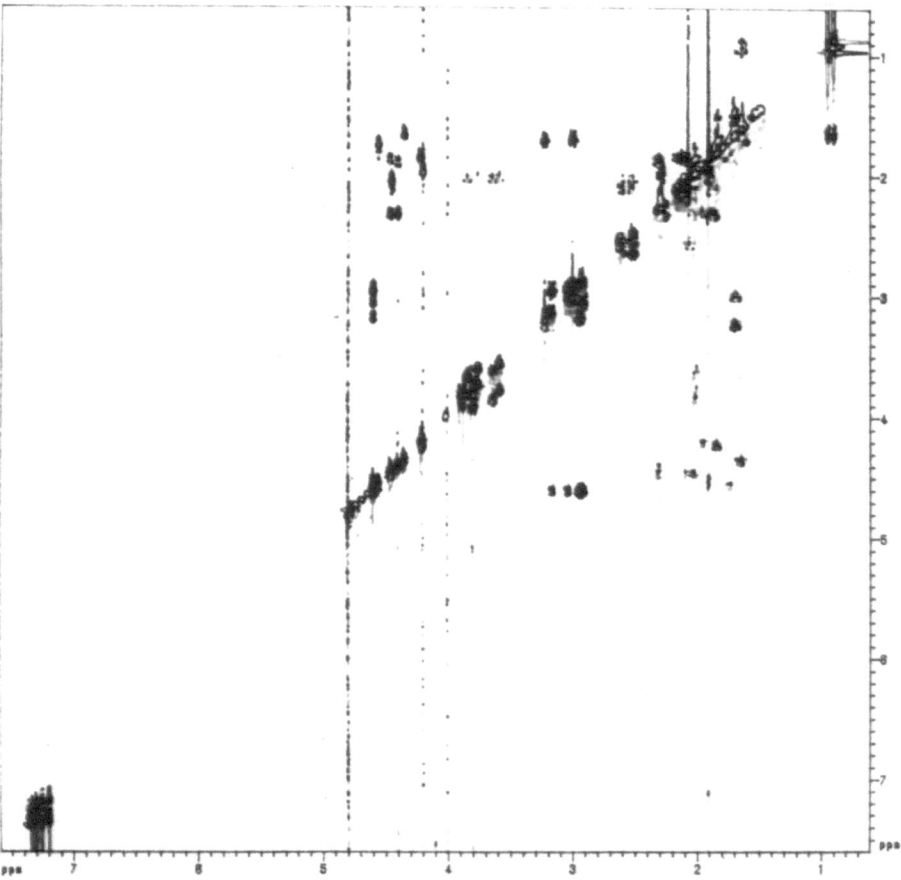

Fig. 7. 2D Phase-sensitive COSY of SP in D₂O.

amide protons which are connected to EH are not detectable. They are detectable below 283 K [4].

A significant difference in chemical shifts was observed in DMSO-d_6 compared to D$_2$O. Particularly the AH-BH protons of Pro[2] and Pro[4] were greatly affected. In DMSO-d_6 doubling of resonance lines is also reduced. This points towards a prominent change in the solution conformation of SP and SP$_{1-4}$.

Table 1 shows the distribution of chemical shifts of undecapeptide Substance P in D$_2$O. Comparing the chemical shifts of Arg[1], Pro[2], Lys[3] and Pro[4] of the undecapeptide with those of the N-terminal fragment, an upfield shift is clearly visible for the former (Figs. 4 and 7). This suggests a stronger shielding of N-terminal residues in the undecapeptide.

Simultaneously we conducted experiments at higher temperature which showed low-field shifts in all protons. Theoretical studies have proposed a stretched conformation of the SP$_{1-4}$ molecule in D$_2$O. The presence of two proline residues in SP$_{1-4}$ hinders formation of any defined structure [7]. In a stretched conformation proximity of protons would be too large. Arg^1 and Lys^3 both are positively charged residues so in this earlier proposed stretched conformation the positions of Arg and Lys will be such that they will repel each other. But the Pro^4 free -CO end will be attracted towards $Lys^3(NH^+)$. The phenomenon of isomerism is governed by electronic factors [11].

Based on our results we may suggest that the conformational contribution of the first four N-terminal residues to the conformation of Substance P is different from the conformation of the N-terminal fragment. In all tachykinin peptides the C-terminal sequence Phe-Xaa-Gly-Leu-Met-NH_2 is identical and it is the heterogeneity of the N-terminal which decides their three-dimensional structure and potency [12].

In the present case *cis/trans* isomerization is not detectable as in the case of SP_{1-4}. Secondly Arg^1 and Lys^3 are not expected to have interaction with the Pro^4-CO moiety as is the case with SP_{1-4}. But the electronic configuration of SP_{1-4}, part of the undecapeptide appears to influence the conformation of the undecapeptide which is clearly observed when compared with data available for the SP_{5-11} fragment. Solvent contribution is also important while considering the influence of the N-terminal end of the undecapeptide. In DMSO-d_6, SP_{1-4} opted for a more rigid conformation. This is important because DMSO-d_6 is an aprotic solvent and does not initiate hydrogen bonding. But it provides almost same the environment as does a lipid. On the basis of our results we may suggest that the role of the N-terminal sequence in determining the undecapeptide conformation depends not only on the electronic configuration of the N-terminal part but also on the nature of the solvent.

At high pH (11.0) various protons resonate at high field except aromatic protons of phenylalanines (Table 3.1). Temperature-dependent measurements show a gradual low-field shift in the peak position of various protons with rise in temperature. This is equally true for aromatic protons. This will help us to detect conformational changes at different solvent conditions. Further studies are in progress in order to gain information about conformations of Substance P and its N-terminal fragment SP_{1-4} in lipid matrix.

Acknowledgements

The authors thank Prof. G. Govil for his contribution in this work. The help rendered by the National facility of 500 MHz at TIFR at Bombay is gratefully acknowledged. This project is sponsored by a grant from the Council of Scientific and Industrial Research, Government of India.

References

1. Regoli, D., Escher, E. and Mizrahi, J. (1984) *J. Pharmacol.* **28**, 301.
2. Bury, R.W., Mashford, M.L. (1976) *J. Med. Chem.* **19**, 854.
3. Couture, R., Fournier, A., Pierre, S. and Regoli, D. (1979) *Can. J. Pharmacol.* **57**, 1427.
4. Otter, A. and Kotovych, G. (1986) *J. Magn. Reson.* **69**, 187.
5. Higa, T., Wood, G. and Deosidrio, D.M. (1989) *Int. J. Peptide Protein Res.* **33**, 446.
6. Otter, A. and Kotovych, G. (1987) *J. Magn. Reson.* **74**, 293.

7. Cotrait, M. (1983) *Int. J. Peptide Protein Res.* **22**, 110.
8. Billeter, M., Braun, W. and Wüthrich, K. (1982) *J. Mol. Biol.* **155**, 321.
9. Bundi, A., Grathwohl, C., Hochman, J. and Keller, R.M. (1975) *J. Magn. Reson.* **18**, 191.
10. Bundi, A. and Wüthrich, K. (1979) *Biopolymers* **18**, 279.
11. Wüthrich, K. (1986) *NMR of Proteins and Nucleic Acids*, Wiley, New York.
12. Kawaki, H., Otter, A., Beierbeek, H. and Kotovych, G. (1986) *J. Biomol. Struct. Dyn.* **3**, 4.
13. Szollosy, S., Otter, A., Stewart, J.M. and Kotovych, G. (1986) *J. Biomol. Struct. Dyn.* **3**, 795.
14. Szollosy, S., Otter, A., Stewart, J.M. and Kotovych, G. (1986) *J. Biomol. Struct. Dyn.* **4**, 501.
15. Waedler, S., Lee, L. and Redfield, A.G. (1976) *J. Am. Chem. Soc.* **98**, 2927.
16. Chassaing, G., Convert, O. and Levielle, S. (1986) *Eur. J. Biochem.* **154**, 77.

Solution conformation of dermorphin in DMSO: 2D NMR studies

Y.U. Sasidhar, M.M. Dhingra and Anil Saran

Chemical Physics Group, Tata Institute of Fundamental Research, Bombay 400 005, India

Abstract

The solution conformation of unprotonated dermorphin (Tyr[1]-D-Ala[2]-Phe[3]-Gly[4]-Tyr[5]-Pro[6]-Ser[7]-NH$_2$), an opioid peptide, in DMSO has been studied by 1D and 2D NMR techniques at 500 MHz. The temperature dependence of amide proton chemical shifts showed that none of the amide protons is involved in an intramolecular hydrogen bonding and that all the amide protons are exposed to the solvent. There are two interconverting conformers one of which is more populated than the other. Exchange crosspeaks are observed between the two conformers at 330 K. The major conformer is *trans* at the Tyr[5]-Pro[6] peptide bond and is in extended conformation.

Introduction

Dermorphin is a heptapeptide (Tyr[1]-D-Ala[2]-Phe[3]-Gly[4]-Tyr[5]-Pro[6]-Ser[7]-NH$_2$) which has been isolated from the skin of South American frogs and shown to possess opiate-like activity much higher than that of endogenous enkephalins [1–3]. The presence of a D-Ala residue in position 2 in peptides isolated from a non-microbial organism was quite unusual. However, its presence was confirmed by synthesis of this peptide and its [L-Ala[2]] analog [3]. It is known that the insertion of a D-amino acid residue in position 2 of several synthetic enkephalins led to enhanced biological activity [4]. From this point of view, it is quite interesting to note that the [L-Ala[2]] analog of dermorphin is virtually devoid of any opioid activity [3]. This is important in considering the structure-activity relationship of dermorphin.

Several NMR investigations have been carried out to determine the conformation of dermorphin in solution [5–9]. Attempts have been made to correlate the NMR parameters such as chemical shifts and ^{1}H-^{1}H coupling constants of dermorphin in DMSO to its biological activity [5]. The results of ^{1}H NMR at 500 MHz as well as theoretical calculations using semi-empirical partioned functions on dermorphin and its [L-Ala[2]] analog indicate the predominance of type I β-turn around Pro[6]-Ser[7] residues at the C terminal end and an extended conformation in the central part involving Phe[3]-Gly[4]-Tyr[5] residues [6,7]. The hydrochloride and trifluoroacetate of dermorphin analog [L-Ala[2]] in DMSO solution were also subjected to ^{1}H NMR investigations at various fields from 80 to 600 MHz and an extended conformation of dermorphin was deduced from the combined use of NMR parameters and 1D NOE measurements [8]. In another study involving ^{1}H and ^{13}C NMR, CD and UV on dermorphin and its [L-Ala[2]] analog, it was concluded that both peptides exist in DMSO as a mixture of rapidly interconverting conformers and an extended β-sheet conformation stabilized by intermolecular association was proposed [9]. A peculiar folded conformation of dermorphin was deduced from a CD study on dermorphin and its [L-Ala^2] analog [10]. A conformation of dermorphin characterized by β-turns and β-sheets was also proposed from molecular modeling, vibrational spectroscopy and CD studies [11].

From the above discussion it is quite clear that there is no concurrence on the conformation of dermorphin in solution. Therefore we have undertaken a study of the conformation of dermorphin in DMSO. Conformational analysis is facilitated by 2D NMR techniques like COSY, NOESY and 2D - J Spectroscopy. Based upon our results we propose that dermorphin is adopting an extended conformation in solution.

Experimental

The sample of dermorphin was obtained from Sigma. 5 mg of sample was dissolved in 0.5 ml of DMSO. 1D and 2D ^{1}H NMR spectra of dermorphin were recorded on a Bruker 500 MHz FT-NMR spectrometer. The 2D NMR techniques used were COSY [12], J-Resolved [13], and phase sensitive NOESY [14]. The DMSO resonance was used as an internal reference with a δ value of 2.49 ppm relative to TMS. A ROESY spectrum was recorded on a Varian 300 MHz spectrometer. The mixing time used was 200 ms.

Results and Discussion

(a) Assignment of resonances

Figure 1 shows the 1D spectrum of dermorphin at 298 K. The assignments indicated are obtained from the COSY experiment. Figure 2 shows the COSY spectrum of dermorphin at 298 K. All the spin systems are indicated in the figure. The assignments

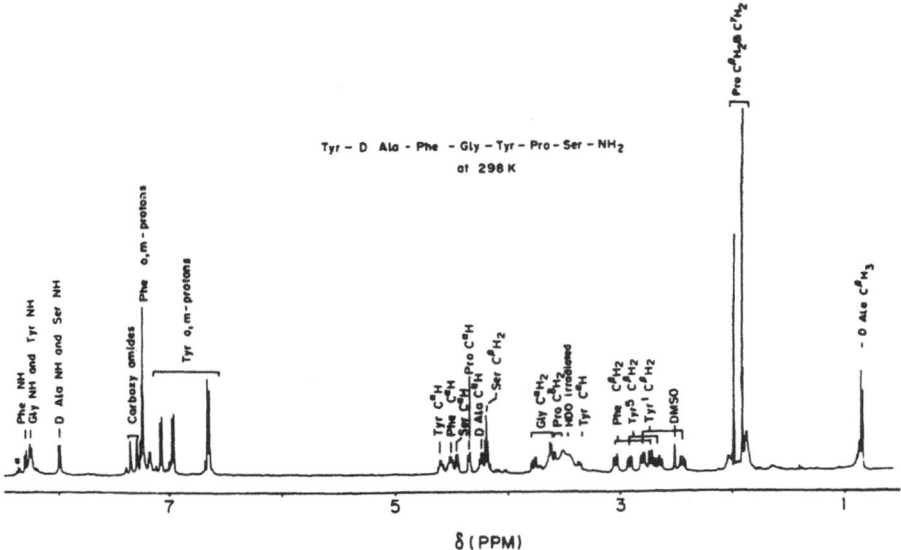

Fig. 1.   1D spectrum of dermorphin at 298 K.

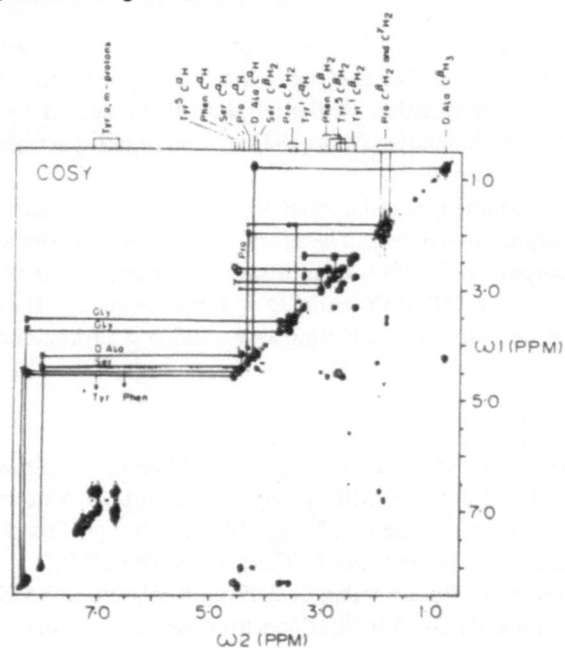

Fig. 2. COSY spectrum of dermorphin at 298 K.

of Gly, D-Ala, Pro and Ser are straightforward. In assigning the Tyr[5]NH and Phe[3]NH resonances use is made of the observation of sequential NOE crosspeaks between Phe NH and D-Ala α-CH. Throughout the 1D spectrum one may observe certain weak resonances like the one indicated by an asterisk in Fig. 1. It looks as though there are two conformers of which one is less populated than the other. This is possible considering the presence of an X-Pro bond that gives rise to *cis* and *trans* isomerism at the Tyr[5]-Pro[6] peptide bond. This point will be further elaborated in the section dealing with nuclear Overhauser spectroscopy.

(b) Temperature dependence of amide proton chemical shifts

Table 1 lists chemical shifts (δ), coupling constants ($^3J_{NH-C\alpha_H}$) and temperature

Table 1   *Chemical shifts (δ), coupling constants ($^3J_{NH-C\alpha_H}$), and temperature coefficients (− dδ/dT) of dermorphin in DMSO at 298 K*

NH resonance	δ/ppm	J/Hz	− dδ/dT 10^{-3} ppm / K	Allowed φ values*
D-Ala[2]/Ser[7]	7.98	7.8	4.1	−150, −90, 60
Phe[3]	8.30	8.8	5.5	−140, −100, 60
Gly[4]	8.26	6.3, 5.2	5.2	−160, −80, 30, 90
				−170, −70, 20, 90
Tyr[5]	8.25	7.8	5.5	−150, −90, 60
Amide 1	7.34		3.3	
Amide 2	7.29		4.4	

* In degrees, derived from J values using Bystrov curves [16].

98

coefficients $d\delta/dT$ of the amide proton resonances (at 298 K). From the values of the temperature coefficients it is quite clear that all the amide protons are exposed to the solvent [15]. They do not seem to be involved in any kind of intramolecular hydrogen bonding. The weak resonances mentioned earlier are observed even at 330 K suggesting that the barrier separating the major and minor conformers is rather high.

(c) Nuclear Overhauser spectroscopy

In the following, the major conformer is discussed. Figure 3 shows the NOESY spectrum of dermorphin at 298 K in the NH to α-CH region. The mixing time, τ_m, used is 200 ms.

The following sequential NOEs along the backbone have been observed.

1. Ser NH to Pro α-CH
2. Tyr⁵NH to Gly α-CHs or Gly NH to Gly α-CHs
3. Gly NH to Phe α-CH
4. Phe NH to D-Ala α-CH

The ambiguity in NOE 2 listed above arises because the Tyr⁵NH and Gly NH resonances are very close to each other. As a consequence, it is not clear whether the crosspeak is a sequential NOE between Tyr⁵NH and Gly α-CHs or an intraresidual NOE between Gly NH and Gly α-CHs. In order to resolve this point, we carried out a NOESY experiment at 310 K (τ_m = 200 ms). At this temperature the NH resonances are more resolved and we found that at least one of the crosspeaks is coming from Gly NH. However, this does not rule out the contribution of a sequential NOE from Tyr NH to Gly C$^\alpha$Hs.

Figure 4 shows the plots of NOESY crosspeak volumes vs mixing time in the NH to α-CH region. This exercise has been done to gain the effects of spin diffusion and to find a mixing time where spin diffusion is minimal. The NOEs seen at a mixing time of 200 ms are also present at a mixing time of 50 ms. Intraresidual NOEs like the one from Phe NH to Phe α-CH are seen, but these are rather weak. No new crosspeaks appeared when the mixing time was varied from 300–600 ms. In this range the intraresidual NOEs excepting that of Gly⁴NH/Tyr⁵NH to Gly⁴α-CHs didn't appear. Crosspeaks from Tyr⁵α-CH to Tyr⁵β-CHs, from Phe α-CH to Phe β-CHs, from D-Ala α-CH to D-Ala β-CHs etc., gained somewhat in intensity.

Similar patterns of crosspeaks were observed in the ROESY spectrum recorded at 300 MHz with a mixing time of 200 ms. This result rules out the possibility of spin diffusion and confirms the inferences drawn from the NOESY data.

(d) Conformation of dermorphin

From the temperature dependence of amide proton chemical shifts it has been concluded that all the NH protons are exposed to the solvent. Sequential NOEs have been observed as detailed in the section on NOE spectroscopy. Intraresidual NOEs are very weak or absent. Of the allowed ϕ values derived from $^3J_{NH-C\alpha_H}$ values (Table 1) positive ϕ values can be ruled out on steric grounds. All the residues listed have entries for ϕ values in the neighbourhood of 180°. These ϕ values would yield somewhat longer intraresidual NOE distances when compared to sequential NOE

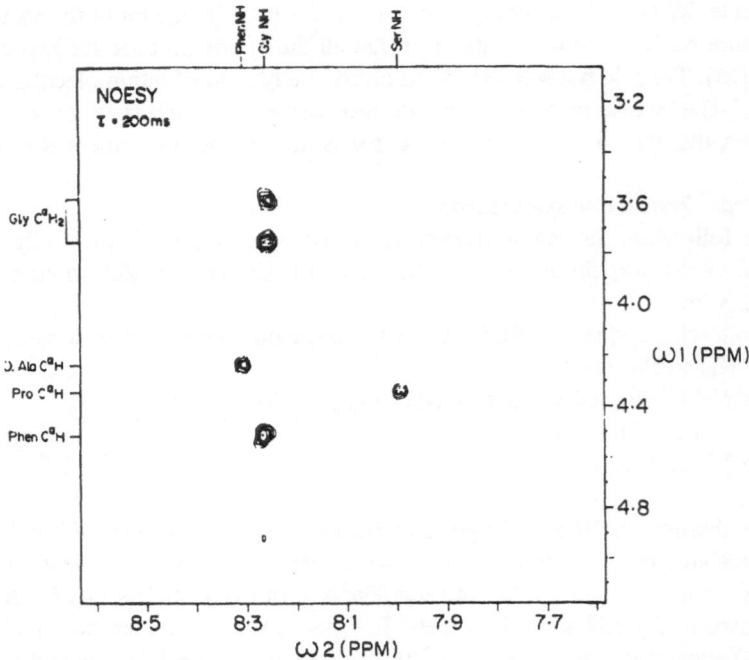

Fig. 3. *NOESY spectrum of dermorphin at 298 K (τ_m = 200 ms). NH to α-CH region shown.*

distances for ψ around 180° and hence intraresidual NOEs might be weaker compared to sequential NOEs.

Taking into consideration all the above mentioned points we propose an extended conformation for dermorphin in DMSO solution.

(e) Cis-trans isomerism

Earlier we mentioned that certain weak resonances are observed throughout the spectrum and speculated on the possibility of *cis* and *trans* isomerism at the Tyr⁵-Pro⁶ peptide bond. We have observed NOESY crosspeaks between Tyr⁵α-CH and Pro δ-CHs. These crosspeaks indicate the proximity of Tyr⁵α-CH and Pro δ-CHs and hence *trans* isomerism [17] at the Tyr⁵-Pro⁶ peptide bond of the major conformer. For *cis* isomerism one would expect a crosspeak between Tyr⁵α-CH and Pro⁶α-CH. We have not observed this crosspeak between main conformer resonances. We do not have any direct evidence that the minor conformer is adopting *cis* isomerism at the Tyr⁵-Pro⁶ bond. The difficulty is that though at certain places the weak resonances from the minor conformer are clearly seen, a majority of them appear to be buried under major conformer resonances. So it can only be said that the major conformer is adopting *trans* isomerism at the Tyr⁵-Pro⁶ peptide bond. No definite comments about the conformation of the minor conformer can be made.

The presence of two interconverting conformers which are slowly exchanging can be confirmed by the observation of exchange crosspeaks in the NOESY spectrum between the resonances of the two conformers. However, at room temperature no

100

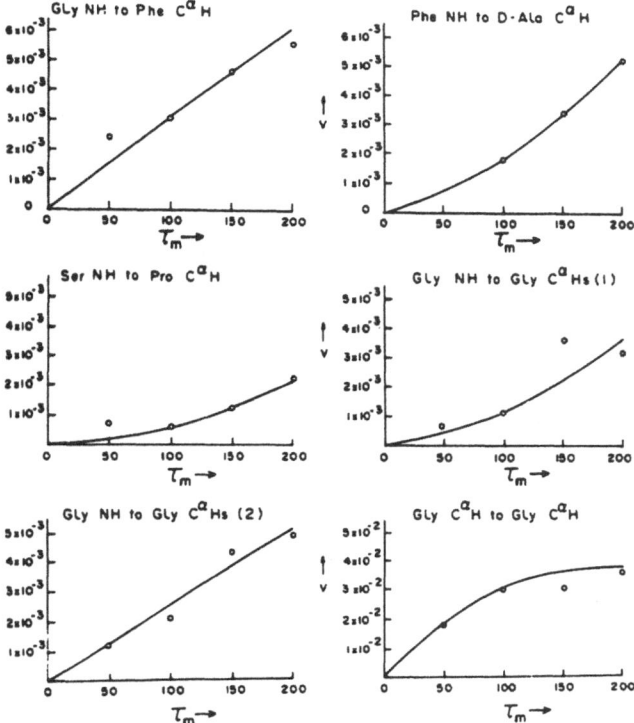

Fig. 4.   Plots of NOESY crosspeak volumes versus mixing time.

exchange crosspeaks are observed, probably because the exchange is too slow to give intense exchange crosspeaks. So we carried out COSY and NOESY experiments at 330 K so that at this higher temperature the increased exchange rate may facilitate the observation of exchange crosspeaks.

In the COSY spectrum at 330 K D-Ala NH does not show a crosspeak to D-Ala α-CH indicating that it has exchanged with the trace of HDO present in DMSO.

Figure 5 shows sections of the NOESY spectrum at 330 K depicting the following exchange crosspeaks.

1. Ser NH (major) to Ser NH (minor)
2. Tyr NH (major) to Tyr NH (minor)
3. Pro α-CH (major) to Pro α-CH (minor)
4. Pro γ-CH (major) to Pro γ-CH (minor)
5. Tyr α-CH (major) to Tyr α-CH (minor)
6. Ser β-CH (major) to Ser β-CH (minor)

These resonances are indicated in the 1D spectrum at 330 K shown in Fig. 6 and also are shown in the sections of 1D spectra plotted on top of the 2D NOESY spectra shown in Fig. 5. The observation of the exchange crosspeaks confirms our supposition that there are two interconverting conformers. We note that the only residues that

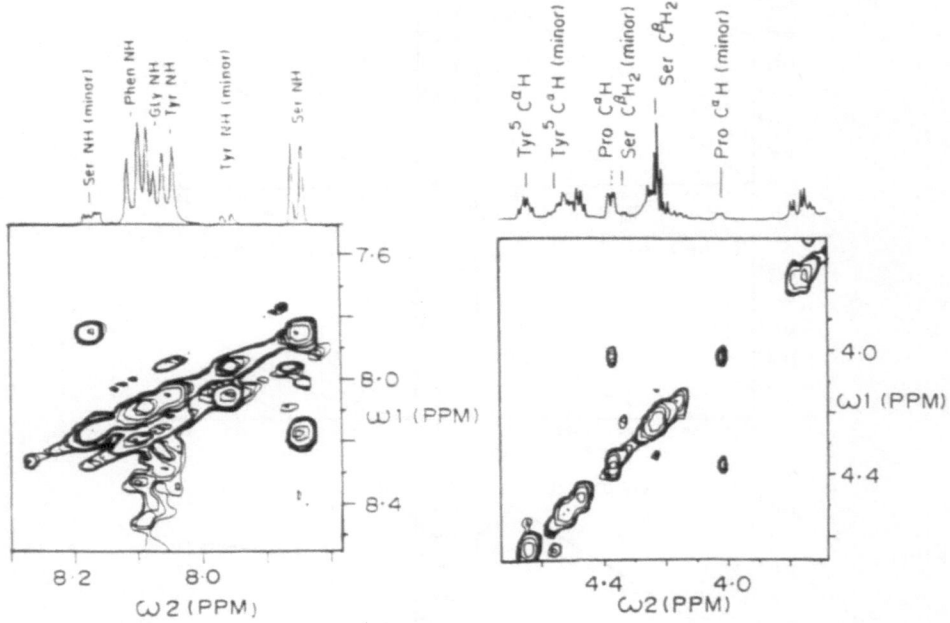

Fig. 5. *NOESY spectrum of dermorphin at 330 K showing exchange crosspeaks. (a) NH region, (b) α-CH region.*

show exchange crosspeaks are adjacent to Pro thus lending further support to the contention of *cis-trans* isomerism at the Tyr^5-Pro^6 peptide bond.

Discussion

In Table 2 we compare our NMR parameters with those available from earlier studies on dermorphin. It appears that in most of the earlier studies, dermorphin used is in its protonated form at its N-terminal while in the present study dermorphin used is in its unprotonated form.

As we can see from Table 2 chemical shifts of NH protons are downfield for the protonated sample when compared to the unprotonated sample (present work) as can be expected. Significant difference is observed for D-Ala NH as it is nearer to the N-terminal. The coupling constants are more or less similar. The temperature coefficients are all characteristic of solvent-exposed NH protons. Pastore et al. [8] proposed an extended conformation for dermorphin (protonated form). They also reported weaker intraresidual NOEs and observation of sequential NOEs. Arlandini et al. [9] proposed a β-sheet structure. It appears to us that all the available NMR data is consistent with an extended conformation for dermorphin in DMSO with solvent-exposed NH protons. Protonation at the N-terminal doesn't seem to affect the conformation. There doesn't seem to be a strong case for a folded conformation. Even

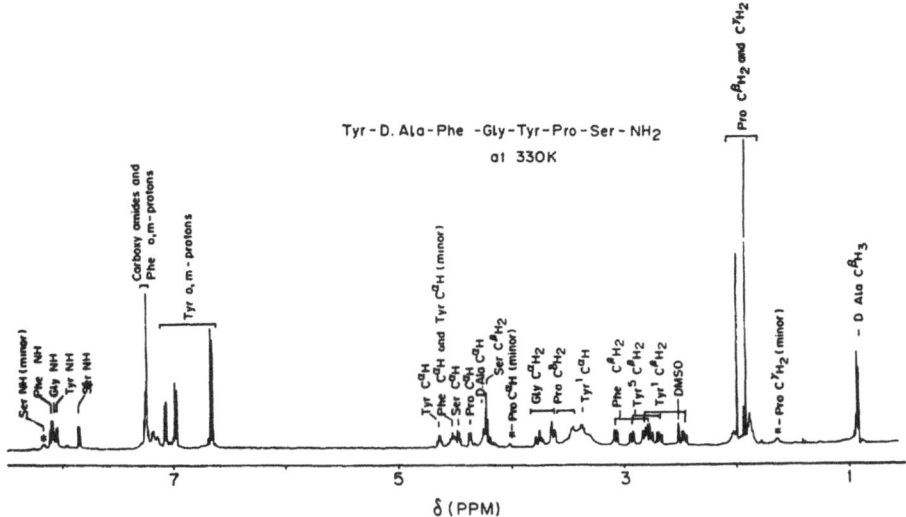

Fig. 6. *1D spectrum of dermorphin at 330 K.*

in the present study no long-range NOEs are observed nor are the NOESY crosspeaks characteristic of β-turn structures. Therefore it is concluded that dermorphin is adopting an extended conformation in DMSO.

Conclusions

The temperature dependence of amide proton chemical shifts suggested that none of the protons is involved in an intramolecular hydrogen bonding and that amide protons are exposed to the solvent.

There are two interconverting conformers, one of which is more populated than the other. Exchange crosspeaks are observed at 330 K between the two conformers. The only residues that showed exchange crosspeaks (besides Pro) viz. Tyr[5] and Ser[7] are adjacent to Pro.

Table 2 *A comparison of chemical shifts for protonated and unprotonated forms of dermorphin*

Ref.		Tyr[1]	D-Ala[2]	Phe[3]	Gly[4]	Tyr[5]	Ser[7]
A. Pastore	δ_{NH}	8.11	8.38	8.39	8.29	8.31	7.70
et al.	J		7.9	8.8	5.7, 6.0	8.1	7.9
	$-d\delta/dT$	2.3	5.2	8.1	10	10	6.2
F. Toma et al.	J		7.3	8.7	6.0	8.0	7.8
	$-d\delta/dT$		2.9	5.1	5.4	5.8	3.8
E. Arlandini	δ_{NH}	8.14	8.40	8.34	8.32	8.42	7.81
et al.	J		7.5	7.5	6.0	9.0	7.6
	$-d\delta/dT$	1.2	3.8	7.3	6.6	5.8	4.8
Present work[a]	δ_{NH}		7.98	8.30	8.26	8.25	7.98
	J		7.8	8.8	6.3, 5.2	7.8	7.8
	$-d\delta/dT$		4.1	5.5	5.2	5.5	4.1

[a] The present work is on the unprotonated form only.

The major conformer is adopting *trans* isomerism at the Tyr^5-Pro^6 bond. This is evidenced by the observation of NOESY crosspeaks between $Tyr^5\alpha$-CH and $Pro^6\delta$-CHs.

The major conformer is adopting an extended conformation. No long-range NOEs or NOEs characteristic of turn structures are observed.

Acknowledgements

The facilities provided by the 500 MHz FT-NMR National Facility supported by the Department of Science and Technology and located at T.I.F.R., Bombay are gratefully acknowledged.

References

1. Erspamer , V. and Melchiorri, P. in *Growth Hormone and Other Biologically Active Peptides*, (Pecile, A. and Muller, E.E., Eds.) Excerpta Medica, Amsterdam, p. 185.
2. De Castiglione, R., Faoro, F., Perseo, G. and Piani, S. (1981) *Int. J. Peptide Protein Res.* **17**, 263.
3. Broccardo, M., Erspamer, V., Falconieri Erspamer, G., Improta, G., Linari, G., Melchiorri, P. and P.C. Montecucchi (1981) *Brit. J. Pharmacol.* **73**, 625.
4. Chipkin, R.E. (1981) *Life Sci.* **28**, 1517.
5. Pastore, A., Temussi, P.A., Tancredi, T., Salvadori, S. and Tomatis, R. (1984) *Biopolymers* **24**, 2349.
6. Toma, F., Sirge, F., Krzsztqf, D. and Zbigniew, G. (1984) *Spectroscopy (Ottawa)* **3**, 465.
7. Toma, F., Dive, V. and Fermandjian, S. (1985) *Biopolymers* **24**, 2417.
8. Pastore, A., Temussi, P.A., Salvadori, S., Tomatis, R. and Mascagni, P. (1985) *Biophys. J.* **48**, 195.
9. Arlandini, E., Ballabio, M., De Castiglione, R., Gioia, B., Lusia Malnati, M., Perseo, G. and Rizzo, V. (1985) *Int. J. Peptide Protein Res.* **25**, 33.
10. Scatturin, A., Salvadori, S., Vertuani, G. and Tomatis, R. (1985) *Farm. Ed. Sci.* **40**, 709.
11. Bhatnagar, R.S., Pattabiraman, N., Sorensen, Keith R., Collette, Timothy W., Carreira, Lionel A., Renugopalakrishnan, V. and Rapaka, R.S. (1985) in *Peptides: Structure and Function* (Proceedings of the 9th American Peptide Symposium), (Deber, C.M., Hruby, V.J. and Kopple, K.D., Eds.), Pierce Chemical Company, Rockford, IL, p. 525.
12. Aue, W.P., Bartholdi, E. and Ernst, R.R. (1976) *J. Chem. Phys.* **64**, 2229.
13. Aue, W.P., Karhan, J. and Ernst, R.R. (1976) *J. Chem. Phys.* **64**, 4226.
14. Bodenhausen, G., Kogler, H. and Ernst, R.R. (1984) *J. Magn. Res.* **58**, 370.
15. Ohnishi, M. and Urry, D.W. (1969) *Biochem. Biophys. Res. Commun.* **36**, 194.
16. Bystrov, V.F., Arseniev, A.S. and Gavrilov, Yu D. (1978) *J. Magn. Res.* **30**, 151.
17. Wüthrich, K. (1986) *NMR of Proteins and Nucleic Acids*, John Wiley & Sons, New York, p. 123.

An ^{1}H NMR study of HTyr-D-Cys-Phe-D-PenOH (JOM-13), a highly selective ligand for the δ opioid receptor

Henry I. Mosberg and Katarzyna Sobczyk-Kojiro

College of Pharmacy, University of Michigan, Ann Arbor, MI 48109–1065, U.S.A.

Introduction

Although the existence of multiple opioid receptors is well established and analogs with high selectivity toward specific opioid receptor types have been developed, the structural and conformational features which confer high affinity and selectivity for specific opioid receptor types remain unclear. Since opioid receptors are large, membrane-bound proteins whose ability to bind opioid ligands is diminished in the absence of the biological membrane [1], direct evidence of bioactive conformations is likely to remain elusive. Consequently, most experimental studies aimed at elucidating these bioactive conformations continue to focus on indirect evidence extrapolated from conformational studies of the free ligand in solution. An obvious requirement to assure the validity of such extrapolations is that flexibility of the ligand be limited. Thus, considerable emphasis has been placed upon the development and conformational analysis of conformationally restricted analogs with high affinity and selectivity for a single receptor type [2,3].

An excellent example of such a conformationally restricted, receptor-selective ligand is [D-Pen^2, D-Pen^5]enkephalin (DPDPE), HTyr-D-Pen-Gly-Phe-D-PenOH [4]. This peptide, conformationally restricted by virtue of the disulfide cyclization and the further rigidizing effects of the gem dimethyl substituents on the Pen residue (Pen, penicillamine is β,β dimethylcysteine), displays good binding affinity and high selectivity for the δ type of opioid receptor and has served for several years as the standard for this type of activity. Conformational features of DPDPE have been extensively studied [5–11], however despite the efforts of many research groups, no consensus and, in fact, rather little agreement has been reached. A likely explanation of this circumstance is that, the rigidizing effects of cyclization and gem dialkyl substituents notwithstanding, DPDPE retains significant flexibility which is manifested in the rather large number of low-energy conformations which have been proposed for this peptide. This flexibility is clearly a liability for the elucidation of the bioactive conformation at δ receptors. We have previously suggested that the central glycine residue in DPDPE and related disulfide-containing enkephalin analogs play an important role by acting as a hinge to allow a receptor-induced fit of each of these ligands to the δ binding site [5]. Thus, while many analogs in this series display conformational differences in solution, particularly in the glycine residue itself, all can adopt the conformation required for receptor binding.

We have recently explored the effect of reducing flexibility in the series of enkephalin analogs that includes DPDPE, by eliminating the central glycine residue [12–14]. The resulting tetrapeptide series, which structurally resembles the dermorphin class of opioid peptides [15], has led to the development of an analog, HTyr-D-Cys-Phe-D-PenOH, (JOM-13) with extremely high selectivity for the δ opioid receptor, comparable to that displayed by DPDPE, and higher δ-receptor binding affinity than DPDPE [12]. This tetrapeptide can be expected to be structurally better defined than DPDPE and thus represents a more tractable ligand for conformational analysis and for providing insights into the binding site of the δ receptor. We present here results of initial ^{1}H NMR studies of this analog. These studies suggest that conformational averaging is indeed less prevalent in this tetrapeptide and demonstrate the existence of two conformers on the NMR time scale.

Results and Discussion

The results of ^{1}H NMR experiments recorded at 500 MHz on a General Electric GN 500 spectrometer are summarized in Tables 1 and 2. Chemical shifts reported in Table 1 are in ppm downfield from 2,2,3,3-tetradeuterio-3-(trimethylsilyl)-propionic acid sodium salt (d$_4$-TSP) as measured from samples of peptide concentration 2–5 mM in either 100% D$_2$O or H$_2$O/D$_2$O (9 : 1). The sample pH was adjusted to 3 (uncorrected meter reading) with CD$_3$COOD. Except for studies exploring the temperature dependence of the chemical shifts, all experiments were performed at 25° C. Assignments shown in Table 1 resulted from a combination of experiments including phase-sensitive DQF-COSY [16], relayed COSY [17] and ROESY [18] spectra, as well as 1D NOE studies.

As indicated by the data presented in Table 1, JOM-13 displays two distinct sets of resonances, indicative of two slowly interconverting conformers on the NMR time scale. These two conformers, designated A and B for the major and minor conformer, respectively, are present in the concentration ratio of approximately 3 : 1, as determined from the integrated intensities of the generally very well resolved resonances. As shown in Table 1, chemical shift differences for like resonances of the two conformers tend to be small but significant (ca. ≤ 0.15 ppm) with some important exceptions observed in the Cys and Pen residues. A moderate chemical shift difference of 0.21 ppm is observed for the D-Cys amide proton, while quite large shift differences of 0.6 ppm and 1.18 ppm are observed for the D-Cys α proton and the D-Pen amide proton, respectively. This latter chemical shift difference can be attributed to a difference in solvent exposure for the D-Pen amide proton in the two conformers as is suggested by the temperature dependence, dδ/dT, of the amide chemical shift, shown in Table 1. In the major conformer, the temperature dependence displayed by the D-Pen amide proton chemical shift, 7.9 ppb/K, is indicative of exposure to the aqueous solvent, while the very low temperature dependence of the minor conformer amide proton, 0.3 ppb/K, is diagnostic of a solvent-inaccessible amide proton [19], which is usually interpreted for small peptides as indicating participation in an intramolecular hydrogen bond. Such inaccessible protons are unavailable for hydrogen bonding to water molecules and

Table 1 ^1H NMR data for major (A) and minor (B) conformers of Tyr-D-C⌐ys-Phe-D-Pen⌐ in aqueous solution at pH 3.0 and t = 25°C

		Tyr	D-Cys	Phe	D-Pen
δ_{NH}	A		8.66	8.69	8.52
	B		8.45	8.54	7.34
δ_α	A	4.17	4.44	4.57	4.53
	B	4.06	3.84	4.48	4.38
δ_β	A	3.17; 3.00	3.29; 3.15	3.18; 3.15	
	B	3.14; 3.00	3.31; 2.99	3.20; 3.15	
δ_γ	A				1.42; 1.33
	B				1.42; 1.33
δ_{ar}	A	7.05; 6.85			
	B	7.02; 6.84			
$J_{NH\alpha CH}$	A		10.0	5.5	8.0
(Hz)	B		5.0	5.5	7.5
$-d\delta/dT^a$	A		8.1	2.8	7.9
(ppb/K)	B		5.8	4.7	0.3

[a]Temperature dependence of amide proton chemical shift.

thus are not subject to the more pronounced deshielding effect of such hydrogen bonding. Consequently, these inaccessible amide protons can be expected to resonate at significantly higher fields, as is indeed observed here for the D-Pen amide of the minor conformer. It is interesting to note that in almost all cases the resonances arising from the minor conformer are upfield of the corresponding major conformer resonances.

The chemical shift data indicate that the D-Cys residue experiences different magnetic environments in the two observed conformers. Additional data suggest that these different environments result at least in part from large conformational differences for the D-Cys residue in the two conformers. For example, the coupling constant, $J_{NH\alpha CH}$, which is dependent upon the backbone dihedral angle ϕ [20], while virtually identical between conformers for the Phe and Pen residues, is quite different for the D-Cys residue in the two conformers (Table 1). Coupling constants, $J_{\alpha\beta}$, between α protons and side-chain β protons, for the D-Cys residue also support this view. Since rotation about the C^α-C^β bond (χ^1) is generally not as hindered as rotations in the peptide backbone, values of $J_{\alpha\beta}$ are usually interpreted in terms of populations of the three lowest-energy (staggered) conformers about this bond: $\chi^1 = -60°$ (g⁻); $\chi^1 = 180°$ (t); and $\chi^1 = +60°$ (g⁺) [21]. These populations for the Tyr, D-Cys, and Phe residues of both conformers are presented in Table 2. As can be seen from this table, rotamer populations in the two conformers are fairly consistent for the Tyr and Phe residues, however major differences are observed for the D-Cys residue. It must be noted that, in the absence of stereospecific assignment of the resonances of the two β protons of a residue, populations calculated for the g⁻ and t rotamers may be reversed. Despite this possibility, it is clear that a much different distribution of rotamers exists for the D-Cys residue in the two conformers since in one case the g⁻ or t rotamer is essentially exclusively populated while in the other conformer both of these rotamers are virtually excluded.

Taken together the results imply that the two conformers observed for JOM-13

Table 2 *Rotamer populations (%) about C^{α}-C^{β} bond (χ^1) for major (A) and minor (B) conformers of Tyr-D-Cys-Phe-D-Pen*

		Tyr	D-Cys	Phe
A	P (g⁻)	35	3	49
	P (t)	58	3	49
	P (g⁺)	7	94	2
B	P (g⁻)	49	0	49
	P (t)	44	94	44
	P (g⁺)	7	6	7

Populations for low-energy staggered conformers, $\chi^1 = -60°$ (g⁻); 180° (t); and +60° (g⁺), calculated by method of Pachler [21].

differ primarily due to a large variation in the conformational features of the D-Cys residue. One possible explanation of this finding would be the presence of a *cis/trans* peptide bond equilibrium, especially involving the D-Cys residue. While observation of both *cis* and *trans* conformers about a peptide bond is usually confined to peptide bonds between residues X-Y in which Y is proline or an N-methyl amino acid, recent examples without this restriction have been reported [22,23]. NMR provides a diagnostic test for the presence of a *cis* peptide bond through the observation of a NOE interaction between sequential α-proton resonances which is indicative of a *cis* peptide bond between the two residues [24]. No such NOE interactions were observed in either conformer of JOM-13. While such a negative result is not definitive, it does argue against the two observed conformers for this peptide being attributable to a *cis/trans* equilibrium.

A second possible explanation is that the large differences between conformers observed for the D-Cys residue reflect a major change in the orientation of the disulfide-containing side chain. Molecular mechanics studies of DPDPE have led to many proposed low-energy conformers, several of which differ in chirality about the disulfide bond [7,8,11] and molecular dynamics studies have been reported to identify a transition between such conformers in DPDPE [7]. A similar transition may occur in JOM-13 between energetically similar conformers separated by a higher energy of activation that arises from the smaller, less flexible 11-membered ring in the tetrapeptide. More extensive NMR studies, complemented by other experimental as well as computational investigations are in progress to more fully elucidate the solution conformations of JOM-13.

Acknowledgements

This work was supported by USPHS grants DA 03910 (HIM) and DA 00118 (HIM) (Research Scientist Development Award) from the National Institute on Drug Abuse. The General Electric GN 500 NMR spectrometer is supported by grant RR-02415 from the USPHS and by funds from the University of Michigan, College of Pharmacy.

References

1. Loh, H. H. and Smith, A. P. (1990) *Annu. Rev. Pharmacol. Toxicol.* **30**, 123.
2. Schiller, P. W. (1986) in *Opioid Peptides: Medicinal Chemistry* (Rapaka, R. S., Barnett, G. and Hawks, R. L., Eds.), National Institute on Drug Abuse, p. 291.
3. Hruby, V. J. (1986) in *Opioid Peptides: Medicinal Chemistry* (Rapaka, R. S., Barnett, G., and Hawks, R. L., Eds.), National Institute on Drug Abuse, p. 128.
4. Mosberg, H.I., Hurst, R., Hruby, V.J., Gee, K., Yamamura, H.I., Galligan, J.J. and Burks, T.F. (1983) *Proc. Natl. Acad. Sci. USA* **80**, 5871.
5. Mosberg, H.I. (1987) *Int. J. Peptide Protein Res.* **29**, 282.
6. Keys, C., Payne, P., Amsterdam, P., Toll, L. and Loew, G. (1988) *Mol. Pharmacol.* **33**, 528.
7. Hruby, V.J., Kao, L.F., Pettitt, B.M. and Karplus, M. (1988) *J. Am. Chem. Soc.* **110**, 3351.
8. Mosberg, H.I., Sobczyk-Kojiro, K., Subramanian, P., Crippen, G.M., Ramalingam, K. and Woodard, R.W. (1990) *J. Am. Chem. Soc.* **112**, 822.
9. Nikiforovich, G. V., Golbraikh, A. A., Shenderovich, M. D. and Balodis, J. (1990) *Int. J. Peptide Protein Res.* **36**, 209.
10. Wilkes, B.C. and Schiller, P.W. (1990) in *Peptides: Chemistry, Structure and Biology*, Proceedings of the 11th American Peptide Symposium (Rivier, J.E. and Marshall, G.R., Eds.) ESCOM, Leiden, p. 341.
11. Froimowitz, M. (1990) *Biopolymers*, in press.
12. Mosberg, H.I., Omnaas, J.R., Smith, C.B. and Medzihradsky, F. (1988) *Life Sci.* **43**, 1013.
13. Heyl, D.L., Omnaas, J.R., Sobczyk-Kojiro, K., Medzihradsky, F., Smith, C.B. and Mosberg, H. I. (1990) *Int. J. Peptide Protein Res.*, in press.
14. Mosberg, H. I., Heyl, D. L., Haaseth, R. C., Omnaas, J. R., Medzihradsky, F. and Smith, C.B. (1990) *Mol. Pharmacol.* **38**, 924.
15. Montecucchi, P.C., deCastiglione, R. and Erspamer, V. (1981) *Int. J. Peptide Protein Res.* **17**, 316.
16. Rance, M., Sorensen, O.W., Bodenhausen, G., Wagner, G., Ernst, R.R. and Wüthrich, K. (1983) *Biochem. Biophys. Res. Commun.* **117**, 479.
17. Wagner, G. (1983) *J. Magn. Reson.* **55**, 151.
18. Bax, A. and Davies, D. G. (1985) *J. Magn. Reson.* **63**, 207.
19. Deslauriers, R. and Smith, I.C.P. (1980) in *Biological Magnetic Resonance*, Vol. 2, (Berliner, L.J. and Reuben, J., Eds.) Plenum Press, New York, p. 243.
20. Bystrov, V.F. (1976) in *Progress in Nuclear Magnetic Resonance Spectroscopy*, Vol. 10, Part 2, Pergamon Press, London, p. 41.
21. Pachler, K.G.R. (1964) *Spectrochim. Acta* **20**, 581.
22. Mierke, D. F., Yamazaki, T., Said-Nejad, O. E., Felder, E. R. and Goodman, M. (1989) *J. Am. Chem. Soc.* **111**, 6847.
23. Kessler, H., Anders, U. and Schudok, M. (1990) *J. Am. Chem. Soc.* **112**, 5908.
24. Wüthrich, K. (1986) *NMR of Proteins and Nucleic Acids*, Wiley, New York.

The precision of protein structures determined from NMR data: Reality or illusion?

Timothy F. Havel

Biophysics Research Division, University of Michigan, Ann Arbor, MI 48109-2099, U.S.A.

Introduction

Distance geometry is the mathematical basis of a general theory of molecular conformation [1]. The fundamental idea behind it is to circumscribe the energy minima of a flexible molecule by means of *distance and chirality constraints*. These are, respectively, lower and upper bounds upon the interatomic distances, and the signs of the oriented volumes of the rigid tetrahedra of atoms. Although this description is necessarily only approximate, the fact that the intramolecular energy can be modelled by a sum of unimodal functions of the distances means that it is usually not a bad one. Compared to a molecular dynamics trajectory, distance geometry provides us with an extremely compact and concise description of the infinite set of conformations that a given molecule can assume under equilibrium conditions.

Applied to experimental data such as NMR, the goal of distance geometry is to describe our *state of knowledge* concerning a molecule's conformational state, and to understand the structural implications of that knowledge. The main technique used here is to sample the space of all conformations consistent with the experimental distance and chirality constraints randomly. Then, provided that this finite ensemble of conformations is sufficiently large, any geometric features that are uniformly present in all its members can safely be inferred to be necessary consequences of the given constraints. This in turn provides us with a crude but effective method of *geometric reasoning*, which can be used to better understand the actual infinite ensemble. Note that it is not the ensemble itself, but rather its invariant geometric features, that are of interest!

To explain these ideas in more detail, let us distinguish three sets of conformations:
(1) The set of all conformations present in significant concentration, i.e. with an energy within about kT of the global minimum.
(2) The set of all conformations that are compatible with the experimental data, to within the attainable signal-to-noise ratio.
(3) The set of all conformations consistent with the distance and chirality constraints derived from the data (along with the known covalent structure of the molecule).

Note that whereas sets (1) and (2) are a little 'fuzzy', in that a complex calculation involving numerous assumptions and approximations must be made in order to determine membership, set (3) is clearly defined. Also, if the experiments have been done properly, set (2) contains set (1). Because the experiments are seldom complete

110

enough to fully define the set of conformations which exists in significant concentration, however, set (2) will not usually be equal to set (1). In a similar way, provided the distance and chirality constraints have been properly defined, set (3) will contain set (2) (as well as set (1)), but is not generally equal to it. Thus, set (3) essentially models what we know for *certain* about the conformational state of the molecule.

What frequently happens in practice is that mistakes are made in the interpretation of the data in terms of distance and chirality constraints. Because of the strong interdependencies present among the constraints, this may cause the constraints to become geometrically inconsistent, so that there exists *no* conformation which satisfies them all. Under these circumstances, the error or total violation of the constraints by a conformation cannot be made to go to zero, so the mistake can be detected. Unfortunately, many investigators have become rather careless about this, and do not even try to drive the error all the way to zero. The reason this is unfortunate is because it results in an *unjustifiable bias* in the sampling obtained, which will essentially always be away from set (1) above.

Similarly, many investigators have tried to arbitrarily force their distances towards values in the center of their allowed ranges, e.g. by driving hydrogen bonds towards their energetically optimal lengths. The rationale behind this is that the middle constitutes a 'best guess' at the actual value of that distance in the equilibrium conformation. Even in those cases in which the molecule has a single well-defined equilibrium conformation, however, this value will not usually be equal to the energetically optimal value within the range that the user has derived from experiment. Thus, the net result again is that the sampling is generally biased *away* from set (1).

An even worse consequence of these practices is that the strong and opposing forces used to enforce the inconsistent constraints or idealized distances will tend to hold the molecule artificially rigid, a phenomenon we call *pinning*. This means that the user will conclude that certain geometric features that appear in all members of the computed ensemble are necessary consequences of the constraints, when in fact they are artifacts of an unjustifiable bias in the sampling. Moreover, they will conclude that the conformation is more precisely defined by the data than it actually is. In many cases, I suspect it will subsequently be found by more careful study that the conformation was precisely wrong!

One can in fact argue that it is preferable to sample near the boundaries of set (3), rather than to sample it uniformly. The reason is that this enables us to better discover what the constraints are telling us for certain about the conformation. Thus, if a 'typical' (i.e. entropically favored) conformation has a right-handed α-helix in it, but a left-handed helix exists which satisfies the constraints, it would be nice to know this, even though a right-handed helix is much more probable on energetic grounds. Then, if one concludes that the helix is actually right-handed, one at least knows *why* one has done so. In addition, it is generally computationally more efficient to try to find conformations satisfying the geometric constraints, and then to modify these as necessary to obtain low energy. For further discussion, the reader is referred to ref. [2].

Results and Discussion

I have recently written a new distance geometry program which combines the best features of the metric matrix method, simulated annealing refinements, and torsion space optimizations. In the process, significant improvements have been made in each of these techniques. Since a more complete account is in preparation, here I will only list a few of the more important improvements:

a. The use of *tetrangle inequality* bound smoothing, to obtain more precise limits on the possible values of the interatomic distances from the relatively incomplete set that is experimentally available.

b. The use of *randomized metrization*, to obtain an improved and unbiased guess at the values of the distances in a possible conformation; this is particularly important in order to obtain good sampling.

c. The use of a technique known as *Guttman transformation* in order to obtain a set of trial coordinates which are a best fit to these distances in a weighted least-squares sense.

d. The use of *very heavy atoms* (with masses between 10 and 1000 Daltons) to drastically improve the convergence obtained during the simulated annealing. This enables us to use a much larger time step without instability, provides the system with more momentum for barrier crossing, and causes the velocities of the atoms in any given step to reflect an average of the gradient over a larger region of conformation space.

e. A highly efficient *regularization algorithm*, which can.correct small deviations from ideal covalent geometry while preserving the overall conformation to a high degree. Although the deviations from ideality are not generally significant after the annealing, ideal geometry is assumed by the next, and final step.

f. A dihedral angle optimization routine based upon the *truncated Newton method*, which can be used to further reduce the total error, to minimize the energy, or to obtain a best fit to the actual data.

I call this program (actually a collection of programs) the *DG-II package*.

Whatever algorithm one uses to calculate random ensembles or whatever bias one prefers, the only way to tell how good the results are (or, more precisely, what they are good for) is by applying the method to *simulated* constraints. The use of simulated constraints is necessary, first in order to be certain of geometric consistency, and second in order to have a reasonably good idea of what the correct answer should be. Unfortunately, such extensive evaluations have so far been performed only with the DISGEO program, so this is all we can compare the DG-II package to. Table 1 shows the results of this comparison, using the simulated data sets I–X for the protein BPTI described in ref. [3]. Each column contains an RMSD averaged among the computed conformations and with the X-ray structure. Data set 0 is a control containing all covalent constraints (including the disulfides) but no other data. The 1985 calculations were made using the DISGEO program, while the 1990 calculations were made using DG-II.

It will be observed that whereas in 1985 the RMSD averages among the computed

Table 1   *Simulated BPTI data sets*

	Data set	1985 alpha	1990 alpha	1985 heavy	1990 heavy	1985 phi/psi	1990 phi/psi
0	among	–	10.75	–	11.89	–	98.8
	X-ray	–	10.68	–	11.71	–	95.3
I	among	1.25	1.18	1.91	1.77	63.1	34.9
	X-ray	1.17	0.99	2.08	1.78	49.7	29.8
II	among	1.06	1.49	1.89	2.17	51.4	46.8
	X-ray	1.19	1.27	2.21	2.07	47.7	38.0
III	among	1.63	2.42	2.53	3.30	74.2	59.6
	X-ray	1.91	2.21	3.05	3.00	63.0	49.7
IV	among	1.65	1.79	2.55	2.52	68.3	47.8
	X-ray	1.51	1.61	2.71	2.38	55.8	39.8
V	among	1.52	2.07	2.27	2.70	72.3	69.7
	X-ray	1.65	1.72	2.88	2.46	71.0	62.5
VI	among	0.94	0.95	1.84	1.61	73.0	50.1
	X-ray	0.79	0.71	2.11	1.54	62.2	38.3
VII	among	0.89	0.78	1.78	1.51	38.5	25.5
	X-ray	0.70	0.68	1.82	1.45	34.9	21.0
VIII	among	0.92	1.20	1.75	1.91	61.2	41.9
	X-ray	0.87	0.97	2.18	1.69	50.3	33.8
IX	among	1.84	2.89	2.72	3.69	84.5	62.4
	X-ray	2.48	2.54	3.32	3.31	74.6	56.5
X	among	1.61	1.88	2.51	2.70	54.1	38.8
	X-ray	2.22	1.97	3.11	2.73	52.9	33.5

conformations tended to be significantly lower than the averages with the X-ray structure, in the 1990 calculations this trend is reversed. This is indicative of the improved sampling obtained with DG-II. The phi/psi RMSD averages in 1990 are distinctly lower than they were in 1985, however, a trend we attribute to the improved convergence obtained with the DG-II package. Since the α-carbon and heavy atom RMSD averages with the X-ray were about the same in 1985 as they are in the 1990 calculations, the general conclusions of the 1985 study are still supported by the 1990 study (see ref. [3]). It is, nevertheless, worthwhile to reiterate the two most important of these conclusions:

1.   Interatomic contacts are a powerful constraint upon the long-range structure of chain molecules; the exact value of these upper bounds, though it can be significant, is not critical.
2.   Short-range interatomic distance constraints and/or direct torsion angle constraints are necessary in order to precisely define the short-range conformation (e.g. the phi/psi map); here the precision of the individual constraints is important.

In recent years a number of investigators have claimed to obtain protein confor-

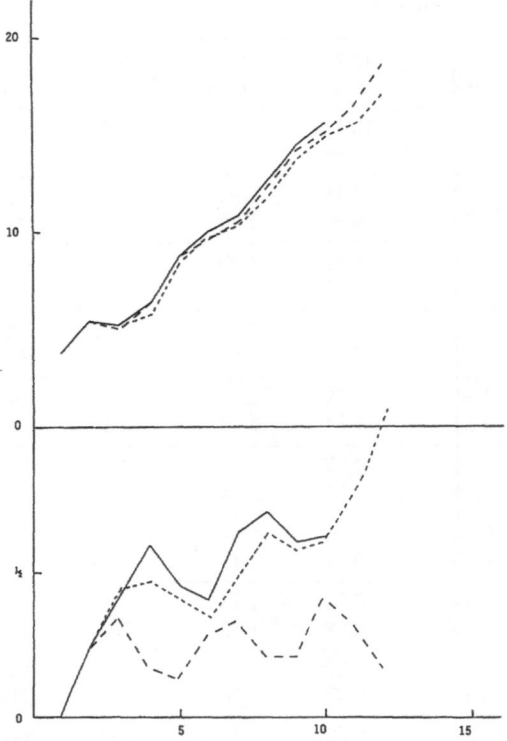

Fig. 1.   The top graph shows the average distance in Å between α-carbons as a function of residue separation; the bottom graph shows the standard deviations in these distances. The solid lines are the average and standard deviation of this function in the α-helices of 14 high-resolution crystal structures; the dashed lines are the average and standard deviation in helix #2 of family #1 of Williamson and Madison's C5a structures; the dotted lines are the same in the five structures obtained from DG-II without a parabolic forcing potential.

mations from NMR data with much higher precision than our evaluation indicates is possible. In order to find out why, we decided to have a careful look at a particular case, namely the 65 residue C5a structure computed by M.P. Williamson and V.S. Madison from only 151 NOE distance constraints, using restrained molecular dynamics [4]. Our reason for singling out this particular paper is not because these investigators have been particularly careless (indeed, quite the contrary!), but rather because they have made their structures and their constraints fully available to the public. There are quite a few other anomalously precise NMR structures available in the literature.

To begin with, we have repeated their calculation using their published lower and upper bounds on the distances, with no parabolic potential forcing distances towards the center of these ranges. The net result was an ensemble of 5 conformations with an average α-carbon RMSD of 2.65 Å (1.67 Å to the mean coordinates), as opposed to the values of 1.8 Å (1.15 Å to the mean) obtained by Williamson and Madison in their best family #1. Part of this increase, of course, may be due to the fact that in this preliminary round of calculations we used neither $1/r^6$ averaging nor floating chiralities, a possibility we plan to look into further. There is, however, another potentially significant source of increased variability.

In Fig. 1, we show plots of the standard deviations in the distances between pairs of α-carbons as a function of residue separation in the second helix of C5a, both in Williamson and Madison's family #1 as well as in the 5 conformations that we have computed using DG-II. Also shown is a plot of the standard deviations in these distances as a function of separation in a sample of helices from 14 highly refined crystal structures. It can immediately be seen that the variability in helix length in Williamson and Madison's NMR conformations is significantly lower than it is in these crystal structures. Thus, unless NMR is now capable of determining helix geometry more precisely than high-resolution crystallography, one is forced to conclude that their helices are a little too perfect to be completely real. Similar plots have been made for helices 3 and 4 in C5a. Helix 1 is underconstrained by the NMR data in both sets of conformations, although much more so in those from DG-II.

The reason for this is undoubtedly the powerful and artificial forces they used to obtain idealized hydrogen-bond lengths, and to drive all their helix phi angles towards exactly 60°. Hydrogen-bond lengths in crystal structures vary between 1.8 Å and 2.8 Å; one of the most slowly exchanging amide protons in BPTI is in fact involved in a 2.8 Å hydrogen bond in its crystal structure (G. Wagner, personal communication). Similarly, the α-helix region of the Ramachandran map spans at least 60° in phi angle, and the Karplus curve has a minimum nearby. Thus, these idealized values are not justified by the data, and it is better to admit one's ignorance than it is to arbitrarily force the protein to concur with our ideals!

The moral is: No bias is more dangerous than operator bias. And the more precisely one tries to determine a structure, the more dangerous the game becomes. The goal of distance geometry, however, has never been to determine a *single* best solution conformation (whatever that is), but rather to characterize as best as can be done the *full range* of structures present in significant concentration. Despite the many difficulties involved in putting this approach into practice, we believe that in the end NMR spectroscopists will be forced to adopt it in order to fully explain their experimental observations.

Acknowledgement

This work was supported by NIH Grant GM-38221.

References

1. Crippen, G.M. and Havel, T.F. (1988) *Distance Geometry and Molecular Conformation*, Research Studies Press, U.K.; Wiley, New York, U.S.A.
2. Havel, T.F. (1990) *Biopolymers* **29**, 1565.
3. Havel, T.F. and Wüthrich, K. (1985) *J. Mol. Biol.* **182**, 281.
4. Williamson, M.P. and Madison, V.S. (1990) *Biochemistry* **29**, 2895.

Distance geometry in distance and torsion angle space: Flexibility, convergence and sampling properties

W. Braun

Institute for Molecular Biology and Biophysics, ETH Zürich, CH-8093 Zürich, Switzerland

Abstract

The variable target function method in torsion angle space, has been frequently used in the determination of polypeptide and protein structures from NMR data. The range of applications was increased by a new graphics tool, GEOM. Modified amino acids or monomeric organic entities can be easily constructed in an interactive way and deposited in the library of the distance geometry program for structure calculations in torsion angle space.

Implementing the variable target function method in a vectorized form, significant increase in computational speed can be obtained on a supercomputer. With the improved program, sampling and convergence properties of this method are studied in detail for α-helical and β-sheet proteins. Problems relating the accuracy of the calculated structures to calculated RMSD values are discussed. Special considerations are given for the different behaviour of distance geometry programs based on the metric matrix method and on the variable target function method. Theoretical considerations and empirical data indicate that the poor sampling property of the metric matrix approach is an intrinsic property of this method.

Introduction

The determination of three-dimensional structures of macromolecules in solution from NMR data relies only on few established calculational methods [1-5]. Distance geometry calculation is currently the most extensively used method when no approximate structure is available. Restrained molecular dynamics [6,7] is most useful as a refinement tool.

Rather than to describe these methods, I want to address several unsolved problems. How does one judge the quality of the NMR structures? Usual parameters are residual violations, RMSD values, conformational energy and the R factor. The problems to be discussed are the sampling properties of different methods, calibration methods and the question if refinement procedures are more than just cosmetics. Another question concerns the comparison of NMR structures with X-ray structures.

Distance geometry methods in distance and torsional angle space

Distance geometry means the characterization of all macromolecular conformations compatible with distance and dihedral angle constraints.

One particular distance geometry approach is based on the metric matrix method [8-11]. Distances are converted to three-dimensional Cartesian coordinates by a partial diagonalization of a certain matrix, the metric matrix. In the first applications of this method for the calculation of molecular structures from NMR data [12-15] new programs were written to interface the embedding procedure with a standard library of amino acid residues (e.g. ECEPP [16]) and additional heuristic data processing led to improved convergence. Treatment of complete covalent polypeptide structures required an efficient way of handling large distance matrices. In practice, this method showed good convergence properties, and it is in widespread use in the software packages DISGEO from T. Havel and DSPACE from Hare Research.

A second method, the variable target function method in torsion angle space, as implemented in the program DISMAN [17] has been successfully applied to determine the tertiary structure of several polypeptides [18–38]. New developments include the program DIANA [39], which is an improved implementation of the variable target function method in torsion angle space, and a new graphics tool GEOM [40], which is used to extend the capabilities of the distance geometry approach in torsion angle space to other covalent structures besides polypeptides and proteins. This tool is especially useful in the field of drug design.

The program DIANA is a new implementation of the variable target function method in torsion angle space. Full use has been made of the short list of nonbonded interactions which is periodically updated. The time-consuming parts have been vectorized. The cpu time to calculate a complete protein structure for a typical protein (e.g. BPTI) through all levels is about 1 minute on a Cray X-MP and about 20 minutes on a Sun 4. This compares favourably to the times quoted for the vectorized version of the metric matrix distance geometry calculation as implemented in the program VEMBED [41]. Details on the times required are given in [39]. A new feature included in DIANA is the treatment of distance constraints for diastereotopic pairs of protons. It involves new distance constraints for the pseudo atom representing the pair of protons. These distance constraints are tighter than using the conventional pseudo atom corrections [42]. Test calculations showed that this procedure gives most pronounced advantages for data sets with no stereospecific assignments.

The program GEOM [40] is a new graphics tool for use of distance geometry programs in torsion angle space with linear or cyclic structures. In designing new drugs, it is important to be flexible in the choice of building blocks so that one is not restricted to a library of standard amino acids or nucleic acids. With GEOM, monomeric organic entities can be sketched by the user on a graphics screen. These are regularized by the program package to a standard geometry. These building blocks can then be deposited into a library which is used by the distance geometry program DISMAN. After the calculation the user can analyze the structures on the basis of residual distance and dihedral angle constraints. The similarity of a set of best structures can be checked by graphical superpositions and RMSD values. This package has been used in the structure determination of a cyclic bouvardin analogue [33] and in a model study of cyclosporin A [40].

In the structure determination of the cyclic bouvardin analogue 212 [33], distance geometry calculations in torsion angle space with the program DISMAN were extended to cyclic structures. Previous methods in calculation of cyclic peptide structures from NMR data were based on a grid search [43,44] which are still limited to rather small ring systems. Two different strategies were compared in their efficiency to give a high number of convergent structures. In method A, first only the distance constraints for the ring closure condition were included in the calculation, starting from 500 random conformations. The best 100 cyclic structures were selected and then minimized in a second step with all NMR distance constraints. In the second method B, all constraints were simultaneously applied. The quality of the best structures, calculated with both methods, were roughly similar, but the second approach yielded a higher number of good structures.

Sampling properties of the metric matrix approach

For trypsin inhibitor [35] and anthopleurin [45], both distance geometry methods, the metric matrix approach and the variable target function method, were compared. For both proteins, the best structures calculated by both programs, and the quality of these structures in terms of residual violations were about the same. A different behaviour was observed in the sampling property for the trypsin inhibitor [35]. In regions with low numbers of distance constraints, the structures calculated by the program DISGEO did not represent a realistic picture of all possible solutions. The same observations have been made by others [4,46,47]. I will sketch some mathematical ideas, which might be the reason for the poor sampling property of the metric matrix method.

I suggest the following explanation. After the bound smoothing, the next step in the EMBED procedure consists in calculating the metric matrix from a set of estimated distances within the given bounds.

$$G_{ii} = \frac{1}{N} \sum_j^N D_{ij}^2 - \frac{1}{2N^2} \sum_{j,k}^N D_{jk}^2 \tag{1}$$

Usually D_{ij} are chosen randomly between the bounds or they are metrized, i.e. treated so that they satisfy the triangle inequality and are still within the given bounds [11,13]. In either case, the dominant part of the summation in Eq. (1) comes from distances, which were derived from insignificant upper and lower bounds in case of a sparse data set. The law of large numbers then suggests that all diagonal elements get similar values for large values of N, if the distances D_{ij} are chosen independently.

$$G_{ii} \rightarrow \frac{1}{2} <d^2> + \varepsilon \tag{2}$$

This is valid for any initial distance distribution. As the distance of the atom i from the centroid is given by $\sqrt{G_{ii}}$, this asymptotic behaviour forces all atoms to be near a surface of a sphere. This explains the similarity of the metric matrix distance geometry structures in unconstrained loop regions of a protein. The observed bias is then a statistical effect by choosing the distances as independent variables. If one introduces in the first phase of the embedding procedure correlations through the triangle inequalities, one must carefully choose the parameters to avoid introducing an additional bias.

These remarks should show that the reason for this strange behaviour might be deeper than a simple implementation problem. Quite recently, a more detailed mathematical and numerical study [48] confirmed the proposed hypothesis and came to similar conclusions.

Refinement of structures

In protein structure determination by X-ray methods, it is usual practice to refine the structure using diffraction intensity values. In the NMR structure determination of macromolecules various refinement procedures have been proposed [5,49–53]. All of these procedures take care of spin diffusion effects. They differ as working in distance space or in real space. In the first case [5,50] the method gives an improved set of distances from the measured NOE data and seems to be quite robust with respect to initial trial structures. However, to improve the coordinates, distance geometry methods or restrained molecular dynamics are needed. The methods working in real space [49,51,53] are impaired in their general use by a small convergence radius. Using more robust global minimizing procedures, these technical problems might be overcome. A more fundamental problem not yet solved is the treatment of the internal flexibility within the molecule.

For a careful structure determination, a comparison of calculated and experimental NOESY spectra is in any case necessary. A qualitative comparison, already showing the major problems was done in the structure determination of the α-amylase inhibitor Tendamistat [23] and rabbit liver metallothionein-2 [54]. In both cases ensemble averages for calculating the expected NOESY cross peaks using the ensemble of distance geometry structures were taken. Overall the calculated and the experimental spectra coincide. All but a few cross peaks in the experimental NOESY spectrum occurred also in the calculated spectrum. This is not surprising as most of the experimental cross peaks have been used as input constraints for the structure calculation, but this test shows that the NOESY spectrum was completely analyzed. The exceptions could be attributed to calibration problems using pseudo-atom corrections. The corresponding cross peaks in the calculated spectrum were always present, but at a lower cutoff. Also some peaks were found in the calculated spectrum, but not in the corresponding experimental spectrum. These differences can in most cases be explained by different flexibility in different parts of the molecule. Using lower limits or refining the structure against the R-factor could lead to a misinterpretation of internal flexibility as a large distance.

Unsolved questions are: What type of average should be taken, $\left\langle \dfrac{1}{r^6} \right\rangle$ or $\left\langle \dfrac{1}{r^3} \right\rangle^2$? Should one use an empirical calibration curve by comparing NOE intensities with distances from calculated structures? What is a good value for the R-factor?

Any refinement of structures should also include energy minimization. For the energy refinement of the distance geometry structures [23] of the α-amylase inhibitor Tendamistat, two energy-refinement programs, AMBER [55] and FANTOM [56] were compared [57]. With both programs, the potential energy values of the distance geometry structures could be significantly reduced without increasing the sum of residual distance constraints violations. The slightly lower value of this sum for the AMBER structures (18 Å as compared to 23 Å for the FANTOM structures) was achieved by distortions of the AMBER structures in their covalent geometry. The FANTOM structures have standard ECEPP/2 geometries [58]. By changing the weight between the experimental distance constraints and the potential energy function, the sum of residual distance constraints violations could be shifted by a factor of six. This showed that with existing empirical force fields it is not possible to decrease the residual constraints violation just by free energy minimization. Also no convergence of the structures was observed.

The same nuclear magnetic resonance data set has also been used to refine the distance geometry structures with the GROMOS package [59]. With restrained energy minimization the sum of residual distance constraints violations increased roughly by a factor of two, but could be reduced by restrained molecular dynamics calculations. A new time-averaged interpretation of the NOE data was proposed in that study, which successfully predicted possible conformational averaging of side-chain motions of Tyr[15]. However, use of the rather limited time scale (30 ps) and the artificial memory function could introduce artefacts such that the fluctuations obtained by these simulations have to be carefully interpreted to avoid wrong interpretations.

Comparison of NMR and X-ray structures

Comparison of the X-ray structure of the α-amylase inhibitor [60] and the solution structure obtained by NMR [22,23] showed a quite close coincidence especially in the interior of the protein [61]. As more structures in single crystals and in solution will be determined it will be interesting to see if this is the general case. Differences between crystal and solution structures are still quite often grossly exaggerated or are simply consequences of errors in the structure determination. ·

Metallothionein is an example of such a case. The NMR structure determinations were done on proteins from three different species, rabbit liver [54,62], rat liver [63] and human liver [64]. All three proteins showed the same global fold and Cd-cysteine connectivities. However, a significant difference between the NMR results and an independent X-ray study [65] was observed. Quite recently, a new X-ray structure analysis has been performed. In this new X-ray structure, all significant differences to the NMR structure vanished. The backbone RMSD values between the best NMR structure of rat liver MT-2 and the new X-ray structure (C.D. Stout, personal communication) are 2.1 Å for the β-domain and 1.8 Å for the α-domain. These values are similar to the average of the pairwise RMSD values between the 10 best rat liver NMR structures [63]. A detailed comparison will be published elsewhere.

References

1. Bax, A. (1989) *Annu. Rev. Biochem.* **58**, 223.
2. Braun, W. (1987) *Quart. Rev. Biophys.* **19**, 115.
3. Clore, G.M. and Gronenborn, A.M. (1989) *Crit. Rev. Biochem. Mol. Biol.* **24**, 479.
4. Kaptein, R., Boelens, R., Scheek, R.M. and van Gunsteren, W.F. (1988) *Biochemistry* **27**, 5389.
5. Keepers, J.W. and James, T.L. (1984) *J. Magn. Res.* **57**, 404.
6. Brünger, A.T., Clore, G.M., Gronenborn, A.M. and Karplus, M. (1986) *Proc. Natl. Acad. Sci. USA* **83**, 3801.

7. Kaptein, R., Zuiderweg, E.R.P., Scheek, R.M., Boelens, R. and van Gunsteren, W.F. (1985) *J. Mol. Biol.* **182**, 179.
8. Crippen, G.M. (1977) *J. Comp. Phys.* **26**, 449.
9. Crippen, G.M. (1981) Distance geometry and conformational calculations, in *Chemometrics Research Studies Series*, Vol. 1 (Bawden, D., Ed.), Research Studies Press, New York.
10. Crippen, G.M. and Havel, T.F. (1978) *Acta Cryst.* A **34**, 282.
11. Havel, T.F., Kuntz, I.D. and Crippen, G.M. (1983) *Bull. Math. Biol.* **45**, 665.
12. Braun, W., Wider, G., Lee, K.H. and Wüthrich, K. (1983) *J. Mol. Biol.* **169**, 921.
13. Braun, W., Bösch, C., Brown, L.R., Gō, N. and Wüthrich, K. (1981) *Biochim. Biophys. Acta* **667**, 377.
14. Crippen, G.M., Oppenheimer, N. and Conolly, M. (1981) *Int. J. Peptide Protein Res.* **17**, 156.
15. Havel, T.F. and Wüthrich, K. (1985) *J. Mol. Biol.* **182**, 281.
16. Momany, F.A., McGuire, R.F., Burgess, A.W. and Scheraga, H.A. (1975) *J. Phys. Chem.* **79**, 2361.
17. Braun, W. and Gō, N. (1985) *J. Mol. Biol.* **186**, 611.
18. Bazzo, R., Tappin, M.J., Pastore, A., Harvey, T.S., Carver, J.A. and Campbell, I.D. (1988) *Eur. J. Biochem.* **173**, 139.
19. Endo, S., Inooka, H., Ishibashi, Y., Kitada, C., Mizuta, E. and Fujiino, M. (1989) *FEBS Lett.* **257**, 149.
20. Haruyama, H. and Wüthrich, K. (1989) *Biochemistry* **28**, 4301.
21. Inagaki, F., Shimada, I., Kawaguchi, K., Hirano, M., Terasawa, I., Ikura, T. and Gō, N. (1989) *Biochemistry* **28**, 5985.
22. Kline, A.D., Braun, W. and Wüthrich, K. (1986) *J. Mol. Biol.* **189**, 377.
23. Kline, A.D., Braun, W. and Wüthrich, K. (1988) *J. Mol. Biol.* **204**, 675.
24. Kobayashi, Y., Ohkubo, T., Kyogoku, Y., Nishiuchi, Y., Sakakibara, S., Braun, W. and Gō, N. (1989) *Biochemistry* **28**, 4853.
25. Kohda, D., Gō, N., Hayashi, K. and Inagaki, F. (1988) *J. Biochem.* **103**, 741.
26. Kohda, D., Shimada, I., Miyake, T., Fuwa, T. and Inagaki, F. (1989) *Biochemistry* **28**, 953.
27. Montelione, G.T., Wüthrich, K., Nice, E.C., Burgess, A.W. and Scheraga, H.A. (1987) *Proc. Natl. Acad. Sci. USA* **84**, 5226.
28. Mulvey, D., King, G.F., Cooke, R.M., Doak, D.G., Harvey, T.S. and Campbell, I.D. (1989) *FEBS Lett.* **257**, 113.
29. Pastore, A., Harvey, T.S., Dempsey, C.E. and Campbell, I.D. (1989) *Eur. Biophys. J.* **16**, 363.
30. Qian, Y.Q., Billeter, M., Otting, G., Müller, M., Gehring, W.J. and Wüthrich, K. (1989) *Cell* **59**, 573.
31. Saudek, V., Atkinson, R.A., Williams, R.J.P. and Ramponi, G. (1989) *J. Mol. Biol.* **205**, 229.
32. Saudek, V., Wormald, M.R., Williams, R.J.P., Boyd, J., Stefani, M. and Ramponi, G. (1989) *J. Mol. Biol.* **207**, 229.
33. Senn, H., Loosli, H.R., Sanner, M. and Braun, W. (1990) *Biopolymers* **29**, 1387.
34. Steinmetz, W.E., Bougis, P., Rochat, H., Redwine, O.D., Braun, W. and Wüthrich, K. (1988) *Eur. J. Biochem.* **172**, 101.
35. Wagner, G., Braun, W., Havel, T.F., Schaumann, T., Gō, N. and Wüthrich, K. (1987) *J. Mol. Biol.* **196**, 611.
36. Widmer, H., Billeter, M. and Wüthrich, K. (1989) *Proteins* **6**, 357.
37. Zuiderweg, E.R.P., Nettesheim, D.G., Mollison, K.W. and Carter, G.W. (1989) *Biochemistry* **28**, 172.
38. Zuiderweg, E.R.P., Henkin, J., Mollison, K.W., Carter, G.W. and Greer, J. (1988) *Proteins* **3**, 139.
39. Güntert, P., Braun, W. and Wüthrich, K. (1990) *J. Mol. Biol.* (in press).
40. Sanner, M., Widmer, A., Senn, H. and Braun, W. (1989) *J. Comput.-Aided Mol. Design* **3**, 195.

41. Kuntz, I.D., Thomason, J.F. and Oshiro, C.M. (1989) in *Methods in Enzymology*, Vol. 177, (Oppenheimer, N.J. and James, T.L., Eds.) p. 159.
42. Wüthrich, K., Billeter, M. and Braun, W. (1983) *J. Mol. Biol.* **169**, 949.
43. Smith, G.M. and Veber, D.F. (1986) *Biochim. Biophys. Res. Commun.* **134**, 907.
44. Taylor, J.S., Garret, D.S. and Wang, M.J. (1988) *Biopolymers* **27**, 1571.
45. Torda, A.W., Mabbutt, B.C., van Gunsteren, W.F. and Norton, R.S. (1988) *FEBS Lett.* **239**, 266.
46. Metzler, W.J., Hare, D.R. and Pardi, A. (1989) *Biochemistry* **28**, 7045.
47. Nilges, M., Clore, G.M. and Gronenborn, A.M. (1988) *FEBS Lett.* **229**, 317.
48. Havel, T.F. (1990) *Biopolymers* **29**, 1565.
49. Baleja, J.D., Moult, J. and Sykes, B.D. (1990) *J. Magn. Res.* **87**, 375.
50. Boelens, R., Koning, T.M.G., van der Marel, G.A., van Boom, J.H. and Kaptein, R. (1989) *J. Magn. Res.* **82**, 290.
51. Borgias, B.A. and James, T.L. (1988) *J. Magn. Res.* **79**, 493.
52. Borgias, B.A., Gochin, M., Kerwood, D.J. and James, T.L. (1990) *Prog. Nucl. Magn. Res. Spec.* **22**, 83.
53. Yip, P. and Case, D.A. (1989) *J. Magn. Res.* **83**, 643.
54. Arseniev, A.S., Schultze, P., Wörgötter, E., Braun, W., Wagner, G., Vašák, M., Kägi, J.H.R. and Wüthrich, K. (1988) *J. Mol. Biol.* **201**, 637.
55. Singh, U.C., Weiner, P.K., Caldwell, J.W. and Kollmann, P.A. (1986) AMBER, University of California, San Francisco.
56. Schaumann, Th., Braun, W. and Wüthrich, K. (1990) *Biopolymers* **29**, 679.
57. Billeter, M., Schaumann, Th., Braun, W. and Wüthrich, K. (1990) *Biopolymers* **29**, 695.
58. Némethy, G., Pottle, M.S. and Scheraga, H.A. (1983) *J. Phys. Chem.* **87**, 1883.
59. Torda, A.W., Scheek, R.M. and van Gunsteren, W.F. (1990) *J. Mol. Biol.* **214**, 223.
60. Pflugrath, J.W., Wiegand, G. and Huber, R. (1986) *J. Mol. Biol.* **189**, 383.
61. Billeter, M., Kline, A.D., Braun, W., Huber, R. and Wüthrich, K. (1989) *J. Mol. Biol.* **206**, 677.
62. Braun, W., Wagner, G., Wörgötter, E., Vašák, M., Kägi, J.H.R. and Wüthrich, K. (1986) *J. Mol. Biol.* **187**, 125.
63. Schultze, P., Wörgötter, E., Braun, W., Wagner, G., Vašák, M., Kägi, J.H.R. and Wüthrich, K. (1988) *J. Mol. Biol.* **203**, 251.
64. Messerle, B., Schäffer, A., Vašák, M., Kägi, J.H.R. and Wüthrich, K. (1990) *J. Mol. Biol.* **214**, 765.
65. Furey, W.F., Robbins, A.H., Clancy, L.L., Winge, D.R., Wang, B.C. and Stout, C.D. (1986) *Science* **231**, 704.

Section II

Optical spectroscopy as a probe for protein structure

Dedicated to the memory of

R.C. Lord

This section is dedicated to the memory of Professor Richard C. Lord, Massachusetts Institute of Technology, Cambridge, MA, U.S.A. for introducing laser excited Raman spectroscopy which has become a major technique to probe protein structures.

Probing protein secondary structure by infrared spectroscopy

Henry H. Mantsch and Witold K. Surewicz
Steacie Institute for Molecular Sciences, National Research Council of Canada, Ottawa, Ontario, Canada K1A 0R6

Introduction

The ability of modern biotechnology to produce new or modified proteins has presently outpaced our understanding of the intricate relationship between protein structure and protein function. It is therefore important to search for new analytical tools that are able to determine the complete structure of a new protein and/or to assess the change in structure of a protein after genetic or chemical manipulation.

The determination of the primary structure of newly cloned or isolated proteins is now relatively straightforward and the procedures are automated. However, the translation of the amino acid sequence of a protein into its secondary structure or higher levels of organization, is not an easy task and constitutes one of the most intriguing problems in biophysical chemistry today.

The theoretical approaches to predicting protein structure from the primary sequence are based on numerous statistical and/or energy minimization methods [1,2], and the search is on for the Holy Grail of all protein designers, the definitive folding algorithm. However, despite undeniable progress, the Holy Grail has remained elusive so far. Thus, while the theoretical approaches are still being refined, it is imperative to exploit all experimental methods available today in order to generate novel experimental data and thus enlarge the basis for understanding the principles of protein structure and folding/unfolding reaction.

Amongst the experimental tools for the determination of protein structure, X-ray crystallography ranks first. However, it has limitations: one needs good crystals, and at present many proteins (e.g., most membrane proteins) are not ameanable to this technique. Furthermore, the question always arises as to whether the crystal structure indeed accurately represents the structure in solution or that in a membrane environment. Techniques that can provide information regarding the solution structure of proteins include nuclear magnetic resonance spectroscopy (NMR), circular dichroism (CD), and vibrational (Raman and infrared) spectroscopy. Of these, 2D NMR is unquestionably the most powerful technique; it provides atomic resolution, just like X-ray diffraction. Yet, it too has its limitations: it is restricted to polypeptides of less than 20 kdaltons and membrane proteins cannot be studied in a straightforward manner. The technique routinely used today for estimating the overall secondary structure of globular proteins in solution is CD; however, while the total amount of α-helical structures can be estimated reasonably accurately by CD analysis, the quantitative determination of other secondary structures, such as β-sheets and strands is less accurate and CD cannot distinguish between different types of helices or between different β-type structures. Also, membrane-bound proteins are difficult to study by CD due to light scattering of the membrane fragments.

In this presentation we shall discuss the analysis of protein secondary structure by Fourier transform infrared (FT-IR) spectroscopy, the most recent addition to the arsenal of biophysical techniques capable to provide information about the overall molecular conformation imposed upon proteins by intra- and intermolecular forces under a variety of experimental conditions.

Methodology

The secondary structures that are found in proteins are regularly repeating conformations of the polypeptide chain. While it is generally accepted that the driving force of protein folding is provided by hydrophobic interactions, a major stabilizing force sustaining a given secondary structure is provided by hydrogen bonding among the secondary amide groups of the polypeptide backbone. Since in a protein each backbone amide group -(C = O)-NH- represents a minute vibrating dipole, and since the nature of the hydrogen bonds is different in the different secondary structures, the vibrational modes occur at different frequencies, which in turn leads to different absorption bands in the infrared spectrum. This is the basis for the infrared spectroscopic analysis of protein secondary structure.

Proteins show a wealth of infrared absorption bands of which many originate from the polypeptide backbone. They are generally referred to as amide bands. Figure 1A illustrates the amide I, amide II and amide III bands in a typical protein. Of these, the amide I mode – which is 80% C = O stretching vibration – is the most conformation-sensitive and most commonly used amide band. There is, however, one serious problem with the conceptual interpretation of the amide I band. The experimentally measured band envelopes typically represent the sum of unresolved components, arising from the different secondary structures. A curve fitting of such a complex (and often featureless) band is practically meaningless as long as both the number of the individual bands and their approximate positions are not known. Therefore, a first step in the analysis of infrared data is the identification of the component bands present under the amide I band envelope. Since the separation between individual bands is small compared to their width, this cannot be achieved by a simple increase in the instrumental resolution. The only feasible approach is to reduce the width of the component bands by numerical data treatment. One possible approach is the calculation of derivative spectra. We have developed a different band-narrowing methodology, known as Fourier self-deconvolution. The deconvolution is performed in the Fourier transform domain and takes advantage of the fact that FT-IR spectra are already obtained interferometrically. The broader a band, the more rapid the decay of its interferogram. Therefore, by slowing down the decay of its interferogram, one can reduce the width of an infrared band. In Fourier self-deconvolution, the composite interferogram (which consists of the sum of the interferograms corresponding to each infrared band), is multiplied by an exponentially increasing weighting function which

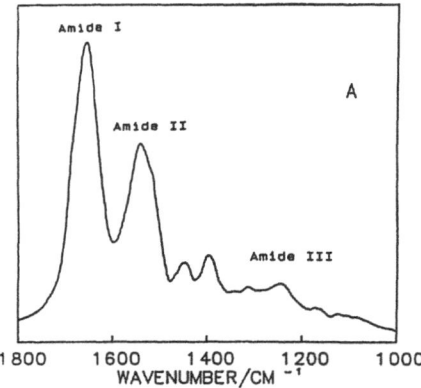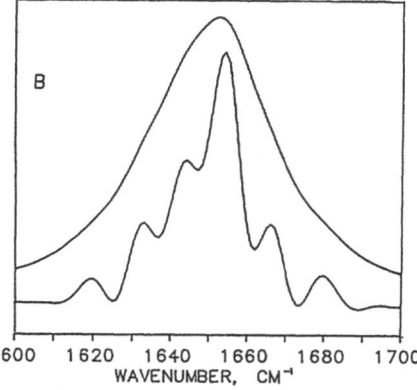

Fig. 1. A: Typical protein infrared spectrum showing the amide I, amide II and amide III bands. B: Demonstration of band narrowing applied to the amide I band. Fourier self-deconvolution was performed with a Lorentzian of 20 cm^{-1} halfbandwidth and a band-narrowing factor of 2.5.

slows down its rate of decay, and by an appropriate apodization function which minimizes the noise introduced by extending the interferogram. The reverse Fourier transformation of this new interferogram then yields a new infrared spectrum with narrower bands, as illustrated in Fig. 1B (for details of this methodology the interested reader may consult references 3-5).

Conformational Analysis of Proteins and Peptides in Solution

Once the number and the frequency position of the different amide I components have been established by band narrowing procedures, curve fitting methods can be applied to quantify the area of the different components of the amide I band envelope, which in turn is directly related to the amount of secondary structure present in that particular protein. The underlying assumptions are that the infrared extinction coefficients of the various amide I component bands are the same or very similar, and that the absorption from glutamate and aspartate side chains, which also have amide groups, is negligible in relation to the large number of backbone amide groups.

A first comprehensive study, based on the methodology outlined above, of the secondary structure of a number of soluble globular proteins was published in 1986 by Byler and Susi [6]. Subsequenly, the infrared spectroscopic method has been applied to many aspects of protein structural research, and we estimate that between 1986 and 1990 over a hundred publications (about 20 from this lab), have been published on this topic.

In order to illustrate the type of questions that can be posed, and the type of answers that one can expect from an infrared spectroscopic analysis, we shall use the example of β-lactoglobulin B. Figure 2 shows the infrared spectrum in the 1600-1700 cm^{-1} region of β-lactoglobulin B after band narrowing by Fourier self-deconvolution (solid trace). The frequency distribution of the amide I mode in proteins ranges from

127

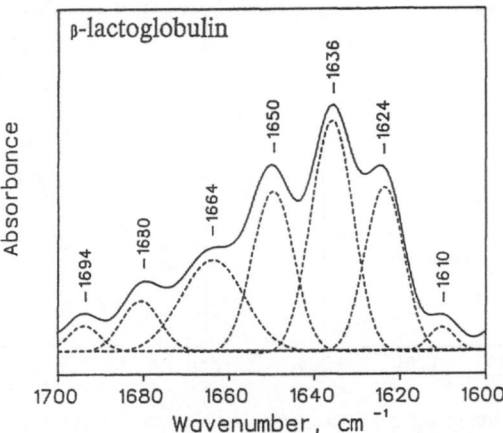

Fig. 2. *Solid trace: Infrared spectrum of β-lactoglobulin B in phosphate buffer (2.5 mM, pH 7.0) after band narrowing by self-deconvolution with a Lorentzian of 18 cm^{-1} halfbandwidth and a band-narrowing factor of two. Broken traces: Best fitted individual component bands of the amide I band envelope.*

1615 to 1695 cm^{-1}. Generally, the stronger the hydrogen bonds, the lower the infrared frequency of the corresponding amide I band. A relationship between the nature of the hydrogen bonding in a given secondary structure and the position of the amide I frequency has been established on the basis of infrared spectra of proteins with known X-ray secondary structures [6–8], as well as on theoretical grounds based on force field calculations [9]. The amide I domains are shown in Table 1. In practice, the indicated limits may show considerable overlap (especially the α-helical and random coil domains).

In the spectrum in Fig. 2 there are seven individual bands of which one (the weak band at 1610 cm^{-1}) is not an amide I band and originates from the aromatic side chain of tyrosine. The amide I bands at 1694 and 1664 cm^{-1} are due to turns, and represent essentially free or very weakly hydrogen-bonded peptide C = O groups [10]. The band at 1650 cm^{-1} is mostly due to random (irregular) peptide segments with a small contribution from the only short α-helix seen in the X-ray structure [11]. We should mention here that one has to distinguish between random (or disordered) structures that are not observed crystallographically, and irregular structures that are observed crystallographically, but do not adopt standard, well recognized, intramolecular hydrogen-bonded conformations. To avoid confusion, we prefer to use the term irregular to refer to conformations which are not classified as α-helix, β-sheet or turns. A totally new observation, coming so far only from infrared experiments, is the recognition that in β-lactoglobulin (as well as in other proteins, e.g., concanavalin A), there are two distinct types of β-structures, documented by the two bands at 1636 and 1624 cm^{-1}. The area of these two bands represents 53% of the total area of all amide I bands, which is close to the content of β-conformation found by X-ray (i.e., 51%). While the band at 1636 cm^{-1} can be assigned to normal β-sheets, the origin of the 1624 cm^{-1}-band is still under cogitation. The lower C = O stretching frequency of the latter indicates that the

128

Table 1 *Amide I domains*

Type of H bonding	Frequency range
β-structures	1615-1638 cm^{-1}
(additional band for antiparallel β strands)	1682-1688 cm^{-1}
Unordered	1638-1645 cm^{-1}
α-helix	1645-1662 cm^{-1}
Turns	1662-1695 cm^{-1}

amide groups are involved in stronger hydrogen bonds, and it has been suggested that they originate from 'exposed' β-strands that can engage in tighter intermolecular or interdomain interactions. The presence of a band at 1680 cm^{-1} reveals that the β-strands are antiparallel.

The question inevitably arises as to how precise is the determination of secondary structures from infrared spectra. The IR procedures described here for the evaluation of the secondary structure of proteins were claimed to yield an estimation of the α-helix and β-sheet content with a standard deviation with respect to the corresponding X-ray structure, of 2-3% (according to reference 6) or 8-9% (see Table 1) (according to reference 12). In our experience, it is better than 10% if done properly. However, we have to emphasize that there are potential pitfalls associated with each of the three steps of an infrared spectroscopic analysis (i.e., band narrowing, band assignment and band quantitation), which have to be recognized in order to avoid misleading interpretations. While these observations apply to the determination of 'absolute' protein conformations, changes as small as 1-2% produced in a given secondary structure can be detected by this technique. It should also be noted, that the secondary structures determined for large proteins are likely to represent families of substructures, rather than a single structure. This is particularly true for the α-helical and irregular structures which yield broad amide I bands, consisting of numerous closely spaced subsidiary bands which cannot be separated by the band-narrowing methods presently in use. Indeed, theoretical work indicates that the frequency of a given α-helix depends on its length, its disposition within the protein, and whether it is distorted or not. Thus, it is not unreasonable to expect that in a complex globular protein, each helix has a unique amide I frequency within a given (narrow) range of such frequencies.

The Study of Protein Folding and Unfolding

One of the more remarkable properties of native globular proteins is their marginal stability relative to the denatured, inactive state. For small proteins this stability margin does not exceed 100 kJ/mol, which is the equivalent of only about 4 hydrogen bonds. This stability margin is very sensitive to environmental factors such as temperature, pressure, pH, ionic strength or organic solvents. Subtle variations of these environmental factors can affect the infrared spectrum in the amide I region considerably, and these changes may be exploited for correlating the particular denaturation with specific changes in the overall secondary structure. Thus, infrared

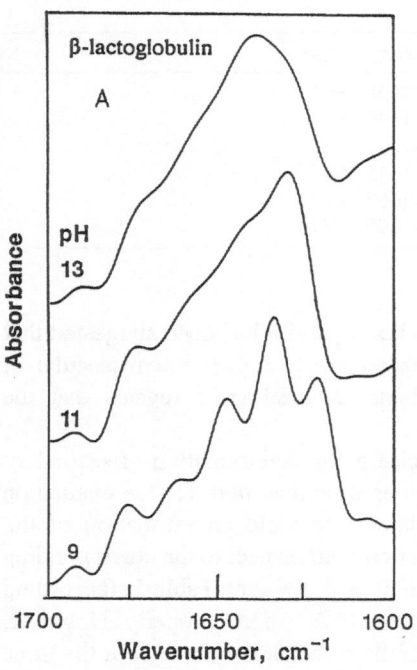
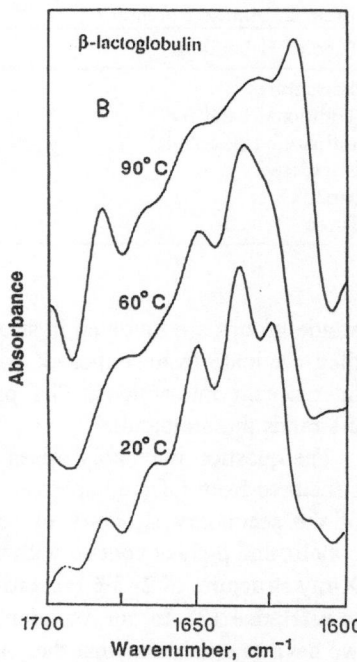

Fig. 3.   Infrared spectra in the amide I region of β-lactoglobulin B in phosphate buffer (2.5 mM) at the indicated pH values (A), and spectra at room temperature at the indicated temperatures (B).

spectra of controlled protein unfolding may shed light on the 'conformational structure' of folding intermediates. There are now many reports that not all denatured proteins are truly random. This is illustrated below.

Figure 3A shows the effect of high pH on the infrared spectrum of β-lactoglobulin B [10]. The irreversible alkaline denaturation of this protein occurs in two stages. The first stage, from pH 9 and 11 affects primarily the bands at 1649 and 1624 cm^{-1}, however the principal β-structure remains intact, as evidenced by the strong amide I band at 1633 cm^{-1}. In the second stage of the alkaline denaturation the remaining β-sheets unfold and a broad band emerges at 1640 cm^{-1}, characteristic of irregular protein segments. However, the spectrum at pH 13 reveals the persistence of some β-structure, indicating that the unfolding is not complete. Figure 3B shows the effect of increased temperature on the infrared spectrum of β-lactoglobulin. There are two temperatures at which the infrared spectra show discontinuous changes, at ~58° C and then at ~70° C. The spectrum at 60° C, which represents the first stage of thermal denaturation, is practically superimposable with the spectrum at pH 11 recorded at room temperature. We recall that pH 11 represents the first stage of the alkaline denaturation, which suggests the existence of a common denaturation intermediate. The final stage of thermal denaturation is exemplified by the spectrum at 90° C with bands at 1682, 1667, 1650, 1632 and 1620 cm^{-1}. The infrared data demonstrate that in the final stage of the thermal denaturation process, β-lactoglobulin aggregates and polymerizes extensively. Such aggregation of the thermally unfolded protein is commonly observed with globular proteins, and is likely to be induced by the exposure to water of originally 'buried' hydrophobic protein segments.

Membrane-bound Proteins/Peptides

Infrared spectroscopy is well-suited for probing the structure of membrane proteins [13]. In particular, the technique can be used to examine the secondary structure of the same protein in aqueous buffer and in its native membrane environment. Similarly, the solution conformation of bioactive peptides can be compared with the conformation after reconstitution into model membranes. This point is made in Fig. 4 which shows infrared spectra of the atrial natriuretic peptide, atriopeptin III, in aqueous solution and after reconstitution with the acidic phospholipid dimyristoyl phosphatidylglycerol [14]. The solution spectrum of this 24-amino acid peptide hormone (bottom spectrum) consists of only one band at $1643 \ cm^{-1}$, pointing to a totally structureless peptide. On the other hand, the conformation of the membrane-bound peptide (top spectrum) consists predominantly of extended β-strands (infrared bands at 1618 and $1637 \ cm^{-1}$) and turns (band at $1666 \ cm^{-1}$), with only a small contribution from peptide segments without well-defined conformations (band at $1647 \ cm^{-1}$). A similar situation was found with the octapeptide angiotensin II, which is also structureless in an aqueous environment, but adopts a β-type structure in a lipid environment [15].

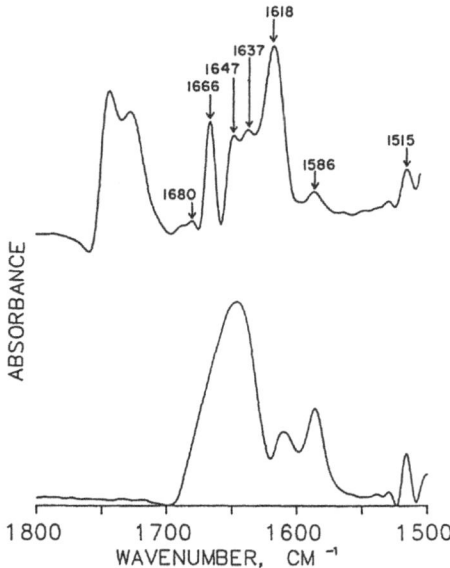

Fig. 4. Bottom: Infrared spectrum of the atrial natriuretic peptide atriopeptin III in Pipes buffer (4 mM, pH 6.0). Top: Infrared spectrum of atriopeptin III in the presence of the lipid dimyristoyl phosphatidylglycerol (peptide:lipid ratio, 1:10). Both spectra are shown after band narrowing by a factor of two.

131

In conclusion, IR spectroscopy, unlike X-ray diffraction or NMR spectroscopy cannot provide detailed knowledge on individual bond lengths and bond angles; it does however provide information on the overall molecular conformation imposed upon proteins by intra and intermolecular interactions. Its forte is the capability to assess subtle variations in protein conformation, brought about by a change in the biomolecular environment, such as changes in pH, temperature, pressure or interaction with other molecules. The particular advantage of the FT-IR method is that it can be used to study proteins not only in solution, but also in a variety of interfacial environments such as in membranes or on solid substrates.

References

1. Chou, P. Y. and Fasman, G. D. (1974) *Biochemistry* **13**, 221.
2. Jaenicke, R. (1987) *Prog. Biophys. Mol. Biol.* **49**, 117.
3. Kauppinen, J. K., Moffatt, D. J., Mantsch, H. H. and Cameron, D. G. (1981) *Appl. Spectrosc.* **35**, 271.
4. Mantsch, H. H., Moffatt, D. J. and Jones, R. N. (1986) in *Spectroscopy of Biological Molecules* (Clark, R. J. H. and Hester, R. E., Eds.) Wiley and Sons, New York, pp. 1–46.
5. Mantsch, H. H., Moffatt, D. J. and Casal, H. L. (1988) *J. Mol. Struct* **173**, 285.
6. Byler, M. and Sus, H. (1986) *Biopolymers* **25**, 469.
7. Harris, P. I., Lee, D. C. and Chapman, D. (1986) *Biochim. Biophys. Acta* **874**, 255.
8. Surewicz, W. K. and Mantsch, H. H. (1988) *Biochim. Biophys. Acta* **952**, 115.
9. Krimm, S. and Bandekar, J. (1986) *Adv. Protein Chem.* **38**, 181.
10. Casal, H. L., Köhler, U. and Mantsch, H. H (1988) *Biochim. Biophys. Acta* **957**, 11.
11. Papiz, M. Z., Sawyer, L., Eliopoulos, E. E., North, A. C. T., Findlay, J. B. C., Sivaprasadarao, R., Jones, T. A., Newcomer, M. E. and Drewry, J. (1986) *Nature* **324**, 383.
12. Goormaghtigh, E., Cabiaux, V. and Ruysschaert, J.M. (1990) *Eur. J. Biochem.* **193**, 409.
13. Surewicz, W. K. and Mantsch, H. H. (1990) in *Protein Engineering: Approaches to the Manipulation of Protein Folding.* (Narang, A. S., Ed.) Butterworths, Boston, pp. 131–157.
14. Surewicz, W. K., Mantsch, H. H., Stahl, G. L. and Epand. R. M. (1987) *Proc. Natl. Acad. Sci. USA* **84**, 7028.
15. Surewicz, W. K. and Mantsch, H. H. (1988) *J. Am. Chem. Soc.* **110**, 4412.

Structural studies of cucumber mosaic virus: Fourier transform infrared spectroscopic studies

P. Piazzolla[a], A.M. Tamburro[b] and V. Renugopalakrishnan[c]

[a]*Dipartimento di Protezione delle Piante Dalle Malattie, Via G. Amendola, 165/A, 70126 Bari, Italy*
[b]*Dipartimento di Chimica, Università degli Studi della Basilicata, Potenza, Italy*
[c]*Harvard Medical School, Boston, MA 02115, U.S.A.*

Introduction

Cucumber mosaic virus (CMV) is an isometric plant virus which occurs in all temperate regions of the world. It manifests an extremely wide host range. The structural integrity of CMV is essentially maintained by strong protein-RNA interactions which are sensitive to changes in ionic strength. Its structural units display hexamer-pentamer clustering in a T = 3 icosahedral surface lattice with an RNA content of 20% [1].

The molecular mechanisms of disassembly of CMV or other viruses into nucleic acid and coat protein are not presently known. The disassembly, essential to virus activity initiation, involves structural modifications of virus coat protein subunits and nucleic acid which in turn influence the protein-protein or protein-RNA interactions, essential to virus architecture. Therefore, structural studies of CMV may facilitate an understanding of the phenomena underlying the virus activity in the host cells.

Recently, cloning and sequencing of the coat protein gene of several strains of CMV has been reported in the literature. Previously, CD studies of two isolates of CMV were reported by Piazzolla et al. [2]. The primary structure of the coat protein of CMV has thus provided renewed impetus to begin detailed structural studies of the intact virus. In this report, circular dichroism (CD) and Fourier-Transform Infrared (FT-IR) studies of CMV-D strain in aqueous solution are utilized to derive the secondary structure of the coat protein. Cloning and sequencing of coat protein gene of this strain have been reported by Cuozzo et al. [3]. A plausible model is proposed based on Chou-Fasman predictions of the secondary structure from its primary structure which is further supported by CD and FT-IR studies.

Materials and Methods

A. Virus Purification: CMV-D strain was a gift from Dr. J.M. Kaper (Agricultural Research Service, U.S. Department of Agriculture, Beltsville, MD 20705, USA). It was propagated in tobacco (*Nicotiana tabacum* L. cv. Xanthi nc) and purified as described by Lot et al. [4].

B. RNA Extraction: Viral RNA was prepared by phenol-SDS treatment of purified virus and ethanol precipitation.

C. *Circular Dichroism (CD) Studies:* CD spectra of CMV-D (0.2 mg/ml) in aqueous solution were obtained on a JASCO 500 A spectropolarimeter. Aqueous background was digitally subtracted. The data were expressed in molar ellipticity as a function of the wavelength.

D. *Fourier-Transform Infrared (FT-IR) Studies:* FT-IR studies of CMV-D were performed by dissolving the virus in aqueous solution at a concentration of 6 mg/ml, pH 7.1. The experimental protocol was identical to previous FT-IR studies of proteins in aqueous solutions [5]. The FT-IR spectrum of CMV-D was recorded on a Digilab FTS-60 FT-IR spectrometer at room temperature. A microcircle cell with GE crystal was used, which was coupled to an MCT detector. The FT-IR spectrum of protein solution was obtained by a standard digital subtraction procedure. The above procedure has been successfully employed in FT-IR studies of polypeptides and proteins [5–7]. Fourier self-deconvolution was performed on the amide I region in accordance with the algorithm developed by Kauppinen et al. [8]. The standard protocol consisted of varying σ, band width at half height, until a slight negative dip appeared in the baseline adjoining the peaks. The resolution enhancement factor, K, was then varied until side lobes could be observed in the baseline. At this stage K was incremented by 0.1 and the results were plotted. The FT-IR spectrum of CMV-D was also recorded after the addition of 0.1% SDS or 2 M LiCl.

Results and Discussion

The CD spectrum of RNA was not subtracted from the CD spectrum of CMV-D. Hence CD studies are useful only in providing information on the conformation of RNA component of CMV-D strain. CD spectra displayed for either encapsidated or isolated RNA the same significant base pairing as revealed by the coincidence of the wave length of the positive band at 268 nm.

FT-IR Studies: The FT-IR spectrum of CMV-D manifested an intense broad amide I band centered at 1638 cm^{-1} in aqueous solution. The mean frequency of 1638 cm^{-1} is typical of an extended polypeptide chain, see Byler et al. [9]. Due to the broad character of the amide I band, we decided to resolve the band by Fourier self-deconvolution [8]. On Fourier self-deconvolution, the broad amide I band at 1638 cm^{-1} splitted into five components at 1638 cm^{-1}, 1657 cm^{-1}, 1666 cm^{-1}, 1676 cm^{-1} and 1688 cm^{-1}. While the intensive band at 1638 cm^{-1} characteristic of extended polypeptide chain persisted on Fourier self-deconvolution, a new less intense band at 1676 cm^{-1} appeared which is also characteristic of extended polypeptide chain [9]. These two amide I components taken together provide strong support of antiparallel β-sheet structure in CMV-D. The amide I component of 1657 cm^{-1} is indicative of α-helical structure as this band has been shown to occur in numerous α-helical polypeptides [10] and proteins (see Byler et al. [9]). The amide I components at 1666 cm^{-1} and 1688 cm^{-1} are generally assigned to β-turns [9]. Therefore, FT-IR studies of CMV-D are suggestive of a number of conformational substates coexisting in the intact CMV. A summary of FT-IR bands and their intensities, qualitatively described, is presented in Table 1. Table 1 also contains the amide I components of CMV-D when 0.1% SDS is added to the aqueous solution

Table 1 *Summary of IR frequencies (in cm^{-1}) observed in CMV-D strain*

		CMV	CMV + 0.1% SDS	Supernatant of CMV/LiCl mixture
Undeconvoluted amide I band		1638	1636.6	1639.5
Fourier self-deconvoluted amide I components	1)	1637.56 (s)	1637.56 (s)	1637.50 (s)
	2)	1656.85 (ms)	1654.92 (sh)	1651.06 (ms)
	3)	1665.49 (sh)	1664.52 (sh)	1666.90 (vw)
	4)	1676.14 (ms)	1676.04 (sh)	1676.98 (vw)
	5)	1687.71 (s)	1687.71 (s)	1687.71 (s)

(s)   = intensity of the band strong.
(ms) = intensity of the band medium strong.
(sh) = shoulder to a major band.
(vw) = intensity weak or very weak.

of virus. SDS induces disassembly of virus into RNA components and protein subunits [1]. Thus, after this treatment, the secondary structure of CMV protein subunits should not be influenced by protein-RNA interactions. 0.1% SDS causes a slight shift of α-helical component from 1657 cm^{-1} to 1655 cm^{-1} and a decrease in intensity of the band. Additionally, the higher-frequency component of the β-sheet band at 1676 cm^{-1} is greatly reduced in the intensity. The influence of 0.1% SDS on the secondary structure of CMV-D is surprisingly small, suggesting that the protein-RNA interactions are not so strong in maintaining the structural integrity of the virus assembly.

2 M LiCl should be also effective in inducing virus disassembly, as it is reported in Kaper [1] that 1.3 M LiCl induced complete disassociation of the virus. To verify the effects of the salt either on the system protein-RNA or on the protein subunits alone, we have recorded the FT-IR spectrum of the supernatant obtained after centrifugation of the mixture virus-LiCl kept 1 hour at 0°C. In these conditions, RNA should be precipitated as Li salt and the supernatant should contain the protein subunits.

Fourier self-deconvolution of the supernatant obtained from LiCl treatment reveals a further decrease in the frequency of α-helical component from 1657 cm^{-1} to 1651 cm^{-1}, the intensity was restored to its original intensity in the intact virus. Amide I component at 1666 cm^{-1}, characteristic of a β-turn structure, and the higher-frequency β-sheet component at 1676 cm^{-1} underwent a decrease in their intensities. The higher-frequency β-turn component at 1688 cm^{-1} was also reduced in its intensity. The effect of LiCl treatment on the secondary structure of CMV-D appears to be negligible.

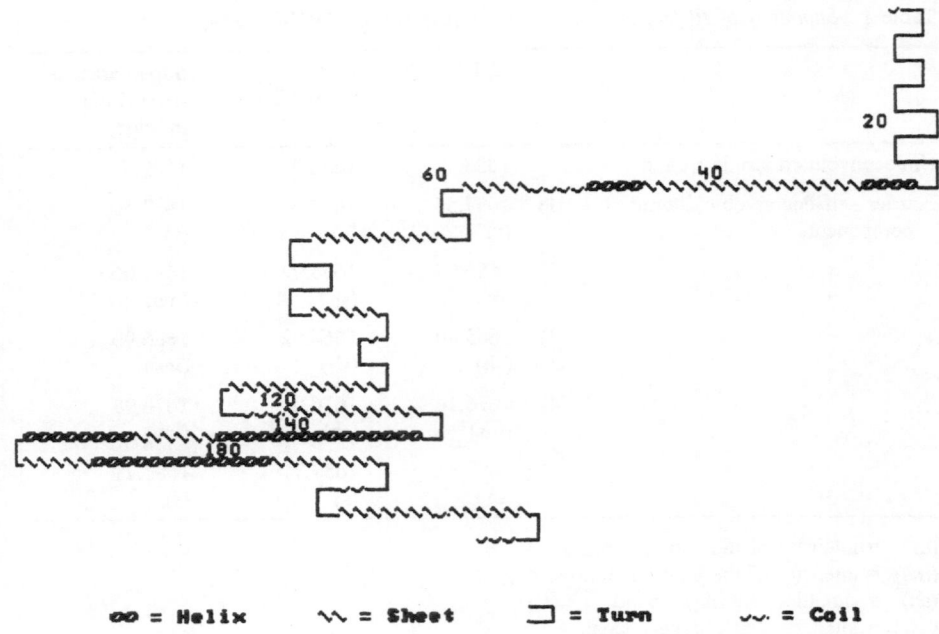

oo = Helix   ʌʌ = Sheet   ☐ = Turn   ᴗᴗ = Coil

Fig. 1.   Tentative secondary structure of coat protein of CMV-D.

Secondary Structure Predictions from Chou-Fasman Algorithm: The Chou-Fasman algorithm [11] predicts approximately 41% β-sheet structure with α-helical and β-turn structures accounting for the remainder of the secondary structure. At the C-terminal region, β-sheet structure with β-turn acting as spacers may fold back and form an anti-parallel β-sheet structure. Detailed structural studies, especially 2D NMR studies, alone can provide complete answers and until that time, the schematic figure, Fig. 1, should be considered tentative. Surprisingly, the general agreement between FT-IR and Chou-Fasman predictions is very good.

Conclusion

FT-IR studies of CMV-D are indicative of a β-sheet structure for the coat protein present in the intact virus. The recently determined primary structure of the coat protein, Cuozzo et al. [2] has ushered in renewed interest in the elucidation of its secondary and tertiary structures. Interestingly, the Chou-Fasman predictions of the secondary structure of the coat protein are in accord with CD and FT-IR studies reported here.

References

1. Kaper, J.M. (1975) in *The Chemical Basis of Virus Structure, Dissociation, and Reassembly,* North-Holland Publishing Company, Amsterdam.
2. Piazzolla, P., Guantieri, V. and Tamburro, A.M. (1986) *J. Gen. Virol.* **67,** 69.

3. Cuozzo, M., O'Connel, K.M., Kaniewski, W., Fang, R.-X., Chura, N.-H. and Turner, N.E. (1988) *Bio-Technology* **6**, 549.
4. Lot, H., Marrou, J., Quiot, J.B. and Esvan, G. (1972) *Ann. Phytopathol.* **4**, 25.
5. Renugopalakrishnan, V., Damle, S.P., Horowitz, P.M., Moore, S., Hutson, T.B. and Gregory, J.D. (1989) *Biopolymers* **28**, 1923.
6. Renugopalakrishnan, V., Horowitz, P.M. and Glimcher, M.J. (1985) *J. Biol. Chem.* **260**, 11406.
7. Renugopalakrishnan, V., Strawich, E.S., Horowitz, P.M. and Glimcher, M.J. (1986) *Biochemistry* **25**, 4879.
8. Kauppinen, J.K., Moffat, D., Mantsch, H.H. and Cameron, D.G. (1981) *Appl. Spectrosc.* **35**, 271.
9. Byler, D.M., Brouillette, J.N. and Susi, H. (1986) *Spectroscopy* **1**, 19.
10. Renugopalakrishnan, V. and Bhatnagar, R.S. (1984) *J. Am. Chem. Soc.* **106**, 2217.
11. Chou, P.Y. and Fasman, G.D. (1978) *Annu. Rev. Biochem.* **47**, 251.

Protein conformation and stability in relation to virus assembly: Investigation of bacteriophage P22 structural proteins by Raman spectroscopy

Renee Becka, Stacy A. Towse and George J. Thomas Jr.*

Division of Cell Biology and Biophysics, School of Basic Life Sciences,
University of Missouri-Kansas City, Kansas City, MO 64110, U.S.A.

Introduction

During the past decade, significant progress has been made in elucidating structural details of plant and animal viruses in the crystal [1]. Recent developments suggest that the X-ray crystal structure of a bacterial virus, the bacteriophage ϕX174, may also be forthcoming [2]. Nevertheless, relatively little is known of the pathways of polypeptide chain folding and subunit association which lead to the proper assembly and activity of viruses. In vitro folding and assembly of viral subunits to form correctly dimensioned capsids and functional capsid substructures continue to represent challenging problems for the structural biochemist. The prospect of meeting these challenges by means of high-resolution structure methods alone is neither practical nor desirable. Structural methods suited to proteins in solution and applicable over wide ranges of sample concentration, temperature, pH, and the like, are required. One such method is laser Raman spectroscopy [3].

We have employed laser Raman spectroscopy to investigate the folding pathways and thermostabilities of viral proteins and their assemblies [4,5], and to identify subunit domains and conformations which can be correlated with viral morphogenesis [6]. In this report, we describe results obtained recently for structural proteins of the *Salmonella* phage P22. This work has been conducted in collaboration with Professor Jonathan King and Dr. Peter Prevelige at the Department of Biology, M.I.T. A crucial element in these studies is the progress made by King and coworkers in identifying protein residues and environmental factors essential to maturation and assembly of P22 proteins. Also important is the capability of isolating viral gene products in highly purified form and in quantities sufficient for spectroscopic analysis.

Subunits occupying discrete positions in a mature virion exist in different and well-defined conformations. Construction of the mature capsid therefore requires that protein subunits in distinct conformations occupy their appropriate locations within the growing shell. Accurate switching between the possible subunit conformations in the context of the growing shell is presumably required in order to build properly dimensioned closed capsid shells. Switching of a protein subunit between unassociable and associable forms at the growing edge of a structure has been termed autosteric

---

* To whom correspondence should be addressed.

switching [7]. It is likely that selection of the proper subunit conformation is coupled to this process. In the double-stranded DNA bacteriophage P22, this process is mediated in vivo by interaction between coat protein subunits (product of viral gene 5, or gp5) and scaffolding protein (gp8), leading to the formation of a metastable procapsid. The transformation from the metastable procapsid to the mature capsid requires packaging of the genomic DNA and an accompanying 10% increase in radius of the shell [8].

In vivo in the absence of scaffolding protein, the coat protein does not polymerize until a high intracellular concentration is reached. This scaffold-free polymerization process produces aberrant structures, consisting of improperly dimensioned shells and spiral assemblies. These structures, which display improper radii of curvature, likely arise from incorrect relative positioning of subunits. Similar results have been obtained in vitro [9]. In order to detect and characterize specific conformational states at the molecular level, we have examined the coat protein subunit assemblies shown in Fig.

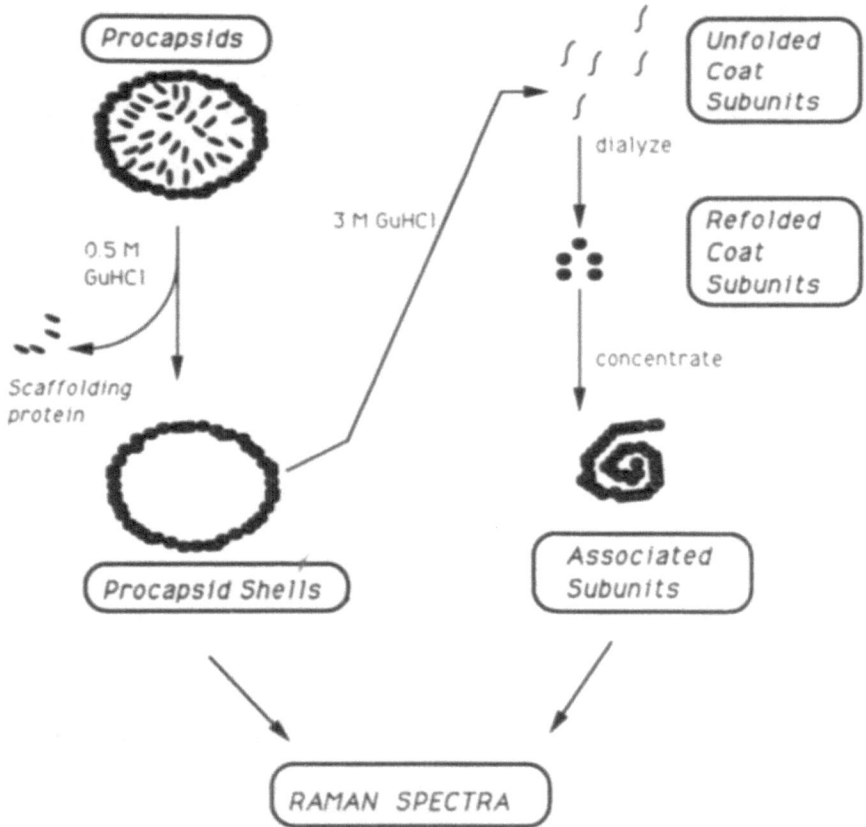

Fig. 1. *Procapsid shells (left) were obtained by treatment of procapsids with 0.5 M GuHCl to remove scaffolding protein. Associated subunits (right) were obtained by treatment of procapsid shells with 3 M GuHCl yielding unfolded subunits, which were refolded and reassociated in solution.*

1 by Raman spectroscopy. The feasibility of Raman spectroscopy as a structural probe of P22 proteins was demonstrated in an earlier study [10]. However, instrumentation and methods employed previously did not permit detection of conformational differences involving less than 10% of the protein secondary structure. The present study, utilizing techniques of higher sensitivity, allows the detection of conformational differences involving as few as 2% of the peptide residues per subunit. The current results demonstrate that the conformation of coat protein subunits in topologically closed procapsid shells (PS) assembled under the control of scaffolding protein is different than the conformation of the same associated subunits (AS) self-assembled in the absence of scaffolding protein.

The thermostable tailspike endorhamnosidase of P22, the product of viral gene 9, provides a model system for comparing the roles of amino acid sequences in determining the intracellular folding pathway with their role in stabilizing the mature structural protein [4,11]. A large set of temperature-sensitive-folding (tsf) mutants of gp9 have been isolated and characterized [12]. When released from the ribosome at permissive temperature (30°C), the mutant polypeptide chains form the native tailspike; when released at higher temperature (40°C) the mutant chains do not achieve the native fold. If maintained at the restrictive temperature, the tsf mutants form aggregates of partially or incorrectly folded chains. However, once correctly folded at permissive temperatures, the native forms of the mutant proteins are as active as wild-type. The proposed maturation pathways of wild-type and mutant tailspikes are presumed identical at permissive temperatures (Fig. 2).

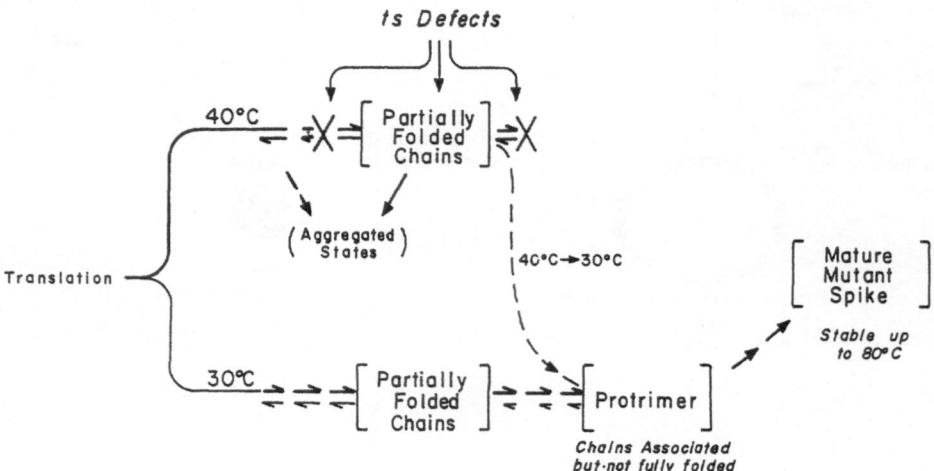

Fig. 2.   *Intracellular pathway of maturation of the bacteriophage P22 tailspike. The polypeptide chain released from the ribosome at 30°C (left) partially folds, and associates to form a protrimer intermediate. The thermostable tailspike results from subsequent folding of the protrimer. Temperature-sensitive folding mutants released at 40°C are blocked at an early stage in chain folding (temperature-sensitive stages) and do not associate to form competent protrimer intermediates at this restrictive temperature. However, if infected cells are shifted to 30°C, the mutant chains continue through the productive pathway.*

We have undertaken a systematic study of the Raman spectra of wild-type and mutant tailspikes in order to determine the conformations in solution and to investigate the loci and nature of tsf defects [4,5]. We have chosen for careful study the wild-type tailspike and three well-characterized tsf mutants: tsU24 (Ile[258] → Leu), tsH302 (Gly[323] → Asp) and tsU38 (Gly[435] → Glu). The tsU24 and tsH302 mutants, unlike tsU38, are as robust as wild-type in their resistance to thermal inactivation [4,13]. In the case of tsH302, a charged residue is substituted for a glycine thought to be at the protein surface [12]. The possibility existed that the mutants were causing local disturbances in the structure of this large protein, but in regions remote from the location involved in the rate-limiting step of the unfolding pathway. The Raman analysis allows us to address this question in greater detail. Experimental methods and procedures in these studies have been described elsewhere [4–6].

Results and Discussion

1. P22 capsid protein

Fig. 3 shows the amide I region (1500–1750 cm^{-1}) of Raman spectra of procapsid

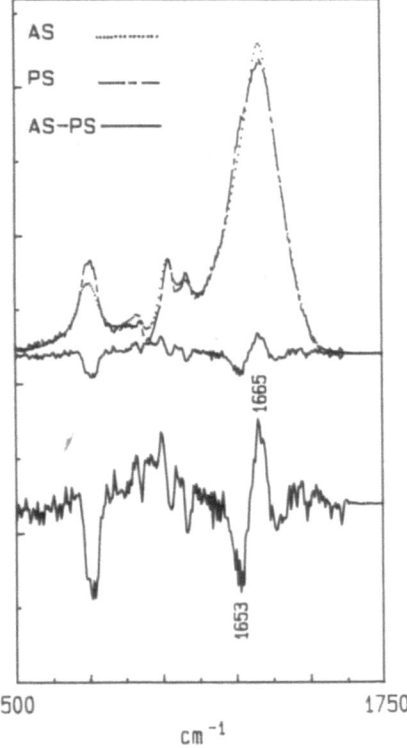

Fig. 3. Raman spectra of PS (— - —) and AS (·····) in the amide I region, and their difference spectrum, AS minus PS (——). The lowermost trace shows a fourfold expansion of the difference spectrum, including labels (cm^{-1} units) for the principal amide I difference peaks and troughs.

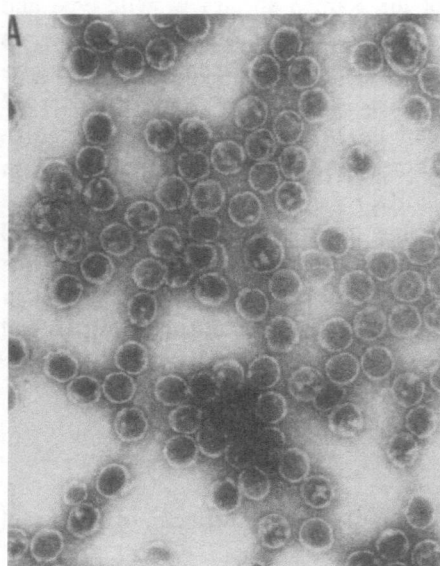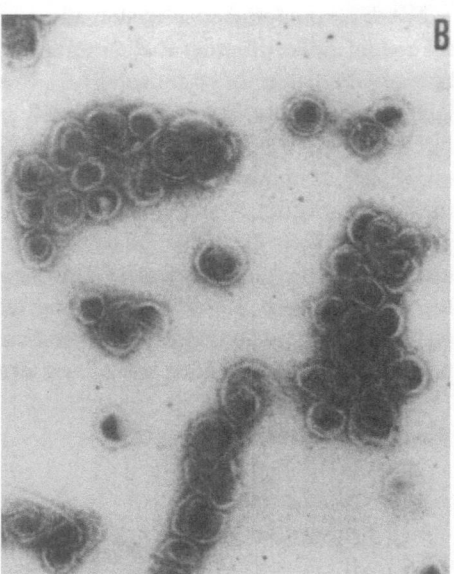

Fig. 4.   Electron micrographs of associated subunits (right panel) and procapsid shells (left panel) which yield the spectra of Fig. 3. Samples were negatively stained with 2% (w/v) uranyl acetate.

shells (PS) and associated subunits (AS). The two samples, prepared as described in Fig. 1, consist exclusively of the major capsid protein (gp5) and yield the electron micrographs of Fig. 4. The difference between Raman amide I bands of PS and AS, amplified in the difference spectrum of Fig. 3, is reproducible and can be interpreted in a straightforward manner by analogy with data from proteins of known structure [6]. The amide I results demonstrate a clearcut shift of Raman intensity from 1653 cm^{-1} (α-helix) in PS particles to 1665 cm^{-1} (β-sheet) in AS particles. A concomitant shift of Raman amide III intensity from ca. 1273 cm^{-1} (α-helix) in PS to ca. 1238 cm^{-1} (β-sheet) in AS is also observed (data not shown). We have confirmed that these spectral differences do not arise from random unfolding of subunits, such as that commonly ascribed to thermal unfolding.

The Raman amide I intensity change involves 4.5% ($\pm$ 0.4%) of the total amide I band area, and consists of a 2.2% intensity loss at 1653 cm^{-1} (α-helix) and 2.3% intensity gain at 1665 cm^{-1} (β-sheet). By assuming that the same amide I intensity is contributed from each peptide group, we interpret our result to indicate that, on average, 10 $\pm$ 1 of the 430 peptide residues of gp5 are converted from α-helix to β-sheet in a conformational switch associated with the PS $\rightarrow$ AS transition.

Other spectral changes indicative of different side-chain environments in PS and AS have been cataloged [6]. In general, the side-chain bands are broader or weaker in AS than in PS particles, consistent with a greater distribution of side-chain geometries and greater variety of side-chain interactions in the associated subunits than in procapsid shells. These results suggest that the protein monomer in the AS

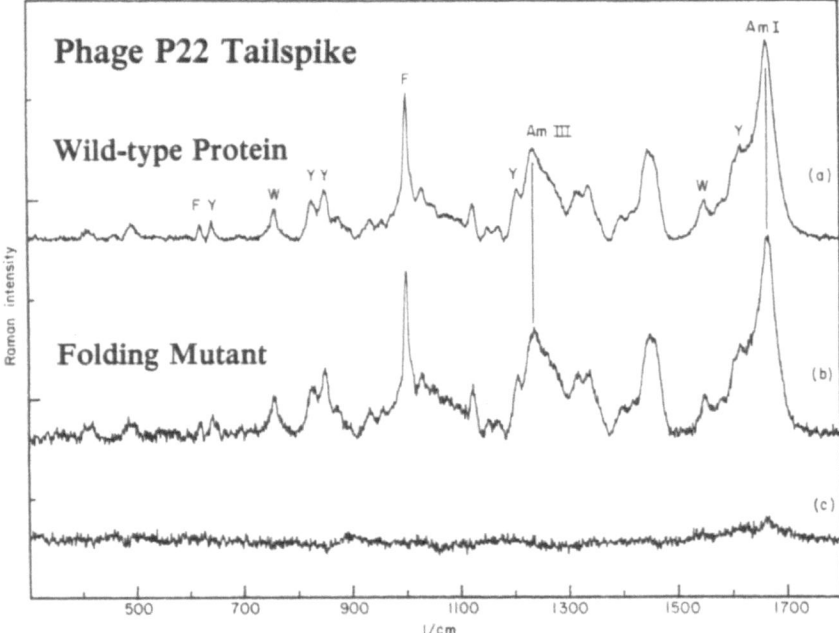

Fig. 5. *Raman spectra in the region 300–1800 cm^{-1} of the wild-type tailspike of P22 (top) and a temperature-sensitive-folding mutant (middle), and their computed difference spectrum (bottom) which illustrates the absence of significant structural differences [4].*

state is typically more exposed to solvent and perhaps less compactly folded than that of the procapsid shell.

2. P22 tailspike protein

The spectra of Fig. 5 show that a typical tsf mutant tailspike, once matured at the permissive temperature (30°C), is structurally identical to the wild-type tailspike. Analysis of this spectrum by Raman amide I and amide III curve fitting and Fourier deconvolution methods shows that the native tailspike consists predominantly of β-sheet (57 ± 4%) and reverse turns (18 ± 3%) [4]. We have also studied the thermostability of this structure and its unfolding by chemical denaturants in some detail [4,5]. The data show that wild-type and tsf mutants can be distinguished by Raman spectroscopy on the basis of their *different* and *characteristic* unfolding patterns, which yield unfolding intermediates of distinguishable secondary and tertiary structures, despite the fact that identical native structures are produced from maturation of tsf mutants at the permissive temperature. Recent efforts to crystallize the P22 tailspike have proved successful and analysis of the X-ray diffraction data is in progress (T. Alber, personal communication). Raman spectra show that the solution and crystal structures are indistinguishable for the spectral region 300–1800 cm^{-1}, consistent with very similar secondary and tertiary structures. On the other hand, the crystal structure incorporates Cd(II) from the mother liquor (containing CdSO$_4$). The Raman signature in low-frequency (0–300 cm^{-1}) and high-frequency regions

(2500–2600 cm^{-1}) demonstrates that Cd(II) is coordinated to cysteine sulfurs by displacement of sulhydryl groups [14].

Acknowledgements

We thank Mr. Tiansheng Li for assistance in data collection and Dr. Peter Prevelige (M.I.T.) for sample preparations. This research was supported by N.I.H. Grant AI11855.

References

1. Jurnak, F.A. and McPherson, A. (Eds.) (1984) *Biological Macromolecules and Assemblies*, Vol. 1, Virus Structures, Wiley, New York.
2. Willingmann, P., Krishnaswamy, S., McKenna, R., Smith, T.J., Olson, N.H., Rossmann, M.G., Stow, P.L. and Incardona, N.L. (1990) *J. Mol. Biol.* **212** (in press).
3. Thomas, G.J., Jr. (1987) in *Biological Applications of Raman Spectroscopy*, Vol. 1 (Spiro, T.G., Ed.) Wiley, New York, pp. 135–201.
4. Sargent, D., Benevides, J.M., Yu, M.-H., King, J. and Thomas, G.J., Jr. (1988) *J. Mol. Biol.* **199**, 491.
5. Thomas, G.J., Jr., Becka, R., Sargent, D., Yu, M.-H. and King, J. (1990) *Biochemistry* **29**, 4181.
6. Prevelige, P., Thomas, D., King, J., Towse, S.A. and Thomas, G.J., Jr. (1990) *Biochemistry* **29**, 5626.
7. Caspar, D.L.D. (1980) *Biophys. J.* **32**, 103.
8. Casjens, S. (1979) *J. Mol. Biol.* **131**, 1.
9. Prevelige, P., Jr., Thomas, D. and King, J. (1988) *J. Mol. Biol.* **202**, 743.
10. Thomas, G.J., Jr., Li, Y., Fuller, M.T. and King, J. (1982) *Biochemistry* **21**, 3866.
11. Goldenberg, D. and King, J. (1981) *J. Mol. Biol.* **145**, 633.
12. Yu, M.-H. and King, J. (1988) *J. Biol. Chem.* **263**, 1424.
13. Sturtevant, J.M., Yu, M.-H., Haase-Pettingell, C. and King, J. (1989) *J. Biol. Chem.* **264**, 10693.
14. Li, T., Alber, T. and Thomas, G.J., Jr., in preparation.

Pursuit of higher order structural changes of proteins by time-resolved ultraviolet resonance Raman spectroscopy

Teizo Kitagawa and Shoji Kaminaka

Institute for Molecular Science, Okazaki National Research Institutes, Myodaiji, Okazaki 444, Japan

Introduction

Higher order structures of proteins, which result from folding of polypeptide chains and their relative arrangements, play an essential role in regulating the activity of various proteins. A typical example is an allosteric effect, in which binding of a ligand such as a substrate or product of the enzymatic reactions to the regulation site causes an activity change. However, little is known about the structural mechanism of an allosteric effect. Hemoglobin (Hb) has served as a test case for studies of general allostery, since the oxygen affinity of Hb exhibits homotropic and heterotropic allosteric effects [1]. The Hb allostery is known to arise from a quaternary structure change, and its dynamical features have been a topic of various spectroscopic studies in the last decade.

Recently it was noticed that ultraviolet resonance Raman (RR) spectroscopy provides essential information on the tertiary and quaternary structures of proteins. When Raman scattering of proteins is excited around 200–240 nm, some Raman bands of aromatic residues are selectively intensity-enhanced [2–5]. Since the time resolution in this technique is much faster than that in NMR spectroscopy, which is currently the most common technique for studying a structure of proteins, RR spectroscopy allows us to investigate dynamical features of a protein conformational change. Accordingly, we pursued a change in the quaternary structure of Hb from the carbon monoxide-bound form (COHb) with the relaxed (R) quaternary structure to the deoxy form (deoxyHb) with the tensed (T) quaternary structure by using time-resolved UV RR spectroscopy [6].

Experimental Procedures

The layout of the observation system is illustrated in Fig. 1. The excitation light for Raman scattering was generated by Raman shifting the fourth harmonic of an Nd : YAG laser. The fundamental, second, and fourth harmonics from the Nd : YAG laser were focused into a 1-m H_2 cell (7.5 atm) and the second anti-Stokes line (218 nm) was selected by a beam separator made of rectangular and Pellin-Broca prisms. The laser power for the 218 nm line was 100–120 μJ/pulse at the sample point. The pump beam for photodissociating the COHb was obtained mostly from an N_2 laser-pumped dye laser, which was operated at 419 nm with a dye of bis-MSB, and its power was

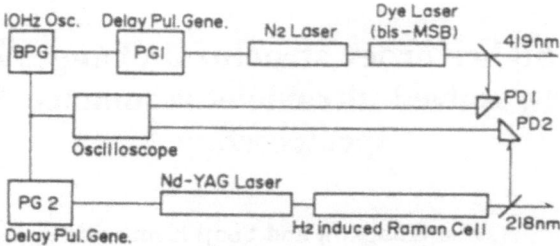

Fig. 1. The time-resolved UV RR measurement system.

180 µJ/pulse at the sample point. The delay time (Δt_d) from the pump pulse to the Raman probe pulse was controlled by a pulse generator (PG) system. BPG in Fig. 1 represents a homemade 10 Hz PG consisting of a 1-MHz quartz oscillator that generates two pulses; one triggers PG1 and the other triggers PG2 to fire each laser at the desired time. The actual delay time was determined by putting the partially reflected light of the 419- and 218-nm pulses into the photodiodes PD1 and PD2 and by monitoring their output with an oscilloscope [6].

The sample was circulated by a peristaltic pump and the illuminating chamber consisting of the wire-guided sample-flow system was constructed. The back-scattered light at 120° was dispersed with a 1.26-m monochromator equipped with a 2400 gr/mm holographic grating and detected with a UV-sensitive photo-multiplier. The grating was used in the second order. The output from the photo-multiplier was averaged by a boxcar integrator and transferred to a microcomputer. The wavenumbers of the Raman bands were calibrated with acetone, acetonitrile, and cyclohexane as standards.

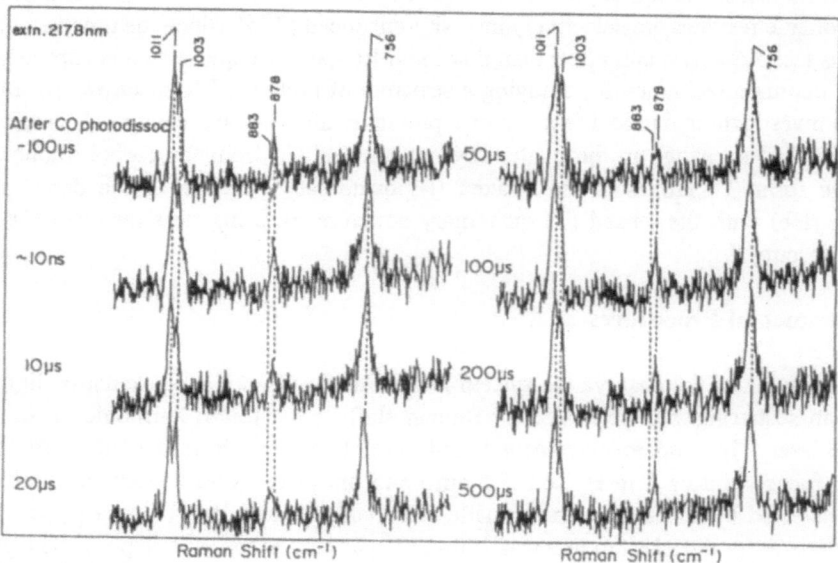

Fig. 2. 217.8-nm excited pump-probe RR spectra of COHb in the 1050–650 cm^{-1} region [6].

Results and Discussion

Time-resolved UV RR spectra of COHb in the $1100-600$ cm^{-1} region are displayed in Fig. 2, where Δt_d are specified on the spectra. Although the pumping system was different for $\Delta t_d = -100$ μs and 10 ns, the two RR spectra were practically the same and they were identical with the spectrum observed in the absence of the pump beam (not shown). The spectrum for $\Delta t_d = 5$ μs (not shown) was also close to that for $\Delta t_d = 10$ μs. This clearly indicates that the protein structural change does not occur within 10 ns after photolysis. The band at 1003 cm^{-1} is assigned to Phe residues but all other bands are attributed to Trp residues. The relative intensity of the two bands at 1011 and 1003 cm^{-1} remained unaltered until $\Delta t_d = 10$ μs, decreased around $\Delta t_d = 10-50$ μs, and restored to the original value at $\Delta t_d = 10$ μs. The 878 cm^{-1} band started to decrease in intensity from $\Delta t_d = 10$ μs and a new band appeared at 883 cm^{-1}, but the original pattern was restored at $\Delta t_d = 100$ μs.

The S/N ratios in these spectra are not always sufficient to draw a general conclusion. Therefore, we carried out completely independent experiments with

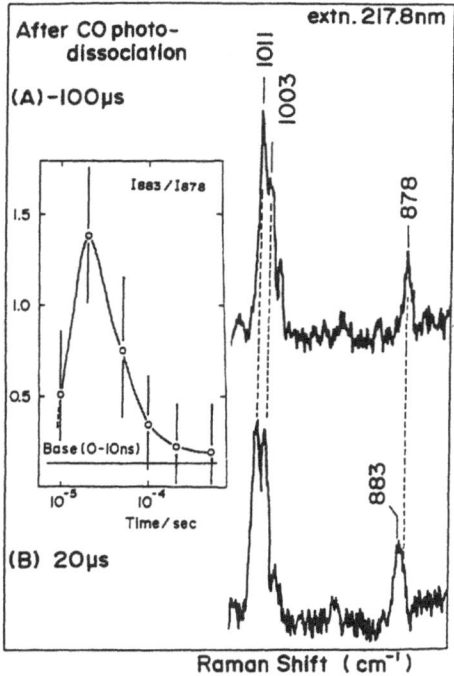

Raman Shift (cm^{-1})

Fig. 3. Completely independent results from the pump-probe experiments for two typical values of Δt_d, $\Delta t_d = -100$ μs (upper) and $\Delta t_d = 20$ μs (lower) with longer accumulation time (average of 16 scans): pump, 439 nm; probe, 217.8 nm; 100–120 μJ/pulse. The inset figure shows the delay-time dependence of the relative peak intensity of the two bands at 883 and 878 cm^{-1} in the spectra displayed in Fig. 2. 'Base' line means the value obtained from the spectrum for the 10-ns delay. Error bars were estimated from the highest and lowest values in the 8 component spectra [6].

different preparation of sample only for two typical values of Δt_d with longer accumulation time. The results are shown in Fig. 3, where the spectra for $\Delta t_d = -100$ μs (A) and $\Delta t_d = 20$ μs (B) are displayed. The S/N ratios are much raised due to longer accumulation times (16 scans are averaged). The intensity reduction of the 1011 cm^{-1} band and the frequency shift of the 878 cm^{-1} band noticed in Fig. 2 are well reproduced in Fig. 3; other spectra in Fig. 2 are also considered to be reliable. In the inset of Fig. 3, the relative intensities of the two bands at 878 and 883 cm^{-1} are plotted against the delay time. The horizontal line indicates the value at $\Delta t_d = -100$ μs and 10 ns. Error bars were estimated from the highest and lowest values in the 8 spectra. The relative intensity reached the maximum at $\Delta t_d = 20$ μs and decreased to its original value. This restoration is due to regeneration of the CO-bound form in the sample-illuminating chamber.

When the sample-illuminating chamber was filled with N_2, the value for $\Delta t_d = 20$ μs was retained even at $\Delta t_d = 100$ μs. The relative intensities of the Trp band at 1011 cm^{-1} to the Phe band at 1003 cm^{-1} obtained under N_2 atmosphere are plotted against Δt_d in Fig. 4, where a broken line indicates the value for $\Delta t_d = -100$ μs. The relative intensity does not change until $\Delta t_d = 5$ μs, then decreases smoothly, and reaches the stationary value at $\Delta t_d = 20$ μs. Therefore, it is likely that there is no intermediate state of the protein structure between the CO-bound R state and the deoxy T state so far as Trp residues are concerned. Consequently, the R to T quaternary structure change starts at ca. 5 μs after cleavage of the FeCO bond and finishes within 20 μs at neutral pH. However, at pH 5.8 where the Bohr proton is attached, the relative intensity changed between the same values as in Fig. 4 within 10 μs. This implies occurrence of a similar structural change but with very fast dynamics.

It is known from X-ray crystallographic analysis that metHb fluoride (HbF) assumes the T structure in the presence of inositol hexaphosphate (IHP) at pH 6.4 and the R structure in its absence at pH 7.4 [7]. Fig. 5 shows the 218 nm-excited RR spectra of HbF without IHP at pH 7.4 (upper) and with IHP at pH 6.8 (lower). The upper and lower spectra in Fig. 5 closely resemble those for $\Delta t_d = -100$ μs and 20 μs, respectively. Thus, it became clear that the most prominent RR spectral changes upon

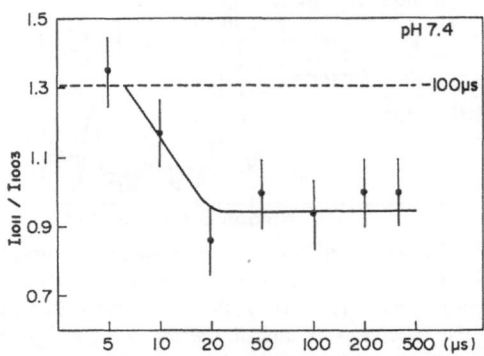

Fig. 4. Delay-time dependence of the relative intensity of the two bands at 1011 (Trp) and 1003 cm^{-1} (Phe) obtained under N_2 atmosphere.

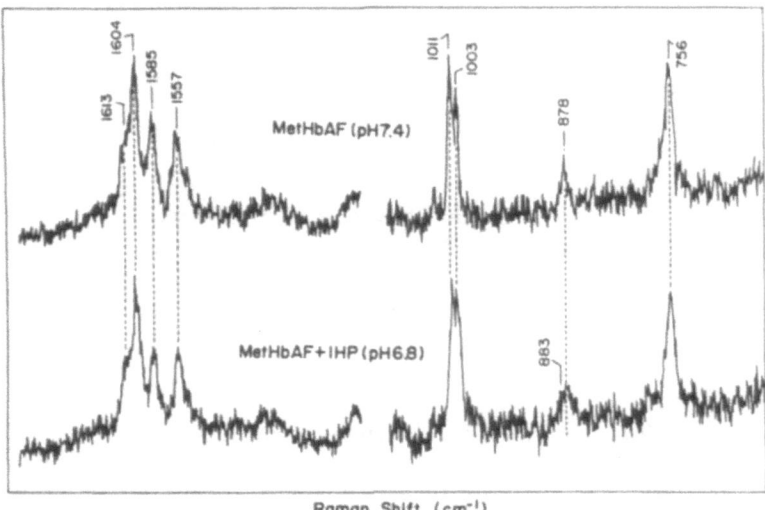

Fig. 5. UV RR spectrum of metHbF excited by the single 217.8 nm line. Upper and lower spectra were obtained for solutions at pH 7.4 without IHP and at pH 6.8 in the presence of IHP, respectively.

the quaternary structure change appeared for the 1011 and 878 cm^{-1} bands of Trp at Δt_d = 10–20 μs. According to normal coordinate calculations [8], the 878 cm^{-1} band of Trp involves motions of the NH group and therefore, its frequency is sensitive to hydrogen bonding. There are three Trp residues per αβ-dimer of Hb A and one of them (β37) undergoes a status change upon the quaternary structure change while the other two residues (α14 and β15) are contained in the α-helix and remain unchanged upon the quaternary structure change. The β37-Trp is located at the α1-β2 subunit interface where this residue is in contact with α140-Tyr in the T structure but is released from it in the R structure. Accordingly, the observed UV RR spectral change is presumably caused by β37-Trp. This implies that the ligand dissociation from the heme iron is first perceived by the Fe-His bond at the *trans* position and the movement of the His is communicated to the subunit interface in 10–20 μs through the movement of the F helix.

References

1.   Perutz, M.F. (1979) *Annu. Rev. Biochem.* **48**, 327.
2.   Rava, R.P. and Spiro, T.G. (1985) *Biochemistry* **24**, 1861.
3.   Asher, S.A., Ludwig, M. and Johnson, C.R. (1986) *J. Am. Chem. Soc.* **108**, 3186.
4.   Bajdor, K., Peticolas, W.L., Wharton, C.W. and Hester, R.E. (1987) *J. Raman Spectrosc.* **18**, 211.
5.   Mayne, L. and Hudson, B. (1987) *J. Phys. Chem.* **91**, 4438.
6.   Kaminaka, S., Ogura, T. and Kitagawa, T. (1990) *J. Am. Chem. Soc.* **112**, 23.
7.   Fermi, G. and Perutz, M.F. (1977) *J. Mol. Biol.* **114**, 421.
8.   T. Miura, H. Takeuchi and I. Harada (1988) *Biochemistry* **27**, 88.

Femtosecond resonance Raman spectroscopy of photochemical and photophysical changes in the chromophores of some proteins

R. van den Berg and M.A. El-Sayed

Department of Chemistry and Biochemistry, University of California, Los Angeles, Los Angeles, CA 90024-1569, U.S.A.

Abstract

We have studied the photochemical and photophysical events that occur upon excitation of two important biological molecules, bacteriorhodopsin (bR) and two differently ligated hemoglobin (Hb) molecules (HbCO and HbO_2) on an 800–900 fs timescale. For bR, spectra were obtained of the J_{625} intermediate in both the fingerprint region (~ 1200 cm^{-1}) and the ethylenic stretch region (~ 1530 cm^{-1}). Furthermore, evidence was found for an even earlier intermediate. In the case of HbCO and HbO_2, the frequencies of the core-size markers (1500–1650 cm^{-1}) gave information on the spin state and structure of the heme chromophore. A distinct down-shift of the ν_4-mode was observed for both molecules with respect to the equilibrium value, which is indicative of an elevated temperature of the heme after photo-dissociation. All spectra are compared with those observed in tens of ps up to ns.

Introduction

Although transient absorption spectra of chromoproteins are obtainable in the short-time domain, they are generally very broad and structureless, and give little or no information on configurational or conformational changes. Time-resolved resonance Raman spectroscopy of picosecond transients was first demonstrated in 1980 by Terner et al. [1] for the carboxyhemoglobin (HbCO) photointermediate and in 1981 by Hsieh et al. [2] for bacteriorhodopsin (bR). The vibrational spectra are greatly simplified when the laser is tuned to the electronic absorption wavelength of the chromophore, since only vibrations which mimic a molecular distortion in the excited state, are enhanced. Using microbeam and flow techniques with ps excitation and optical multichannel analyzer (OMA) detection, it has proven possible to detect the resonance Raman spectra of picosecond transients.

bR is a protein found in the purple membrane of *Halobacterium halobium*. It has a retinal chromophore, which is covalently linked to the protein backbone. Upon light absorption, it goes through a cycle of different intermediates on timescales varying from fs up to ms [3], which have been extensively studied with optical transient absorption techniques. During the cycle protons are pumped from the inside of the cell to the outside, leading to an electrochemical gradient that drives the bacteria's metabolic processes [4].

The primary photochemical event in the photocycle has been a subject of considerable interest. A number of experiments showed a redshifted primary photoproduct which forms in ~ 500 fs and relaxes to K_{610} on a 3 ps timescale [5–9].

This precursor was labeled J_{625}, although there was some uncertainty about its absorption maximum.

bR has been the subject of a great number of resonance Raman studies on timescales from ms to ps. It has been shown that upon light absorption an *all-trans* to *13-cis* isomerization occurs in the retinal. This isomerization has already taken place in the K_{610} intermediate, i.e. before 10–40 ps [2,10,11]. Whether isomerization has already taken place in the J_{625} intermediate has been an open question.

Hemoglobin (Hb) is probably one of the most thoroughly investigated biological molecules. It is a tetrameric protein, which exists in two different quaternary structures, one of which, designated T, is stable in the absence of ligands like O_2, CO and NO, and the other, designated R, is stable when ligands are bound [12]. The existence of these two quaternary structures is the basis for the cooperativity of ligand binding: the affinity for oxygen and other ligands increases as the number of ligand molecules already bound increases.

The study of the dynamics of the photodissociation of ligated hemoglobins has yielded a wealth of information regarding the process of ligand binding to deoxy-hemoglobin (deoxy-Hb). The availability of lasers and spectrometers providing time-resolution in the picosecond and femtosecond domain has enabled the studies of the primary photophysical processes occurring immediately after ligand photodissociation. The complicated photochemistry and photophysics of ligated hemoglobins have mostly been studied by transient optical absorption spectroscopy (see [13] and references therein). The initial photodissociation is assumed to occur in less than 50 fs and results in two excited-state species, Hb_I^* and Hb_{II}^* [13]. Whereas the former decays with a time constant of 350 fs into a deoxy-like photoproduct ($Hb^\dagger$), the latter lives much longer (~ 2.5 ps) and exhibits a much higher reactivity towards the photodissociated ligand. This was assumed to result from a planar heme in this species. Moreover, it was found that the population of Hb_{II}^* from HbO_2 was much larger than that resulting from HbCO (75% vs. 14%). This would partly explain the much lower quantum yield of photodissociation for HbO_2 (~ 0.05 on a microsecond timescale) compared to that of HbCO (~ 0.5). The deoxy-like species, $Hb^\dagger$, was assumed to be in a high-spin state ($S = 2$), although no direct evidence for this was obtained. Furthermore, a distortion of the spectrum of the deoxy-like Hb in the Soret-region was interpreted to be indicative of doming of the heme [13].

Materials and Methods

Sample preparation

The purple membrane was purified from the ETI-001 strain of *Halobacterium halobium* according to the method of Oesterhelt and Stoeckenius [3] and Becher and Cassim [14]. The purified purple membrane was suspended in doubly deionized water at a concentration of ~ 60 µM as measured by the retinal absorption at 570 nm.

Human Hb A was isolated in the oxygenated form (HbO_2) from packed red blood cells, according to the method of Antonini and Brunori [15]. Carbonmonoxy-hemoglobin (HbCO) was obtained by stirring a deoxygenated solution under CO overnight. Before every experiment the stock solution was diluted to a concentration of 0.6–0.8 mM with 0.01 M phosphate buffer (pH 7.4).

Experimental set-up

The experimental arrangement used for obtaining resonance Raman spectra is similar to that used by Terner et al. [1,16] and Nagumo et al. [17]. However, a different laser system was used: a cavity-dumped dye laser (Spectra Physics 375B) was synchronously pumped by the compressed, frequency-doubled output of a Nd:YAG laser (Spectra Physics model 3800). The dye Rhodamine 6G in ethylene glycol provided an output power of 20–25 mW at a dump rate of 800 kHz (25–31 nJ/pulse) at 587 nm (bR) and 578 nm (Hb). The bandwidth of the laser was found to be ~ 16–18 cm^{-1} (bR) and ~ 22–25 cm^{-1} (Hb). This is approximately two times the transform-limited bandwidth of an 800–900 fs pulse, as was measured by autocorrelation (Inrad model 5–14A) assuming a sech^2 pulseshape.

The beam was directed via a microscope objective (Zeiss Neofluar 40 ×) onto a vertically flowing free jet of bR suspension, and was thus tightly focused to a ~ 5–10 μm spot. The solutions were pumped from the sample-vial through a syringe needle with a 110 μm diameter by a peristaltic pump (Cole-Parmer Masterflex) at a flow rate of 25 m/s. Because of the low peak energy of the pulse, tight focusing is necessary in order to obtain a reasonable amount of photolysis and to eliminate multiple excitation of the sample. Each pulse simultaneously serves as a photolysis and a Raman excitation light source. Difference spectra were obtained by subtracting an arbitrarily scaled spectrum of the unphotolyzed molecule from a tight focus/high power spectrum, which also contains contributions from photointermediates.

Scattered light was collected at 90° and focused onto the slit of a 0.5 m spectrograph (SPEX 1870) with a 1200 grooves per mm ruled grating. Resonant scattering was blocked by appropriate cut-off filters. The resolution of the monochromator was ~ 3 cm^{-1}. The detection system consisted of a diode array detector, thermoelectrically cooled to −20°C (EG&G Princeton Applied Research Corporation model 1455) and an optical multichannel analyzer (OMA) model 1460, with a 1462 detector controller and 1462/99 14-bit A/D controller (EG&G Princeton Applied Research Corporation).

Results and Discussion

Bacteriorhodopsin

Resonance Raman spectrum of bR$_{570}$

The bR$_{570}$ spectrum obtained by us was in agreement with spectra reported previously [10,11,18–20]. Moreover, a complete vibrational assignment has been given by Smith et al. [21]. Two different spectral regions can be distinguished: 1450–1600 cm^{-1} and 900–1400 cm^{-1}. The former contains vibrations which characterize the isomeric configuration of the retinal, and the latter those which are sensitive to conformational distortions along the C$_7$-C$_{15}$ hydrocarbon chain [21]. The region around 1500 cm^{-1} has been assigned to a C = C stretching mode. Its frequency has been shown to correlate linearly with the electronic absorption maximum of the chromophore [22,23].

Resonance Raman spectrum in the ethylenic stretch region

The difference spectrum obtained in the region around 1530 cm^{-1} is shown in Fig. 1. An intense residual band around 1518 cm^{-1} was found by subtracting the bR_{570} from a high-power spectrum.

In Fig. 2 the relation is shown between the frequency of the C = C stretch vibration and the absorption maximum of the photointermediate for the different bR intermediates. From this figure the C = C stretch frequency of 1518 cm^{-1} as found from the difference spectrum of Fig. 1, would correspond to an intermediate with an absorption maximum of 625 nm and might therefore be attributed to J_{625}. Furthermore, it should be noted that the 1518 cm^{-1} band is asymmetric on the low-frequency side with a shoulder ~ 1510 cm^{-1}. This frequency corresponds to an intermediate with an absorption maximum at 660 nm and might indicate that during the 800–900 fs photolysis pulse either another intermediate is formed, which has an even more redshifted absorption spectrum than J_{625}, or that an intermediate is formed corresponding to J_{625} from another type of bR.

The absorption maximum which has been attributed to J_{625} seems to be shifted to the red when it is probed at shorter timescales. At times varying between 1–3 ps the absorption maximum seems to be around 625 nm, the value commonly assumed for the J intermediate. The C = C stretch frequency of 1518 cm^{-1} would then correspond to J_{625}, which would be in complete accord with the linear relation between λ_{max} and $v_{C=C}$ (Fig. 2) [22,23]. At earlier times, however, as shown by the differential transmittance spectra of Mathies et al. [8] there seems to be a redshift. It is therefore

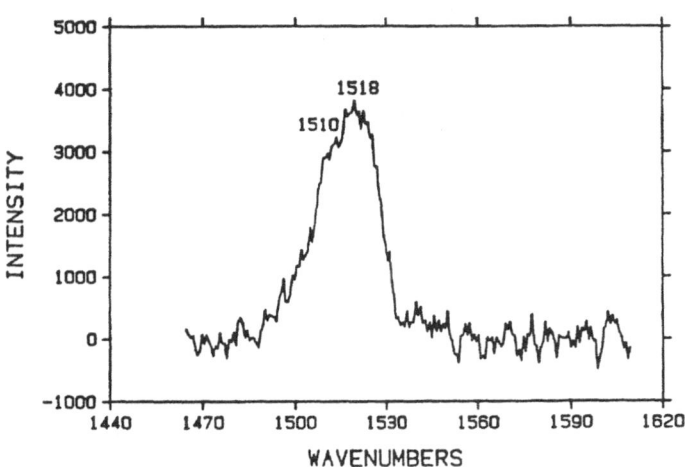

Fig. 1. Sub-picosecond resonance Raman spectra of the earliest intermediates of bR in the ethylenic stretch region. Two frequencies can be distinguished: a strong band at 1518 cm^{-1}, corresponding to the J_{625} intermediate, and a shoulder at 1510 cm^{-1}, due to an even earlier photocycle intermediate.

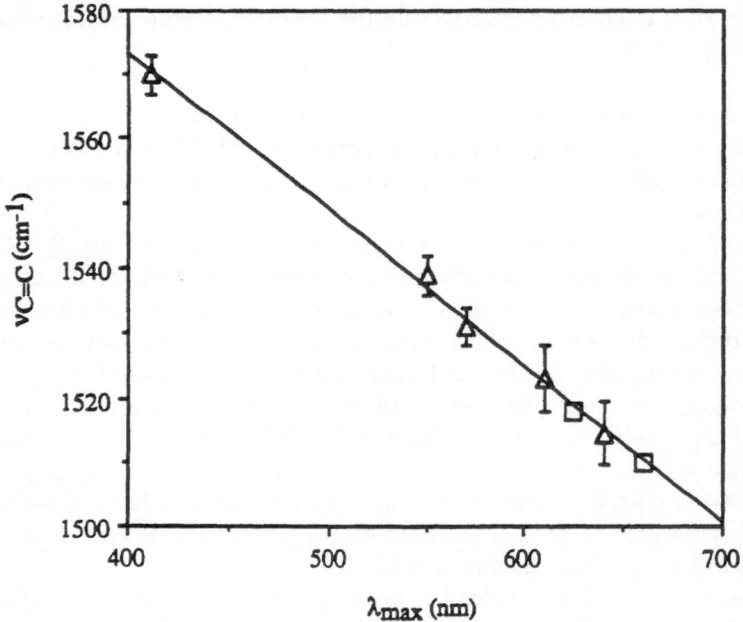

Fig. 2. Plot of the dependence between the frequency of the ethylenic stretch mode ($\nu_{C=C}$) and the absorption maximum (λ_{max}) of the different bR photocycle intermediates. The intermediates that have been plotted are: M_{412} (1570 cm^{-1}), L_{550} (1539 cm^{-1}), bR_{570} (1531 cm^{-1}), K_{610} (1523 cm^{-1}) and O_{640} (1514 cm^{-1}). The $C = C$ stretch frequencies of 1510 and 1518 cm^{-1} would correspond to absorption maxima of 660 and 625 nm, respectively. They have been put in the figure with different symbols. A linear relation is obtained.

possible that the 1510 cm^{-1} band is due to an absorption of an intermediate prior to J_{625}, which absorbs around 660 nm and is present on the 200–800 fs timescale.

Resonance Raman spectrum in the fingerprint region

It is this vibrational region which is sensitive to the isomeric state of the retinal chromophore. In the difference spectrum of Fig. 3 a band is clearly resolved around 1195 cm^{-1} and so is another band at a lower frequency of 1164 cm^{-1}. The latter confirms the observation of Atkinson et al. [11] of a similar vibration appearing in their 8 ps resonance Raman spectrum. The spectrum presented here is the first that can be exclusively attributed to J_{625}, since the 800–900 fs pulsewidth does not allow for appreciable K_{610}-formation.

The occurrence of the 1195 cm^{-1} band is indicative of isomerization at the $C_{13} = C_{14}$ position, i.e. from *all-trans* to *13-cis* retinal. This band was also observed in spectra obtained on a 40 ps [2,24], a 200 ps [10] and a 20–30 ns timescale [10,24,25]. It is also observed for the K_{610} intermediate which Braiman and Mathies [26] have shown to have a *13-cis* configuration. The 1164 cm^{-1} vibration on the other hand was not observed in any of the spectra mentioned above, and therefore does show that the retinal isomerization at these very early times (< 1 ps) is of a different nature than at much longer times.

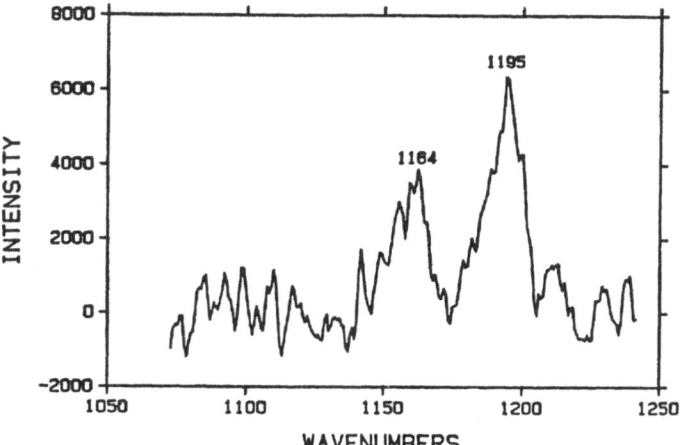

Fig. 3. Sub-picosecond resonance Raman spectra of earliest intermediates of bR in the fingerprint region. The all-trans to 13-cis isomerization of the retinal is evident from the peak at 1195 cm^{-1}. The 1164 cm^{-1} vibration, which is unique to the J_{625} spectrum, shows that the isomeric configuration at these early times (800–900 fs) is clearly different from that of the K_{610} intermediate.

Hemoglobin

A vibrational mode which has been shown to be sensitive to the ligation and oxidation state of the heme is the v_4-mode. It corresponds to a polarization dependent, totally symmetric stretching mode of the $C_\alpha N$ bond of the heme. Upon deligation it has been shown to shift from ~ 1370 cm^{-1} to ~ 1350 cm^{-1}. The frequency of the v_4-mode of the CO-photointermediate was carefully studied as a function of delay time between 0.2 ps and 90 ps [27]. It was shown to be down-shifted with respect to the equilibrium value of unligated Hb. The time-evolution of this downshift was interpreted in terms of a vibrational cooling of vibrationally hot heme.

We have been able to see the same effect on the 800–900 fs timescale. In Fig. 4 difference spectra are shown of the vibrational region around 1350 cm^{-1} of HbCO. When a diffusely focused spectrum (solid line) is subtracted from a tightly focused one (dotted line), a clear intensity decrease around 1370 cm^{-1}, and an increase around 1350 cm^{-1} can be observed. A fit with two Gaussian envelopes is also shown. The equilibrium value of the v_4-mode (as found in deoxy-Hb) on a long timescale is 1357 cm^{-1}. The downshift with respect to this value was calculated to be 8-9 cm^{-1} for HbCO, which is in excellent agreement with the value obtained by Petrich et al. [27] on this timescale. Similar spectra to those presented for HbCO were obtained for HbO_2. We find a value for the downshift of 7-8 cm^{-1}, which is slightly lower than that observed for HbCO on the same timescale, but which shows that a similar process might take place in the two Hb molecules.

In Fig. 5 (upper part) difference spectra are presented of the initial photoproduct after photodissociation of HbCO in the vibrational region between 1450 and 1750 cm^{-1}, which contains the core-size markers v_{11}, v_{19} and v_{10}. These Raman bands

155

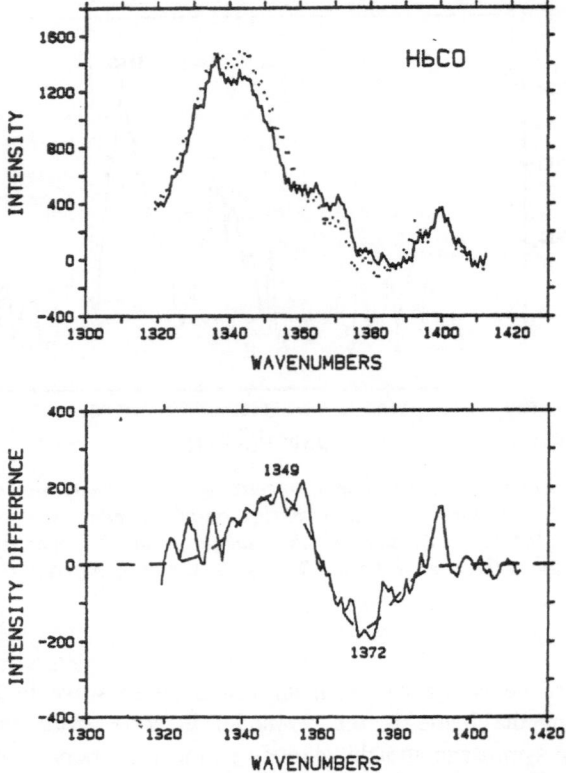

Fig. 4. Spectra of the v_4-region for HbCO obtained at tight (dotted line) and diffuse focus (solid line) respectively (upper part). The lower part shows a difference spectrum. Photolysis is found to occur indicated by a shift of the v_4-mode from ~ 1370 cm^{-1} to ~ 1350 cm^{-1}.

are found at 1545 cm^{-1}, 1556 cm^{-1} and 1605 cm^{-1} for deoxy-Hb [1,16]. In the spectrum of the photointermediate, the vibrational frequencies of a deoxy-like photointermediate are revealed. They appear at 1543, 1553 and 1602 cm^{-1}.

On a timescale of 30–40 ps similar spectra of a deoxy-like photoproduct have been reported in the literature [1,16]. The frequencies of the core-size markers depend on the ligation of the heme and on the spin state of the iron. Since deoxy-Hb is known to have a high-spin (S = 2), the photointermediate must also be in the S = 2 state. The further downshift of the core-size markers with respect to deoxy-Hb can be explained by the fact that the Fe atom is apparently more constrained in the heme plane following photoexcitation than in deoxy-Hb, in which it has been shown to be fully out-of-plane.

For HbO_2 the difference spectrum (Fig. 5, lower part) shows even more down-shifted bands than those found for HbCO, which were attributed to $Hb^{\dagger}$. This additional downshift can not be attributed to a further expansion of the porphyrin core. Alternatively, the larger downshifts might be explained by assuming that the molecule is in an excited state. The v_{10} and v_{11} modes, found in the HbO_2 photopro-

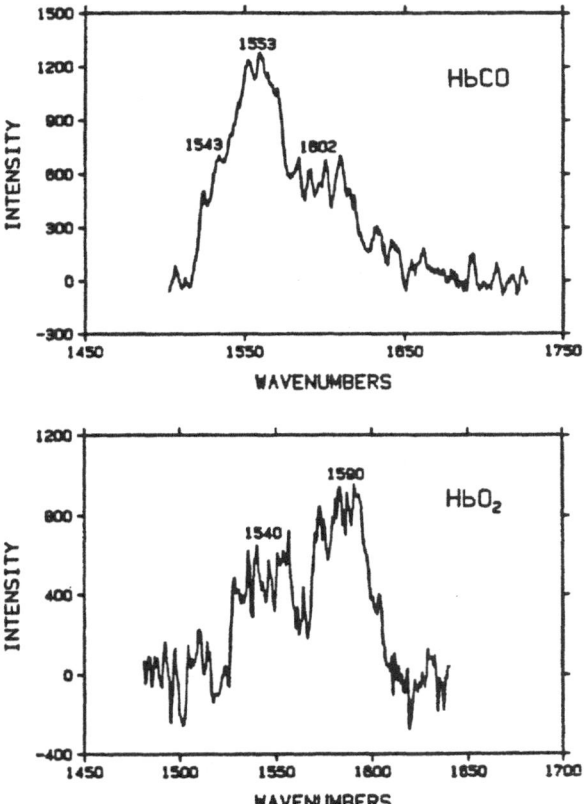

Fig. 5. Upper part: sub-picosecond resonance Raman spectra of the photoproducts of HbCO in the core-size region. The spectrum is a difference spectrum, and shows the Raman modes of a deoxy-like photoproduct, Hb†. Lower part: sub-picosecond resonance Raman spectra of the photoproducts of HbO₂ in the core-size region. The spectrum is a difference spectrum, and shows the Raman modes of an excited-state photoproduct, Hb_{II}. Note the much larger downshift of the frequencies in this spectrum, compared to those obtained for HbCO.*

duct spectrum at ~ 1540 cm^{-1} and ~ 1590 cm^{-1} respectively, have been shown to decrease when the porphyrin π^* orbital is occupied, i.e. in an excited state [28]. A similar spectrum has been obtained by Terner et al. [29] using 25 ps pulses. They also attributed the spectrum to that of an electronically excited high-spin heme. The spectra of the HbO₂ photoproduct of Nagumo et al. [17], obtained with ~ 50 ps pulses, were similar however to the HbCO photoproduct. Terner et al. [29] speculated that the excited-state species they observed was due to electronically excited deoxy-Hb, directly formed from the photodissociation of photoexcited HbO₂. They concluded on the basis of the rather broad spectra they observed that on the timescale of their experiment they were indeed observing a mixture of prompt and delayed photoproducts. Instead of being sharper, our spectrum presented in Fig. 5 (lower part) taken at shorter times is broader than the spectra obtained by Nagumo et al. [17] and by Terner et al. [29]. This, and the difference in timescale between their and our experiments (30-50 ps vs. 800 fs), suggests that the broadening in our spectrum is most likely a result of scattering from vibrationally unrelaxed species, in accordance with the observations made for the v_4-mode.

We attribute the spectrum of Fig. 5 (lower part) to that of Hb_{II}^*. It was noted by Petrich et al. [13] that the reactivity of Hb_{II}^* towards ligands should imply a planar heme, since a partially domed heme might present a sizable barrier towards ligand recombination. If so, the resonance Raman spectra presented here might be those of the planar heme in Hb_{II}^*. Both spectra shown in Fig. 5 therefore give structural information on the earliest photointermediates of both HbCO and HbO_2 on this timescale.

Acknowledgements

This work was supported by the Department of Energy (Office of Basic Energy Sciences) under Grant DE-FG03-88ER13828. R. van den Berg gratefully acknowledges the SHELL Research Laboratory, Amsterdam (K.S.L.A.) for financial support and the Netherlands Organization for Scientific Research (N.W.O.) for a research fellowship.

References

1. Terner, J., Spiro, T.G., Nagumo, M., Nicol, M.F. and El-Sayed, M.A. (1980) *J. Am. Chem. Soc.* **102**, 3238.
2. Hsieh, C.-L., Nagumo, M., Nicol, M. and El-Sayed, M.A. (1981) *J. Phys. Chem.* **85**, 2714.
3. Oesterhelt, D. and Stoeckenius, W. (1974) *Methods Enzymol.* **31**, 667.
4. Stoeckenius, W. and Bogomolni, R.A. (1982) *Annu. Rev. Biochem.* **52**, 587.
5. Sharkov, A.V., Pakulev, A.V., Chekalin, S.V. and Matveetz, Y.A. (1985) *Biochim. Biophys. Acta* **808**, 94.
6. Nuss, M.C., Zinth, W., Kaiser, W., Kölling, E. and Oesterhelt, D. (1985) *Chem. Phys. Lett.* **117**, 1.
7. Polland, H.J., Franz, M.A., Zinth, W., Kaiser, W., Kölling, E. and Oesterhelt, D. (1986) *Biophys. J.* **49**, 651.
8. Mathies, R.A., Brito Cruz, C.H., Pollard, W.T. and Shank, C.V. (1988) *Science* **240**, 777.
9. Dobler, J., Zinth, W., Kaiser, W. and Oesterhelt, D. (1988) *Chem. Phys. Lett.* **144**, 215.
10. Stern, D. and Mathies, R.A. (1985) in *Time-resolved Vibrational Spectroscopy*, (Laubereau, A. and Stockburger, M., Eds.), Springer Verlag, Berlin, pp. 250-254.
11. Atkinson, G.H., Brack, T.L., Blanchard, D. and Rumbles, G. (1989) *Chem. Phys.* **131**, 1.
12. Perutz, M.F. (1979) *Annu. Rev. Biochem.* **48**, 327.
13. Petrich, J.W., Poyart, C. and Martin, J.L. (1988) *Biochemistry* **27**, 4049.
14. Becher, B.M. and Cassim, J.Y. (1975) *Prep. Biochem.* **5**, 161.
15. Antonini, E. and Brunori, M. (1971) *Frontiers of Biology* **21**, pp 1-12.
16. Terner, J., Stong, J.D., Spiro, T.G., Nagumo, M., Nicol, M. and El-Sayed, M.A. (1981) *Proc. Natl. Acad. Sci. USA* **78**, 1313.
17. Nagumo, M., Nicol, M. and El-Sayed, M.A. (1981) *J. Phys. Chem.* **85**, 2435.
18. Terner, J., Hsieh, C.-L., Burns, A.R. and El-Sayed, M.A. (1979) *Proc. Natl. Acad. Sci. USA* **76**, 3046.
19. Terner, J., Hsieh, C.-L., Burns, A.R. and El-Sayed, M.A. (1979) *Biochemistry* **18**, 3629.
20. Atkinson, G.H., Brack, T.L., Grieger, I., Rumbles, G., Blanchard, D. and Siemanowski, L.M. (1985) in *Time-resolved Vibrational Spectroscopy*, (Atkinson, G.H., Ed.), Gordon and Breach, New York, pp. 55-82.

21. Smith, S.O., Braiman, M.S., Meyers, A.B., Pardoen, J.A., Courtin, J.M.L., Winkel, C., Lugtenburg, J. and Mathies, R.A. (1987) *J. Am. Chem. Soc.* **109**, 3108.
22. Rimai, L., Heyde, M.E. and Gill, D. (1973) *J. Am. Chem. Soc.* **95**, 4493.
23. Aton, B., Doukas, A.G., Callender, R.H., Becher, B. and Ebrey, T.G. (1977) *Biochemistry* **16**, 2995.
24. Hsieh, C.-L., El-Sayed, M.A., Nicol, M., Nagumo, M. and Lee, J.H. (1983) *Photochem. Photobiol.* **38**, 83.
25. Smith, S.O., Braiman, M. and Mathies, R.A. (1983) in *Time-resolved Vibrational Spectroscopy*, (Atkinson, G.H., Ed.) Academic Press, New York, pp. 219–230.
26. Braiman, M. and Mathies, R.A. (1982) *Proc. Natl. Acad. Sci. USA* **79**, 403.
27. Petrich, J.W., Martin, J.L., Houde, D., Poyart, C. and Orszag, A. (1987) *Biochemistry* **26**, 7914.
28. Spiro, T.G. and Strekas, T.C. (1975) *J. Am. Chem. Soc.* **96**, 338.
29. Terner, J., Voss, D.F., Paddock, C., Miles, R.B. and Spiro, T.G. (1982) *J. Phys. Chem.* **86**, 859.

Oriented fluorescent streptavidin conjugated phycoerythrin protein on biotinylated lipid LB monolayer films

Lynne A. Samuelson[a], Pascal Miller[c], Dianne M. Galotti[b], Kenneth A. Marx[b], Jayant Kumar[c], Sukant K. Tripathy[b] and David L. Kaplan[a]

[a]*Biotechnology Branch, U.S. Army Labs, Natick, MA 01760, U.S.A.*
Departments of [b]Chemistry and [c]Physics, University of Lowell, Lowell, MA 01854, U.S.A.

Introduction

A methodology which can be employed to rationally orient and couple biological molecules into well-defined, two-dimensional arrays would be extremely valuable for potential bioelectronic, optical, biomedical and protein research applications. In this way, one could, on the one hand, exploit the inherent and unique intelligent material properties of the biological assembly in a hybrid system to create novel properties. On the other hand, these hybrid materials may well serve as both unique biomimetic environments or simple environments for the study of protein structure. For example, crystallographic TEM structure studies on ordered two-dimensional protein crystals would be made easier in many instances where the two-dimensional or large three-dimensional protein crystals are difficult or impossible to obtain [1]. Also, totally oriented two-dimensional protein films are advantageous, allowing for easier interpretation of high-resolution TEM images where surface tertiary structure is well-preserved by a methodology such as the freeze-fracture, deep-etch technique [2,3].

The Langmuir-Blodgett (LB) technique has recently been used to prepare such oriented and spatially organized protein molecular assemblies [4,5] making it the method of choice because following formation, one may subsequently transfer ultrathin monolayer films a single monolayer at a time. Our approach has involved the highly specific recognition of biotin on the LB trough subphase surface of biotinylated LB lipid monolayers by streptavidin conjugated phycoerythrin (Str-PE) and avidin conjugated phycoerythrin (Av-PE). The binding affinity of biotin to the tetramer proteins (four binding sites) avidin and streptavidin is well-known (10^{15} M) and once formed the complex is essentially irreversible [6–8] with a stability comparable to that of a covalent bond. Also, the multiple biotin binding sites of these tetramer proteins should allow for the subsequent attachment of other biotin derivatized biomolecules such that organized two-dimensional biological arrays may be fabricated. These desirable properties have led to the use of the biotin-streptavidin interaction in many biomedical research and biotechnology applications [9].

Phycoerythrin is the outermost phycobiliprotein of the phycobilisome 'Light Harvesting System' found in red algae [10,11]. Individual phycoerythrin phycobiliproteins are highly and characteristically fluorescent, i.e., twenty-fold more fluorescent than the chromophore fluorescein on a molar basis. These proteins also possess an unusually large Stoke's shift, 81 nm (495 nm excitation and 576 nm emission), which is approximately 2.7 times that of fluorescein [12]. In addition, the time-resolved fluorescence properties and molecular environment of the chromophores of this much studied antennae pigment are reasonably well understood [13,14]. These properties, coupled with the protein's stability and its ability to function efficiently in low light level situations, suggest promising new biomedical research, biotechnology and biosensor applications.

Av-PE and Str-PE two-dimensional assemblies were prepared by injecting the conjugated protein system underneath a biotinylated phospholipid monolayer. Protein binding to the monolayer was observed through pressure-area isotherms and fluorescence spectroscopy.

Methods and Materials

The biotinylated phospholipid, N-(biotinoyl)dipalmitoyl-L-α-phosphatidylethanol-amine, triethylammonium salt, (B-DPPE), was purchased from Molecular Probes (Eugene, Oregon) and used as received. L-α-dipalmitoyl phosphatidylethanolamine (DPPE) was purchased from Avanti Polar Lipids (Pelham, Alabama) and was used as received. The unconjugated phycoerythrin (PE) and avidin and streptavidin conjugated R-phycoerythrin proteins were purchased from Biomeda Corporation (Foster City, California).

Monolayer studies were carried out on Lauda MGW Filmwaag troughs with a surface area of approximately 930 cm^2. The subphase was composed of an aqueous solution of 0.1 mM sodium phosphate, 0.1 M NaCl, at pH 6.8. In the case of pressure-area isotherms, the lipid was spread from a 0.5 mM chloroform solution and 0.1 mg of the protein in 5 ml of the buffered subphase was injected under the spread film and left to incubate for 2 hours at 30°C. Compression was then carried out at a speed of approximately 2 mm^2/min until collapse of the film was observed. For transfer studies, the lipid was spread, followed by protein introduction and incubation in the expanded state for two hours and then compressed to an annealing surface pressure of approximately 15 mN/m for deposition. Monolayer films were then transferred onto glass solid supports for fluorescence spectroscopy.

Results and Discussion

Pressure-area isotherms of the protein-injected monolayers may be compared to that of the pure phospholipid to establish the attachment of the proteins to the monolayer films subsequent to incubation. Figure 1 shows the structure of the biotinylated phospholipid (B-DPPE), and the pressure-area isotherms of B-DPPE and protein-injected B-DPPE (PE, Av-PE and Str-PE). As shown, all four isotherms give a relatively steep slope after 15 mN/m which corresponds to an area per lipid molecule of approximately 100 Å^2 prior to film collapse.

It is interesting to note that for the injected Av-PE or Str-PE monolayers, a significant increase in surface pressure of the lipid in the gas-expanded phase was observed. This suggests that the conjugated protein systems are adsorbing by binding

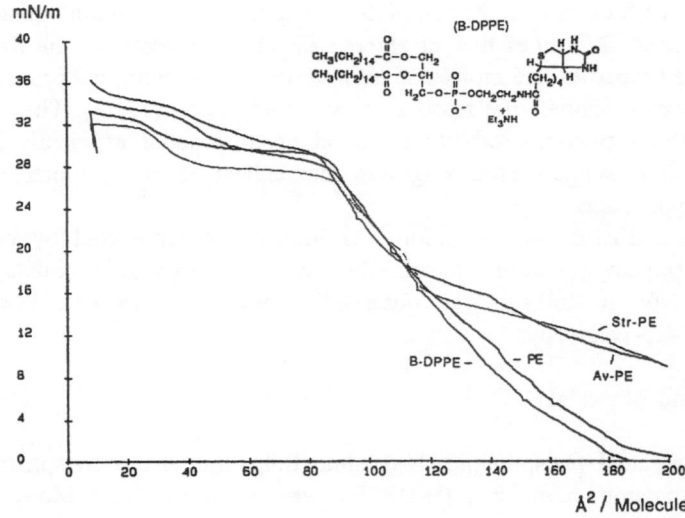

Fig. 1. *Structure of the biotinylated phospholipid (B-DPPE) and pressure-area isotherms of B-DPPE and protein-injected B-DPPE (PE, Av-PE and Str-PE).*

biotin moieties on the B-DPPE monolayer as shown in Fig. 2. Upon further compression, the lipids and/or bound protein conjugates begin to preferentially orient and the isotherms become identical to the pure B-DPPE. This indicates that the bulky protein systems may be 'swinging' down into the aqueous subphase, thus not causing any expansion of the monolayer at higher surface pressure. In comparison, the unconjugated phycoerythrin showed very little change in the isotherm when compared to the pure B-DPPE. This control provides further evidence that the biotin binding sites of the avidin and streptavidin tetramer proteins are likely to be directly involved in the adsorption process.

The monolayer films were transferred to hydrophilic solid glass supports using the vertical dipping technique at an annealing pressure of 15 mN/m, with transfer ratios of the films ranging from 100 to 150%. The presence of the phycoerythrin is probed by virtue of its intense and characteristic fluorescence. A schematic of the fluorescence set-up is given in Fig. 3. The measurements were carried out by exciting the samples with 496 nm light from an Argon Ion laser and scanning the emission from 515 to 670 nm.

The fluorescence spectra of a B-DPPE monolayer with Str-PE, a B-DPPE monolayer with PE and a monolayer of DPPE (not biotinylated) with Str-PE are compared in Fig. 4. As shown, the B-DPPE Str-PE monolayer gives a strong emission at 576 nm which corresponds to the fluorescence spectrum of the native phycoerythrin protein. This is direct evidence that the protein has adsorbed to the B-DPPE monolayer. In comparison, the unconjugated phycoerythrin monolayer shows no fluorescence signal at 576 nm suggesting that the streptavidin is, in fact, necessary for binding of the phycoerythrin to the monolayer. In addition, the monolayer where Str-PE was injected under a monolayer of the parent lipid (DPPE) containing no biotin

162

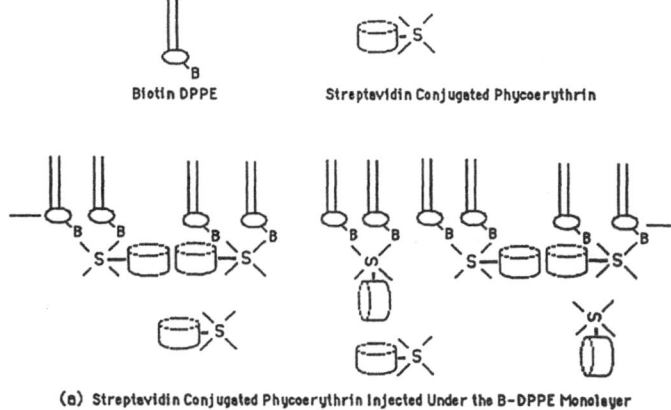

Biotin DPPE   Streptavidin Conjugated Phycoerythrin

(a) Streptavidin Conjugated Phycoerythrin Injected Under the B-DPPE Monolayer

COMPRESSION

(b) Hypothetical Oriented 2-D Protein Lattice

Fig. 2. Idealized schematic of the two-dimensional ordering of derivatized protein monolayers onto a biotinylated lipid LB film.

showed no fluorescence emission at 576 nm, providing further evidence that the biotin-streptavidin complexation is essential in the protein-binding mechanism. Films of Av-PE bound to a monolayer of B-DPPE exhibited similar fluorescence properties (data not shown).

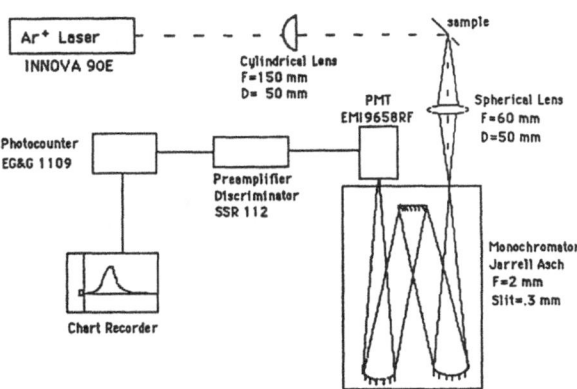

Fig. 3. Schematic diagram of fluorescence set-up.

163

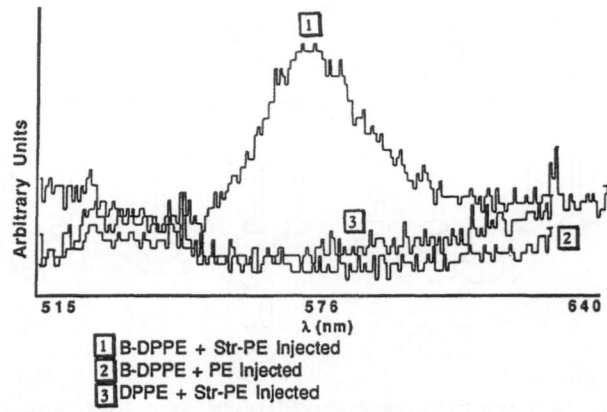

Fig. 4.   Fluorescence spectra of protein-adsorbed LB films.

Conclusion

We have shown that avidin and streptavidin conjugated phycoerythrin, injected under a monolayer, preferentially adsorb to only biotinylated monolayers at the air-water interface. Fluorescence measurements confirmed the presence of the conjugated phycoerythrin protein. These results establish a general technique for the two-dimensional ordering, in a monolayer form, of any biomolecular system which can be derivatized with biotin or avidin/streptavidin. A significant advantage of this attachment methodology lies in the multiple biotin binding sites of the tetramer proteins, which would allow the incorporation of multiple interacting biomolecules in an organized two-dimensional hierarchy. For the first time the monomolecular organization of the photodynamic phycobiliprotein, phycoerythrin, has been fabricated using this general technique.

References

1. Kornberg, R.D. and Ribi, H.O. (1987) *Protein Str., Folding and Design* **2**, 175.
2. Ruben, G.C. and Marx, K.A. (1984) *J. Electron Microscopy Tech.* **1**, 373.
3. Ruben, G.C. (1989) *J. Electron Microscopy Tech.* **13**, 335.
4. Blankenburg, R., Meller, P., Ringsdorf, P. and Salesse, C. (1989) *Biochemistry* **28**, 8214.
5. Haas, H. and Mohald, H. (1989) *Thin Solid Films* **180**, 101.
6. Green, N.M. (1975) *Adv. Protein Chem.* **29**, 85.
7. Gimlick, R.K. and Giese, R.W. (1988) *J. Biol. Chem.* **263**, 210.
8. Weber, P.C., Ohlendorf, P.C., Wendoloski, J.J. and Salemme, F.R. (1989) *Science* **243**, 85.
9. Wilchek, M. and Bayer, E.A. (1989) *Trends Biochem. Sci.* **14**, 409.
10. Gantt, E. (1975) *Bioscience* **25**, 781.
11. Glazer, A.N. (1985) *Annu. Rev. Biophys. Chem.* **14**, 47.
12. Glazer, A.N. and Stryer, L. (1984) *Trends Biochem. Sci.* October, 423.
13. Holzwarth, A. (1989) *Quart. Rev. Biophys.* **22**, 3, 239.
14. Schirmer, T., Huber, R., Schneider, M., Bode, W., Miller, M. and Hackert, M.L. (1986) *J. Mol. Biol.* **188**, 651.

Conformational studies of proteins using vibrational circular dichroism

Timothy A. Keiderling[a], Petr Pancoska[b], Sritana C. Yasui[a], Marie Urbanova[b] and Rina K. Dukor[a]

[a]Department of Chemistry, University of Illinois at Chicago, Box 4348, Chicago, IL 60680, U.S.A.

[b]Department of Chemical Physics, Faculty of Mathematics and Physics, Charles University, Prague 2, Czechoslovakia

Introduction

Spectroscopic studies of proteins have a long history in the biochemical and biophysical sciences. In large part they were originally directed at determining structural information about large-molecule systems that were difficult to probe using conventional structural techniques. Spectral data has always been important in protein studies for preliminary, qualitative estimations of structure. In particular, circular dichroism in the ultraviolet (UV-CD) has become an indispensable tool for qualitative characterization of proteins in solution [1]. However the number of transitions available in the near-UV for analysis of the spectra-structure relationship is very limited. In proteins, the most useful have proven to be the n-π* and π-π* transitions of the amide groups. Interactions among these transitions in a protein yield information about the polymeric backbone, but since such amide electronic excitations are relatively delocalized, the resulting UV spectral bands are often affected by other environmental or local perturbations and, additionally, are broad and overlapping.

This set of circumstances has stimulated the development of vibrational CD (VCD) as a tool for conformational analyses [2]. VCD has the usual spectral advantages found in infrared (IR) and Raman spectroscopies of a large number of resolved, relatively localized transitions and, in addition, is a property of the ground electronic state of the molecule. Thus it has characteristics that offer the potential of avoiding some of the limitations of UV-CD while maintaining the benefits of making an optical activity measurement. The differences in the origins of CD measured in the two spectral regions suggest that they would bear a complementary relationship that could enhance the quality and quantity of structural information derivable from either one alone.

Empirical correlation of spectral features with secondary structure has historically been the most profitable route for stereochemical utilization of both parent techniques of VCD, i.e. electronic CD and vibrational (IR and Raman) spectroscopies. Until a reliable data set and an understanding of environmental effects are available for biopolymer VCD, detailed theoretical analyses will be premature, and successful application to real biomolecular systems will come faster via an empirical approach.

Unlike electronic CD, VCD can be used to correlate data for several different spectrally resolved transitions involving different localized vibrations of the molecule; and, unlike IR and Raman spectroscopies, each of these features will have a sharp and possibly different dependence on stereochemistry. Most importantly, the combination of these techniques proves to be more powerful than each individually. In some sense, they compensate for each other. How best to carry out such a coupled analysis is a main topic of our ongoing studies. Here we review our first efforts to obtain a *uniform systematic analysis* of the data available from each technique.

We have previously carried out a series of studies on polypeptides and oligo-peptides that has given us some insight into the regularities of and difficulties in obtaining such VCD spectra on amide vibrational modes [3]. The amide I band (amide I' in D_2O), mainly $C = O$ stretch, is the most characteristic transition and is the easiest to study with this infrared technique. Exceptional stability of the VCD for right-handed α-helices was found, with deuteration of the amide N-H changing the shape of the amide I VCD from a *positive* couplet to a $(-,+,-)$ three-peaked pattern (amide I') and shifting the negative α-helical amide II VCD from 1550 cm^{-1} to 1450 cm^{-1} (amide II') [4]. Standard test cases for the study of multiple polypeptide conformations in aqueous environments are poly-L-lysine and poly-L-glutamic acid which are in a 'random-coil' form in D_2O and can be transformed to β-sheet or α-helical forms by pH variation. The α-helical result is maintained while the 'random-coil' form in many polypeptides and proteins gives rise to a large *negative* couplet [5,6]. Such a pattern implies that, in these 'random-coils', substantial local ordering exists similar in nature to that of *left-handed helical* poly-L-proline II, as supported by their having a VCD band shape close to that of the oligomers, $(L-Pro)_n$, n = 3 or 4 [7]. If additional bands are studied, even subtle structural variations can be determined with VCD such as the difference between α-helices and 3_{10}-helices [8]. This library of experience with VCD has enabled us to have some understanding of the VCD of proteins now being analyzed [6,9]. In particular, our studies of polypeptides strongly indicate that their VCD depends on relatively short-range effects [7,8] in contrast to their UV-CD which is dependent on longer-range interactions.

Experimentally, VCD spectra are routinely measured at UIC on either a dispersive or an FT-IR-based instrument, both assembled at UIC and described in detail elsewhere in the literature [2,10,11]. The dispersive data are generally obtained at ~ 10 cm^{-1} resolution by averaging of several repetitive scans over the band of interest. The FT-IR measured VCD spectra are obtained at ~ 4 cm^{-1} resolution by averaging several blocks of scans, but the signal-to-noise ratio (S/N) is typically not as good as for dispersive spectra measured over the band of interest in the same time span [6]. Our spectra are usually obtained on solutions in short-path-length sample cells with CaF_2 windows; but, with care, it is possible to measure spectra of films and *interpret their relative changes* on the basis of a comparison of like structures [12]. For comparison and deconvolution purposes, higher resolution and better S/N IR absorption spectra are obtained using an FT-IR (Digilab FTS-60). UV-CD spectra are obtained with a spectropolarimeter (JASCO J-600). For systematization of the data handling from different instruments, all spectra are transferred to an off-line personal computer using SpectraCalc (Galactic Industries, Nashua, NH) for data manipulation and analysis.

Results and Discussion

In extending the application of VCD to proteins, we have obtained consistent protein VCD spectra and profitably compared it to UV-CD and IR data on the same proteins. As compared to IR spectra, the sign variation inherent in VCD gives it a tremendous advantage in differentiation between proteins. Furthermore, since the different types of secondary structure contribute on a comparable basis to VCD, it has more variation in band shape than UV-CD, which is dominated by the α-helical contribution. For example, hemoglobin is in the class of proteins whose secondary structure is dominated by the α-helix, while concanavalin A is dominated by its β-sheet components. Each of these has similarly shaped UV-CD with the main difference being in terms of intensity. In the VCD spectra, they have oppositely signed band shapes and significant frequency shifts, thereby evidencing more sensitivity to the structural variation [6,9]. Furthermore, globular proteins with a mixture of α- and β-components have VCD spectra often resembling a linear combination of these two more limiting types [6,9]. It should be noted that all of our UV-CD data penetrate into the vacuum UV to 180 nm, an extent effectively limited by the solvent; thus extension of the UV-CD wavelength range is unlikely to alter its analysis. However, VCD can be extended to include other bands such as the amide II for proteins dissolved in H_2O [13].

The basis for our first systematic method of empirical VCD analysis [9] is the principal component method of factor analysis whereby the spectra are reduced to relatively small vectors of coefficients that serve to characterize the spectral shapes [14]. Our initial results [9] show that 20 protein amide I' VCD spectra can be fit to their error limit with linear combinations of six orthogonal subspectra. The first subspectrum represents the most common elements of all the experimental spectra used in the analysis. The second represents the major deviations from the first. Each successive one then becomes less significant, eventually representing the noise contributions.

The coefficients of the subspectra calculated for the proteins in the set can be correlated at a statistically significant level with the α-helical, β-sheet, bend and 'other' contributions to the secondary structure as determined from analyses of the X-ray structures of 13 of the 20 proteins [15]. Via a multicomponent regression analysis, the relative contribution of these subspectra, which are determined directly from experimental VCD without any prior assumptions, can be used to give a measure of the fractional secondary structure composition [9]. An exactly parallel analysis has been carried out on the UV-CD data and used to show that the information content of both types of spectra is similar but that the VCD is slightly better quantitatively [16]. This is particularly true for determination of the β-sheet content. The method was then used to make predictions about the other proteins. Those proteins which are well-grouped with the 13 standard ones, as judged by cluster analysis [17], gave the best predictions compared to other determinations of their secondary structure contents.

We have found it most useful to characterize the crystal structure results using the Kabsch-Sander algorithm [15] for generating fractional contributions to the secondary structure. Initially, a 5-vector characterization was used including α-helix, β-sheet, bend, turn and 'other' which encompasses all conformational types not enumerated. This classification is not unique, nor is it obvious that it is the best for any of the spectral techniques to which we wish to apply it. Test studies using the Levitt and Greer [15] 'low resolution' analysis of crystal structures led to similar results [9].

Clearly there remains a question of suitability of solid-state crystal structure data for solution-phase spectral analysis. Based on VCD analysis, we have identified a protein whose solid-state structure and solution spectral parameters are in distinct disagreement. The α-lactalbumin crystal structure shows the same folding pattern as found for lysozyme [18] but the VCD spectra of the two in D_2O are significantly different [19]. If the solvent for α-lactalbumin, known to be solvent and ion sensitive, is altered to contain 33% propanol, a reasonable match with the lysozyme VCD is found. Our data re-emphasize that, in this case, the crystal-line is not the same as the aqueous solvent environment.

Additionally, our studies can isolate those proteins which have spectral behavior far from the norm based on standard crystal structure-spectra correlations. This has already been possible in our study of the pH dependence of the secondary structure of phosvitin, a highly charged protein [20]. The VCD spectra of phosvitin at low pH did not cluster with those of our training set proteins, thus any derivation of the secondary structures for these forms of phosvitin from the spectra are likely to have more error than found with the training set proteins.

Another approach to statistical analysis of spectra is the partial-least-squares (PLS) method which treats the 13 proteins as a calibration set in the language of quantitative analysis, but weights contributions of different spectral components [21]. The regression approach gave about the same average result as the PLS one but since the former was more highly constrained, its 'bad' predictions were more predictable via cluster analysis. The PLS method has already been applied to protein FT-IR [22].

Our approach is different from other efforts using factor analysis for spectral interpretation [23,24] in that we first evaluate the degree of reliability of individual and multiple correlations before finalizing the regression analysis. Prior UV-CD analysis using the principal component method used a total inversion or target transformation [14] of the coefficient matrix to determine secondary structure [24]. The PLS approach [21] shares some of the problems of target transformations. Our factor analysis and limited regression approach is more conservative and predicts fewer fractional contributions, but it allows determination of those aspects of the structure to which the analysis is sensitive and what degree of confidence one should place on the results. Target transformation methods neglect other environmental contributions to the spectra or implicitly assume that they cancel. By first evaluating correlations of the coefficients with structure and then doing a regression analysis for only those shown to be significant, this problem is minimized at the cost of some precision in fitting the data.

For purposes of understanding protein structure, *accuracy* is more important than precision. Accuracy is enhanced by our method in that non-relevant factors have diminished influence in our selective parameter method for interpretation of 'unknown' structures. We are now extending this systematic method of analysis to other spectral methods including IR and Raman spectroscopy and to a broader range of proteins.

Acknowledgements

We wish to thank the National Institutes of Health for support of this research (GM 30147) and NSF, NIH and the University of Illinois for support for purchase of instrumentation used.

References

1. Chang, C.T., Wu, C.S.C. and Yang, J.T. (1978) *Anal. Biochem.* **91**, 13; Johnson, W.C. (1985) *Methods Biochem. Anal.* **31**, 61.
2. Keiderling, T.A. (1990) in *Practical Fourier Transform Infrared Spectroscopy* (Ferraro, J.R. and Krishnan, K., Eds.) Academic, San Diego, pp. 203–284 and references.
3. Keiderling, T.A., Yasui, S.C., Dukor, R.K. and Yang, L. (1989) *Polymer Preprints* **30**, 423; Keiderling, T.A., Pancoska, P., Dukor, R.K. and Yang, L. (1989) *Biomolecular Spectroscopy* (Birge, R.R. and Mantsch, H.H., Eds.), *Proc. SPIE* **1057**, 7; Keiderling, T.A. (1986) *Nature* **322**, 851.
4. Lal, B.B. and Nafie, L.A. (1982) *Biopolymers* **21**, 2161; Sen, A.C. and Keiderling, T.A. (1984) *Biopolymers* **23**, 1519; Yasui, S.C., Keiderling, T.A. and Katakai, R. (1987) *Biopolymers* **26**, 1407.
5. Yasui, S.C. and Keiderling, T.A. (1986) *J. Am. Chem. Soc.* **108**, 5576; Paterlini, M.G., Freedman, T.B. and Nafie, L.A. (1986) *Biopolymers* **25**, 1751; Dukor, R.K. and Keiderling, T.A. (1989) in *Peptides 1988* (Proceedings of the 20th European Peptide Symposium) (Bayer, E. and Jung, G., Eds.) pp. 519–521.
6. Pancoska, P., Yasui, S.C. and Keiderling, T.A. (1989) *Biochemistry* **28**, 5917.
7. Dukor, R.K., Keiderling, T.A. and Gut, V. (1990) *Int. J. Pept. Prot. Res.* (submitted for publication); Dukor, R.K. and Keiderling, T.A. (1991) *Biopolymers* (submitted for publication); Kobrinskaya, R., Yasui, S.C. and Keiderling, T.A. (1988) in *Peptides, Chemistry and Biology* (Proceedings of the 10th American Peptide Symposium) (Marshall, G.R., Ed.) ESCOM, Leiden, pp. 65–66.
8. Yasui, S.C., Keiderling, T.A., Bonora, G.M. and Toniolo, C. (1986) *Biopolymers* **25**, 79. Yasui, S.C., Keiderling, T.A., Formaggio, F., Bonora, G.M. and Toniolo, C. (1986) *J. Am. Chem. Soc.* **108**, 4988.
9. Pancoska, P., Yasui, S.C. and Keiderling, T.A. (1991) *Biochemistry* (in press).
10. Malon, P. and Keiderling, T.A. (1988) *Appl. Spectr.* **42**, 32; Keiderling, T.A., Yasui, S.C., Malon, P., Pancoska, P., Dukor, R.K., Croatto, P.V. and Yang, L. (1989) *7th International Conference on Fourier Transform Spectroscopy* (Cameron, D.G., Ed.), *Proc. SPIE* **1145**, 57.
11. Keiderling, T.A. (1981) *Appl. Spectr. Rev.* **17**, 189.
12. Sen, A.C. and Keiderling, T.A. (1984) *Biopolymers* **23**, 1533; Narayanan, U., Keiderling, T.A., Bonora, G.M. and Toniolo, C. (1985) *Biopolymers* **24**, 1257; (1985) *J. Am. Chem. Soc.* **108**, 2431.
13. Gupta, V.P. and Keiderling, T.A. (to be submitted).
14. Malinowski, E.R. and Howery, D.G. (1980) in *Factor Analysis in Chemistry*, Wiley, New York; Pancoska, P., Fric, I. and Blaha, K. (1979) *Coll. Czech. Chem. Comm.* **44**, 1296, 1298.

15. Kabsch, W. and Sander, C. (1983) *Biopolymers* **22**, 2577; Levitt, M. and Greer, J. (1977) *J. Mol. Biol.* **114**, 181.
16. Pancoska, P. and Keiderling, T.A. (1991) *Biochemistry* (submitted for publication).
17. Massart, D.L. and Kaufman, L. (1983) in *The Interpretation of Analytical Chemical Data by the Use of Cluster Analysis*, Chemical Analysis, Vol. 65, J. Wiley and Sons, New York.
18. Acharya, K.R., Stuart, D.I., Walker, N.P.C., Lewis, M. and Phillips, D.C. (1989) *J. Mol. Biol.* **208**, 99.
19. Urbanova, M., Dukor, R.K. and Keiderling, T.A. (1991) (to be submitted).
20. Yasui, S.C., Pancoska, P., Dukor, R.K., Keiderling, T.A., Renugopalakrishnan, V., Glimcher, M.J. and Clark, R.C. (1990) *J. Biol. Chem.* **265**, 3780.
21. Haaland, D.M. and Thomas, E.V. (1988) *Anal. Chem.* **60**, 1193, 1202.
22. Dousseau, F. and Pezolet, M. (1990) *Biochemistry* **29**, 8771.
23. Lee, D.C., Haris, P.I., Chapman, D. and Mitchell, R.C. (1990) *Biochemistry* **29**, 9185.
24. Hennessey, J.P. and Johnson, W.C., (1981) *Biochemistry* **20**, 1085.

Protein dynamics: Theory and experiment

Section III

Cluster dynamics: Theory and experiment

Protein dynamics: A brief overview

Hans Frauenfelder

Departments of Physics, Chemistry and Biophysics,
University of Illinois at Urbana-Champaign, Urbana, IL 61801, U.S.A.

Introduction

One goal of protein research is the description of function in terms of structure. This goal cannot be reached by considering only the static structure; proteins are dynamic systems and motions are crucial to their function.

Four properties of proteins must be known and understood for fully comprehending function: the protein structure, the conformational energy landscape, the motions within that landscape (dynamics), and the effect of the structure and dynamics on the function. Of these four areas, the first one is best known. Ever since the classic work of Kendrew and Perutz, an enormous amount of energy has been devoted to the elucidation and description of the structure of biomolecules. In addition to the standard tool, X-ray diffraction with conventional X-ray sources, new approaches such as the use of synchrotron radiation and NMR, are greatly adding to our knowledge of biomolecular structures. This effort is important; without structure, the exploration is like a long hike in the beautiful mountains around Whistler, but at night and without a map. Without dynamics, however, the exploration of the function of proteins is like watching a play where the actors are silent and immobile. The exploration of the dynamics of proteins and of the connection of dynamics and function is only at a beginning, about at the point where the first X-ray structure showed myoglobin as a dim and sausage-like shape. A small number of papers presented here give some glimpses of protein motions. I will try to sketch some of the essential concepts of protein dynamics and point out the contribution of these papers.

The Energy Landscape

In an atom like hydrogen, the 'energy landscape' consists of the well-known energy levels, exhaustively described in textbooks. In a diatomic molecule, the energy landscape is already more involved: vibrational and rotational levels are added to the electronic states. In a biomolecule, the energy landscape becomes truly complex. A protein can exist in a very large number of different conformational substates (CS) [1,2].

CS perform the same function, but possibly with different rates; they have the same overall structure, but differ in detail. CS can be pictured as energy valleys in a high-dimensional conformation space. Experimental evidence for CS comes from the nonexponential time dependence of protein reactions [1] and motions [3], from the

173

characteristic Debye-Waller factor [4,5], and from the existence of inhomogeneously broadened spectral lines [6].

Evidence for conformational substates is presented here, directly or indirectly, in the papers by Loncharich and Brooks [7], Cusak and Doster [8], Teeter et al. [9], Cheung [10] and Madison et al. [11].

While the existence of CS is established and some features of the conformational energy landscape are known, much (or nearly everything) remains to be studied. What is the organization of the landscape? There is good evidence for a hierarchical arrangement (valleys within valleys within valleys). How are these valleys arranged and how high are the ridges between the valleys? Extensive experimental and theoretical work will be needed before reliable maps are available [12]. Moreover, most of the detailed work done so far has used myoglobin as prototype. Myoglobin is a simple protein, presumed to have a simple function, and it consists mainly of α-helices. The paper of Prabhakaran and co-workers [13] suggests that amelogenin, which is rich in β-sheets and turns, may have a different energy landscape. It is likely that general features of substates and the landscape are similar in all proteins, but characteristic differences may well be crucial for function.

Conformational Motions

A working protein moves and breathes. These motions correspond to jumps between the energy valleys. To explore these motions, the transitions among substates must be characterized. It is customary to employ simple theoretical tools and preferably assume a monoexponential time dependence and an Arrhenius temperature dependence. When a time course is found to be nonexponential, two or more exponentials are used and the biphasic nature is explained by more than one process [10]. Studies of protein motions over wide ranges in time and temperature, particularly with relaxation experiments, present a different picture [1,3,14]: Protein motions are often truly nonexponential in time and must be described by a distribution rather than a sum over a few exponentials. A convenient form for fitting nonexponential time courses is the stretched exponential (Kohlrausch function)

$$\Phi(t) = \exp[-(kt)^\beta] \tag{1}$$

with $0 < \beta \leq 1$. The rate coefficient k often displays a non-Arrhenius temperature dependence that can be approximated for instance by the relation

$$k(T) = A \exp \{-(E/RT)^2\} \tag{2}$$

Equations (1) and (2) express the fact that protein motions are not simply transitions over a fixed potential barrier, but involve major parts of the protein.

To study conformational motions, a wide variety of techniques is being used. Theoretically, motions are studied by molecular dynamics computations [7,9,15] and by Monte Carlo calculations [16]. Experimentally, one approach is to start at low temperatures where a given protein remains frozen into a particular substate. As the temperature increases, the protein starts to move. First, small-scale motions become activated. Near 180 K, a glass-like transition can be observed [3,7,8] and large parts of the protein are involved. Finally, the entire protein becomes a dynamic system. Both experimental and theoretical studies have severe limitations. At present, the molecular dynamics calculations are restricted to short times, of the order of 100 ps. Since protein function may involve motions that occur as slowly as milliseconds, this limitation severely hampers the theoretical understanding. Experimentally, it is equally difficult to observe all time scales, from picoseconds to seconds, and to look at all parts of a protein. Consequently, while the exploration of the energy landscape of proteins is still at an early stage, the quantitative understanding of protein motions is even farther behind and constitutes a major challenge.

Dynamics and Function

One major goal of the studies of protein dynamics is the quantitative explanation of protein function in terms of protein motions. The goal is close only when the model or theories are powerful enough to predict for instance the catalytic efficiency of a newly designed mutant. While this goal is still far in the future, progress has been made for some simple processes. Consider ligand binding to myoglobin. A CO molecule entering a protein passes through the protein matrix into the pocket and then binds at the heme iron. The crucial steps, passage through the matrix and forming the covalent bond, are dominated by dynamics. The motion can be described surprisingly well with Eqs. (1) and (2) and these relations lead to a quantitative understanding of the association process [17]. Structural, spectroscopic, energetic, and dynamic aspects all come together to form a coherent picture of a simple biological reaction which may be a paradigm for more complex biological phenomena.

The binding of small ligands to a heme protein is, of course, a very simple process. Nevertheless it is encouraging that a quantitative description can be given. We expect that further studies of protein reactions with all the tools of theoretical and experimental sciences may indeed lead to a theory in which structure, dynamics, and function are related.

References

1. Austin, R.H., Beeson, K.W., Eisenstein, L., Frauenfelder, H. and Gunsalus, I.C. (1975) *Biochemistry* **14**, 5355.
2. Frauenfelder, H., Parak, F. and Young, R.D. (1988) *Annu. Rev. Biophys. Biophys. Chem.* **17**, 451.
3. Iben, I.E.T. et al. (1989) *Phys. Rev. Lett.* **62**, 1916.
4. Frauenfelder, H., Petsko, G.A. and Tsernoglou, D. (1979) *Nature* **280**, 558.
5. Petsko, G.A. and Ringe, D. (1984) *Annu. Rev. Biophys. Bioeng.* **13**, 331.
6. Agmon, N. (1988) *Biochemistry* **27**, 3507.
7. Loncharich, R.J. and Brooks, B.R. (1991) in *Proteins: Structure, Dynamics and Design* (Renugopalakrishnan, V. et al., Eds.) ESCOM, Leiden, pp. 177–183.

8. Cusak, S. and Doster, W. (1991) in *Proteins: Structure, Dynamics and Design* (Renugopalakrishnan, V. et al., Eds.) ESCOM, Leiden, pp. 193-201.
9. Teeter, M.M., Rao, U. and Case, D. (1991) in *Proteins: Structure, Dynamics and Design* (Renugopalakrishnan, V. et al., Eds.) ESCOM, Leiden, pp. 220-228.
10. Cheung, H.C. (1991) in *Proteins: Structure, Dynamics and Design* (Renugopalakrishnan, V. et al., Eds.) ESCOM, Leiden, pp. 184-192.
11. Madison, V.S. et al. (1991) in *Proteins: Structure, Dynamics and Design* (Renugopalakrishnan, V. et al., Eds.) ESCOM, Leiden, pp. 234-239.
12. Pastor, R.W. (1991) in *Proteins: Structure, Dynamics and Design* (Renugopalakrishnan, V. et al., Eds.) ESCOM, Leiden, pp. 229-233.
13. Prabhakaran, M. et al. (1991) in *Proteins: Structure, Dynamics and Design* (Renugopalakrishnan, V. et al., Eds.) ESCOM, Leiden, pp. 202-207.
14. Frauenfelder, H. et al., *J. Phys. Chem.* **94**, 1024.
15. Brooks, C.L., Karplus, M. and Pettitt, B.M. (1988) *Proteins*, John Wiley, New York.
16. Gō, N. and Noguti, T. (1989) *Chemica Scripta* **29A**, 151.
17. Steinbach, P.J. et al. (1991) *Biochemistry*, in press.

Temperature and phase dependence of protein dynamics: A simulation study of myoglobin

Richard J. Loncharich and Bernard R. Brooks*

Molecular Graphics and Simulation Laboratory, Division of Computer Research and Technology, National Institutes of Health, Bethesda, MD 20892, U.S.A.

Abstract

Molecular dynamics is used to study the detailed behavior of carboxy-myoglobin as a function of temperature and environment. The experimentally observed glass transition temperature is accurately reproduced with molecular dynamics simulations. The magnitude of the low-temperature behavior for the powder state is more accurately reproduced if clusters of myoglobin are simulated. The high-temperature behavior can be reproduced with a single myoglobin, and is strongly correlated with dihedral transitions. It is found that the glass transition involves all classes of atoms, and is not caused by a subset of the system. At 300 K, 42% of the behavior is attributed to harmonic motion, and 58% is attributed to substate jumping and other anharmonic motion for an isolated hydrated myoglobin. Anisotropy and skewness increase with temperature. Kurtosis increases with temperature below the glass transition temperature, and then becomes negative at higher temperature reflecting multiple state behavior. Work continues on examining environmental effects. For partially solvated systems, water rapidly reorganizes to make the system more spherical and to hydrate exposed charged groups. This can be done with little deformation to the protein structure. In vacuum, proteins become spherical at the expense of the protein structure.

Introduction

The temperature dependence of structure and dynamics of proteins is important in the study of biological function [1,2]. The dynamics of proteins as a function of temperature has been studied by many experiments. The dynamical behavior of metmyoglobin has been evidenced by X-ray structure analysis [3,4]. Frauenfelder and collaborators [3] carried out an X-ray diffraction study of sperm whale metmyoglobin at 220, 250, 275, and 300 K. The data from this study was later compared with structure determination of metmyoglobin at 80 K [4]. The protein is more rigid at 80 K; the overall structure is similar to that at 300 K, but cell dimensions and unit cell volume are smaller at 80 K. Mössbauer experiments on the heme iron of myoglobin have also been used to investigate dynamic properties [5-7]. Below 210 K the mean-square displacements of the iron atom are indicative of normal harmonic vibrational motion. Above this temperature the mean-square displacements of the iron atom become anomalously large. The additional motion has been ascribed to internal large scale motions of the protein [6,7]. Inelastic neutron scattering has also been used to characterize atomic motions of myoglobin as a function of temperature [8,9]. This

---

* To whom correspondence should be addressed.

experimental technique explores atomic motions on a timescale which is accessible by molecular dynamics simulations. Experiments indicate that below 180 K the behavior of myoglobin resembles a harmonic solid, possessing primarily vibrational motion. Above this temperature, non-vibrational types of motion are excited. The authors suggested that the transition in motion corresponds to torsional jumps between states of different energy.

In this paper we summarize the results of molecular dynamics simulations of hydrated carboxy-myoglobin at 11 different temperatures published elsewhere [10], and we present an overview of some new results on cluster simulations and for dehydrated myoglobin. The neutron scattering experiments were done on the fully hydrated protein, a 400 mg sample of sperm-whale myoglobin powder hydrated to 0.38 g D_2O per g of protein [8]. We attempt to mimic experimental conditions very closely by including a partial hydration shell around the protein. The results of these simulations provide information about dynamics of atomic motions which aid in the interpretation of experimental observations. In addition, this work addresses the following points. What is the nature of water in the dynamics of a partially hydrated protein? How well does a hydrated protein in vacuum mimic the powder state of proteins? What is the best theoretical model for the powder state? How does the detailed atomic behavior change as a function of temperature? What portion of atomic fluctuations can be attributed to local oscillations (e.g., harmonic limit behavior)? Does the mobility transition of the protein reflect the experimental glass transition temperature of water and can it be observed theoretically? What are the effects of the environment on protein behavior? How good are the current simulation techniques, and which techniques should be avoided?

The model system for the partially hydrated simulations of carboxy-myoglobin and the simulation methods used are described in detail elsewhere [10]. The model system for the simulations is a coordinate set which includes all hydrogen atoms generated from the X-ray structure [11] using the HBUILD routine [12] of CHARMm [13]. There are 2536 atoms in this representation of carboxy-myoglobin. All simulations are performed in double precision on the Star Technologies ST-50 and ST-100 array processors using the program GEMM [14]. Data transfer to the array processors is accomplished with Apollo DN-590 host computers. The Polygen Corporation all atom parameter set [15], with a few modifications [10] to the heme group, is used for all calculations.

Results and Discussion

Molecular dynamics simulations of a partially hydrated carboxy-myoglobin show that water tends to cluster about charged amino acid residues while leaving nonpolar residues unsolvated. The water molecules move rapidly on the protein surface in a short simulation time to hydrate the hydrophilic groups and make the entire system more spherical.

As shown in Table 1, at each temperature the simulation time-averaged structures are compared to the starting structure and to the X-ray structure (in parentheses) by least-squares fitting of all selected atoms. The rms deviations from the starting

Table 1   *RMS difference (Å), and average fluctuations (Å²) for the last 100 ps of each simulation*

Temperature (K)	RMS difference[a]			Average fluctuations		
	protein	backbone	neh[b]	protein	backbone	neh[b]
20	0.56 (1.69)	0.47 (1.49)	0.66	0.011	0.0074	0.019
60	0.54 (1.68)	0.45 (1.47)	0.63	0.036	0.024	0.060
100	0.62 (1.66)	0.53 (1.45)	0.72	0.065	0.044	0.11
180	0.56 (1.73)	0.46 (1.55)	0.66	0.13	0.081	0.21
220	0.53 (1.70)	0.41 (1.49)	0.59	0.23	0.15	0.34
240	0.60 (1.64)	0.39 (1.39)	0.71	0.27	0.18	0.42
260	0.65 (1.82)	0.48 (1.62)	0.84	0.30	0.19	0.44
280	0.72 (1.73)	0.55 (1.55)	0.94	0.41	0.25	0.62
300	0.98 (1.70)	0.75 (1.39)	1.17	0.63	0.39	0.89
320	1.04 (1.75)	0.83 (1.46)	1.23	0.54	0.33	0.80
340	0.98 (1.73)	0.65 (1.36)	1.27	0.76	0.46	1.18
100[c,e]	(1.70)	(1.43)		0.10	0.067	0.15
100[c,f]	(2.24)	(1.83)		0.088	0.059	0.14
100[d,e]	(1.49)	(1.18)		0.045	0.025	0.082
300[c,e]	(2.82)	(2.23)		0.70	0.43	1.09
300[c,f]	(2.95)	(2.47)		0.47	0.28	0.84
300[c,e]	(1.96)	(1.54)		0.34	0.19	0.64
300[g]	0.77 (1.69)	0.60 (1.37)	0.93	0.48	0.29	0.73
300[g]	0.85 (1.81)	0.62 (1.47)	1.02	0.50	0.32	0.78
300[g]	0.57 (1.73)	0.31 (1.33)	0.78	0.41	0.24	0.64
ave	0.73 (1.74)	0.51 (1.39)	0.91	0.46	0.29	0.72

[a] Values are for the difference between time-average structure and starting structure, and in parentheses the difference between the time-average structure and X-ray structure.
[b] The abbreviation neh stands for nonexchangeable hydrogen atoms experimentally observed.
[c] Simulation in vacuum using a shifted potential with a cutoff distance of 12.0 Å.
[d] Simulation in vacuum using an 11-12 Å switching function.
[e] Distance-dependent dielectric is used.
[f] Constant dielectric is used.
[g] Results for a single myoglobin in the cluster simulation.

structure are relatively small in magnitude. The rms differences of protein atoms are nearly constant from 20 to 280 K within the standard error of molecular dynamics simulations [16]. At room temperature and above there is a slightly larger rms difference. A similar trend is seen for rms differences of the backbone atoms from the starting structure. The nonexchangeable hydrogens have larger rms differences than the protein as a whole. The rms difference for the vacuum simulations is generally larger than the value for hydrated simulation at the corresponding temperature. In vacuum and hydrated systems, the structures become more spherical. In vacuum this occurs by deforming the protein whereas for the hydrated system this occurs primarily by reorganization of bound waters. The rms difference of any single myoglobin in the cluster simulation is smaller than the value for the isolated partially hydrated myoglobin simulation.

A detailed analysis of atomic behavior is performed to further understand protein mobility as a function of temperature. In Table 1 we list the average fluctuations, $<< \Delta R^2 >>$, of the protein, backbone, and nonexchangeable hydrogen atoms as a function of temperature. Comparison of the average fluctuations of each class of protein structure shows that the fluctuations of the nonexchangeable hydrogen atoms are largest. Figure 1 shows the temperature dependence of the $<< \Delta R^2 >>/3$ and $<< \Delta x^2 >>$ values for all nonexchangeable hydrogens of carboxy-myoglobin for molecular dynamics simulations and experiment [8], respectively. The atomic fluctuations increase linearly from 0 to 180 K. From extrapolation of harmonic limit behavior to high temperatures, shown by the solid line for simulation, we estimate that approximately 40% of atomic fluctuations is attributed to local oscillations, while the remaining 60% is mainly due to jumping between substates (*vide supra*). The most striking feature of Fig. 1 is that protein mobility deviates from linearity at 210–220 K. An estimate of the glass-like transition temperature can be determined by least-squares fit of the data points above 220 K and intersection with the straight line harmonic behavior. The harmonic intersection is 218 and 219 K, respectively for simulation and experimental data. The glass transition is predicted in the atomic fluctuations in accord with experimental observations.

Table 1 also shows the atomic fluctuations for several vacuum simulations of dehydrated myoglobin. The fluctuations are smallest in the vacuum simulation which

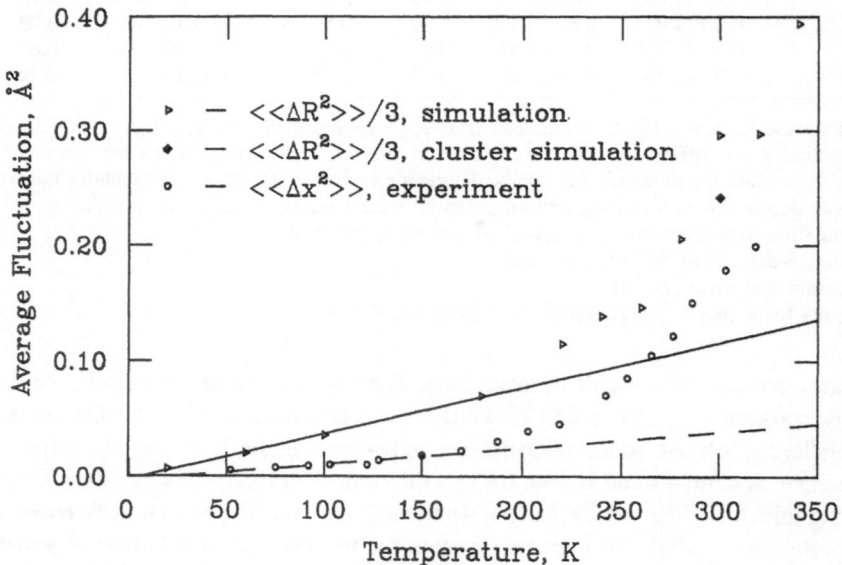

Fig. 1. Temperature dependence of average fluctuations ($\mathring{A}^2$) of the nonexchangeable hydrogen atoms for the last 100 ps of each simulation. Triangles represent simulation results whereby the average fluctuations are obtained from an isotropic average of the mean-square displacements, $1/3 << \Delta R^2 >>$. Circles represent mean-square displacements, $<< \Delta x^2 >>$, from neutron scattering experiments in (Doster et al., 1989). Diamonds represent the average fluctuations for a cluster simulation of three myoglobins. The solid and dashed straight lines are the vibrational contribution extrapolated linearly from low temperatures to 350 K for the simulation data and experiment, respectively.

uses a switching function to truncate nonbonded interactions. The average fluctuations determined from the vacuum simulations using the shifted potential with either a distance-dependent dielectric or a constant dielectric are larger than the fluctuations predicted in the hydrated simulation. The fluctuations for the cluster simulation are also shown in Table 1. The fluctuations of any single myoglobin in the cluster are reduced as compared to the isolated partially hydrated myoglobin simulation.

Further comparison of the experimental $<< \Delta x^2 >>$ data and the simulation $<< \Delta R^2 >>/3$ fluctuation data is presented in Fig. 2. Here we plot the difference in the straight line harmonic limit behavior (given by the solid line for simulation data and dashed line for experimental data in Fig. 1) and the corresponding fluctuation value at each data point. This gives the anharmonic contribution to the atomic fluctuations at each temperature. The most striking result of this plot is that above the glass-like transition temperature the theoretical and experimental anharmonic behavior is remarkably similar. Protein mobility is highly anharmonic in character above the transition temperature. The simulation-derived atomic fluctuations are too large in the harmonic limit behavior, but agree well above the glass transition temperature. One possible explanation for this discrepancy is that simulations of the hydrated protein in vacuum do not mimic the powder state of proteins very well since the amplitude of harmonic motion is too large without intermolecular contacts.

Another method to explore protein behavior is by the analysis of dihedral transitions. A dihedral transition is monitored as described previously [10]. Figure 3 shows a plot of the total number of heavy atom dihedral transitions with temperature; the resemblance to Fig. 2 is striking. Once the temperature is above the

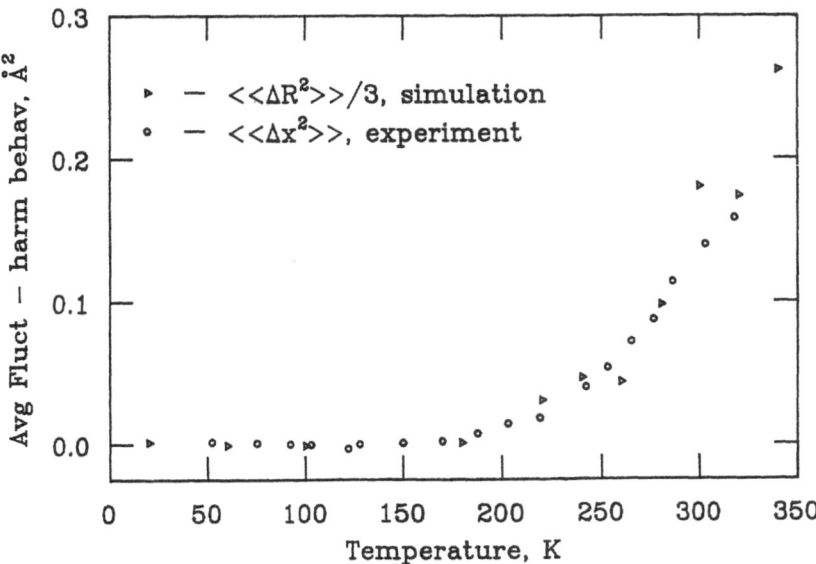

Fig. 2. Plot of the average fluctuation in Fig. 1 minus the straight line harmonic contribution for simulation data (triangles) and experiment (circles). The harmonic contribution is determined by linear extrapolation of the low-temperature data to 350 K.

Table 2 *Anisotropy and anharmonicity of atomic motions of all protein atoms at 20, 100, 180, 240, 300, and 340 K*

Temperature (K)	A_1	A_2	σ_x^2	α_{4x}	$|\alpha_{3x}|$	$|\alpha_{4x}|$
20	0.52	0.12	0.01	0.01	0.11	0.17
100	0.56	0.12	0.05	0.01	0.17	0.23
180	0.57	0.12	0.09	0.04	0.20	0.31
240	0.58	0.13	0.19	−0.01	0.22	0.36
300	0.70	0.14	0.47	−0.06	0.33	0.49
340	0.71	0.14	0.60	−0.04	0.34	0.51

glass transition temperature, there is a large increase in the total number of transitions. The harmonic intercept of 214 K is in accord with the value of the glass transition temperature derived from the average fluctuations as a function of temperature. We believe that the glass transition and increased fluctuations are primarily due to torsional jumping, rather than single state anharmonicity or other structural changes in the protein.

To assess the shape of the atomic motions we have also calculated the anisotropies and anharmonicities of the atomic motion as a function of temperature, shown in Table 2. The anisotropy (A_1 and A_2) of the atomic position distributions is large at all temperatures with a gradual monotonic increase as a function of temperature. The skewness (α_3) and the absolute value of the kurtosis (α_4) of the atomic position distribution also show a monotonic increase with temperature. The kurtosis of the whole protein (Table 2) below the glass transition temperature is positive, and goes to zero as the temperature approaches 0 K. However, above the glass transition

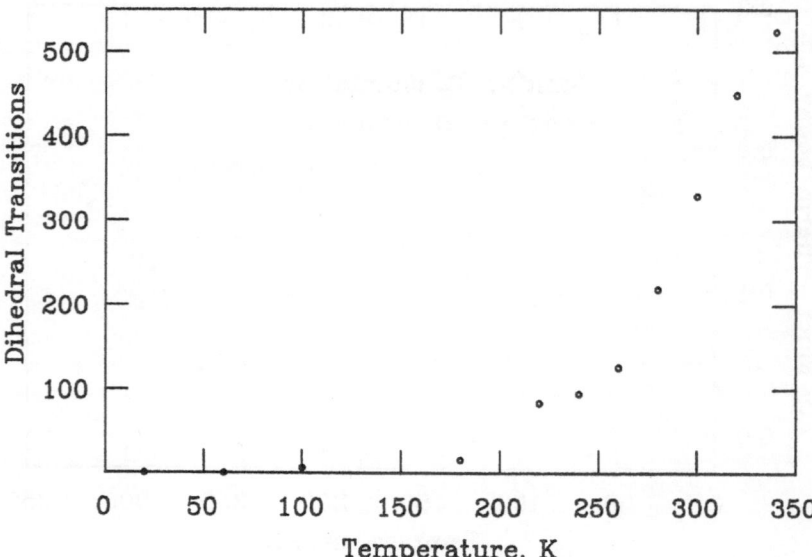

Fig. 3. *Plot of the total number of heavy atom dihedral transitions as a function of temperature.*

temperature the kurtosis is negative suggesting the crossing of double potential behavior. The absolute value of the kurtosis for the whole protein increases monotonically as the temperature increases.

Future investigations in our laboratory will include additional calculations of a cluster of hydrated myoglobins and the crystal structure of myoglobin to understand how the atomic fluctuations change with the introduction of intermolecular protein-protein contacts as probably found in the powder state. The protein in bulk solvent will also be studied.

Conclusion

The simulations of the single myoglobin in vacuum do well in predicting the glass transition in the atomic fluctuations of the protein. The simulations suggest that the glass transition is highly correlated with the occurrence of dihedral transitions. The simulation results at high temperature agree well with the experimental data, but there remains a large discrepancy (factor of 3) for the amplitude of the fluctuations at low temperature. Work continues to try to further identify the nature and causes of this discrepancy and to more fully examine the environmental effects on protein dynamics.

References

1. McCammon, J.A. and Harvey, S.C. (1987) in *Dynamics of Proteins and Nucleic Acids*, Cambridge University, New York, pp. 79–150.
2. Brooks, C.L., III, Karplus, M. and Pettitt, B.M. (1988) in *Proteins: A Theoretical Perspective of Dynamics, Structure, and Thermodynamics*, Advances in Chemical Physics, Vol. 71, (Prigogine, I. and Rice, S.A., Eds.) John Wiley and Sons, New York, pp. 7–21.
3. Frauenfelder, H., Petsko, G.A. and Tsernoglou, D. (1979) *Nature* **280**, 558.
4. Hartmann, H., Parak, F., Steigemann, W., Petsko, G.A., Ringe Ponzi, D. and Frauenfelder, H. (1982) *Proc. Natl. Acad. Sci. USA* **79**, 4967.
5. Parak, F., Frolov, E.N., Mössbauer, R.L. and Goldanskii, V.I. (1981) *J. Mol. Biol.* **145**, 825.
6. Parak, F., Knapp, E.W. and Kucheida, D. (1982) *J. Mol. Biol.* **161**, 177.
7. Bauminger, E.R., Cohen, S.G., Nowik, I., Ofer, S. and Yariv, J. (1983) *Proc. Natl. Acad. Sci. USA* **80**, 736.
8. Doster, W., Cusack, S. and Petry, W. (1989) *Nature* **337**, 754.
9. Cusack, S. and Doster, W., unpublished results.
10. Loncharich, R.J. and Brooks, B.R. (1990) *J. Mol. Biol.* **215**, 439.
11. Kuriyan, J., Wilz, S., Karplus, M. and Petsko, G. (1986) *J. Mol. Biol.* **192**, 133.
12. Brunger, A.T. and Karplus, M. (1988) *Proteins* **4**, 148.
13. Brooks, B.R., Bruccoleri, R.E., Olafson, B.D., States, D.J., Swaminathan, S. and Karplus, M. (1983) *J. Comp. Chem.* **4**, 187.
14. Brooks, B.R. (1987) in *Supercomputer Research in Chemistry and Chemical Engineering*, ACS Symposium Series 353 (Jensen, K.F. and Truhlar, D.G., Eds.) American Chemical Society, Washington DC, pp. 123–145.
15. Polygen Corporation Parameter file for CHARMm version 20. Copyright 1986. Released August 1988.
16. Loncharich, R.J. and Brooks, B.R. (1989) *Proteins* **6**, 32.

Protein dynamics from fluorescence resonance energy transfer

Herbert C. Cheung

Department of Biochemistry, University of Alabama at Birmingham,
Birmingham, AL 35294, U.S.A.

Introduction

Nonradiative fluorescence resonance energy transfer (FRET) has been widely used to determine distances between specific sites in macromolecules. The technique is based on measurements of the transfer efficiency of excitation energy between a donor (D) and an acceptor (A) that are either intrinsic chromophores or extrinsic probes located at specific sites, and calculation of the D-A separation from an observed transfer efficiency as first described by Förster [1]. The phenomenon in question arises from dipolar interactions between the donor emission and acceptor absorption dipoles, and the transfer efficiency (E) is proportional to an inverse sixth power of the separation between the dipoles and the angle between them (orientation factor). The useful range of distances that can be recovered from dipolar interactions is about 15–60 Å. Because of the $1/r^6$ dependence of the transfer efficiency, FRET can be a sensitive measure of structural perturbations. Up to recently a single D-A distance has been assumed in most FRET studies and this single distance is used to gain insight into structure/function relationships. Typically the D-A separation is determined before and after functional perturbation, and a difference between the two measured distances is usually taken as evidence of structural perturbations that accompany biological function. Such an approach, while widely used, provides no information on the kinetics of the distance change and, therefore, no clue as to whether the distance change may be kinetically competent to be related to functional events.

A second problem with FRET studies of macromolecules is the implicit assumption of a unique D-A separation. Clearly this assumption is not always valid because of structural fluctuations that can occur in many macromolecules. These fluctuations may result in a range of D-A distances, and the transfer efficiency is different for each of the distances. Thus there is a distribution of transfer efficiencies for a given D-A pair, and from this information a distribution of the D-A distances, P(r), can be recovered. The possibility of recovering P(r) from measurements of steady state donor fluorescence intensity was first suggested in 1971 [2]. Although such measurements are easy to make, the suggested experimental protocol is difficult to implement and has not been successfully used because it requires labeling of a given macromolecule with several different D-A pairs at the same donor and acceptor sites. This procedure would yield a range of the Förster critical distances (R_0). Each D-A pair between the same two sites has a different transfer efficiency, and from these data an estimate of

P(r) can be obtained. An alternate procedure to obtain a range of R_0 is from measurements of collisional quenching of the donor steady-state intensity [2], and has been sucessfully implemented only recently [3].

A more direct approach to the determination of P(r) is from measurements of the time-dependent decays of the donor emission using either the time-domain [4,5] or frequency-domain [6] data. Recent instrumental developments [7,8] have enabled us to recover complex decays down to subnanoseconds with high accuracy. This capability in turn allows recovery of the distribution of distances with tryptophan as the energy donor and other donor fluorophores showing complex decay patterns. We summarize here our recent results obtained from three types of novel FRET studies of muscle proteins: (a) recovery of P(r) from lifetime measurements with multifrequency phase fluorometry, (b) determination of P(r) from steady-state intensity measurements, and (c) time-resolved (stopped-flow) measurements of changes in molecular distances.

Results

Recovery of P(r) from lifetime measurements
The details for recovery of the distribution of D-A distances have been published in our previous work [9]. A summary of the theory for FRET in the presence of a range of D-A distances is given here for the case where the donor emission decays monoexponentially with a decay time τ_D in the absence of energy transfer. If a single acceptor is located at a unique distance r from the donor, the donor decay of the D-A pair, $I_{DA}(r,t)$, will remain monoexponential with decay time τ_{DA}: $I_{DA}(r,t) = I_D^0 \exp(-t/\tau_{DA})$, where $1/\tau_{DA} = (1/\tau_D) [1 + (R_0/r)^6]$, and R_0 is Förster's critical distance at which the transfer efficiency is 50%. If the acceptor is not rigidly held in space with respect to the donor coordinates, there will be a range of D-A distances. The decay of the donor intensity at each D-A distance will still be single-exponential, but the overall observed decay will contain contributions reflecting the range of D-A distances and is more complex than a single exponential. It is this deviation from monoexponentiality that provides a measure of the distribution of the D-A distances. The energy transfer efficiency and the intensity decay of the donor in the presence of energy transfer are each given by the average of the individual transfer efficiencies and individual decays weighted by a normalized distance probability distribution P(r) of the D-A pairs:

$$E = \int \frac{P(r) \; R_0^6}{R_0^6 + r^6} \; dr \tag{1}$$

$$I_{DA}(t) = I_D^0 \int P(r) \exp(-t/\tau_{DA}) \; dr \tag{2}$$

In the present studies we assume that the probability distribution is a Gaussian:

$$P(r) = \frac{r^n}{\sigma\sqrt{2\pi}} \exp \left[-\frac{1}{2}(\frac{r - \bar{r}}{\sigma})^2 \right] \tag{3}$$

185

where $\bar{r}$ is the average, σ the standard deviation of the distribution, and $n = 0$, 1, or 2. The standard deviation is related to the half-width by $hw = 2.354\sigma$.

We recovered the decay times of the donor and the D-A pair (τ_D and τ_{DA}) from the measured phase (ϕ_ω) and modulation (m_ω) of the donor emission that were determined at 20–22 frequencies (ω) over the range of 5–300 MHz. The phase and modulation are related to the intensity decay function I(t) through a sine (N_ω) and cosine (D_ω) transformation [9]. For a distribution of D-A distances these transforms are weighted averages that depend on P(r) and the decay times, and are calculated numerically. The parameters describing the intensity decay and the distribution function (τ_D, τ_{DA}, $\bar{r}$, and σ) are calculated using nonlinear least squares as previously described [9]. If the donor emission decay in the absence of energy transfer is multiexponential, Eq. 2 and the expressions for N_ω and D_ω will be sums of exponentials.

Determination of P(r) from steady-state measurements

We used collisional quenching of the donor steady-state intensity to obtain a range of R_0, R_0^Q. For this case Eq. 1 becomes

$$E = \int_{r=r_m}^{\infty} \frac{P(r) \ (R_0^Q)^6}{(R_0^Q)^6 + r^6} \ dr \tag{4}$$

where r_m is some minimum distance for the D-A pair, and R_0^Q is the R_0 value determined in the presence of collisional quenching. The range of R_0^Q can be determined from the decrease in donor quantum yield due to collision quenching using a quencher agent such as acrylamide [3]:

$$R_0^Q = R_0 \ (\phi_D^Q / \phi_D^0)^{1/6} \tag{5}$$

where ϕ_D^0 and ϕ_D^Q are the donor quantum yield determined in the absence and presence of quencher, respectively. From a series of quantum yields determined in the presence of different quencher concentrations, a range of R_0^Q is readily obtained. The determination of ϕ_D^Q is less straightforward because it requires knowledge of the dynamic quenching constant (K_D). In general the quenched donor intensity may contain contributions from both static and dynamic components. A modified Stern-Volmer equation for this case [3] can be used to extract two quenching constants from steady-state measurements, one corresponding to the static constant (K_S) and the other to K_D. These measurements alone cannot distinguish between K_S and K_D, but assignment of the correct value to K_D can be accomplished either by auxiliary decay measurements or from knowledge of the expected K_D values based on the quenching efficiency and/or solvent viscosity [3].

Recovered distributions of FRET distances

In a previous study [10] we determined six single D-A distances in the complex formed between skeletal troponin I(TnI) and troponin C(TnC). Two of these distances, one between Trp[158] and Cys[133] and TnI, and the other between Met[25] and Cys[98] on TnC, were subsequently studied by multifrequency phase fluorometry and

186

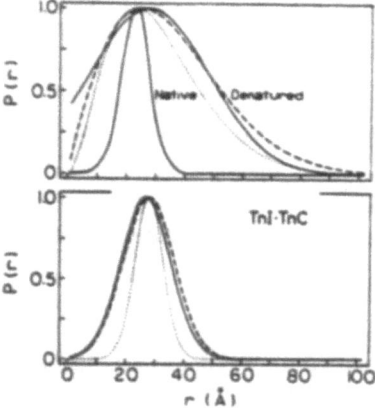

Fig. 1. *Peak-normalized distributions of FRET distances between Trp[158] and Cys[133] for isolated TnI, TnI in the TnI•TnC complex, and denatured TnI. Upper panel: (—) distributions for native and denatured TnI recovered with n = 0 in Eq. 3; the distributions recovered with n = 1 (---) and n = 2 (...) for denatured TnI. Lower panel: the distributions for TnI•TnC in the presence of EGTA (—), Mg^{2+} (---), and Ca^{2+} (...).*

the steady-state method to recover the distributions of the D-A distances. For TnI the single Trp served as the energy donor and N-(iodoacetyl)-N'-(1-sulfo-5-naphthyl)-ethylenediamine (AEDANS) attached to Cys[133] was the acceptor; for TnC dansylaziridine (DNZ) covalently linked to Met[25] was the donor and 5-(iodoaceta-mido)-eosin (IAE) attached to Cys[98] served as acceptor. Figure 1 shows the recovered distributions for native and denatured TnI and TnI complexed to TnC. Upon denaturation the mean distance ($\bar{r}$) between Trp[158] and Cys[133] increases from 23 to 27 Å and the half-width (hw) of the distribution increases by a factor of 4. The

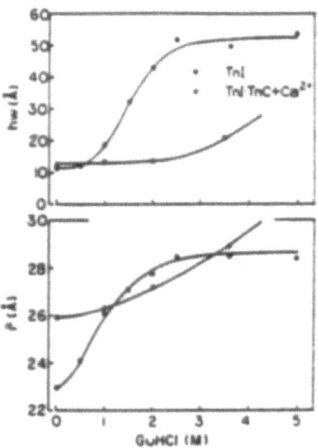

Fig. 2. *Effects of guanidine hydrochloride on $\bar{r}$ and hw of the D-A distribution for TnI (•) and TnI complexed to unmodified TnC + Ca^{2+} (o).*

187

effects of increasing concentrations of quanidine hydrochloride (Gu•HCl) on both r and hw are shown in Fig. 2. TnI is known to be stabilized against denaturation by complexation with TnC in the presence of Ca^{2+}. This protection is evident in Fig. 2. The distribution parameters for TnI, TnC, and their complexes are listed in Table 1. Figure 3 shows that the distributions recovered for TnI from steady-state and frequency-domain data are in good agreement.

Myosin subfragment 1 (S1), which is the cross-bridge moiety of myosin, has been extensively studied in relation to energy transduction because its heavy chain contains both the ATPase site and actin-binding sites. The short segment of the heavy chain between Cys^{697} (SH_2) and Cys^{707} (SH_1) is of particular interest because it may play a role in modulating the interaction between myosin and actin. The distribution of the SH_1-SH_2 distances was recovered from frequency-domain lifetime data [11] with AEDANS attached to SH_1 as energy donor and N-(4-dimethylamino-3,5-dinitrophenyl)-maleimide (DDPM) linked to SH_2 as acceptor [12]. The distribution parameters are also listed in Table 1. Addition of MgADP decreases the mean distance (increasing energy transfer) from 27.3 to 14.1 Å. In contrast to TnI, denaturation of S1 increases the hw by only 40%.

Time-resolved studies of changes in FRET

The MgADP-induced decrease of the distances between SH_1 and SH_2 on S1 was

Table 1 *Distribution of FRET distances for troponin I, troponin C and myosin subfragment 1[a]*

Parameter	\multicolumn{4}{c}{Troponin I and its complex with troponin C[b]}
	TnI*	Tn*I+Gu•HCl	TnI*•TnC+Mg^{2+}	TnI*•TnC+Ca^{2+}
$\bar{r}$ (Å)	24.3 (23.8)[c]	27.0	27.5	27.4
hw (Å)	11.5 (12.1)[c]	55.9	21.4	11.5

Parameter	\multicolumn{6}{c}{Troponin C and its complex with troponin I}
	TnC*	TnC*+Gu•HCl	TnC*+Mg^{2+}	TnC*+Ca^{2+}	TnC*•TnI+Mg^{2+}	TnC*•TnI+Ca^{2+}
$\bar{r}$ (Å)	22.4	18.0	15.2	22.1	16.1	22.2
hw (Å)	13.3	34.3	15.0	10.9	17.2	15.1

Parameter	\multicolumn{3}{c}{Myosin subfragment 1}
	S1	S1•MgADP	S1+Gu•HCl
$\bar{r}$ (Å)	27.3	20.7	17.2
hw (Å)	14.1	13.7	19.7

[a] The distance distribution parameters $\bar{r}$ and hw were recovered from frequency-domain lifetime data.
[b] TnI* and TnC* denote the proteins TnI and TnC containing a donor-acceptor pair, and TnI and TnC denote unmodified proteins.
[c] The values of $\bar{r}$ and hw in parentheses were recovered from steady-state quenching data.

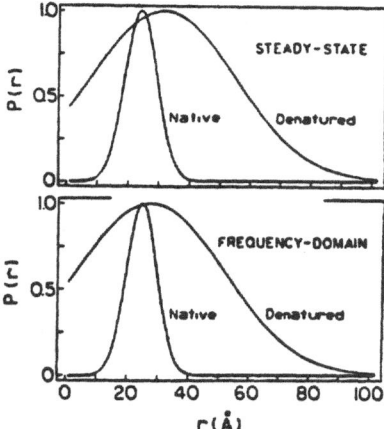

Fig. 3. Comparison of distributions of distances recovered from frequency-domain lifetime data and steady-state quenching data for TnI.

studied kinetically in stopped-flow experiments in which S1 labeled with AEDANS and DDPM was rapidly mixed with varying concentrations of ADP [13]. The reaction was monitored by the decrease of the steady-state donor intensity (increase in energy transfer). The kinetic tracings were fitted to a biexponential function yielding two observed first-order rate constants, k_1 and k_2. Figure 4 shows the change of the two rate constants with increasing [ADP]. At the highest [ADP] used k_1 is 31 s^{-1} and k_2 is 2.6 s^{-1}. These results suggest that the structural change in S1 induced by ADP binding to its active site occurs in two kinetic steps.

Discussion

The results shown here for the distribution of FRET distances are the first results obtained for proteins from frequency-domain and steady-state data. While other distribution functions could be used, we approximated P(r) by a Gaussian as our starting point. The choice of n in Eq. 3 depends upon whether one assumes the D-A distance distribution is along a line, in a plane, or in three dimensions. For a structured protein the choice of n is not obvious, but the recovered distributions are visually similar irrespectively of the choice of n (Fig. 1).

We [14] have previously shown, on the basis of free energy coupling, that the TnI-TnC linkage within the three-subunit troponin serves as the main Ca^{2+} signal transmitter in muscle. The distributions of FRET distances obtained from isolated TnC and the complex TnI•TnC provide a simple model to relate structural perturbation to calcium activation. In relaxed muscle (in the absence of activator Ca^{2+}) a segment of TnI residues 104–115 (inhibitory region) interacts with TnC only weakly, and this same segment of TnI interacts strongly with actin-tropomyosin. Upon activation, activator Ca^{2+} binds to the Ca^{2+}-specific sites of TnC; this binding switches the weak interaction between TnI and TnC to a strong interaction and concomitantly suppresses the TnI interaction with actin-tropomyosin. Thus during a contractile cycle TnI alternately interacts with TnC and actin-tropomyosin. The

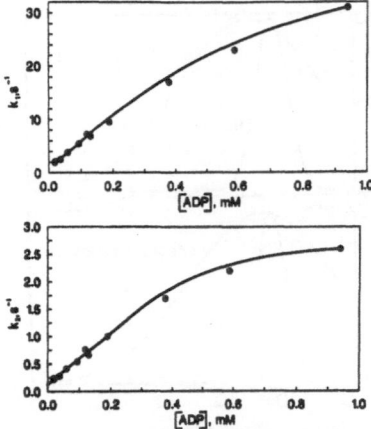

Fig. 4. Dependence of k_1 and k_2 on [ADP] for the change of FRET in myosin subfragment 1 between Cys^{707} and Cys^{697}. The rate constants were obtained from stopped-flow experiments in which 2 μM of donor and acceptor labeled S1 was mixed with ADP in the presence of MgCl$_2$. The labeled S1 was identical to the S1 preparation described in Table 1.

origin of this Ca^{2+}-induced alternate interaction of the inhibitory region of TnI with actin-tropomyosin and TnC may be related to altered conformational fluctuations in TnI and TnC. Under ionic conditions that mimic Ca^{2+} activation (change from Mg^{2+} to Ca^{2+}), the mean distance of TnI in TnI•TnC is not affected, but the half-width of the P(r) of the Trp^{158}-Cys^{133} distances decreases from 21 to 11 Å. A similar, but less pronounced, decrease is observed for the half-width of the P(r) of the Met^{25}-Cys^{98} distances of isolated TnC and TnC in the complex TnI•TnC. At this time we do not fully understand the various factors that can contribute to the width of the distribution of distances [9], but expect to see a narrow width if the donor and acceptor are fixed in space and the D-A pair can be characterized by a unique separation. An appreciable width of the distribution can be ascribed to conformational fluctuations. These and other preliminary considerations suggest that calcium activation may be accompanied by decreased fluctuations in both TnI and TnC. In relaxed muscle conformational fluctuations may prevent strong interaction of the TnI inhibitory region with TnC, thus promoting interaction between the same inhibitory region and actin-tropomyosin. The TnI-TnC interaction is strengthened upon Ca^{2+} activation possibly because of decreased conformational fluctuations, and these structural changes may be the basis of the stabilization (a free energy coupling of -11.6 kJ mol^{-1}) of the TnI-TnC linkage by Ca^{2+} binding to its specific sites on TnC [15].

The decrease of the SH_1-SH_2 distance in S1 induced by MgADP was previously demonstrated on the basis of a single D-A distance [12]. In spite of a decrease of 6–7 Å, the hw remains relatively unaffected by ADP binding. Upon denaturation the protein is expected to approach a random coil and the segment in question exists in a large number of conformations. Consequently, in denatured protein the D-A segment should exist in a wide range of D-A distances and the half-width of the distribution should increase manifold relative to native protein. This general expectation has been confirmed by the results on TnI and TnC, but not S1. The absence of a large increase in the hw upon denaturation suggests that the SH_1-SH_2 segment of the native protein is likely already in an overall conformation approaching that of a random coil, and its conformational fluctuations are relatively unconstrained. The small increase in the hw may be attributed to disruption of a weak secondary or tertiary structure upon denaturation. These results are direct physical evidence that the SH_1-SH_2 segment of myosin has an extraordinary flexibility comparable to a random coil. While the flexible nature of this segment has been inferred from a variety of biochemical studies showing cross-linking of the two sulfhydryl groups by bifunctional reagents with different spans, previous physical studies have not yielded information to demonstrate the flexibility.

The ADP-induced increase in energy transfer between SH_1-SH_2 occurs in two kinetic steps with time constants that can be physiologically relevant. This change can be caused by either a decrease in the D-A distance or a change in the orientation factor. The latter parameter, which cannot be unequivocally determined in solution, can be qualitatively assessed from the mobility of the donor and acceptor through measurements of their anisotropies. We implemented a two-channel stopped-flow apparatus [13] to monitor the anisotropy of the attached donor over the same time interval used in the total donor intensity experiments. The anisotropy remained essentially constant during the change in donor intensity. This result suggests that there is no change in donor mobility during the course of the reaction between MgADP and donor-acceptor labeled S1. Thus the time-dependent increase in energy transfer efficiency is due largely to a decrease in the D-A separation from 29 to 22 Å. The fractional amplitudes of the two kinetic phases suggest that the fast phase reflects a decrease in D-A distance of 4–5 Å and the slow phase an additional decrease of about 2 Å. The use of FRET in conjunction with anisotropy signals in stopped-flow studies to investigate the kinetics of structural perturbations has advantages over the use of signals from single chromophores. The amplitudes of the FRET changes can in principle be related to physical changes. In a previous kinetic study on the S1-nucleotide interaction, we proposed a 3-step mechanism for the binding of nucleotides to the active site [16]. The present time-resolved FRET results are compatible with the 3-step model of nucleotide binding, and suggest a structural basis for the previous kinetic mechanism.

Acknowledgements

This work was supported in part by grants (AR25193 and AR31239) from the U.S. National Institutes of Health.

References

1. Förster, Th. (1948) *Ann. Phys.* **2**, 55.
2. Cantor, C.R. and Pechukas, P. (1971) *Proc. Natl. Acad. Sci. USA* **68**, 2099.
3. Gryczynski, I., Wiczk, W., Johnson, M.L., Cheung, H.C., Wang, C.-K. and Lakowicz, J.R. (1988) *Biophys. J.* **54**, 557.
4. Grinvald, A., Haas, E. and Steinberg, I.Z. (1971) *Proc. Natl. Acad. Sci. USA* **69**, 2237.

5. Amir, D. and Haas, E. (1987) *Biochemistry* **26**, 2162.
6. Gryczynski, I., Wiczk, W., Johnson, M.L. and Lakowicz, J.R. (1988) *Chem. Phys. Lett.* **145**, 439.
7. Gratton, E. and Limkeman, M. (1983) *Biophys. J.* **44**, 315.
8. Lakowicz, J.R., Laczko, G. and Gryczynski, I. (1986) *Rev. Sci. Instrum.* **57**, 2499.
9. Lakowicz, J.R., Gryczynski, I., Cheung, H.C., Wang, C.-K., Johnson, M.L. and Joshi, N. (1988) *Biochemistry* **27**, 9149.
10. Wang, C.-K. and Cheung, H.C. (1986) *J. Mol. Biol.* **191**, 509.
11. Cheung, H.C., Gryczynski, I., Wiczk, W., Johnson, M.L. and Lakowicz, J.R. (1991) *Biophys. Chem.* (in press).
12. Cheung, H.C., Gonsoulin, F. and Garland, F. (1985) *Biochim. Biophys. Acta* **832**, 52.
13. Garland, F., Gonsoulin, F. and Cheung, H.C. (1988) *J. Biol. Chem.* **263**, 11261.
14. Cheung, H.C., Wang, C.-K. and Malik, N.A. (1987) *Biochemistry* **26**, 5904.
15. Wang, C.-K. and Cheung, H.C. (1985) *Biophys. J.* **48**, 727.
16. Garland, F. and Cheung, H.C. (1979) *Biochemistry* **18**, 5281.

Low-frequency dynamics of proteins: Comparison of experiment with theory

Stephen Cusack[a] and Wolfgang Doster[b]

[a]European Molecular Biology Laboratory, c/o ILL, 156X, F-38042 Grenoble, France
[b]Physik Department E13, Technische Universität München, D-8046 Garching, Germany

Introduction

In order to assess the biological importance of picosecond structural fluctuations in proteins it is important that they be well characterised both experimentally and theoretically. Inelastic neutron scattering (INS) is a spectroscopic technique [1,2] which can be used to study internal protein motions on exactly the same timescale (0.1–100 ps) that is now widely accessible by computer simulation [3]. Here we show that INS measurements on myoglobin over a wide temperature range (4–350 K) lead to a global picture of the nature of fast internal motions in a globular protein which can be used to test models of protein dynamics. According to this picture, proteins behave as quasiharmonic systems below a temperature of about 180 K but above this temperature there is a dynamic transition to a state of enhanced mobility due to the excitation of fast non-harmonic motions which are most likely local segmental transitions [4,5]. It has recently been shown that this dynamic behaviour can be reproduced by molecular dynamics simulations of myoglobin as a function of temperature [6,7]. This freezing phenomenon (which in the case of myoglobin is biologically significant as it correlates with cessation of oxygen diffusion through the protein matrix [8]), has a resemblance to the liquid-glass transition occurring in many dense systems. This has been shown by a recent detailed application of modern theories of the glass transition to the INS data obtained on myoglobin [9].

Inelastic Neutron Scattering

Due to the anomalously large incoherent neutron scattering cross section of the ^{1}H nucleus and the abundance of such atoms in proteins, INS measurements give a global view of protein dynamics as sensed via the motions of the hydrogen atoms [1]. Incoherent scattering implies that the measured intensity is a sum of individual atomic contributions; the lack of interference means that no direct structural information is obtained but only information on the timescales, amplitudes and geometry of motion. This contrasts with the dynamic information obtainable from X-ray crystallography, where temperature factors (B-factors) give a measure of the amplitude and structural localisation of motion, but no indication of timescale [10]. The quantity measured experimentally is the incoherent dynamic structure factor $S_{inc}(q,\omega)$ where $\hbar q$ and $\hbar\omega$ are respectively the momentum and energy transfers between system and incident neutron [1,2,11]. $S_{inc}(q,\omega)$ is the Fourier transform of the self-correlation function in a system as given by:

$$S_{inc}(q,\omega) = \frac{1}{2\pi} \int_{-\infty}^{\infty} dt \ e^{-i\omega t} \sum_{L=1}^{N} \frac{\sigma_L^{inc}}{4\pi} <e^{-iq.\hat{R}_L(0)} \ e^{iq.\hat{R}_L(t)} >_T \qquad (1)$$

where $\hat{R}_L(t)$ is the time-dependent Heisenberg position operator and σ_L^{inc} the incoherent cross section of nucleus L and the brackets $< >_T$ indicate an ensemble average at temperature T. This quantity can be directly calculated from any theoretical model yielding atomic trajectories such as molecular dynamics simulations or a normal mode (vibrational) analysis. In general, three features may be distinguishable in an INS spectrum of a protein (Fig. 1): (a) the *elastic peak*, $S_{inc}(q,\omega \approx 0)$ which arises from neutrons which are scattered with no change in energy and is generally the most intense feature in the spectrum in the case of solid samples. Analysis of the elastic intensity as a function of q and say temperature can give information on the geometry and energetics of the motion [5,11]. (b) The *quasielastic scattering* [5,9,11] which appears as a more or less broad, continuous feature centred at $\omega = 0$. This can arise from rotational or translational diffusive motion (e.g. for a protein in solution) or activated motions, for example translational or rotational motion of an atom or group between discrete sites requiring the surmounting of an energy barrier. Transitions between conformational substates [8,12] of proteins should manifest themselves in the quasielastic spectrum. (c) The *inelastic scattering* which arises from transitions of the system between well-defined energy levels and thus gives rise to features shifted from $\omega = 0$. This is the case for a quasiharmonic system undergoing motion which can be decomposed into independent normal modes of vibration. In such systems an incident neutron can excite or de-excite a normal mode of angular frequency ω, giving rise to peaks in the inelastic neutron spectrum at $+ \omega$ or $- \omega$ respectively. The amplitude-weighted vibrational frequency distribution $G(\omega)$, which is very similar in form to the unweighted density of states [13], can be obtained from measurements of $S_{inc}(q,\omega)$ by extrapolating to $q = 0$ according to the definition (for a spherically averaged system) [1,2,4,13]:

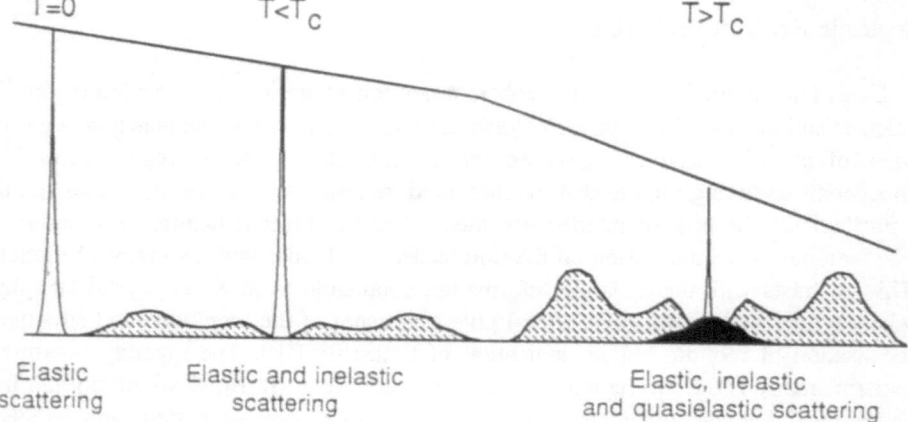

Fig. 1. Schematic diagram of the temperature variation of the inelastic neutron scattering of myoglobin showing an extra decrease in the elastic intensity above a transition temperature due to the excitation of new motions which scatter quasielastically.

$$G(\omega) = \lim_{q \to 0} \frac{6\omega}{\hbar q^2} (e^{\hbar\omega/k_B T} - 1) \; S_{inc}(q,\omega) = \sum_{\lambda=1}^{3N-6} \sum_{L=1}^{N} \frac{\sigma_L^{inc}}{4\pi m_L} |c_L^\lambda|^2 \delta(\omega - \omega_\lambda) \quad (2)$$

where the second term corresponds to the result for a harmonic system of N atoms of mass m_L, with $3N - 6$ normal modes of frequency ω_λ and eigenvectors c_L^λ. Due to the low intensity of neutron beams, samples for INS need to be of several hundred milligrams. In order to focus on the protein dynamics rather than that of the solvent, the sample is usually hydrated with D_2O. The signal thus arises principally from $-CH_3$, $-CH_2$, $-CH$ and a few unexchanged $-NH$ groups in the protein. In neutron scattering, a variety of energy/frequency units is employed, 1 meV $= 8.07$ cm^{-1} $= 0.24$ THz $= 11.61$ K.

The Temperature Dependence of the Dynamics of Myoglobin

The temperature dependence of a number of properties of myoglobin have been studied over a wide range of temperatures. The concept of conformational substates arose from the analysis of low-temperature flash photolysis experiments which measure ligand (e.g. oxygen) binding rates to the heme iron [8,12]. Mössbauer experiments have revealed a striking dynamic transition at about 200 K, above which temperature the mean-square displacement of the heme iron shows a much faster increase with temperature than expected for a harmonic system [14,15]. This has been interpreted as due to the excitation of new slow modes of motion above the transition temperature and correlates with the ability of small ligands to cross the protein matrix between the heme pocket and the solvent [8]. X-ray analysis however shows no dramatic structural changes in myoglobin as a function of temperature [16]. Recently it has been shown that the dynamic transition in myoglobin near 200 K can also be clearly observed in INS experiments and analysis of the data gives some new insights into the nature of picosecond motions in proteins [1,4,5,9].

A schematic view of the observed incoherent dynamic structure factor $S_{inc}(q,\omega)$ of myoglobin at fixed q and three different temperatures is shown in Fig. 1. At very low temperatures (Fig. 1, left), there is essentially no motion (apart from quantum effects) and the spectrum consists only of the sharp elastic peak. As the temperature is raised (to say 150 K, Fig. 1, centre), the amplitude of harmonic vibrations increases giving rise to an inelastic vibrational spectrum and a corresponding decrease of the elastic peak. Figure 1 (right) shows the situation above the transition temperature at say 300 K. In addition to the increased scattering from the vibrations, new non-vibrational modes of motion have been excited and give rise to a quasielastic spectrum, with a corresponding sharper decrease in the elastic peak. Note that this discussion illustrates well the *zeroth-order sum rule* [1,2] which is an exact result and shows that the overall scattered intensity is conserved but can be redistributed between elastic, quasielastic and inelastic scattering.

Experimental measurements have been made on a powder of sperm-whale myoglobin hydrated to 0.33 gD_2O/gprotein. Under these conditions the protein is essentially fully hydrated, whereas the D_2O only contributes about 5–10% of the scattering. Figure 2 shows the temperature variation between 4 and 320 K of the

195

elastic intensity $S_{inc}(q,\omega \approx 0)$ at a number of different values of q measured on the backscattering instrument IN13 at the Institut Laue-Langevin in Grenoble, France [17]. This instrument has a resolution of about 10 μeV, so that motions slower than 10^{-10}s are not resolved. A transition in behaviour, corresponding to a sharper decrease in the elastic intensity, is clearly seen at about 200 K. Figure 3 shows quasielastic and inelastic spectra obtained on instrument IN6 (resolution 50 μeV) at temperatures between 100 and 350 K. These show that at low temperatures (180 K and below) there is a well-resolved but fairly broad maximum in the structure factor at 25 cm^{-1} and no quasielastic broadening. Above 180 K, the maximum becomes less and less resolved due to the rapid increase with temperature of a broad quasielastic band extending to about 30 cm^{-1}. This broad line accounts for the extra decrease observed in the elastic intensity.

(a) Vibrational frequency distribution of myoglobin [4]

The measurements provide strong evidence that myoglobin behaves quasi-harmonically below 200 K. Figure 4 shows G(ω) for the hydrated myoglobin powder derived from the data of Fig. 3 independently at 180 K and 100 K using the extrapolation implied by equation 1. These are the same within experimental error suggesting that the dynamics of myoglobin below 200 K and above a few cm^{-1} is characterised by a distribution of underdamped vibrational modes and has a temperature dependence consistent with quasiharmonic motion. This is confirmed by the low-temperature linear temperature dependence of the atomic mean-squared displacements observed by neutrons (Fig. 5 and [5]) and by Mössbauer spectroscopy [14]. The inset log/log plot in Fig. 4 shows that at very low frequencies the density of states is Debye-like i.e. $G(\omega) \propto \omega^2$, but that there is a distinct enhancement at about 25 cm^{-1} corresponding to the peak in $S_{inc}(q,\omega)$ at 25 cm^{-1} (Fig. 3). It should be stressed that although the enhancement in numbers of modes at about 25 cm^{-1} is not very striking, the real significance of modes of this frequency is evident from the

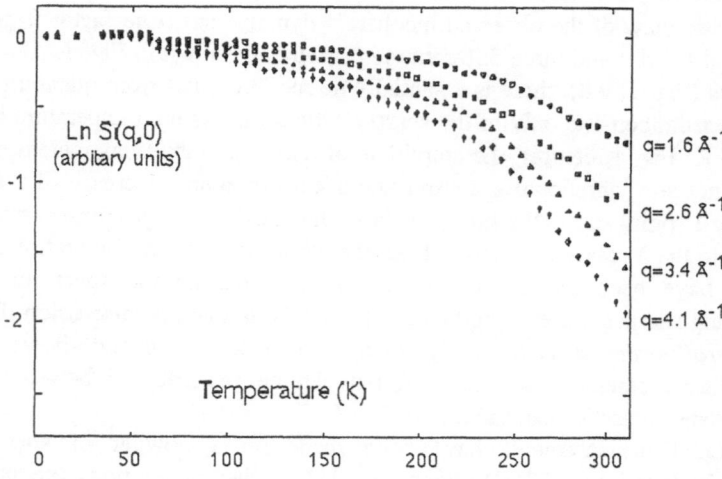

Fig. 2. Normalised elastic intensity $S(q,\omega \approx 0)$ for hydrated myoglobin as a function of temperature at a number of different values of q.

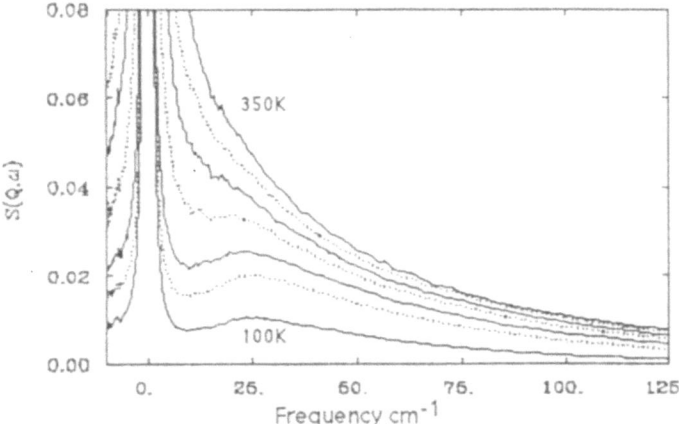

Fig. 3. Inelastic neutron scattering spectra of hydrated myoglobin for a fixed scattering angle at temperatures of 100, 180, 220, 270, 300, 320 and 350 K.

appearance of the corresponding peak in $S_{inc}(q,\omega)$, where modes are weighted by their amplitude. This peak thus tells us that the dominant mean-square displacements are given by vibrational motions with a frequency of about 25 cm^{-1}.

No theoretical normal-mode analysis has yet been published for myoglobin so the experimental results cannot be compared to theory. However this has been done in detail in the case of bovine pancreatic trypsin inhibitor (BPTI) [19–21]. For this protein, comparison of the experimental vibrational density of states with the results of normal-mode analyses using different empirical force fields [22–24] has shown

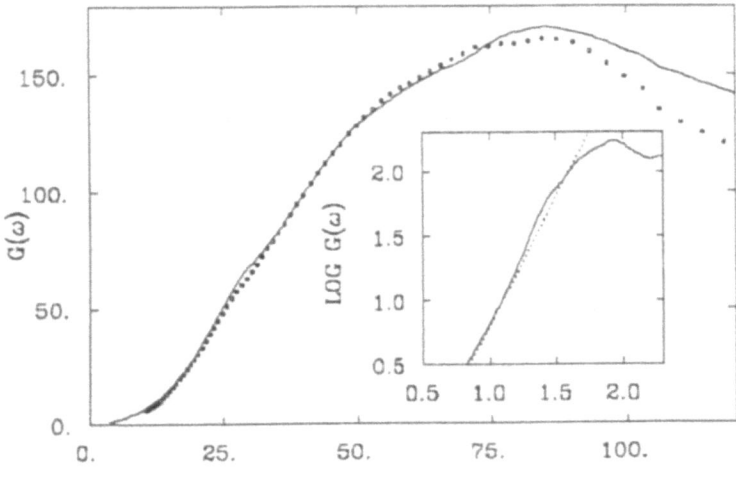

FREQUENCY ω cm^{-1}

Fig. 4. Experimental vibrational distribution of hydrated myoglobin determined independently at 180 K (—) and 100 K (ooo). Inset: log/log plot of G(ω) at 180 K showing deviations from Debye behaviour (the dotted line has slope 2).

the importance of the treatment of long-range electrostatic interactions in correctly predicting the number of very-low-frequency modes. Experimentally derived frequency distributions of BPTI, myoglobin and lysozyme are rather similar, being smooth and without notable structure even at low temperatures. It is furthermore remarkable that very similar curves are obtained with simple amorphous materials [25] and glasses [26]. These observations suggest that simpler models of close-packed but irregular systems may be sufficient to explain the density of states of proteins [27,28].

(b) Elastic and quasielastic neutron scattering in myoglobin [5,9]

Using the vibrational density of states measured at low temperatures and the assumption that it is valid at higher temperatures as well, enables the vibrational part to be subtracted from the spectra above 180 K to obtain the quasielastic scattering [6]. The resulting quasielastic spectra can be Fourier transformed to obtain the corresponding correlation functions which show that there are at least two components, a very fast component ($\tau \sim 0.35$ps) and a slower component ($\tau \sim$ several ps). However, a much more detailed analysis of the quasielastic scattering has now been made in terms of the mode-coupling theory of the glass transition (see below) which predicts the occurrence of two power law regions in the relaxation spectrum with related exponents [9]. The question arises as to what is the nature of the fluctuations giving rise to the quasielastic scattering observed above 180 K? It has been shown that the elastic scattering from myoglobin as a function of temperature can be accounted for by modelling the extra motion as jumps between distinct sites of energy asymmetry about $\Delta H = 12$ kJ/mol and separation 1.5 Å [5]. This model enables the mean-square displacements to be derived as a function of temperature (Fig. 5) and shows that there is a steep increase in these displacements above 180 K, as also deduced from Mössbauer scattering experiments which monitor the motion of the heme iron [18,19].

(c) Comparison of neutron results with molecular dynamics simulations [6,7]

Support for the interpretation of the neutron results has come from the analysis of molecular dynamics simulations on myoglobin where the details of the calculations may give further insight into the real nature of the motion observed. Fast (picosecond) transitions between conformational substates have been revealed by an early molecular dynamics simulation of myoglobin in vacuum [29]. More recently, a detailed comparison of the neutron results has been made with simulations of myoglobin as a function of temperature using the CHARMm force field, one study being made in vacuum with only polar hydrogens explicitly included [6,30] and one with an all-hydrogen atomic model and partial hydration (with 350 water molecules) to mimic the neutron experimental conditions [7]. The latter results are described in more detail in an accompanying contribution to these proceedings [31]. Both studies have shown that the experimental neutron results on myoglobin can be semi-quantitatively reproduced, in particular the change in form of the inelastic spectra as seen in Fig. 3, and the corresponding sharp rise in the mean-square hydrogen displacements above 200 K. They also both show that the empirical force fields result in a somewhat softer protein

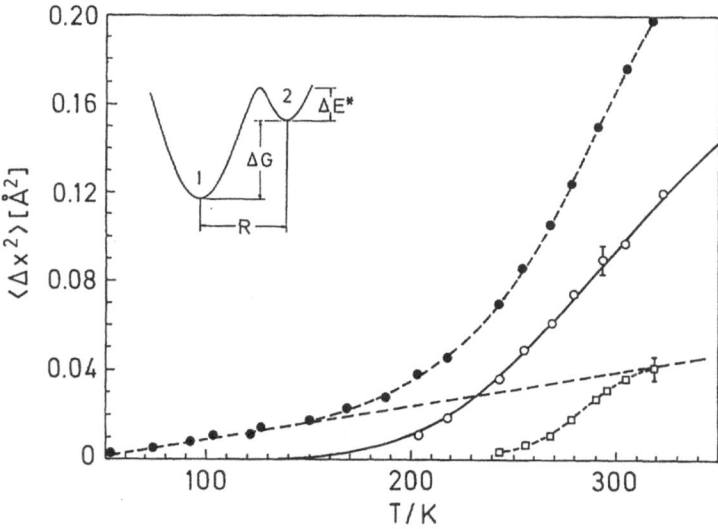

Fig. 5. *Hydrogen mean-square displacements in hydrated myoglobin as a function of temperature derived by fitting an asymmetric double potential well model (inset) to the data of Fig. 2. Full circles: total mean-square displacement; open circles: contribution from jumping; dashed straight line: vibrational contribution extrapolated from low temperatures; open squares: additional Gaussian process above 240 K.*

as the main peak in the calculated dynamic structure factor at low temperatures, in the harmonic regime, is below the observed 25 cm^{-1}. This is particularly true of the Loncharich and Brooks calculations [7] (where the peak is around 10 cm^{-1}) with the result that the observed mean-square vibrational fluctuations are uniformly too large, although the temperature dependence is correct. An interesting result is that the number of dihedral transitions (defined as jumps in the torsional angles involving heavy atoms) shows a steep rise above the transition temperature in line with the increase in mean-square displacements [7]. The authors have estimated that at room temperature about 40% of the fluctuations are due to local oscillations (vibrations) and 60% due to torsional jumps.

(d) The glass analogy

In the last years several authors have drawn attention to certain similarities between the behaviour of hydrated proteins and glasses [32–34]. To appreciate this one has to have some understanding of the physics of glasses [35] which is not an easy subject. Glassification is solidification without crystallisation. At the liquid-glass transition, the shear viscosity diverges (of the order 10^{13} poise at T_g) and structural relaxation (diffusion) is arrested. The exact point of transition is cooling-rate dependent; the glass transition is therefore not a true phase transition, but a dynamic transition. The glassy state is non-ergodic in that only a restricted region of allowable configurational phase space is explored; the corollary to this is that the ground state is highly degenerate. Many different kinds of material can be made to form glasses e.g. simple oxides (SiO_2), ionic compounds ($Ca(NO_3)_2 \cdot KNO_3$), organic compounds (glycerol) and polymers (polybutadiene). Proteins have several features in common with glasses but there also exist important differences. They have no true liquid state, due to the constraints which maintain the 3D structure. But many experiments indicate a glass-like transition at around 200 K. Above this temperature internal mobility rapidly increases; below this temperature individual molecules are frozen into particular conformational substates ('degenerate ground state'). Mode-coupling theory is a very powerful microscopic theory of dense fluids that appears to describe the critical phenomena occurring in glassy materials around the transition point [36]. It makes specific predictions about the temperature and time dependence of the density correlation function $S(q,t)$ and has been tested on a number of systems [37] using INS which can probe the relevant timescales (10^{-7}-10^{-11}s). In simple terms, the theory predicts that the correlation function is separable into two time regions with different scaling laws. This corresponds to the α- and β-relaxations in glasses. The β-relaxation corresponds to fast local motions. e.g. the breathing motions of a cage for a particular atom or group and it survives the glass transition but with decreasing amplitude. The α-relaxation corresponds to escape of the particle out of the cage and ultimately at long times in the liquid state to diffusion and becomes indefinitely slow at the glass transition. The theory describes the coupling of these local and global motions via a non-linear equation. An analysis in terms of mode-coupling theory seems to account for the temperature dependence of the INS results on myoglobin over several orders of magnitude of time [9]. This description of protein dynamics (in terms of a few parameters) could be useful to understand certain features of protein function [38].

Conclusions

Finally the biological importance of the kind of motions described in this paper need to be considered. It is still an open question whether picosecond motions in proteins are simply a general manifestation of the mobility necessary for protein activity or whether certain of these motions are specifically coupled to protein function. It may be rather difficult to test the latter hypothesis by the technique of INS, but the method has already been shown to provide unique information on the nature of fast motions in proteins, against which theoretical models can be tested.

Acknowledgements

The work reviewed in this paper has benefited from numerous discussions and assistance from Jeremy Smith (Saclay), Martin Karplus (Harvard), Albert Dianoux (ILL Grenoble) and Winfried Petry (ILL, Grenoble).

References

1. Cusack, S. (1989) in *The Enzyme Catalysis Process: Energetics, Mechanism and Dynamics*, (Cooper, A. and Houben, J.L., Eds.) *NATO ASI Ser. A. Life. Sci.* **178**, pp. 103–122.

2. Lovesey, S.W. (1984) in *Theory of Neutron Scattering from Condensed Matter,* Vol 1, Clarendon Press, Oxford.
3. McCammon, J.A. and Harvey, S.C. (1987) in *Dynamics of Proteins and Nucleic Acids,* Cambridge University Press, Cambridge.
4. Cusack, S. and Doster, W. (1990) *Biophys. J.* **58**, 243.
5. Doster, W., Cusack, S. and Petry, W. (1989) *Nature* **337**, 754.
6. Smith, J., Kuczera, K. and Karplus, M. (1990) *PNAS* **87**, 1601.
7. Loncharich, R.J. and Brooks, B.R. (1990) *J. Mol. Biol.* **215**, 439.
8. Austin, R.H., Beeson, K.W., Eisenstein, L., Frauenfelder, H. and Gunsalus, I.C. (1975) *Biochemistry* **14**, 5355.
9. Doster, W., Cusack, S. and Petry, W. (1990) *Phys. Rev. Lett.* **65**, 1080.
10. Ringe, D. and Petsko, G.A. (1985) *Prog. Biophys. Mol. Biol.* **45**, 197.
11. Bee, M. (1988) in *Applications of Quasielastic Neutron Scattering to Solid-State Chemistry, Biology and Material Science,* Adam Hilger, London.
12. Ansari, A., Berendson, J., Bowne, S.F., Frauenfelder, H., Iben, I.E.T., Sauke, T.B. Shyamsunder, E. and Young, R.D. (1985) *PNAS* **82**, 5000.
13. Smith, J., Cusack, S., Pezzeca, U., Brooks, B.R. and Karplus, M. (1986) *J. Chem. Phys.* **85**, 3636.
14. Parak, F., Knapp, E.W. and Kucheida, D. (1982) *J. Mol. Biol.* **161**, 177.
15. Parak, F., Heidemeier, J. and Nienhaus, G.U. (1988) *Hyperfine Interactions* **40**, 147.
16. Parak, F., Hartmann, H., Aumann, K.D., Reuscher, H., Rennekamp, G., Bartunik, H. and Steigemann, W. (1987) *Eur. Biophys. J.* **15**, 237.
17. Guide to Neutron research facilities at the ILL high flux reactor (1988) Institut Laue-Langevin, Grenoble, France.
19. Cusack, S., Smith, J., Finney, J., Tidor, B. and Karplus, M. (1988) *J. Mol. Biol.* **202**, 903.
20. Cusack, S. (1989) *Chemica Scripta* **29A**, 103.
21. Smith, J., Cusack, S., Tidor, B. and Karplus, M. (1990) *J. Chem. Phys.* **93**, 2974.
22. Go, N., Noguti, T. and Nishikawa, T. (1983) *PNAS* **80**, 3696.
23. Brooks, B.R. and Karplus, M. (1983) *PNAS* **80**, 6571.
24. Levitt, M., Sander, C. and Stern, P.S. (1985) *J. Mol. Biol.* **181**, 423.
25. Dianoux, A.J., Page, J.N. and Rosenberg, H.M. (1987) *Phys. Rev. Lett.* **58**, 886.
26. Buchenau, U., Prager, M., Nücker, N., Dianoux, A.J., Ahmad, N. and Phillips, W.A., (1986) *Phys. Rev.* **B34**, 5665.
27. Elber, R. and Karplus, M. (1986) *Phys. Rev. Lett.* **56**, 394.
28. Schirmacher, W. and Wagener, M. (1989) in *Dynamics of Disordered Materials,* Springer Proceedings in Physics, Vol. 37 (Richter, D., Dianoux, A.J., Petry, W. and Teixeira, J., Eds.) Springer-Verlag, Berlin, p. 120.
29. Elber, R. and Karplus, M. (1987) *Science* **235**, 318.
30. Kuczera, K., Kuriyan, J. and Karplus, M. (1990) *J. Mol. Biol.* **213**, 351.
31. Loncharich, R.J. and Brooks, B.R. (1991) in *Proteins: Structure, Dynamics and Design* (Renugopalakrishnan, V. et al., Eds.) ESCOM, Leiden, pp. 177–183.
32. Morozov, V.N. and Gevorkian, S.G. (1985) *Biopolymers* **24**, 1785.
33. Doster, W., Bachleitner, A., Dunau, R., Hieble, M. and Lüscher, E. (1986) *Biophys. J.* **50**, 213.
34. Iben, I.E.T., Braunstein, D., Doster, W., Frauenfelder, H., Hong, M.K., Johnson, J.B., Luck, S., Ormos, P. and Schulte, A. (1989) *Phys. Rev. Lett.* **62**, 1916.
35. Jäckle, J. (1986) *Rep. Prog. Phys.* **49**, 171.
36. Gotze, W. and Sjögren, L. (1989) in *Dynamics of Disordered Materials,* Springer Proceedings in Physics, Vol. 37, (Richter, D., Dianoux, A.J., Petry, W. and Teixeira, J., Eds.) Springer-Verlag, Berlin, p. 18.
37. Mezei, F., Knaak, W. and Farago, B. (1987) *Phys. Rev. Lett.* **58**, 57.
38. Frauenfelder, H. (1991) in *Proteins: Structure, Dynamics and Design* (Renugopalakrishnan, V. et al., Eds.) ESCOM, Leiden, pp. 173–176.

Three-dimensional structure of amelogenin from 2D NMR: A molecular dynamics approach

M. Prabhakaran[a], V. Renugopalakrishnan[a], B. Wilson[a], H.C. Cheung[b], E. Strawich[a] and M.J. Glimcher[a]

[a]*Harvard Medical School, Boston, MA 02115, U.S.A.*
[b]*University of Alabama at Birmingham, Birmingham, AL 35294, U.S.A.*

Amelogenin, a hydrophobic 170 residue extracellular protein from developing bovine tooth enamel is characterized by the presence of a large proportion of Pro, Gln, His, and Leu residues and is sparse in Asp and Glu residues [1]. Amelogenin takes part in the critical early events of enamel mineralization [2,3]. Amelogenin is rapidly degraded after initiating mineralization by proteolysis. It appears to be an extremely fast process in which amelogenin is called upon to sequester Ca^{++} ions quite rapidly. In view of this, the interactions of amelogenin with Ca^{++} ions are quite important in understanding the molecular mechanisms of enamel mineralization.

The primary structure of amelogenin has been derived by Edman degradation [1] and more recently from its cDNA sequence [4]. The primary structure does not bear resemblance to any other known protein sequence [5]. Structural studies on amelogenin were initiated in our Harvard laboratory before the primary structure was reported in the literature [1]. Initial CD studies of amelogenin at acid and neutral pH indicated amelogenin to be lacking in α-helical structure and to be representative of a protein rich in β-sheets and β-turns. Conclusions from CD studies were further reaffirmed by FT-IR [6] and Raman [7] spectroscopic studies. When the primary structure of amelogenin was reported by Takagi et al., 1984 [1] the above observations from CD, FT-IR, and Raman studies became apparent and the first glimpse of the secondary structure of amelogenin began to emerge revealing unusual secondary structural features. The major observation was a tandem repeat of β-turns resulting from the tandem repeating segment, Gln^{112} through His^{139} near the middle of the sequence. While tandem repeats are known to occur in other proteins, for example RNA polymerases [8], the cluster of β-turns occurring in amelogenin seems to be unique among protein structures.

Our attention was focused at this stage on predicting the structure of the tandem repeating segment, Gln^{112} through His^{139}, using initially a molecular mechanics approach and the derived structures were later refined by molecular dynamics studies both at normal and high temperatures [9]. The initial structures were built using the AMBER program and later refined using the CHARMm program. These studies clearly demonstrated the low-energy structures [9] were characterized by a staircase-like arrangement of β-turns where the β-turns were repeated periodically along a helical axis. The β-turns were stabilized by inter-molecular hydrogen bonds and the Gln side chains occurring at i and i + 4 positions were further tied up in an unusual

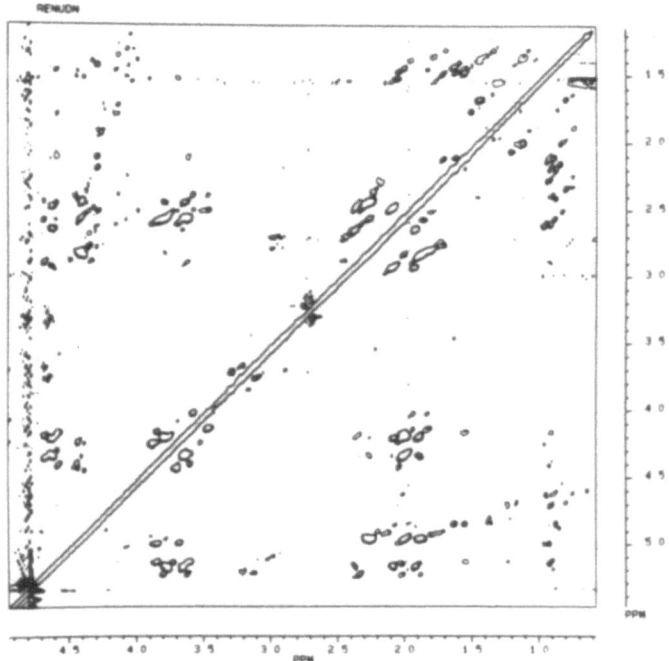

Fig. 1. *Upfield domain of NOESY spectrum of amelogenin, $\tau_m = 400$ ms, in acetate buffer, ~3 mM concentration, pD ~ 4.5.*

side-chain-to-backbone hydrogen bonding scheme. The low-energy β-spiral structure was characterized by the presence of hydrophobic side chains on the exterior with the C = O groups lining the interior of the structure. In addition the low-energy β-spiral structure was characterized by a skewed channel running across the helical axis providing the right dimensions for the passage of Ca^{++} ions. The tandem repeating structure was speculated to occur in the interior of the protein where the hydrophobic side chains on its exterior will most likely be encapsulated by other hydrophilic regions surrounding it. The above model was therefore chosen as the template for folding the polypeptide chain around it as the nucleus. Recently the β-spiral structure has been found to occur in the synthetic tandem repeat polypeptide from 2D NMR studies in dimethyl sulfoxide (Renugopalakrishnan et al., in preparation). The above observation was useful in further studies of amelogenin. The tandem repeat segment bears resemblance to similar structures proposed for tropoelastin polypeptides by Urry and co-workers [10].

2D NMR studies of amelogenin were started in our laboratory in 1986. They have been pursued over the last four years and a partial assignment of proton resonances was completed recently in 1990. From the long-range NOE contacts observed and using the β-spiral polypeptide as a template, molecular mechanics studies and molecular dynamics simulation of intact amelogenin comprising 170 residues were started. In this paper, some of the results from 2D NMR, molecular mechanics studies and molecular dynamics simulation will be presented.

Results and Discussion

2D NMR studies

A typical upfield domain of the NOESY spectrum of amelogenin in deuterated acetate buffer at 600 MHz resolution, pD 4.5, at a concentration of ~3 mM, mixing time τ_m = 400 ms is shown in Fig. 1. The task of assignment of proton resonances was extremely complicated due to the large size of amelogenin. The task was simplified by using numerous fragments of amelogenin obtained by solid-phase synthesis, a sequential assignment strategy [11] based on the primary structure of amelogenin [1], COSY, selected NOE, and TOCSY experiments. NMR studies also included some isotopically enriched amelogenin and fragments. Fortunately, a combination of the above experiments led to the assignments of a number of proton resonances. From the presently confirmed proton assignments, a partial list of NOE contacts, shown in Table 1, has been derived. Complete assignment of proton resonances and results from NOESY studies will be published subsequently (Wilson et al.)

Protein classification

The conformational characteristics of the protein amelogenin are derived from CD, laser Raman, and FT-IR studies [6,7]. From these studies the α-helical content is found to vary from 0% to 16%. These results indicate that this protein should be mainly classified as β-type protein. We adopted the Chou [12] method to classify the protein by finding the percentage difference between the expected and actual

Table 1 *Partial list of NOE contacts presently known (June, 1990)*

Residue number	Residue name	Atom	Residue number	Residue name	Atom
112	Gln	N	114	His	N
118	Gln	N	120	His	N
124	Gln	N	126	Met	N
165	Leu	N	167	Leu	N
40	Gln	N	88	Val	N
112	Gln	CA	115	Gln	N
18	Gln	CB	30	Ile	N
15	Phe	CB	29	Met	N
37	Tyr	CB	46	Leu	N
4	Pro	CB	5	Pro	N
66	Pro	CB	67	Pro	N
151	Pro	CB	152	Pro	N
154	Pro	CB	155	Pro	N
159	Pro	CB	160	Pro	N
168	Gln	CB	165	Leu	N
158	Leu	N	43	Gly	N
169	Ala	N	70	Ile	N

Table 2   *Prediction scheme for amelogenin*

Sequence	Conformational sub-structure	Secondary structure prediction scheme
13-21	β-sheet	N, G1, G2, B, L, Ne
22-29	α-helix	G2, F, L, Co, AS
30-34	β-sheet	G2, G1, N, L, Co, NE
37-40	coil	N, Co, G1, G2, L
42-49	α-helix	F, B, G2, AS
51-58	β-sheet	F, G2, G1, Co, B, L, N
60-65	α-helix	F, B, Co, AS
71-77	β-sheet	G2, G1, C, N
79-88	β-sheet	N, G2, G1, G, B, Ne, Co
93-100	β-sheet	Co
100-105	β-sheet/coil	G2, Co, G1, F
107-112	coil	Ne, N, Co, G1, F
112-138	β-turn spiral	experimental studies on synthetic polypeptide Gln[112]- His[138]
142-153	β-sheet	Ne, G2, G1, F, B, N, Co
155-161	β-sheet	Co, G1, G2
163-170	α-helix	F, N, G2, Ne, AS

F: Chou and Fasman [13], G1: Garnier [14] (0% α-helix), G2: Garnier (16% α-helix), B: Burgess [15], N: Nagano [17], L: Lim [16], Ne: Neural network [18], Co: Cohen [20,21], AS: Strong helix Cohen [20,21].

distribution in each class. The total difference in each class is very high and this protein can not be classified in any class from this method. Along with this result, the high percentage of proline residues in amelogenin led us to infer a special character for this protein. The correspondence rule of Chou classified the protein either as a β or αβ protein. We applied the rules of protein folding for these two classes to amelogenin. Further modeling of amelogenin was carried out on the basis of these two classifications.

Secondary structure prediction scheme

We have attempted to predict the secondary structure of this protein from the sequence. Due to the peculiar nature of its primary structure, we adopted many structural prediction schemes available in the literature [13-18]. From the Garnier method [14] we have derived two prediction schemes: (1) with 0% α-helix and (2) with 16% α-helix, by varying the decision constants. The hydropathy plot [19] was also used to locate β-turns. We have independently used the basic strategy described by Cohen and co-workers [20,21] of first determining the most probable β-turn positions, and then assigning secondary structure between high-probability β-turns. The final prediction scheme derived from a combination of the above approaches is given in Table 2.

Tertiary structure prediction scheme

2D NMR studies have revealed certain unambiguous NOE constraints for amelogenin (see Table 1). Hydrodynamic studies revealed an approximate axial ratio of 1.5. Using spatial dimensions of 21.0 Å, 21.0 Å, and 14.0 Å we built a tertiary structure from the secondary structure derived earlier. The following protein folding principles were used in folding the structure:

(1) Proline backbone constraint on itself and the previous residue; (2) Dense packing as observed in globular proteins; (3) Accepted backbone dihedral angles; (4) Side-chain positions determined from the side-chain rotamer library [22]; (5) Spatial constraints; (6) Handedness of β-structures [23]; (7) Van der Waals overlap; (8) Nonpolar/polar exposure and amphipathic nature; (9) Thornton's rules for β-turn formation [24]; (10) For coil structures dihedrals were scanned from tripeptide to pentapeptide level from the Protein Data Bank; (11) Encapsulation of β-spiral within the protein [9]; (12) Optimal pairing of oppositely charged residues, and hydrogen-bonded residues; (13) Trp residues were assumed to be 10 Å apart and on the surface of the protein (Cheung, H.C., personal communication).

Molecular dynamics simulation

Molecular dynamics studies, using the program CHARMm [25], were effectively used to build up the structure using the available structural information. Initial structures were annealed both by steepest descent and conjugate gradient methods. Both low-temperature and high-temperature dynamics simulation were used to develop this model. Periodically the NOE constraints were added and withdrawn during the MD studies. We have incorporated the 17 tertiary NOE constraints in a unique way. The impact of these constraints is considered not only on the specific atoms derived from 2D NMR studies, but also extended to other nearby atoms as a constraint but with a higher distance limit. In this way, the MD studies pulled not only the specific atoms but also the specific strands nearby. Many structures were modified using computer graphics when the MD simulations led to structures that

Fig. 2. Tentative model of amelogenin showing the β-spiral channel in the middle.

were unable to satisfy NOE constraints due to blockage by other strands. The graphics program FRODO was extensively used for building the starting model and further modifications.

The annealed structures were further examined in terms of their conformational energetics, polar/nonpolar exposures, satisfaction of NOE constraints, spatial volume and the effective formation of pore for passage of Ca^{++} ions. The high mobility of Trp residues as shown by fluorescence studies, was also monitored during the simulation. The flexibility of amelogenin was derived from the trajectory studies. The low flexibility of β-spiral structure further supports our hypothesis. A tentative model satisfying all these constraints is shown in Fig. 2.

References

1. Takagi, T., Suzuki, M., Baba, T., Minegishi, K. and Sasaki, S. (1984) *Biochem. Biophys. Res. Commun.* **121**, 592.
2. Fincham, A.G. and Belcourt, A.B. (1985) in *The Chemistry and Biology of Mineralized Tissues* (Butler, W.T., Ed.) EBSCO Media Services, Inc., Birmingham, AL, pp. 240–247.
3. Deutsch, D. (1989) *The Anatomical Record* **224**, 189.
4. Shimokawa, H., Ogata, Y., Sasaki, S., Sobel, M.E., MacQuillan, C.I., Termine, J.D. and Young, M.F. (1987) *Adv. Dental Res.* **20**, 2.
5. Hunt, L., *Personal Communication, Protein Identification Resource,* National Biomedical Research Foundation, Washington, DC 20007, 1989.
6. Renugopalakrishnan, V., Strawich, E.S., Horowitz, P.M. and Glimcher, M.J. (1986) *Biochemistry* **25**, 4879.
7. Zheng, S., Tu, A.T., Renugopalakrishnan, V., Strawich, E. and Glimcher, M.J. (1987) *Biopolymers* **26**, 1809.
8. Allison, L.A., Moyle, M., Shales, M. and Ingles, C.J. (1985) *Cell* **42**, 599.
9. Renugopalakrishnan, V., Pattabiraman, N., Prabhakaran, M., Strawich, E. and Glimcher, M.J. (1989) *Biopolymers* **26**, 297.
10. Urry, D.W. (1991) in *Proteins: Structure, Dynamics and Design* (Renugopalakrishnan, V. et al. Eds.) ESCOM. Leiden, pp. 352–360.
11. Wüthrich, K. (1986) *NMR of Proteins and Nucleic Acids,* John Wiley & Sons, New York, NY.
12. Chou, P.Y. (1990) *Prediction of Protein Structure and the Principles of Protein Conformation* (Fasman, G.D., Ed.) pp. 549–586.
13. Chou, Y. and Fasman, G.D. (1974) *Biochemistry* **13**, 224.
14. Garnier, J., Osguthorpe, D.J. and Robson, B. (1978) *J. Mol. Biol.* **120**, 97.
15. Burgess, A.W., Ponnuswamy, P.K. and Scheraga, H.A. (1974) *Isr. J. Chem.* **12**, 239.
16. Lim, V.I. (1974) *J. Mol. Biol.* **88**, 873.
17. Nagano, K. (1973) *J. Mol. Biol.* **75**, 401.
18. Qian, N. and Sijnowski, T.J. (1989) *J. Mol. Biol.* **202**, 865.
19. Kyte, J. and Doolittle, R.F. (1982) *J. Mol. Biol.* **157**, 105.
20. Cohen, F.E., Abarbanel, R.A., Kuntz, I.D. and Fletterick, R. (1983) *Biochemistry* **22**, 4894.
21. Cohen, F.E., Abarbanel, R.A., Kuntz, I.D. and Fletterick, R. (1986) *Biochemistry* **25**, 266.
22. Ponder, J.W. and Richards, F.M. (1987) *J. Mol. Biol.* **193**, 775.
23. Richardson, J.S. (1981) *Adv. Protein Chem.* **34**, 168.
24. Wilmot, C.M. and Thornton, J.M. (1988) *J. Mol. Biol.* **203**, 221.
25. Brooks, B.R., Bruccoleri, R.E., Olafson, B.D., States, D.J., Swaminathan, S. and Karplus, M. (1983) *J. Comp. Chem.* **4**, 187.

Analysis and interpretation of tryptophan fluorescence intensity decays in proteins

F.G. Prendergast, Z. Bajzer, P.H. Axelsen and C. Haydock

Department of Biochemistry and Molecular Biology, Mayo Clinic/Foundation, Rochester, MN 55905, U.S.A.

Introduction

In principle, fluorescence spectroscopy is a powerful tool for studying protein structure and function. There are two different approaches to the use of fluorescence data for such purposes. First, fluorescence may be used simply as a signal of an event. In this instance, the investigator need only detect and quantify a *change* in the fluorescence. Such measurements can be conducted with great accuracy and sensitivity, especially if steady-state fluorescence intensity is the property being quantified. Second, the investigator may wish to use fluorescence data to explore the physicochemical character (i.e., the structure and dynamics) of the fluorophores environs within the protein matrix and solvent. The predicate for using fluorescence data in the latter manner is the well documented sensitivity of the fluorescence process to the physicochemical properties of the environment. However, if fluorescence data are to be used for this second purpose, *interpretation* of the data is intrinsically more problematic, especially if one is studying typical biologically derived fluorophores such as tryptophan or tyrosine. In part, this is because the photophysics of these fluorophores is intrinsically so complex, and their fluorescence is so sensitive to a wide variety of environmental factors. Inferences regarding structure drawn from fluorescence intensity decay data are difficult at best and, in fact, may be almost impossible given our present state of knowledge. Yet, under some circumstances, careful measurements of the fluorescence properties of a protein *can* yield quantitative data interpretable in terms of the protein's structure and dynamics, albeit within the bounds of well constructed physical (conceptual) models.

Of all the measurable fluorescence properties, the determination of the fluorescence intensity decay is potentially one of the most useful for studies of proteins. Intensity decay data may be employed either for the measurement of fluorescence lifetimes or for the determination of time-dependent fluorescence anisotropy decay from the decays of the polarized emission intensities. The latter affords direct assessment of fluorophore dynamics. The parameters calculated from such measurements should provide insight into structure, *per se*, and into the nature and extent of time-dependent processes in terms of both the amplitude and the rate of motion of the fluorophore itself and indirectly of the protein matrix in which the fluorophore is embedded. However, it is now clear that even in proteins with a single fluorophore, such as a protein with a single tryptophan residue, the measured fluorescence intensity decay, traditionally analyzed, frequently suggests multiexponential processes. The most straightforward interpretation of such a result is the existence of one conformer of the Trp ring, hence of the protein matrix, for each apparent fluorescence lifetime. Obviously, confidence in such an inference, especially where it has clear implications with respect to the physical and chemical behavior of the protein, is unavoidably influenced, and markedly so, by confidence in the accuracy of the data analysis. And, it does not matter which experimental technique is used in the determination of the intensity decay profiles; either time-correlated single photon counting or phase/modulation fluorometry may be used confidently for data acquisition [1-3]. The much more important consideration lies in the approach to data analysis.

There is another facet to the problem. The assumption that a single fluorescence decay implies a unique conformation of the Trp residue, carries with it distinct implications regarding the factors affecting the fluorescence decay process itself. The burden of proof is then on the investigator to justify the assumption of multiple conformers and to do so beyond mere assertion by demonstrating that more than one conformation indeed can exist, and by showing a rational structural basis for the fluorescence decay assumed to be from a particular conformer.

Our overall goal in this chapter is to examine the validity of inferences drawn from fluorescence intensity decay data about protein structure and dynamics. One objective is to show that there are clear and unavoidable limits to the procedures we use for data analysis. A second objective is to illustrate the usefulness of both molecular graphics and molecular dynamics simulations in the study of protein fluorescence especially when these techniques are used conjointly with experimental data.

The Problem

Our focus is on the fluorescence of tryptophan in proteins, in particular, proteins bearing a single Trp residue and of known crystal structure. The data given in Table 1 for four such proteins illustrate the problem: of the three proteins for which fluorescence lifetimes are given, only RNAse T_1 at pH 5.5 apparently exhibits a single exponential decay for the fluorescence of its single Trp residue (Trp[59]). For the other proteins, the tryptophan fluorescence intensity decay yields two or more exponential terms when analyzed assuming a discrete exponential model. The extent of heterogeneity in lifetimes is especially striking for scorpion neurotoxin variant 3 (SN_3) and phospholipase A_2 (PLA_2). Interpreted by commonly used methods, these results would argue that there are two conformers for RNAse T_1 at pH ≥ 7.0 probably three for PLA_2 and as many as four for SN_3. We must ask: can these inferences linking individual lifetimes with unique protein conformations be justified?

Analytical Considerations

The first issue that needs to be settled is whether the methods of analysis used can reliably detect and quantify both the number of components and the actual parameters (pre-exponential factors and lifetimes) of the intensity decay. In this matter, the problem exists for both time-correlated single photon counting (TCSPC) and multifrequency phase fluorometry (MPF) data, albeit with some differences in analysis based on the peculiarities of the particular method. For the sake of this chapter, we focus on TCSPC data only. A detailed discussion of the overall problem of analysis of TCSPC data is given in Bajzer and Prendergast [4].

The extraction of fluorescence decay lifetimes from time-correlated single photon counting (TCSPC) measurements is a difficult problem of long standing. The issue centers on two numerically ill-conditioned problems: deconvolution with respect to the instrument response function (IRF), and parameter estimation of multiexponential functions. Favorable characteristics of the problem relate to the possibility that the level of noise in TCSPC data can be controlled, and to the fact that one can obtain dense sampling of the fluorescence intensity decay curve.

An important consideration in the analysis of experimentally determined multicomponent decay curves contaminated by noise, is the ability to separate the individual components (separability, or resolvability) in terms of individual pre-exponential factors (or fractions) and lifetimes. For a given number of sampled data, a given level of noise, and a particular algorithm, there is a *critical lifetime ratio* and a *critical fraction ratio* beyond which the required parameters cannot be resolved. In general, if the lifetime ratio is smaller than the critical lifetime ratio, or likewise, if the fraction ratio is greater than the critical fraction ratio, the corresponding two decay components are almost certain to be detected by the algorithm as a single component.

The method of analysis should also be *accurate*, i.e., the values obtained from the analysis for the parameters must be close to the 'true' values. *A priori*, in a protein system one cannot *know* what the 'true' values are, but in ideal circumstances the latter can be known from experiments in which the contribution of each component has been determined prior to creation of a mixture of the two components. For example, this may be possible given a wild-type protein bearing two Trp residues and two mutant forms of the protein each bearing one or other of the Trp moieties, with the proviso that some structural signature, e.g. an ^{1}H NMR fingerprint, shows no major difference in the tertiary structure of the mutated forms when compared to the wild-type protein. But more realistically, one must assume generally that the true values of parameters are those obtained by the analysis of the *measured* data. Thereafter, using these values, one can generate synthetic data, i.e., do simulations. Subsequently, applying the same mathematical analysis to the simulated data, one should be able, *a posteriori*, to conclude whether the method of analysis was sufficiently accurate.

Finally, the statistical criteria which one uses to accept or reject values of pre-exponential factors and lifetimes are important. Traditionally, most investigators have relied almost exclusively on the use of the reduced χ^2 statistic and the Durbin-Watson factor to evaluate the quality of fit for TCSPC data irrespective of the precise algorithm employed in the analysis [3]. Bajzer et al. [5] have also described an approach using Padè approximants and Laplace transformations for the analysis of fluorescence intensity decays (this method has recently been championed for analysis of multiexponential functions [6,7]). The Padè-Laplace method does not rely on statistical criteria and was shown to recover fluorescence parameters well. More recently, Bajzer et al. [8] have re-examined the use of the maximum likelihood method for these analyses comparing it to the more commonly employed least-squares approach. Because noise in TCSPC data is characterized by Poisson statistics, the MLM method is the more appropriate choice for analysis since use of the LS approach is predicated on an assumption that noise in the data is Gaussian. The distinction between methods is not significant when the ratio of lifetimes is large (say > 2.5) but becomes progressively more marked as that ratio falls. Bajzer et al. [8] have proposed that given the ill-conditioned nature of the data and hence the complexity of the analysis, other statistical criteria should be considered. They introduced a likelihood ratio test, and detectability and separability (resolvability) indices to be used in addition to the reduced χ^2 statistic. The value of using these several criteria lies in the realization that if *all* conditions are satisfied the accuracy of the derived fluorescence parameters may be assumed with confidence. Data for the fluorescence lifetimes of scorpion neurotoxin variant 3 and Ribonuclease T_1 (RNAse T_1) from Aspergillus oryzae analyzed by the MLM method [8], are given in Table 2.

An important feature of the data in Table 2 is that the simulations show that there is enough information in the experimental data to recover four discrete lifetime components for SN_3 and two lifetime components in RNAse T_1 at *both pH 5.5 and pH 7.5* (cf. Table 1). The data on RNAse T_1 results are particularly intriguing because they suggest that a species contributing (in terms of intensity weight) as little as 1% of the signal is both detectable and quantifiable. At this juncture we need to pause and ask ourselves: what do these data mean?

Interpretive Considerations

Let us assume for the moment the validity of the parameters derived. How should they be interpreted in terms of physical models of the protein in question? A trivial inference would be that the tryptophan side chain in SN_3 must adapt at least four distinct conformations and that the extent of quenching of the Trp fluorescence is different for each conformer. With RNAse T_1 a similar argument may be proffered

Table 1 *Fluorescence lifetimes of proteins bearing a single tryptophan residue*

| Protein | Fluorescence lifetimes | | | | | |
	τ_1	f_1	τ_2	f_2	τ_3	f_3
RNAse T_1(pH 5.5)	3.0	1.0				
RNAse T_1(pH 7.5)	3.7	0.94	1.6	0.06		
SN_3	0.08	0.08	0.44	0.6	2.0	0.32
PLA_2	6.62	0.66	2.29	0.25	0.66	0.09

Lifetimes, denoted τ, are in nanoseconds. The intensity-weighted fractions of the fluorescence signal contributed by each species (according to a discrete exponential model) are denoted f_1, f_2 and f_3.

Table 2   *Fluorescence lifetimes of proteins bearing a single tryptophan residue analyzed by use of the maximum likelihood method of Bajzer et al. [8]*

Protein	Lifetimes							
	τ_1	f_1	τ_2	f_2	τ_3	f_3	τ_4	f_4
RNAse T_1(pH 5.5)	3.99	0.99	0.9	0.001				
RNAse T_1(pH 7.5)	3.66	0.94	1.6	0.06				
SN_3	0.035	0.02	0.25	0.26	0.66	0.58	3.1	0.14

Lifetimes are denoted by τ and intensity weighted fractional contributions of each putative species by f.

except that there would be only a small contribution (approximately 1% at pH 5.5, approximately 6% at pH $\geq$ 7.0) from one of the species to the fluorescence signal. Such interpretations are commonplace, but how plausible are they?

From first principles, one may assume that a shortened fluorescence lifetime for Trp embedded in a protein is determined primarily by dynamic fluorescence quenching mediated by moieties in the protein matrix. Because the mechanisms for deactivation of the Trp excited state are not fully understood, we cannot even define what we mean by dynamic quenching in this microscopic environment. Certainly, Smoluchowski-Stern-Volmer formalisms cannot be meaningfully applied. However, if we assume, for the sake of argument, that electron transfer from the Trp to nearby proton acceptors – primarily in the protein matrix although bulk solvent (water) could also be an effective electron sink – then the fluorescence lifetime evinced will depend on the efficiency and kinetics of the electron transfer process. Petrich et al. [9,10] have suggested that electron transfer from the indole ring is the principal mechanism for deactivation of the excited state of Trp. The relative orientations of the donor and acceptor could also be important. Since we may reasonably assume

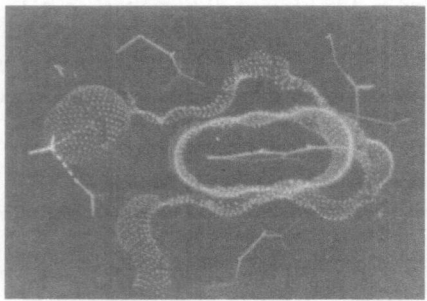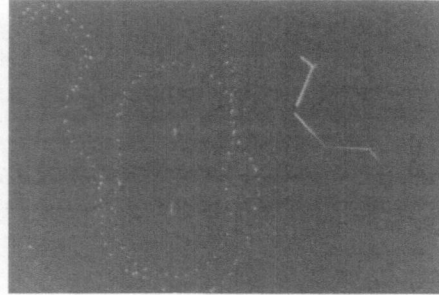

Fig. 1. *Molecular graphics depictions of the 'pocket' for the indole side chain of Trp[59] in RNAse T_1. The thin section through the Trp ring and its immediate environs in the protein matrix shows that the indole moiety is tightly packed with little room for motion. There is also a narrow channel from the surface of the protein with dimensions sufficient for a single water molecule to move through and contact the indole ring. A water molecule is shown posed on the lip of the channel. (Details are provided in Axelsen and Prendergast [12]).*

that the protein matrix exhibits a broad spectrum of dynamics, some on the same time scale as the fluorescence intensity decay, the relative orientations of the electron donor (the excited indole) and the putative electron acceptors cannot be constant. Given these considerations, the fluorescence lifetimes detected will depend on: (i) how quickly the environment of the fluorophore is averaged through molecular motion subsequent to the instant of photon absorption; (ii) the rate of interconversion between the putative conformers; (iii) whether long-range (fluorescence) quenching interactions occur.

From an experimental standpoint, the questions posed implicitly above are very difficult to answer. Our approach has been to turn to molecular graphics depictions and to molecular dynamics simulations for help. For example, details of molecular dynamics simulations of RNAse T_1 have been published [11–14]; the reader is referred to these papers for methodologic information and for the complete data set. For our immediate purposes the molecular graphics depiction in Fig. 1 suffices. This shows that the indole side chain of Trp[59] in RNAse T_1 is embedded in a tightly packed region of the protein with little room for ring motion. Moreover, water has minimal access to the indole ring both in the 'static' crystal structure *and* in a molecular dynamics simulation [12]. Apart from a very limited contact with water, the only other polar contact made by the indole moiety of Trp[59] in RNAse T_1 is with the carbonyl oxygen of a proline residue in the pocket in which the indole is found. This means that the Trp 'pocket' is largely apolar within 4.5 Å of the aromatic ring. Even a simple-minded analysis of this structure is sufficient for us to infer that with restricted motion of the Trp side chain within a predominantly non-polar pocket, the probability is high that the environment of the excited Trp[59] side chain would be 'averaged' very rapidly after photon absorption, i.e., there is a high probability that the fluorescence intensity decay would be mono-exponential. This is essentially supported by the

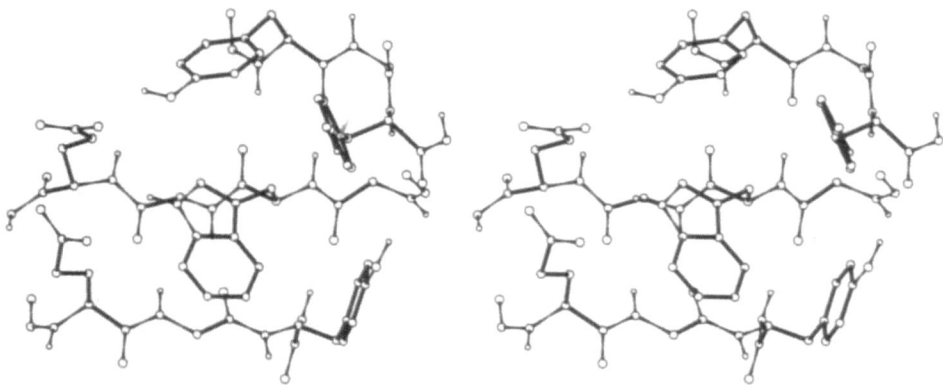

Fig. 2. Stereo drawings of the tryptophan residue of SN₃. The view is from 'above' and shows the indole ring to be exposed to solvent and to be encircled by both carboxylates from glutamic acid residues and tyrosine side chains. A detailed description is given in Haydock et al. [17].

fluorescence lifetime data for RNAse T_1 at pH 5.5. But what then is the significance of the 1% short lifetime component at pH 5.5 or the 6% component at pH 7.5? We cannot say, but there is a nagging consideration, namely, is the protein 'pure'? This problem will be addressed in more detail, below.

Another problem is generated by the possibility of long-range interactions. For example, bound 2'-GMP significantly quenches the fluorescence of Trp[59] in RNAse T_1 [15]. These authors have ascribed this to electron transfer from the indole to the guanine moiety, a plausible explanation but one which will be difficult to prove. However, we have found (Hedstrom and Prendergast, unpublished data) that phosphate ion by itself quenches the Trp fluorescence in RNAse T_1 at concentrations at which we assume the ion should be bound at the active site of the enzyme, i.e., at about 9 Å from the indole ring. It is improbable that phosphate would also act as an electron acceptor, but we can find no evidence from molecular graphics depictions that the bound ion (phosphate) would induce changes in the environment of Trp[59] in RNAse T_1 which might mediate the changes in Trp lifetime. Thus, for the moment, the molecular basis for the heterogeneity of RNAse T_1 Trp lifetimes with or without ligands is unclear. Certainly, the existence of two distinct rotameric conformations for the Trp side chain in RNAse T_1 seems highly implausible [16].

The situation with SN_3 is even more perplexing. The molecular graphics depiction in Fig. 2 shows the principal species in the protein matrix within 5 Å of the indole ring. Bulk water has clear access to the indole moiety. Thus, overall the potential is high for a wide variety of interactions of the Trp side chain with charged groups, with aromatic side chains and with several other polar moieties including water molecules. Again the issue comes down to whether multiple conformers exist. Handwaving arguments based on the location of the Trp side chain on the surface of the protein are temptingly simple, but clearly not good enough. It is obvious that we need to have some quantitation of the ability of the Trp ring to 'flip' into different orientations, i.e., into different (microscopic) environments and that we need to know what the energy barriers might be. With a knowledge of the latter, one can calculate what might be the rate of thermally driven interconversion between conformers. Unfortunately, there is no simple way to determine the likelihood of multiple conformations.

Haydock et al. [17] have used molecular dynamics simulations coupled with thermodynamic perturbation techniques to show that one *can* identify two energetically reasonable configurations of the Trp side chain in SN_3. One of these corresponds to the X-ray crystal structure of the protein [18] and to a structure deduced from NMR data [19]. The other configuration has not yet been specifically sought through experiment. Figure 3 shows the result of the thermodynamic perturbation calculations and Fig. 4 is a schematic showing how a Trp side chain might move within a hypothetical double potential well. This is an admittedly early, but highly encouraging result, particularly as the pattern of molecular interactions is distinctly different in the two conformers. Moreover, although the calculated potential energy barrier (approximately 8.5 kcal/mol) between the two conformers makes barrier crossing relatively infrequent at 25°C, the temperature at which the fluorescence data given in Table 2 were gathered, the model suggested by these

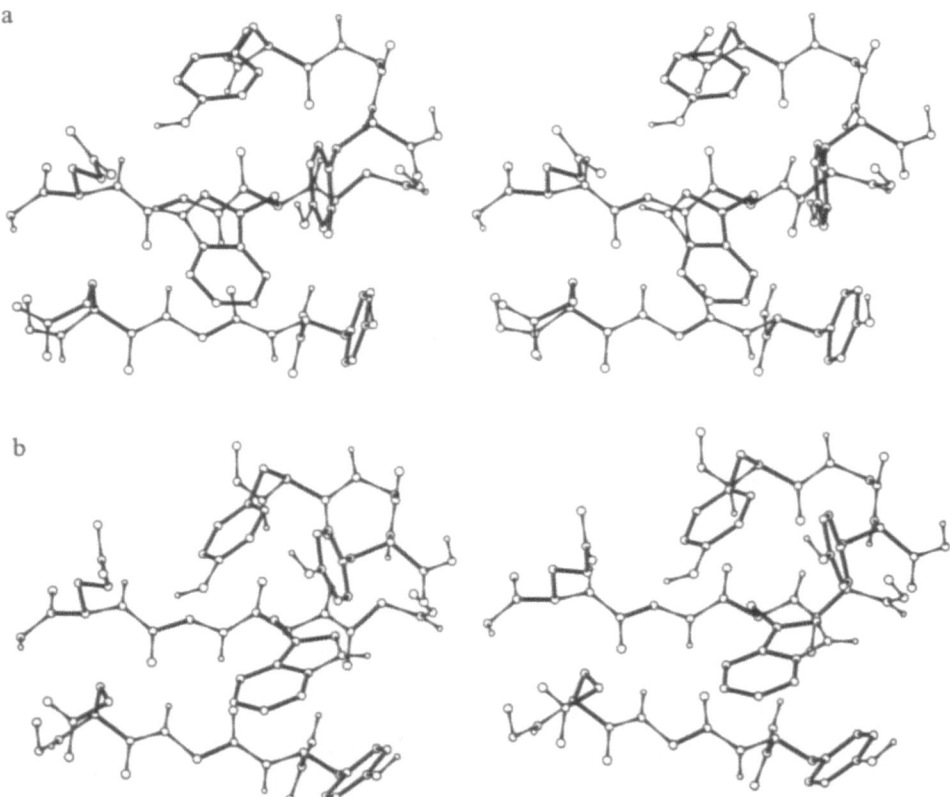

Fig. 3. Stereo drawings of the tryptophan residue in SN₃ subsequent to molecular dynamics simulations and umbrella sampling [14]. Figure 3a shows the fit to the average umbrella sample structure in the crystallographic well. Figure 3b depicts the average umbrella sample structure of the new rotational isomer predicted by the mathematical simulations. The simulations suggest that the tryptophan side chain can plausibly occupy two potential wells even though thermodynamically the crystallographically identified well dominates.

simulations could explain partially the origin of the multi-exponential decay. This is, of course, only conjecture at this stage because we do not know the pathways or mechanisms for de-excitation of the excited indole in SN₃. Until these dissipative mechanisms *are* identified we will have to continue to speculate, but the thermodynamic perturbation calculations done on SN₃ simulations show clearly how difficult it is to 'prove', even through *simulation*, the feasibility of multiple Trp conformers.

The final example we wish to consider is PLA₂. The heterogeneity of porcine pancreatic PLA₂ Trp fluorescence lifetimes is well documented [20,21]. One assumption has been that because the single Trp is on the surface of the protein it would be intrinsically more mobile and therefore be more subject to interactions with the protein matrix and with bulk solvent. At first blush this seems a reasonable proposition; upon closer examination, it is not. From an extensive series of time-

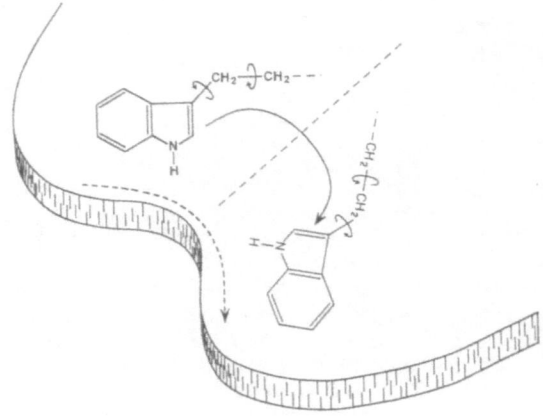

Fig. 4. Schematic drawing depicting rotation of the side chain of a tryptophan residue within a double well. The hypothesis is that the side chain can traverse the barrier during the fluorescence intensity decay time and that molecular interactions with the excited state of the indole moiety differ in the two wells.

resolved anisotropy data determined from modulation ratio measurements through use of multifrequency phase fluorometry, Axelson et al. [11] have shown convincingly that the Trp residue in PLA$_2$ exhibits little librational freedom. The experimental measurements have been prompted by results from molecular dynamics simulations which had suggested that the Trp ring was not free to 'lift' off the surface and rotate. Rather, the χ^1 and χ^2 dihedral angles, non-bonded interactions of the indole with the protein matrix and interactions with water molecules were such as to severely constrain motions of the aromatic ring. Further, umbrella sampling calculations on PLA$_2$, akin to those described earlier for SN$_3$, suggested that the Trp ring remained in a single, narrow, potential well. These data still permit the Trp ring in PLA$_2$ to sample a fair number of local environments during its excursions along the protein surface, but there is no way to determine which interactions might give rise to the fluorescence lifetimes noted in Table 1.

Practical Considerations

At this stage we need to pause and re-consider the issue of protein 'purity'. That word assumes new significance given the apparent sensitivity and probable accuracy of analytic techniques for analysis of TCSPC data. For example, the results with RNAse T$_1$ given above are disquieting because we cannot think of an experiment to detect 1% of a protein which may be simply *configurationally different* from the other 99% of the sample. Thus, a protein may be adjudged 'pure' by all the usual experimental criteria and yet not be 'spectroscopically' pure because a small percentage of the protein molecules have altered tertiary structure, e.g., are partially denatured. This is a completely feasible scenario, one that could obviously have profound influence on interpretation of the measured fluorescence intensity decays. A 1% 'contamination' could have marked effects depending on the relative quantum yields of the (putative) 'properly' and 'improperly' folded species. There is no simple way of avoiding this problem. For enzymes one might use active-site titration and hope that a 100% recovery of active sites is achieved. For protein enzyme inhibitors one might likewise hope for a 100% inhibition at the appropriate molar ratios for proper stoichiometry. To minimize the problem, one must also exercise great care with respect to solvent conditions from one set of measurements to the next. Finally, one should where possible use ¹H-NMR signatures to ensure batch-to-batch identity in protein samples or to determine the stability of protein preparations over time. In the final analysis, however, it will always be difficult to interpret the physical significance of an apparent component contributing only a small fraction to the total fluorescence signal.

Fluorescence Lifetime Distributions

In principle, even a single exponential decay may be mathematically described in the form of a distribution of states albeit a very narrow distribution centered on the value of the single exponential. A number of authors have in recent years suggested that fluorescence lifetimes in proteins and other systems might be better described in terms of distributions rather than as discrete values [16,20,22–24]. The foregoing discussion shows that there is an unavoidable uncertainty inherent in the analysis of presumably multi-exponential decays which is expressed as the resolvability. This means that even when a single exponential is derived from an intensity decay curve the uncertainty in the analysis requires at least an implicit assumption that a narrow distribution of states could exist and be undetectable. From a practical standpoint such an evaluation may seem like quibbling; indeed, such a complaint may be valid if a single exponential decay is observed for the Trp fluorescence of a protein. However, with a clear display of heterogeneous lifetimes the physical models implied by assuming discrete exponentials and by asserting that the lifetimes are distributed are really quite different.

The model assuming discrete lifetimes requires quite distinct conformational states – two states for two lifetimes, three for three and so on. Nothing can be said or even implied about interconversion between states, and one must posit, in fact, that the individual states are stable at least on the time scale of the fluorescence intensity decay. As noted above, when there are as many as four apparent exponential terms, the existence of four unique conformational states stretches credibility somewhat. By asserting instead that there is a continuous or quasi-continuous distribution of lifetimes, one is implying a broad distribution of conformational substates. The width of the distribution would, in this construct, speak to the probable rate of interconversion between substates. Therefore, the temperature dependence of the width would be related to the activation energy of those interconversions. Thinking photophysically, one could propose that at the moment of excitation, the tryptophan ring might be located in any number of orientations even though a particular configuration might dominate because of preferential non-bonded interactions of the aromatic ring with the protein matrix. The form of the subsequent fluorescence decay would then depend on: 1) the number of dissipative mechanisms existing in that particular protein; 2) the efficiency and rate of the interactions with the excited indole moiety – both short (e.g., < 4.5 Å) and long range; and 3) how such interactions evolve in the course of thermally driven protein dynamics. We would expect that interaction geometries between excited fluorophore and dissipative sinks are continuously changing on the time scale of the fluorescence intensity decay. Described in this fashion, the fluorescence intensity decay may well be non-exponential and intuitively a distribution of lifetimes would seem a more plausible way to describe the intensity decay.

The data of Alcala et al. [23] show that such decay data can be fit well by use of distributions, but it is important to note that by χ^2 statistical criteria the fits *per se* are not necessarily better when a distribution of states is assumed compared to when a discrete model is assumed. An investigator's bias will in most instances determine how the data are presented and that bias will in turn be determined by the measure of confidence one has in the method of analyzing the data and by the validity of the statistical criteria employed in determining quality of fit. One weakness of the analytic approach of Alcala et al. [23] was the focus on use of simple distribution models to describe protein Trp fluorescence, i.e., models assuming primarily either Lorentzian or Gaussian distributions. One cannot assume, *a priori*, that either form should exist. For this reason, approaches such as the maximum entropy method [25] or the exponential series method proposed by Siemiarczuk et al. [24] should also be applied to the analysis of data. In the final analysis, however, how an investigator chooses to interpret fluorescence intensity decays should rest on how convincing is the physical model presented to rationalize the data from a particular protein.

Conclusions

Throughout this chapter, we have tried to illustrate contemporary problems in the analysis and interpretation of tryptophan fluorescence intensity decays in proteins. We have deliberately focused on this one issue and have omitted discussion of fluorescence spectra or anisotropy. We have also not discussed the physical meaning of the actual fluorescence lifetime(s) a protein might exhibit – such an evaluation would require an extensive consideration of indole photophysics including the complexity of the overlap of the ^1L_a and ^1L_b absorption transitions and the character of the resulting excited state of the indole. What we have presented shows that the analysis of fluorescence intensity decay data must be treated rigorously with due consideration given to the resolvability of the analytical method and the statistical criteria one employs to determine how well data have been fit. Molecular dynamics simulations and molecular graphics depictions are likely to be invaluable when one is attempting to justify a particular experimental or theoretical model. In the final analysis, however, the correlation between protein structure and fluorescence lifetime will be difficult to make until more is understood about the details of Trp photophysics and of the mechanisms whereby moieties in the protein matrix influence the photophysics of the embedded indole ring.

Acknowledgements

This research was supported by GM34847 of the PHS. We thank Ms. Jill Kappers for typing the manuscript and Mr. Peter Callahan for preparation of the figures.

References

1. Hedstrom, J., Sedarous, S.S. and Prendergast, F.G. (1988) *Biochemistry* **27**, 6203.
2. Gratton, E. and Limkeman, M. (1983) *Biophys. J.* **44**, 315.
3. O'Conner, D.V. and Phillips, D. (1984) Academic Press, New York.
4. Bajzer, Z. and Prendergast, F.G. (1991) in *Methods in Enzymology* (Brand, L. and Johnson, M., Eds.), in press.
5. Bajzer, Z., Sharp, J.C., Sedarous, S.S. and Prendergast, F.G. (1990) *Eur. Biophys. J.*, **18**, 101.
6. Yeramian, E. and Claverie, P. (1987) *Nature (Lond.)* **326**, 169.
7. Aubard, J., Levoir, P., Denis, A. and Claverie, P. (1987) *Comput. Chem.* **11**, 163.
8. Bajzer, Z., Therneau, T.M., Sharp, J.C. and Prendergast, F.G. (1991) Submitted for publication to the *Eur. Biophys. J.*
9. Petrich, J.W., Chang, M.C., McDonald, D.B. and Fleming, G.R. (1983) *J. Am. Chem. Soc.* **105**, 3824.
10. Petrich, J.W., Longworth, J.W. and Fleming, G.R. (1987) *Biochemistry* **26**, 2711.
11. Axelsen, P.H., Gratton, E. and Prendergast, F.G. (1991) *Biochemistry* **30**, 1173.
12. Axelsen, P.H. and Prendergast, F.G. (1989) *Biophys. J.* **56**, 43.
13. MacKerrell, A.D. Jr., Rigler, R., Nilsson, L., Hahn, U. and Saenger, W. (1987) *Biophys. Chem.* **26**, 247.
14. MacKerrell, A.D. Jr., Nilsson, A.D., Rigler, R. and Saenger, W. (1988) *Biochemistry* **27**, 4547.
15. Chen, L.X.-Q., Longworth, J.W. and Fleming, G.R. (1987) *Biophys. J.* **51**, 865.
16. James, D.R., Demmer, D.R., Steer, R.P. and Verrall, R.E. (1985) *Biochemistry* **24**, 5517.
17. Haydock, C., Sharp, J.C. and Prendergast, F.G. (1990) *Biophys. J.* **57**, 1269.
18. Almassy, R.J., Fontecilla-Camps, J.C., Suddath, F.L. and Bugg, C.E. (1983) *J. Mol. Biol.* **170**, 497.
19. Nettesheim, D.G., Klevit, R.E., Drobny, G., Bugg, C.E., Watt, D.D. and Krishna, N.R. (1989) *Biochemistry* **24**, 1548.
20. Alcala, J.R., Gratton, E. and Prendergast, F.G. (1987) *Biophys. J.* **51**, 925.
21. Ludescher, R.D., Wolwerk, J.J., de Hass, G.H. and Hudson, B.S. (1985) *Biochemistry* **24**, 7240.
22. Alcala, J.R., Gratton, E. and Prendergast, F.G. (1987) *Biophys. J.* **51**, 587.
23. Alcala, J.R., Gratton, E. and Prendergast, F.G. (1987) *Biophys. J.* **51**, 597.
24. Siemiarczuk, A., Wagner, B.D. and Ware, W.R. (1990) *J. Phys. Chem.* **94**, 1661.
25. Livesey, A.K. and Brochon, J.-C. (1987) *Biophys. J.* **52**, 693.

Normal modes of crambin and molecular dynamics for structure prediction

Martha M. Teeter[a], Usha Rao[a] and David Case[b]

[a]Department of Chemistry, Boston College, Chestnut Hill, MA 02167, U.S.A.
[b]Department of Molecular Biology, Research Institute of Scripps Clinic,
La Jolla, CA 92037, U.S.A.

Introduction

Molecular dynamics methods have been useful in studying the structure and function of proteins. We consider here a comparison for proteins of the dynamical properties of different empirical potential functions [1] and the usefulness of molecular dynamics in improving prediction of homologous proteins that have sequence deletions (Rao and Teeter, unpublished).

Both studies resulted from analysis of crystals of the plant seed protein crambin, which are exceptionally well ordered and diffract farther than any protein that has yet been crystallized (0.83 Å resolution [2]). Crambin is homologous to 14 plant toxins [3], and the structures of two of the toxins have been predicted from it [4]. Crambin's small size (46 amino acids), neutrality (two positive and two negative charged side chains) and the fact that nearly all the water molecules are located in the crystal at 130 K make it ideal for use in simulations and comparisons of different potential functions [1,5].

Calculation of normal modes provides an effective method to compare the dynamical properties of different empirical potential functions. Various quantities such as the frequency distribution and the magnitude of protein atom fluctuations along the chain at 300 K can be compared for these potentials. Normal modes assume harmonic atomic motion [6]. Anharmonic motions can be included as well if normal modes are back-calculated from the atomic fluctuations in a molecular dynamics simulation [7].

Molecular mechanics [4] and dynamics can also aid in structure prediction of proteins. A particularly difficult problem encountered in prediction of homologous proteins is modeling deletions and insertions from the starting protein. We have successfully employed molecular dynamics in such a situation to improve a structure prediction (Teeter and Rao, unpublished), where the crystal structure disagreed with the predicted model used to solve the structure [8].

Results and Discussion

Comparison of dynamical properties of potential functions

Eight different variations of potential functions were compared using normal mode analysis. These were CHARMm 17,18 [9] and 19 [10], AMBER united atom [11] with the dielectric constant (ε) = r or 4r, AMBER all atom [12] with ε = r or ε = a sigmoidal function of r [13], AMBER/OPLS [14] and DISCOVER [15]. ε is important in Coulombic interactions, where the energy is $q_1q_2/\varepsilon r$. r Is the distance between charges q_1 and q_2. An additional calculation was done with AMBER/OPLS using a 5 Å shell of water. Normal modes were calculated in vacuo from an extensively minimized crambin structure for each potential. A 100 picosecond (ps) trajectory was also calculated with CHARMm 17,18 and quasiharmonic analysis was done.

The starting model in each case was the Pro[22], Leu[25] form of crambin, as refined at 300 K and 0.945 Å resolution (Teeter and Hendrickson, unpublished). The root-mean-square (rms) deviations from the X-ray structure for the energy-minimized models were 0.7 Å or more. However, minimization was done to an energy gradient of less than 2×10^{-4} kcal/(mol Å) which is much smaller than the usual stopping point of 0.1 kcal/(mol Å).

The largest deviations of the energy-minimized structure from the X-ray were for the CHARMm and AMBER/OPLS force fields (1 Å vs 0.7 for the others). The latter force field was developed for solution and may be inappropriately applied in vacuo. Also, the nonbonding cutoff procedure was different in the various force fields used [1], and it has been recently shown that this can substantially influence the rms deviation from the starting model [16]. We do not feel the rms deviations are the best criteria to evaluate the potentials [5]. Deviations among the results are greatest in the most mobile region of the molecule (Fig. 1, upper right-hand corner). This region has few intramolecular hydrogen bonds to the rest of the protein and is mainly connected by a disulfide bridge.

For the various potential functions, the frequency distributions are remarkably similar below 140 cm^{-1} (Fig. 2). These are the frequencies that determine the

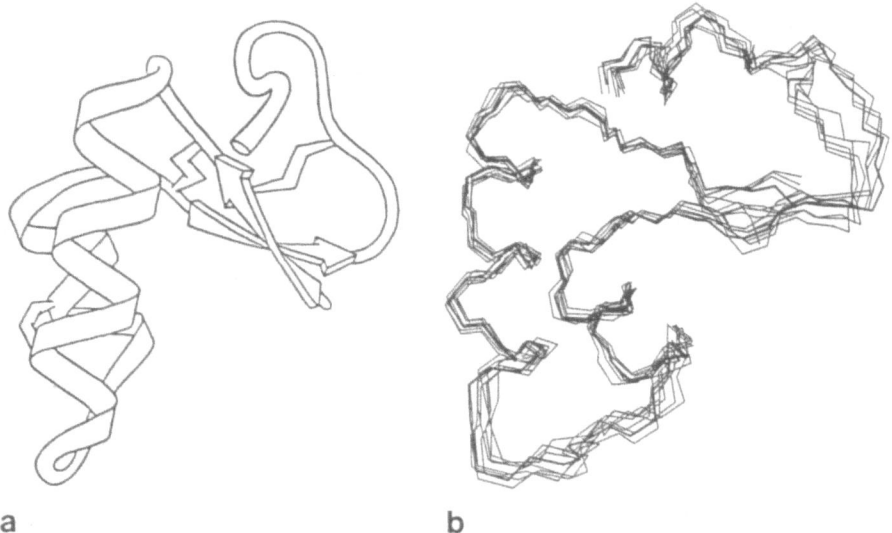

a   b

Fig. 1. Backbone of crambin. a. Ribbon diagram of crambin from Jane Richardson, with modifications by Marc Whitlow. β-sheets are arrows and disulfides are lightning rods. b. Superimposed backbones of energy-minimized and X-ray (bold line) structures.

primary vibrational motion of the molecule (low frequency, high amplitude). Two results that differ from most are the one with ε = 4r (top line – dotted) and normal modes from Levitt et al. [17], that are based only on dihedral angles motion (connected filled circles). The former has more modes than the average due to decreased electrostatic interaction energies at long distances. This leads to larger fluctuations than most force fields, as we shall see below. The latter dihedral-based normal mode treatment has fewer modes below 140 cm^{-1} than other calculations. Above 120 cm^{-1}, where bond and angle fluctuations contribute, the distribution flattens off.

The quasiharmonic frequency distribution calculated for the entire 100 ps trajectory (not shown) is very similar to the harmonic ones. The only difference is that there are frequencies in the 1700–2600 cm^{-1} range in the quasiharmonic but not the harmonic normal modes.

There is a strong dependence of the quasiharmonic frequency distribution on the sampling time for the trajectory (Fig. 3). Shorter sampling times (5–10 ps) give many fewer 20–140 cm^{-1} frequency modes than in the harmonic case. For 'windows' of 50 ps, the harmonic values are approached. But at 75 and 100 ps, the number of modes below 140 cm^{-1} is even larger than for 50 ps and may not have even converged in 100 ps.

The backbone fluctuations at 300 K from the harmonic calculations are in

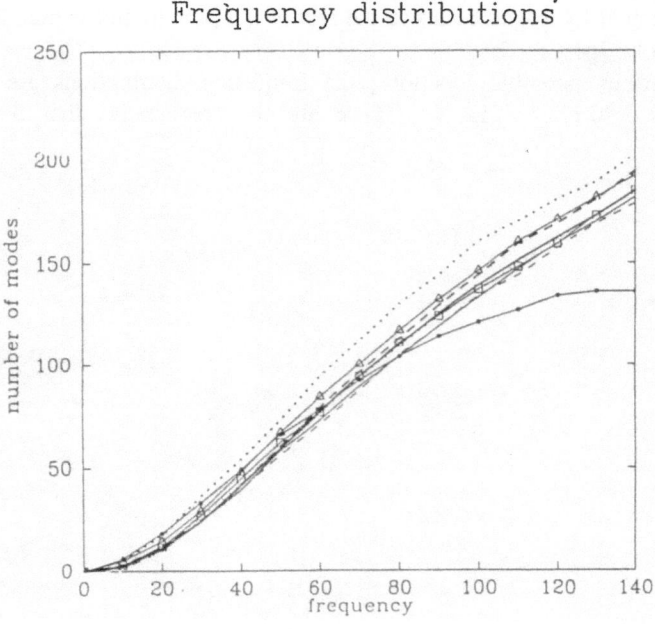

Fig. 2. Harmonic frequency distributions. CHARMm 17,18 United atom (UA), ε = 1: heavy solid line; CHARMm 19 UA, ε = 1: heavy dashed line; AMBER UA, ε = r: light dashed line; AMBER all atom (AA) ε = r: light dots; AMBER UA ε = 4r: heavy dots; AMBER AA sigmoidal dielectric: squares; AMBER/OPLS UA ε = 1: light solid line; DISCOVER AA, ε = 1: triangles. The filled connected circles are from ref. 16.

remarkable agreement (Fig. 4). The X-ray results from vibrational factors, shown for comparison, include not only the internal vibration modeled by the normal modes but also the libration and translation of the molecule in the crystal. Thus the X-ray average value should be and is larger than the fluctuations from normal mode analysis.

Two harmonic normal mode calculations had fluctuations that deviated considerably from the average (Fig. 5). The first employed $\varepsilon = 4r$. It had more low frequency modes than most (see above) and reduced hydrogen bond interactions, which are principally electrostatic. Figure 5 shows that its fluctuations (dashed line) are considerably greater than the other normal-mode-derived fluctuations (AMBER/OPLS which is the solid line). The second showed much less fluctuation for the AMBER/OPLS model with a shell of water around it (dotted line). Here the minimized water has no thermal disorder and fluctuations must increase the local strain energy. Both calculations show that the dynamical properties are sensitive to major changes in application of potential functions, much more than to the differences in the potential function parameterizations.

The quasiharmonic fluctuations (not shown) are generally similar to those for harmonic normal mode calculations (Fig. 4). However, in the most mobile region of the structure (residues 36–44, see also Fig. 1), the fluctuations are approximately 50% greater for the quasiharmonic calculations and are also larger here in the X-ray structure (Fig. 4). This region may include anharmonic contributions to the motion. Still, the differences are not overwhelming and the most mobile regions of the structure agree in the two approaches, as well as with the X-ray structure.

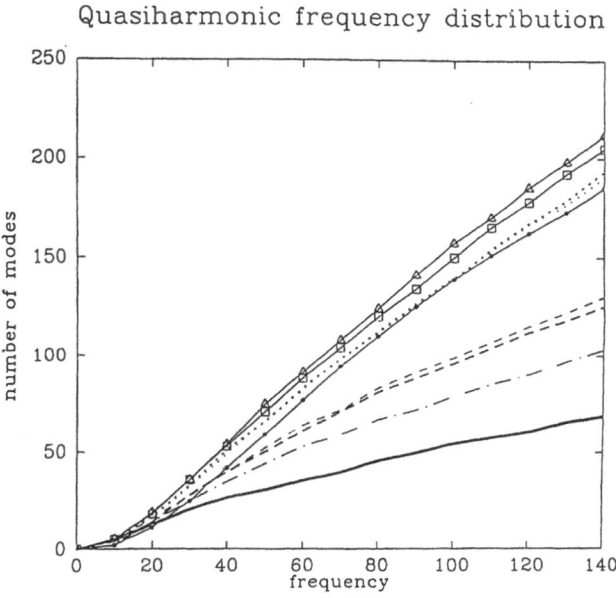

Fig. 3. Quasiharmonic frequency distributions for different simulation lengths. Solid line for 5 ps modes; dot-dash for 10 ps; dashed for 15 ps; dotted line for 50 ps; squares for 75 ps; and triangles for 100 ps. Solid circles give harmonic results for CHARMm 17,18.

Harmonic backbone fluctuations

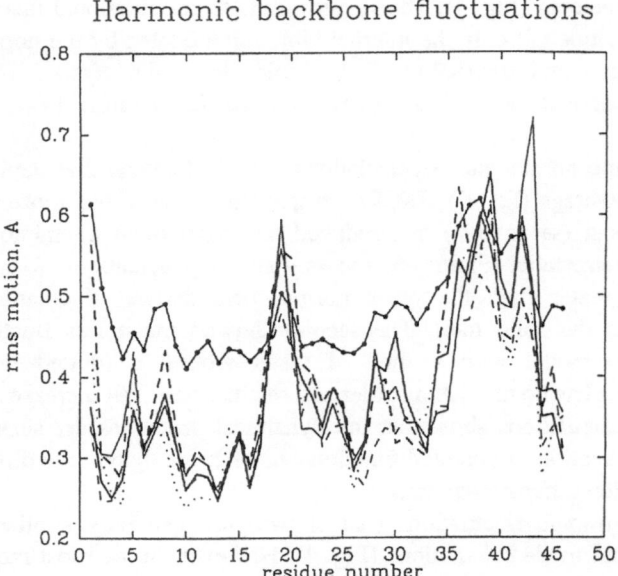

Fig. 4. Root-mean-square fluctuations in backbone atoms versus residue number. Lines coded as in Fig. 2. Solid line connecting circles are derived from the X-ray vibrational factors (B values).

Harmonic backbone fluctuations

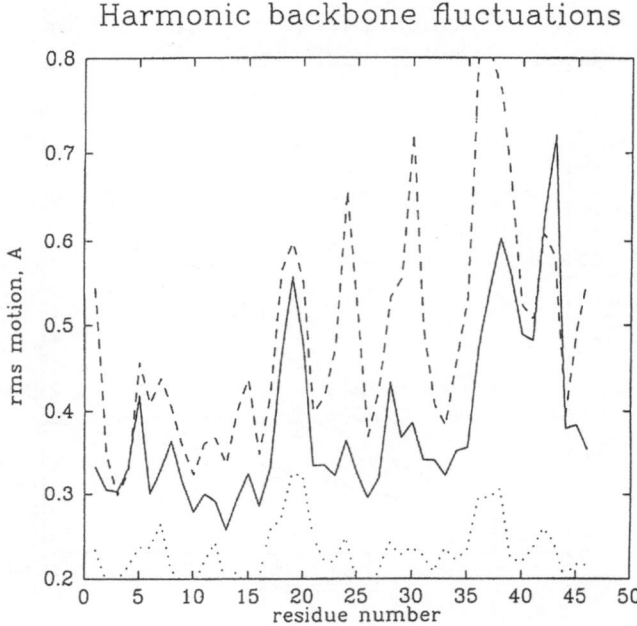

Fig. 5. Same as Fig. 4. AMBER UA with $\varepsilon = 4r$: dotted line; AMBER/OPLS: solid line; and AMBER/OPLS with water shell: dotted line.

In conclusion, we have seen that there is considerable agreement between dynamical properties of proteins from normal mode calculations and from molecular dynamics simulations. Three points should be made about these calculations, however. First, decreasing the strength of hydrogen bonds with $\varepsilon = 4r$ greatly influences the dynamics of the molecule (Fig. 5). Calculation with a sigmoidal dielectric which does not reduce these interactions shows much more comparable dynamical properties. Second, if bond and angle fluctuation are not allowed, there are many fewer modes above 60 cm^{-1} (Fig. 2). Third, the correspondence between normal modes and quasiharmonic calculations from molecular dynamics holds for sampling times longer than 50 ps. Shorter windows lead to significant loss of density of states in the 20–140 cm^{-1} region.

Improving predictions with molecular dynamics

The 2.5 Å resolution crystal structure of α_1-purothionin, a small (MW 4812), basic (pI ~ 11) wheat toxin, was solved with a model derived from crambin [8]. The model was built by graphical replacement of nonhomologous residues using the program FRODO [18] and global potential energy minimization with the program AMBER [11], under in vacuo conditions first optimized for the crambin structure [5].

However, the turn between two helices (residues 17–24), where there is a deletion relative to crambin, did not agree [8] with the X-ray structure of the protein (Fig. 6). We have now undertaken a molecular dynamics (MD) simulation of the turn where the error occurred, in order to explore whether the predicted model would convert into the turn from the refined X-ray structure. The AMBER program [11] was used in these calculations.

The turn in question from the α_1-purothionin model was first energy minimized. Then it was subjected to molecular dynamics at 300 K (heating for 3 ps, equilibration

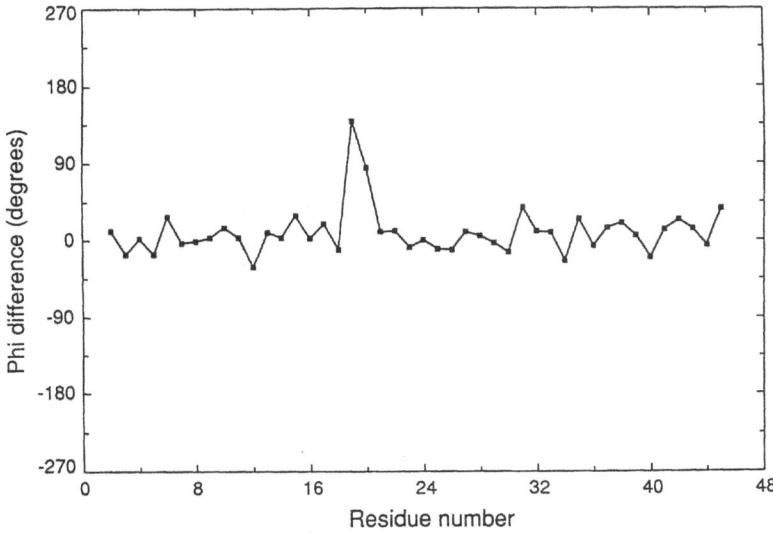

Fig. 6. Difference in backbone Φ angle between the predicted model and refined X-ray structure. Residues 19 and 20 deviate most.

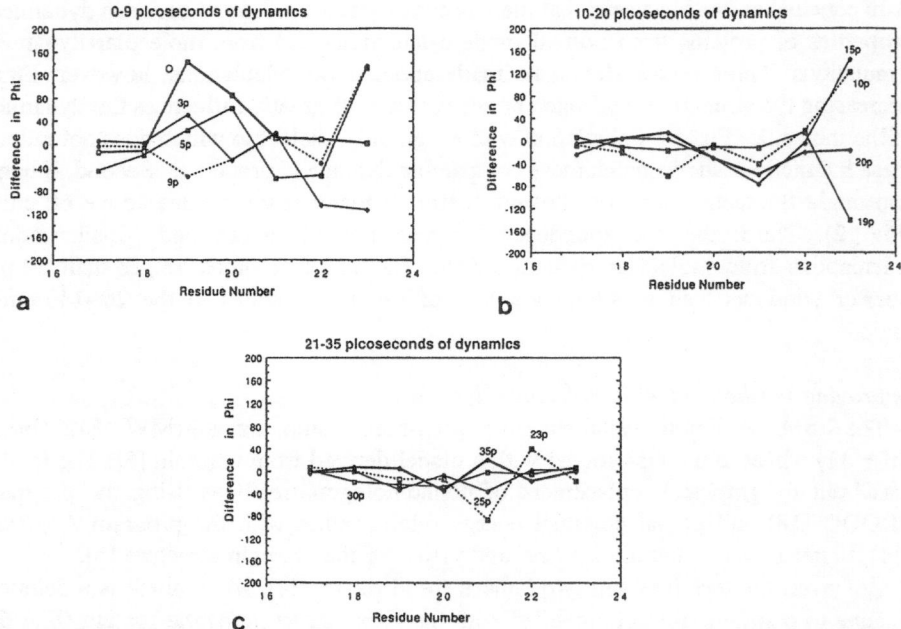

Fig. 7. *The turn backbone* Φ *difference between molecular dynamics and X-ray refined structures in* α$_1$*-purothionin. a. 0–9 ps; b. 10–20 ps; c. 21–35 ps. 0–5 ps is for heating and equilibration.*

for 2 ps and 30 ps of production dynamics). During minimization and dynamics of the turn, the non-turn atoms were constrained to their original position with 100 kcal restraints. Charged groups were neutralized with small ions to compensate for the missing shielding effect of ·solvent. The same conditions were used as for the optimized minimization of crambin [5], i.e. ε = 4r, Jorgensen's van der Waals radius, and nonbonding atom-based cutoff of 10 Å.

Figure 7a shows the turn Φ backbone angle differences between the molecular dynamics model and the X-ray model during the 35 ps of dynamics. During heating and equilibration (the first 5 ps), the difference in the Φ for residues 19 and 20 is transferred to residue 23, a residue close to the end of the turn, like a wave motion in a tethered string. At 19 ps or 14 ps of production dynamics (Fig. 7b), residue 23 Φ difference flips from 160 to −150°, and at 20 ps (15 ps of dynamics), the difference is close to zero. The deviations remain small in the last 15 ps of dynamics (Fig. ·7c).

The rms difference between the model backbone and the X-ray structure illustrates well the better agreement of the predicted model with the X-ray structure after MD. The rms deviation started at 0.76 Å for all residues (1.59 Å for the turn and 0.52 Å for the non-turn residues). After MD, the deviation for all residues was 0.54 Å (0.94 Å for the turn and 0.51 Å for the non-turn residues). Note the value for all residues is close to that for non-turn residues. In contrast, the rms deviation of the MD model from the starting model has increased from 0 to 2.1 Å for the turn residues and 0.06 for non-turn residues. After molecular dynamics, Fig. 8 shows that the model and

226

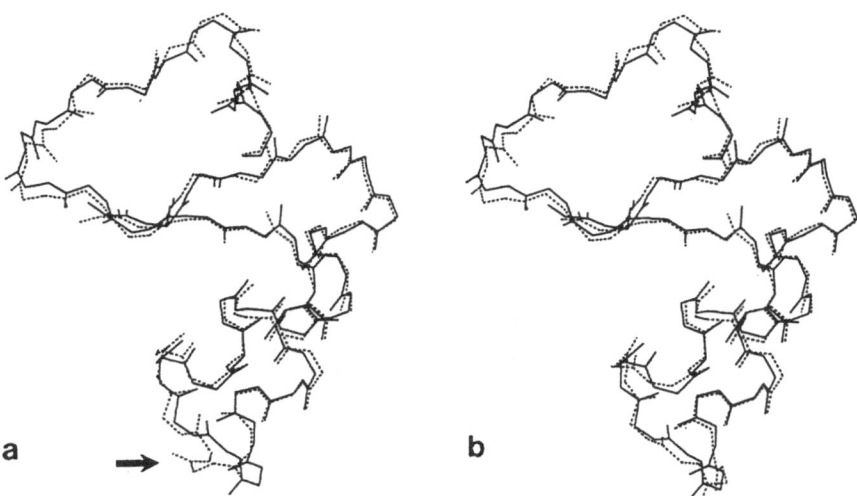

Fig. 8. Superimposed backbones for the predicted model (dotted) and refined X-ray model (solid) for α₁-purothionin. a. Before molecular dynamics; b. After molecular dynamics.

X-ray structure are considerably closer.

From these results, we would recommend that in cases where there is a deletion between homologous structures (and perhaps an insertion as well), it is valuable to subject the region in question to molecular dynamics. This removes the arbitrariness of deciding which residues correspond best between the two structures [4]. We recommend the combined use of energy minimization and limited molecular dynamics because it has been observed that molecular dynamics calculations can diverge from the crystal structure [19]. It should be noted that the low deviations both of our initial prediction and the improved prediction with molecular dynamics may be related to the fact that we started with an excellent model.

It is somewhat surprising that the X-ray conformation of the turn seems favored by the molecular dynamics (i.e. is lower in energy). In the crystal, this turn participates in a crystal contact and we surmised that the conformation we saw in the crystal structure might be due to crystal packing forces. We plan to continue the molecular dynamics longer to see whether the turn flips back to the starting modeled structure. We will also use this improved model with the crystal data to see if refinement can be further improved.

Acknowledgements

The authors wish to thank Dr. Bernard Brooks for considerable help in setting up the molecular dynamics of crambin with CHARMm and the quasiharmonic calculations. Dr. Marc Whitlow contributed to the α₁-purothionin project since its inception. The crambin dynamics research was supported by NSF grant DMB86-06636 and DMB89-04337 (M.M.T.) and NIH grant GM39266 (D.A.C.). The structure prediction was supported by NIH grant GM38114 (M.M.T.). Usha Rao acknowledges graduate support from this grant as well.

References

1. Teeter, M.M. and Case, D.A. (1990) *J. Phys. Chem.*, in press.
2. Teeter, M.M. and Hope, H. (1986) *Trans. New York Acad. Sci.* **482**, 163.
3. Teeter, M.M., Mazer, J.A. and L'Italien, J.J. (1981) *Biochemistry* **20**, 5437.
4. Whitlow, M. and Teeter, M.M. (1985) *J. Biomol. Struct. Dyn.* **2**, 831.
5. Whitlow, M. and Teeter, M.M. (1986) *J. Am. Chem. Soc.* **108**, 7163.
6. Brooks, B. and Karplus, M. (1983) *Proc. Natl. Acad. Sci. USA* **80**, 6571.
7. Karplus, M. and Kushick, J.N. (1981) *Macromolecules* **14**, 325.
8. Teeter, M.M., Ma, Xing-qi, Rao, U. and Whitlow, M. (1990) *Proteins* **8**, 118.
9. Brooks, B.R., Bruccoleri, R.E., Olafson, B.D., States, D.J., Swaminathan, S. and Karplus, M. (1983) *J. Comput. Chem.* **4**, 187.
10. Reiher, W.E., Ph.D. Thesis, Harvard University, 1985.
11. Weiner, S.J., Kollman, P.A., Case, D.A., Singh, U.C., Ghio, C., Algona, G. and Weiner, P. (1984) *J. Am. Chem. Soc.* **106**, 765.
12. Weiner, S.J., Kollman, P.A., Nguyen, D.T. and Case, D.A. (1986) *J. Comput. Chem.* **7**, 230.
13. Hingerty, B.E., Richie, R.H., Ferrell, T.L. and Turner, J.E. (1985) *Biochemistry* **24**, 427.
14. Jorgensen, W.L. and Tirado-Rives, J. (1988) *J. Am. Chem. Soc.* **110**, 1657.
15. Hagler, A.T. (1985) in *The Peptides Vol. 7 (Analysis, Synthesis, Biology)* (Udenfriend, S. and Meienhofer, J., Eds.) Academic Press, Orlando, FL, pp. 213-299.
16. Loncharich, R.J. and Brooks, B.R., 1989, *Proteins.* **6**, 32.
17. Levitt, M. Sander, C. and Stern, P.S. (1985) *J. Mol. Biol.* **181**, 423.
18. Jones, T.A. (1982) in *Computational Crystallography*, (D. Sayre, Ed.) Clarendon Press, Oxford, p. 303.
19. van Gunsteren, W.F., Berendsen, H.J.C., Hermans, J., Hol, W.G.J. and Postma, J.P.M. (1983) *Proc. Natl. Acad. Sci. USA* **80**, 4315.

Analyses of statistical errors in dynamics simulations

Richard W. Pastor

Biophysics Laboratory, Center for Biologics Evaluation and Research,
Food and Drug Administration, 8800 Rockville Pike, Bethesda, MD 20892, U.S.A.

Introduction

The estimation of statistical errors in molecular dynamics (MD) simulations is relevant in any application where an attempt is made to determine an ensemble average. In parameter development it is critical. This paper reviews a useful approach to the analysis of statistical variance involving the correlation function of the calculated property, and includes several illustrations.

Theory

The variance, $\sigma^2[A]$, of an average $< A >$ of N independent samples is usually evaluated as follows

$$\sigma^2[A] = (1/N - 1) \sum_{k=1}^{N} (A_n - < A >)^2 \tag{1}$$

where A_n is the value (which may itself be an average) of the nth sample. The standard deviation, $\sigma[A]$, is the square root of $\sigma^2[A]$. If the distribution of samples is normal (generally a reasonable assumption), σ can be used in standard statistical tests [1].

For a variety of reasons, it is common in simulations to have a single long trajectory from which it is not clear how to estimate the statistical error. It is intuitively obvious, nevertheless, that the correlation function of the property, $< A(0) A(t) >$, should be important; i.e., a trajectory contains fewer independent samples of an equilibrium property with a long correlation time, and consequently larger variance in the calculated average, than of a property with a short correlation time. Additionally, the average of a property with large fluctuations will have a larger variance than one with small fluctuations, even if the correlation times are the same. The preceding relationships may be written as follows for a trajectory of length T_{run} [2]:

$$\sigma^2[A]_{run} = (2/T_{run}) \int_0^{T_{run}} (1 - t/T_{run}) < \delta A(0) \, \delta A(t) >_{exact} dt \tag{2}$$

where $\delta A(t) = A(t) - < A >$; the subscript *run* will be included in references to variances calculated using the correlation function, as distinct from Eq. (1). Recalling that $< \delta A^2 > = < A^2 > - < A >^2$, we define the relaxation time τ_A by

$$\tau_A = \int\limits_0^\infty \left[\frac{< A(0)\ A(t) > - < A >^2}{< A^2 > - < A >^2} \right] dt \tag{3}$$

(i.e., τ_A is the decay constant when $< A(0)\ A(t) >$ is single exponential). Then, for $T_{run} \gg \tau_A$, Eq. (2) simplifies to

$$\sigma^2[A]_{run} = (2/T_{run}) \int\limits_0^\infty < \delta A(0)\ \delta A(t) >_{exact} dt \tag{4}$$

or

$$\sigma^2[A]_{run} = (2/T_{run}) (< \delta A^2 >\tau_A)_{exact} \tag{5}$$

Equation (5) is the principle result of this section. Note that exact quantities appear on the right-hand side, while in most cases of practical interest the correlation function must be obtained from the same trajectory as the average in question. However, because the variance in the correlation time is also of the order (τ/T_{run}) [3], this approach provides a useful guide to the level of error even when it is approximate. A detailed presentation of these and related topics is contained in the excellent monograph by Allen and Tildesley [2].

Results

Positional averages for a diffusive harmonic oscillator

The average position, $< x >$, and the mean squared position, $< x^2 >$, are 0 and $kT/m\omega^2$ for a harmonic oscillator, where k is Boltzmann's constant, T is the temperature, m is the mass, and ω is the frequency. For our first example we will use exact values in Eq. (5) and determine $\sigma^2[x]_{run}$ and $\sigma^2[x^2]_{run}$ for an oscillator undergoing diffusive motion. (It is also easy, and a good exercise, to do this by simulation; the appropriate algorithms are discussed in ref. [4]).

The required correlation functions are:

$$< x(0)\ x(t) > = < x^2 > \exp(-\omega^2\ t/\gamma) \tag{6a}$$

$$< x^2(0)\ x^2(t) > = < x^2 >^2 + 2 < x(0)\ x(t) >^2 \tag{6b}$$

where γ is the collision frequency, which equals the friction constant divided by the mass [4]. Equation (6b) is derived from Eq. (6a) using the pairwise decomposition of Gaussian random variables [3]; $< A^2(0)A^2(t) >$ does not in general have such a simple relationship to $< A(0)A(t) >$. Combining Eqs. (5) and (6),

$$\sigma^2[x]_{run} = (kT/m\omega^2)\ (2\gamma/\omega^2 T_{run}) \tag{7a}$$

$$\sigma^2[x^2]_{run} = (kT/m\omega^2)\ \sigma^2[x]_{run} \tag{7b}$$

Now consider a 100 ps simulation of an oscillator with $m = 11$ a.u., $\gamma = 53$ ps^{-1}, $\omega = 23$ ps^{-1} and $T = 300$ K ($\tau_x = \gamma/\omega^2 \approx 0.1$ ps); these parameters approximate the torsional oscillation of *trans* butane immersed in a solvent with a viscosity of water [5]. From Eq. (7), $\sigma[x] = 0.00927$ Å and $\sigma[x^2] = 0.00192$ Å^2. Hence, the relative error $\sigma[x^2] / < x^2 > \approx 4.5\%$. A 2 ns simulation is required to reduce the relative error to 1%.

The trans/gauche equilibrium constant for glycol

The dynamics of a particle in pseudo one-dimensional potential provides a good example of the statistical error associated with determining an equilibrium constant from a trajectory average. Here we consider the equilibrium fraction of the *trans* conformation, $< N_t >$, of the O-C-C-O torsion angle of glycol, utilizing preliminary results of MD simulations of a single glycol in water (G. Widmalm and R.W. Pastor, unpublished data). The correlation function is obtained from the trajectory, and is therefore approximate. For this barrier-crossing process, $\sigma^2[N_t]_{run}$ is given by:

$$\sigma^2[N_t]_{run} = (2/T_{run}) \int_0^\infty < \delta N_t(t) \; \delta N_t(0) > \, dt \tag{8a}$$

$$= (2 \, \tau_N/T_{run}) < \delta N_t^2 > \tag{8b}$$

where $N_t(t) = 1$ if the conformation is in the *trans* state, and 0 if not. $< \delta N_t(t) \; \delta N_t(0) >$ is the number correlation function, and is useful for calculating rate constants in isomerizations of polymers [5,6]. The relaxation time, τ_N, is inversely proportional to the rate constant; thus, the variance in the equilibrium constant will decrease as the rate constant increases, which is reasonable. Note that $< \delta N_t^2 > = < N_t > - < N_t >^2$.

Table 1 shows the equilibrium data and relaxation times obtained from 5 simulations, each of 1 ns; $\sigma[N_t] = 0.061$. The final column of Table 1 lists $\sigma[N_t]_{run}$ estimated from $< \delta N_t^2 >$ and τ_N from each sample (i.e., $T_{run} = 1$ ns). Even though there is spread (the average equals 0.046), $\sigma[N_t]_{run}$ and $\sigma[N_t]$ are in reasonable agreement, considering the limited number of samples and the level of uncertainty in $< \delta N_t(t) \; \delta N_t(0) >$ (e.g., $(2\tau_N/T_{run})^{1/2} = 0.12$). The issue of using the $\sigma[N_t]_{run}$ to establish confidence intervals [1] is separate; this example only demonstrates that $\sigma[N_t]_{run}$ provides a good measure of the standard deviation even when an approximate correlation function is used.

Table 1 *Fraction* trans *and* $< \delta N_t(t) \; \delta N_t(0) >$ *from 5 molecular dynamics simulations of glycol in solvent.* $\sigma[N_t]$ *and* $\sigma[N_t]_{run}$ *are calculated from Eqs. (1) and (8), respectively*

Run	$< N_t >$	$< \delta N_t^2 >$	τ_N (ps)	$\sigma[N_t]_{run}$
1	0.177	0.146	8.2	0.049
2	0.167	0.139	8.6	0.049
3	0.138	0.119	5.7	0.037
4	0.285	0.204	8.9	0.060
5	0.134	0.116	5.8	0.037
Average	0.180	0.145	7.4	0.046
$\sigma[N_t]$	0.061			

The deuterium order parameters of a lipid chain

The final example involves the deuterium order parameter, S_{CD}, for carbons in a lipid chain determined from a Brownian dynamics simulation [7]. If we assume axial averaging of the lipid about the bilayer normal S_{CD} is given by

$$S_{CD} = (1/2) < 3 \cos^2 \theta - 1 > \tag{9}$$

where θ is the angle between the CD vector and the normal. Hence, S_{CD} is a measure of the average orientation of the chain in the bilayer.

The trajectory described in ref. 7 was carried out in 90 segments of 2.46 ns each (when scaled for $\gamma = 50$ ps^{-1}). Because data from additional 90 segment trajectories are not available, we determine averages and the standard deviations of 10 segment (i.e., 24.6 ns) blocks; these are listed in Table 2 for the two hydrogens of carbon 2 (nearest to the headgroup), carbon 9 and carbon 15 (nearest to the middle of the bilayer).

As described in ref. [8], Eq. (9) is simply related to the correlation function of the second rank spherical harmonic:

$$1/4 < (3 \cos^2 \theta(0) - 1)(3 \cos^2 \theta(t) - 1) > = (4\pi/5) < Y_2^{0*}(t) \, Y_2^0(0) > \tag{10}$$

Hence, we may use the amplitudes and decay constants of $< Y_2^{0*}(t) \, Y_2^0(0) >$ listed in ref. 8 to estimate the variance in the calculated order parameter. Even though the dynamics underlying the averaging is complex, Table 2 shows that $\sigma[S_{CD}]$ and $\sigma[S_{CD}]_{run}$ are in good agreement. This implies that the standard deviation associated with the entire set of 90 segments is 3 times smaller; for example, many trials of such

Table 2 *Averages and standard deviations of deuterium order parameters (S_{CD}) determined from a Brownian dynamics simulation [7] of a lipid chain. Each set of 10 runs is 24.6 ns, assuming $\gamma = 50$ ps^{-1}; correlation times are from ref. [8]. $\sigma[S_{CD}]$ is calculated from Eq. (1) and $\sigma[S_{CD}]_{run}$ from Eqs. (10) and (5)*

Runs	C2–H1	C2–H2	C9–H1	C9–H2	C15–H1	C15–H2
1–10	−0.222	−0.247	−0.180	−0.164	−0.0803	−0.1059
11–20	−0.236	−0.256	−0.171	−0.161	−0.0654	−0.0586
21–30	−0.151	−0.249	−0.146	−0.155	−0.0984	−0.0743
31–40	−0.060	−0.252	−0.152	−0.165	−0.1008	−0.0862
41–50	−0.303	−0.187	−0.198	−0.189	−0.0928	−0.0979
51–60	−0.079	−0.193	−0.190	−0.203	−0.1012	−0.0889
61–70	−0.173	−0.266	−0.149	−0.175	−0.0710	−0.1081
71–80	−0.270	−0.116	−0.213	−0.140	−0.0783	−0.0790
81–90	−0.174	−0.240	−0.215	−0.160	−0.0682	−0.0853
Average	−0.185	−0.223	−0.179	−0.168	−0.0841	−0.0871
$\sigma[S_{CD}]$	0.082	0.049	0.027	0.019	0.0145	0.0156
τ (ps)	394.7	447.2	68.13	65.42	9.268	10.58
$< \delta S_{CD}^2 >$	0.1543	0.1094	0.1096	0.1106	0.1656	0.1669
$\sigma[S_{CD}]_{run}$	0.070	0.063	0.025	0.024	0.011	0.012

trajectories would show that $\sigma[S_{CD}]$ for carbon 15 $\approx$ 0.004, or that the relative error is $\approx$ 5%.

Summary

The examples included here make clear that the variance in dynamics averages is uncomfortably large for trajectories of the usual length; this should certainly be considered when drawing conclusions from simulations. A simple formula for estimating the variance from single trajectories was illustrated. The formula is most accurate when the exact correlation function is known, but is also useful when the correlation function itself has statistical error.

References

1. See, for example, DeGroot, M.H. (1975) *Probability and Statistics*, Addison-Wesley, Menlo Park.
2. Allen, M.P. and Tildesley, D.J. (1987) *Computer Simulation of Liquids*, Clarendon Press, Oxford, and references therein.
3. Zwanzig, R. and Ailawadi, N.K. (1969) *Phys. Rev.* **182**, 280.
4. Pastor, R.W., Brooks, B.R. and Szabo, A. (1988) *Mol. Phys.* **65**, 1409.
5. Pastor, R.W. and Karplus, M. (1989) *J. Chem. Phys.* **91**, 211.
6. Chandler, D. (1978) *J. Chem. Phys.* **68**, 2959.
7. Pastor, R.W., Venable, R.M. and Karplus, M. (1988) *J. Chem. Phys.* **89**, 1112.
8. Pastor, R.W., Venable, R.M., Karplus, M. and Szabo, A. (1988) *J. Chem. Phys.* **89**, 1128.

Derivation of solution conformers of peptide hormones via constrained molecular dynamics based on 2D NMR data

Vincent S. Madison[a], David C. Fry[a], Bogda B. Wegrzynski[a],
Michael P. Williamson[b], Robert M. Campbell[a], Waleed Danho[a],
Edgar P. Heimer[a] and Arthur M. Felix[a]

[a]Roche Research Center, Hoffmann-La Roche Inc., Nutley, NJ 07110, U.S.A.
[b]Roche Products Ltd., Welwyn Garden City, Hertsfordshire AL7 3AY, U.K.

Introduction

An optimization procedure based on molecular dynamics and energy minimization constrained by distances from nuclear Overhauser effect (NOE) measurements has been developed using the CHARMM program package [1] to derive and analyze peptide conformations [2,3]. This procedure readily folds a peptide from its fully-extended conformation to an ensemble of conformers satisfying the distance constraints and, in the process, makes self-consistent chiral assignments of pro-chiral protons with split resonances. Statistical analysis of the ensemble yields families with similar overall backbone conformations and reveals which segments of the peptide chain have well-defined single conformations. Results will be presented for a series of analogs of growth hormone releasing factor (GRF), an internal peptide from interleukin-2 (IL-2), and the blood complement protein porcine C5a(des-Arg).

Results and Discussion

Our previous analysis of GRF analogs [4] has been extended. Circular dichroism (CD) spectra show that all but one of the GRF analogs (See Table 1 for abbreviated names of the analogs) have an α-helical content of about 90% in 75% methanol: water, pH 6. The one exception, tricycloGRF, has about 75% helix (Table 2). In aqueous solution, pH 3, the helix content increases from 20–25% for the linear

Table 1 *Abbreviations for peptides*

Peptide	Abbreviation
GRF(1–29)NH$_2$	GRF
[Ala^{15}]GRF(1–29)NH$_2$	[Ala^{15}]GRF
cyclo(8–12)[Asp^8,Ala^{15}]GRF(1–29)NH$_2$	cyclo(8–12)GRF
cyclo(12–16)[Ala^{15},Glu^{16})GRF(1–29)NH$_2$	cyclo(12–16)GRF
cyclo(16–20)[Ala^{15},Glu^{16},Lys^{20}]GRF(1–29)NH$_2$	cyclo(16–20)GRF
cyclo(21–25)[Ala^{15}]GRF(1–29)NH$_2$	cyclo(21–25)GRF
dicyclo(8–12;21–25)[Asp^8,Ala^{15}]GRF(1–29)NH$_2$	dicycloGRF
tricyclo(8–12;16–20;21–25)[Asp^8,Ala^{15},Glu^{16},Lys^{20}]GRF(1–29)NH$_2$	tricycloGRF

Table 2   *Helix content (from CD spectra) and bioactivity*

Peptide	Helix content (%)		Relative bioactivity
	Water, pH 3	75% Methanol, pH 6	
GRF(1-44)NH$_2$	–	–	1.0
GRF	20	90	0.7
[Ala^{15}]GRF	25	95	3.8
cyclo(8-12)GRF	35	90	0.8
cyclo (12-16)GRF	30	95	0.8
cyclo (16-20)GRF	40	90	0.2
cyclo (21-25)GRF	35	95	1.3
dicycloGRF	40	90	0.7
tricycloGRF	40	75	0.0
IL-2(36-56)Cys		20[a]	

[a] In 100% methanol.

peptides, to *ca.* 35% for the monocyclic analogs, and to *ca.* 40% for the dicyclic and tricyclic analogs.

Helical segments were apparent for the GRF analogs from the pattern of consecutive i → (i + 3) NOEs in 2D spectra. Further analysis depended on the optimized structures constrained by the experimental distances. Pairwise interproton distances in the optimized structures agreed well with the experimental ones. In only two cases, there were more than two distances which exceeded the experimental range (generally, ± 0.5Å) by more than an additional 0.5 Å in more than half the members of the ensemble. Over the ensemble of 30 optimized structures for each peptide, the average root-mean-square deviation (RMSD) exceeded the estimated error by 0.044 to 0.173 Å. Each member of the ensemble fits the experimental data as shown by the small standard deviation (SD) about the average RMSD (0.003–0.006 Å). The interproton distances

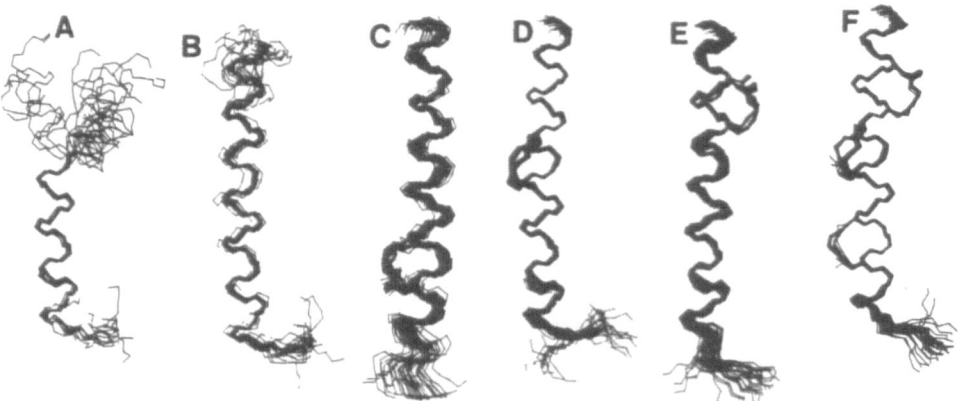

Fig. 1.   *Superposition of peptide backbones for all members of a family for GRF analogs in 75% methanol : water. The N-termini are at the bottom. The side-chain lactam bridges are also shown for the cyclic peptides. The straight helical segments (residues 4:18, 4:26 or 4:29, Table 3) were overlapped. The analogs and number of members (out of 30) in the family are: A. [Ala^{15}]GRF – family 2, 10; B. [Ala^{15}]GRF – family 1, 20; C. cyclo(8–12)GRF, 30; D. cyclo(16–20)GRF, 30; E. cyclo(21–25)GRF, 30; F. tricycloGRF, 26.*

in the optimized structures were distributed throughout the estimated range of the experimental data as shown by the full RMSDs of 0.332 to 0.522 Å, comparable to the estimated error.

Structural details were obtained from analysis of the optimized structures. In 75% methanol, the consensus conformation is a single α-helical segment with fraying of up to three residues at each end (Fig. 1, Table 3). The linear peptides are well ordered for residues 4:26, and the cyclic analogs for residues 4:29. The three disordered residues at the amino terminus tend to continue in the direction of the helix axis (GRF, cyclo(8–12)GRF, cyclo(12–16)GRF, and dicycloGRF) or to extend toward the right, perpendicular to the helix axis ([Ala^{15}]GRF, cyclo(16–20)GRF, cyclo(21–25)GRF, and tricycloGRF). In addition to the conformers with a single, straight helical segment, the linear peptides also have a tendency to form helices with kinks near residues 16 and/or 25 (Fig. 1A). The fraction of structures with kinks is 6/30 for GRF, 10/30 for [Ala^{15}]GRF, and 13/30 for cyclo(12–16)GRF. None of the other monocyclic structures have significant kinks, 2/49 of the dicyclic structures have large kinks, and 4/30 of the tricyclic structures are somewhat kinked, but still

Table 3 *Fit of ordered segments to average structure*[a]

Peptide	Solvent	No. of structures[b]	Residue range	RMSD ± SD (Å)
GRF, Family 1	75% MeOH	24	4:26	0.71 ± 0.29
GRF, Family 2	75% MeOH	6	4:18	0.90 ± 0.21
[Ala^{15}]GRF, Family 1	75% MeOH	20	4:26	0.58 ± 0.33
[Ala^{15}]GRF, Family 2	75% MeOH	10	4:18	0.54 ± 0.12
cyclo(8–12)GRF	75% MeOH	30	4:29	0.64 ± 0.24
cyclo(12–16)GRF, Family 1	75% MeOH	17	4:26	0.96 ± 0.41
cyclo(12–16)GRF, Family 2	75% MeOH	13	4:20	0.52 ± 0.28
cyclo(16–20)GRF	75% MeOH	30	4:29	0.42 ± 0.14
cyclo(21–25)GRF	75% MeOH	30	4:29	0.72 ± 0.42
dicycloGRF	75% MeOH	47	4:29	0.55 ± 0.17
tricycloGRF, Family 1	75% MeOH	26	4:29	0.28 ± 0.07
tricycloGRF, Family 2	75% MeOH	4	4:29	0.51 ± 0.77
IL-2(36–56)Cys	Methanol	25	40:54	0.81 ± 0.31
GRF	Water	30	9:14	0.23 ± 0.16
[Ala^{15}]GRF	Water	30	7:14	0.53 ± 0.16
cyclo(8–12)GRF	Water	30	7:17	0.28 ± 0.11
cyclo(21–25)GRF, Family 1	Water	16	9:14	0.41 ± 0.15
cyclo(21–25)GRF, Family 2	Water	14	9:14	0.86 ± 0.37
dicycloGRF	Water	30	7:20	0.59 ± 0.29
GRF	Water	30	24:28	0.24 ± 0.13
[Ala^{15}]GRF	Water	30	21:28	0.34 ± 0.14
cyclo(12–16)GRF	Water	29	22:28	0.97 ± 0.42
cyclo(8–12)GRF	Water	30	21:26	0.71 ± 0.29
cyclo(21–25)GRF	Water	25	19:28	0.91 ± 0.43
dicycloGRF	Water	28	21:28	1.09 ± 0.21

[a] Fits of α-carbons to the average structure for the family within the specified segment were performed. The average RMS deviation and its standard deviation are listed. Structures whose RMSD was greater than 1.6 Å were placed in another family.

[b] For each peptide, a total of 30 optimized structures were analyzed; for dicycloGRF in 75% methanol 49 structures were used, and for cyclo(12–16)GRF in water 58 structures were used.

have essentially a single helical segment.

In aqueous solution, the GRF analogs have two helical segments with flexible regions between them and at the termini of the peptides. One helical region is at residues 9:14 for GRF and progressively increases in length for [Ala^{15}]GRF, cyclo(8–12)GRF and dicycloGRF to include residues 7:20 at the maximum. For these peptides, all 30 of the optimized structures agree in this helical region (Fig. 2, Table 3). For cyclo(16–20)GRF and cyclo(21–25)GRF, the helix in this region is destabilized. In cyclo(16–20)GRF, residues 9:14 are disordered. In 16/30 cyclo-(21–25)GRF structures, residues 9:14 are helical as for GRF (Fig. 2B and 2C), but this helical segment is shorter than for the parent, linear [Ala^{15}]GRF. Furthermore, for the remaining 14/30 structures the residue 9:14 region is essentially fully-extended (Fig. 2A).

The second helical segment in aqueous solution is near the carboxyl terminus. It increases in length from residues 24:28 for GRF, to 21:28 for [Ala^{15}]GRF, to 19:28 for cyclo(21–25)GRF. For cyclo(8–12)GRF, this segment is residues 21:26, shorter than for the linear parent. Only half the structures are helical for cyclo(12–16)GRF residues 22:28. In dicycloGRF, this segment covers residues 21:28 as in the linear

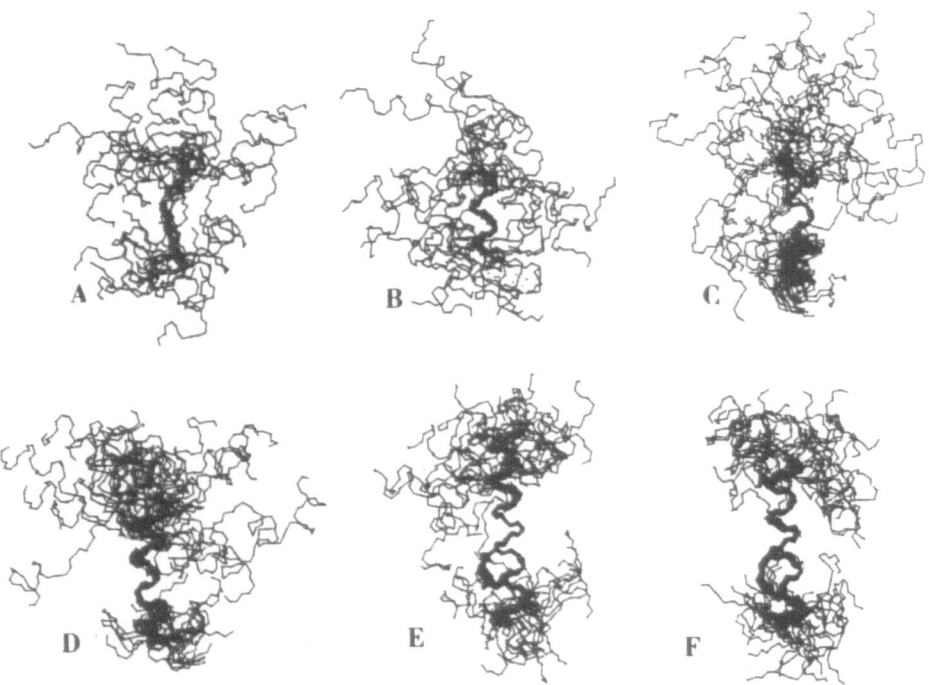

Fig. 2.   Overlapped peptide backbones for all members of a family for GRF analogs in aqueous solution. The N-termini are at the bottom. The side-chain lactam bridges are also shown for the cyclic peptides. The overlapped segments ranged from 9:14 to 7:20 (Table 3). The analogs and number of members (out of 30) in the family are: A. cyclo(21–25)GRF – family 2, 14; B. cyclo(21–25)GRF – family 1, 16; C. GRF, 30; D. [Ala^{15}]GRF, 30; E. cyclo(8–12)GRF, 30; F. dicycloGRF, 30.

parent, but is less well ordered (Table 3).

For [Ala^{15}]GRF in 75% methanol, both the backbone and side chains are well ordered for all 30 structures for residues 4:17. One face of the helix is hydrophobic with residues Phe^{6}, Tyr^{10}, Val^{13}, Leu^{14} and Leu^{17} aligned (Fig. 3). In aqueous solution, only the central two turns of this segment remain reasonably well ordered including the side chains of Tyr^{10}, Val^{13} and Leu^{14}.

In the crystal structure of IL-2, residues 36:56 comprise helices B and B', two helical segments with a kink at the central Pro^{47} residue [5]. For the peptide IL-2(36–56)Cys, an overall helix content of 20% was derived from CD spectra in methanol solution. The agreement of experimental interproton distances with those in optimized structures for this peptide is comparable to that for the GRF analogs. The consensus optimized structure is an α-helix kinked at Pro^{47} spanning residues 40:54 with considerable fraying of about four residues at each end of the helix. Within the helical segment, the degree of order is comparable to that in the GRF analogs (Table 3). About 70% of the residues are within the helical segment, compared to 20% overall helix content from CD spectra. This latter value is lower due to dynamic fluctuations of the structure and because there are two short helical segments separated by Pro^{47} rather than one continuous helix.

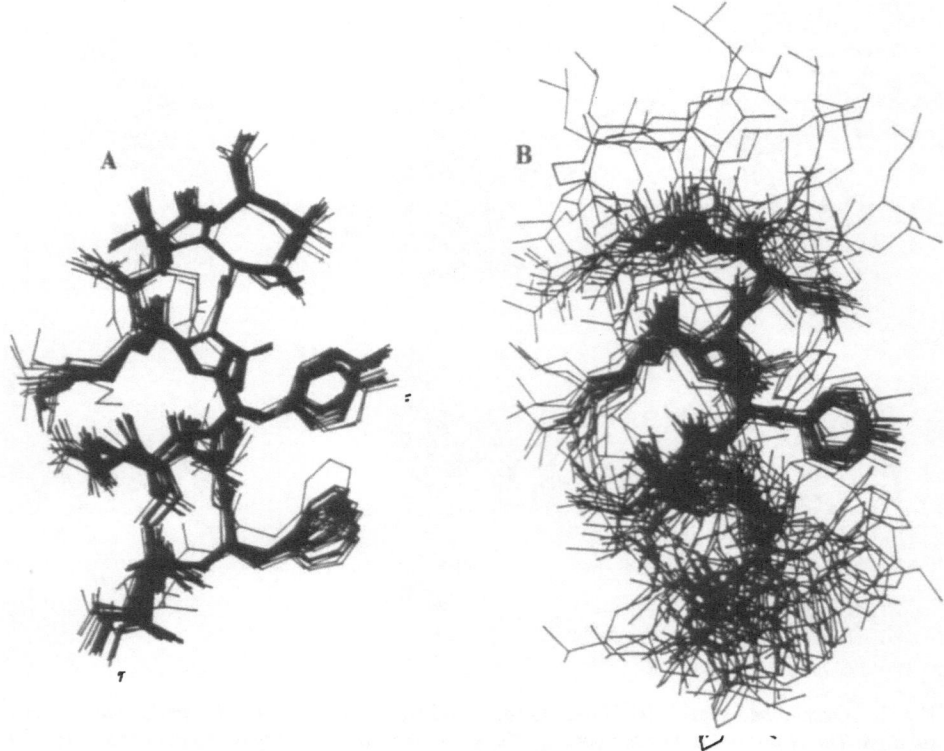

Fig. 3. Overlapped structures including side chains for all 30 optimized [Ala^{15}]GRF conformations for residues 4:17. The N-termini are at the bottom. Solvent for the constraining data and overlapped residues: A. 75% methanol : water, 4:17; B. water, 8:14.

C5a is a small protein (73 residues) which forms a four-helix bundle in solution. Interproton distances in optimized structures for this protein fit experimental ones as well as for the smaller peptides. We have presented our structural results for C5a in detail elsewhere [3]. Here, we emphasize the fact that the optimization protocol, starting from a fully-extended conformation, readily folded C5a into the four-helix bundle. This folding proceeded with no constraints for disulfide bonds. In fact, early introduction of covalent disulfide bonds tended to lock the protein into incorrectly folded conformations.

References

1. Brooks, B.R., Bruccoleri, R.E., Olafson, B.D., States, D.J., Swaminathan, S. and Karplus, M. (1983) *J. Comput. Chem.* **4**, 187.
2. Fry, D.C., Madison, V.S., Bolin, D.R., Greeley, D.N., Toome, V. and Wegrzynski, B.B. (1989) *Biochemistry* **28**, 2399.
3. Williamson, M.P. and Madison, V.S. (1990) *Biochemistry* **29**, 2895.
4. Madison, V.S., Fry, D.C., Greeley, D.N., Toome, V., Wegrzynski, B.B., Heimer, E.P. and Felix, A.M. (1990) in *Peptides: Chemistry, Structure and Biology*, Proceedings of the 11th American Peptide Symposium (Rivier, J.E. and Marshall, G.R., Eds.) ESCOM, Leiden, pp. 575–577.
5. Brandhuber, B.J., Boone, T., Kenney, W.C. and McKay, D.B. (1987) *Science* **238**, 1707.

Parallel processing and computational chemistry

Roberto Gomperts and Paul Weiner*

Alliant Computer Systems Corporation, One Monarch Drive, Littleton, MA 01460, U.S.A.

With the increasing use of such methods as dynamic simulations, free energy perturbation and large molecule ab initio calculations, as well as the use of very elaborate graphical display programs, chemical researchers find increasing frustration in the use of conventional computers to solve their problems. These machines are either too slow or too expensive to be practical for everyday use. However, the advent of parallel computer architectures and parallel algorithms offers a solution to this problem.

Before discussing the types of parallel computers and the programming requirements to run codes efficiently on them, it is important to define the basic idea behind parallel computation. In Fig. 1, an example of a loop that can be run in parallel is given. The fact that all tasks are independent and can be executed in any order implies that there can be no data dependencies between any of the tasks. Any data dependency will result in the need for a particular task to be run before another one and the tasks can no longer be sent independently to different processors. Thus a major programming problem of converting code from a single to a multiple processor computer is the removal of data dependencies from the computationally

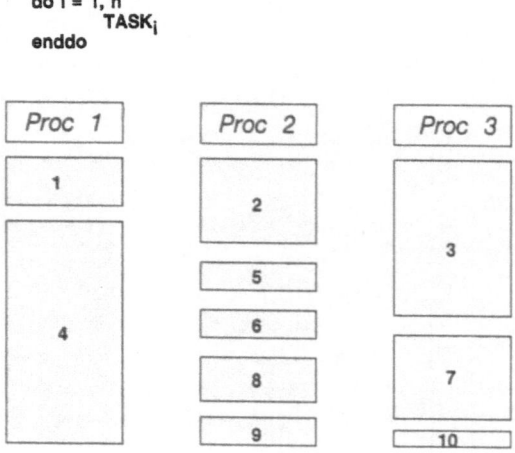

Fig. 1. *The basic unit of parallel computation.*

---

*To whom correspondence should be addressed.

intensive loops. There are many tools, which will be discussed later, to assist in this process.

There are two common types of parallel computers – shared memory and distributed memory machines. The shared memory computers have a central memory that is shared by all processors. This is the easiest type to program since the data is available to all processors. However, the programmer must worry about the data integrity. If there is a loop of code that has different iterations executed across different processors and there is an array that uses values calculated in a previous iteration, the code may give incorrect answers. This is due to data dependencies in the loop. The programmer must find a way to avoid this problem.

The distributed memory computers, such as the hypercube, have several nodes loosely coupled. Each node has a processor, memory, and some communication hardware and operates independently. The nodes work together by sending and receiving messages. The systems can scale up relatively cheaply to hundreds or thousands of processors. These systems can achieve very impressive speeds. However, if there is a need for the nodes to frequently communicate, the performance drops quickly. In addition, these machines can be difficult to program when there is a need for frequent communication between nodes. The program must be modified to include code both to send and to receive messages. After this has been done, the code will often need to be modified to be run on a serial or a shared memory parallel computer.

There are many opportunities for both high-level or coarse-grain parallelism and loop-level or fine-grain parallelism in chemistry codes. In molecular mechanics codes, these opportunities occur in the computation of the nonbonded pairwise interactions. In ab initio codes, the integral and derivative calculations, the Fock matrix construction, as well as some parts of the eigenvalue problem may be computed concurrently. Also, most of the post-SCF calculations involve inherently parallel matrix operations. The algorithms to render atoms, bonds, and orbitals are highly parallelizable.

To exploit the parallelism that exists in these codes, special parallel code must be written by the user or a compiler must be used that recognizes opportunities for parallelism in the serial code. In a shared memory computer, fine-grained parallelism can be handled by the compiler automatically. Simple data dependencies in a loop will be detected by the compiler and do not prevent parallelism. Figure 2 gives an example of a loop that has a simple data dependency, but which can still be run concurrently simply by waiting for the required values of $F(i)$ to be computed before going on with the calculation. All codes can take advantage of their fine-grain parallelism merely by being compiled.

Medium-grain parallelism involves several loops or many operations within a single loop. Since many operations are carried out, it is often too difficult for the compiler to analyze the code and to locate data dependencies. However, the programmer understands the algorithm and often knows whether or not data dependencies occur. In this case, it is possible to give directives to the compiler that tell it to go ahead and make the loop run concurrently. In Fig. 3, the programmer knows that the three calls to molecular mechanics routines are all independent and can run on separate

R. Gomperts and P. Weiner

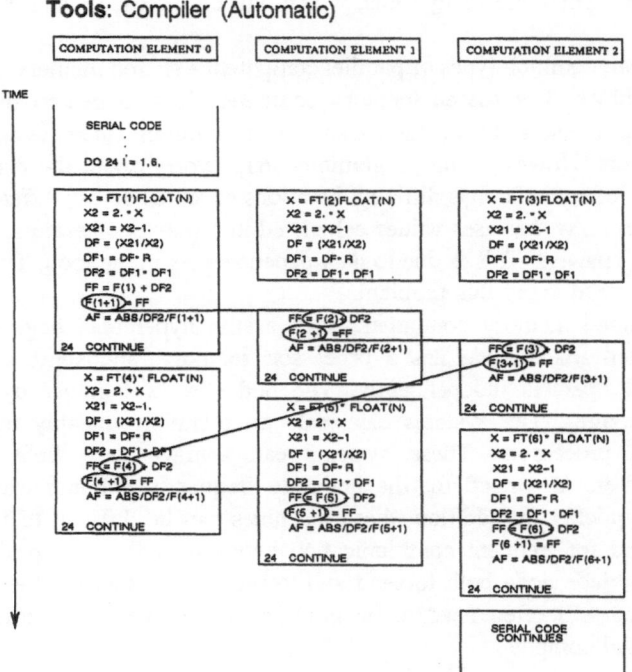

Fig. 2. Fine-grain parallelism.

processors. The CVD$CNCALL directive tells the compiler to run this loop concurrently. The second example uses a directive to tell the compiler to ignore dependency checking because the index j(i) has different values for all values of i in the do loop.

Coarse-grain parallelism involves very high level parallelism and often occurs in nested loops that contain complex subroutine calls. This type of parallelism usually involves human intervention; however, it is the most effective form of parallelism. There are two powerful tools developed by Alliant to help the user take full advantage of this form of parallelism. Both tools must address the problem of data dependencies in the code.

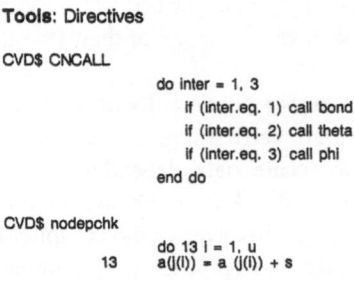

Fig. 3. Medium-grain parallelism.

242

The first tool is called FAST and is illustrated in Fig. 4. It performs an inter-procedural analysis to detect dependencies in the code and then automatically rewrites the code to remove the dependencies. Each processor is given a unique copy of any array that has a data dependency and the results of each processor are combined after the last iteration is completed.

This method has the disadvantage of requiring a complex interprocedural analysis and the algorithm is difficult to generalize for arbitrary FORTRAN constructs. There can also be load-balancing problems when the loops do not require approximately the same amount of work. In this case, it is possible for several processors on a complex to remain idle while the remaining processors finish up their computations. In an effort to correct these deficiencies, a new algorithm, FASTf, was developed at Alliant. This algorithm is based on forking of processes and does not require restructuring of code to protect the global data. The shared data and communications between the processes occur via the shared memory regions. The user only needs to add directives to the code (for the compiler) in order to run a section of code concurrently. Figure 5 shows an example of the directives required for a typical molecular mechanics kernal.

The directives are used to mark the beginning and end of the code segments that are to be executed in parallel by the different processes and are replaced by FASTf with calls to subroutines that handle forking and joining of processes. (These directives are explained in detail in the Alliant *Focus on Chemistry* brochure by the developer of the technique - Roberto Gomperts.)

The advantages of FASTf are the ability of a job to use multiple processors only when it needs to and the fact that processors are available for other tasks when they are not needed. The disadvantage of FASTf is its inability to handle very fine-grain parallelism well. In this case, forking and joining overhead can offset gains from running a parallel execution of the code.

Method

– Perform inter-procedural analysis to detect dependencies

– Rewrite code to remove dependencies

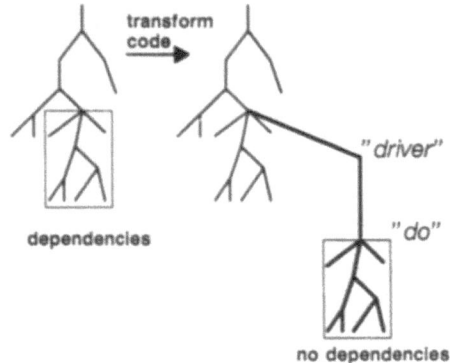

Fig. 4. Tools for parallelization: Fast.

243

Example for a Typical Molecular Mechanics Kernel:

```
CVD$  DATA_DEFS                         CVD$  PDO
   preliminary work; initialization, etc.        here comes the bulk of the work
                                                  to compute gradients and energy
CVD$  PARALLEL REGION                   CVD$  END PDO
c  non-bonded terms                        enddo
   do i = 1,nat                          c  bending-terms
      a little work done here               some preparation
      do j = 1,i                            do i = 1,list_bending
CVD$  PDO                                CVD$  PDO
         here comes the bulk of the work          here comes the bulk of the work
          to compute gradients and energy          to compute gradients and energy
CVD$  END PDO                            CVD$  END PDO
      enddo                                 enddo
   enddo                                 c  torsional-terms
c  coulombic-terms                          some preparation
   some preparation                         do i = 1,list_torsional
   do i = 1,nat                          CVD$  PDO
      a little work done here                      here comes the bulk of the work
      do j = 1,i                                    to compute gradients and energy
CVD$  PDO                                CVD$  END PDO
         here comes the bulk of the work       enddo
          to compute gradients and energy  CVD$  END PARALLEL REGION
CVD$  END PDO                            CVD$  JOIN gradients $DIM 3,natom
      enddo                              CVD$  JOIN etotal
   enddo
c  bonded terms
   some preparation
   do i = 1,list_bonded
      a little work done here
```

Fig. 5. Tools for parallelization: Fastf.

A combination of all of these methods is used on computational chemistry codes to decrease the time-to-solution of a single job, as well as to increase the throughput or to decrease the time-to-solution of multiple jobs.

Examples of several types of parallelism will now be given on the Alliant FX/2800. This computer is a shared memory parallel computer with a maximum of 28 I860 processors. The processors can be combined to form a complex, for optimum time-to-solution for a particular job, or run individually in detached mode, for high throughput. Combinations of these are also allowed. Figure 6 lists some of the capabilities of the Alliant.

Figure 7 shows the speedup achieved on a molecular mechanics code, AMBER 3+ (developed by Chandra Singh and Amber Systems, Inc., and derived from AMBER 3.0 by Chandra Singh, Paul Weiner and Peter Kollman), by taking advantage of medium-grain parallelism. Directives were used to make the non-bonded routine run concurrently. The rest of the code exploited the fine-grain parallelism that the compiler detected. The code achieved a speedup of 4.2X in going from 1 to 6 processors.

Figure 8 shows the speedup achieved on an ab initio code. These codes are much more difficult to parallelize than the molecular mechanics codes and typically require the use of coarse-grain parallelism tools. This code has been parallelized with FASTf, using forking, and achieved a 4.1X speedup in going from 1 to 8 processors.

These examples both show how parallelism can help achieve a faster time-to-solution for a single job. Another use of parallelism is to help achieve a faster throughput or time-to-solution of multiple jobs. Figures 9, 10 and 11 give a mix of

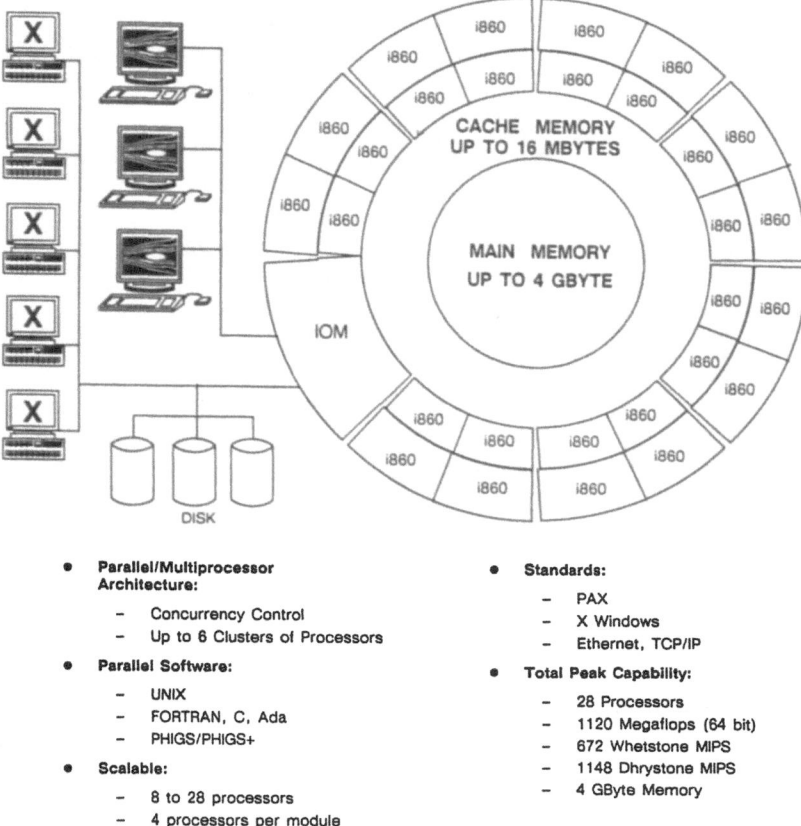

- **Parallel/Multiprocessor Architecture:**
  - Concurrency Control
  - Up to 6 Clusters of Processors
- **Parallel Software:**
  - UNIX
  - FORTRAN, C, Ada
  - PHIGS/PHIGS+
- **Scalable:**
  - 8 to 28 processors
  - 4 processors per module
- **Standards:**
  - PAX
  - X Windows
  - Ethernet, TCP/IP
- **Total Peak Capability:**
  - 28 Processors
  - 1120 Megaflops (64 bit)
  - 672 Whetstone MIPS
  - 1148 Dhrystone MIPS
  - 4 GByte Memory

Fig. 6. FX/2800 architecture, shared memory parallel computer.

computational chemistry jobs. Table 1 shows the results of running each job individually ('Time to solution' column) and the time for running each job when all jobs are run at once ('Under load' column). If all jobs were run one after another, they

# of Processors	Time (sec.)
1	707.5
2	455.0
4	237.8
6	167.4

1121 Atoms, 10Å NB cutoff, 100 steps of minimization.

Fig. 7. Single job comparison of parallel and non-parallel Amber 3+, FX/2800.

245

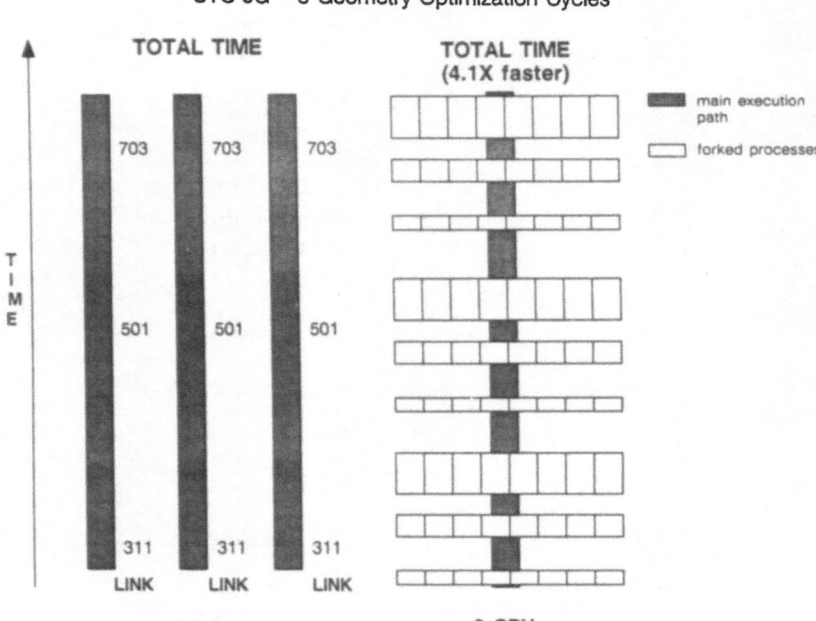

Fig. 8. *Single job comparison of parallel and non-parallel ab initio, FX/2800.*

would take 4 h and 50 min. By running all jobs simultaneously, they only take a total of 1 h to run.

Both forms of parallelism can also be combined to run multiple jobs, all working on different aspects of the same research problem, at the same time. An example of this, illustrated in Fig. 12 on an Alliant FX/80, is a set of tasks consisting of AMBER 3+, AMBER 3+ analysis, and the GRAMPS graphics display program. These programs are traditionally run in serial mode – batch processing of AMBER 3+, followed by analysis and graphics post-processing. At Alliant a program was written* that uses UNIX shared memory calls for synchronization and data passing between the applications. The use of intermediate disk files, which slows execution, is avoided. AMBER 3+ passes the results of the dynamics simulation to the analysis program. After analyzing the results, this program passes the results to GRAMPS, which then renders the molecule in colors that represent values computed in the analysis program. AMBER 3+ can be computing the next dynamics point while the analysis and graphics programs are working.

For this process to work, it is important for all steps to take approximately the same time. This is normally not true and requires that the computationally intensive steps be run in parallel. The computer must also be able to efficiently run several different jobs at the same time to maintain interactive performance. By achieving these goals,

---

*By Paul Weiner, Steve Gallion, Rosario Caltabiano and T.J. O'Donnell.

Molecular Mechanics – Discover 2.60

Test1	–	1876 atoms
		150 steps of Minimization
		225 steps of Dynamics
Test2	–	2073 atoms
		800 steps of Minimization
Test3	–	2073 atoms
		300 steps of Minimization
Test4	–	3082 atoms
		600 steps of Dynamics

Fig. 9. Computational chemistry load test: simultaneous jobs molecular mechanics, Discover 2.60.

Semi-Empirical – MOPAC 5.0

Test1	–	96 atoms (C37 H48 N4 07)
		1 Cycle of Nllsq using AM1
Test2	–	20 atoms (C6 H9 N 03 S)
		300 Cycles of Nllsq using AM1
Test3	–	24 atoms (C14 H10)
		90 Cycles of Nllsq using MNDO

Fig. 10. Computational chemistry load test: simultaneous jobs semi-empirical, MOPAC 5.0.

Ab-Initio

Test1	–	18 atoms (C5 H10 N O CL)
		6-31G*, 144 basis functions
		600 MB disk
		Single point RHF
Test2	–	10 atoms (C2 H6 O S)
		6-31G*, 76 basis functions
		7 MB disk
		Direct SCF (RHF) Geometry Opt.

Fig. 11. Computational chemistry load test: simultaneous jobs ab initio.

the user can interactively explore conformational space for a small molecule by using a combination of minimization and dynamics steps.

In the future, there is a possibility of combining multiple forms of parallelism within a single program. Both FAST and FASTf could be combined and have different subroutines run concurrently across different numbers of processors. This will help

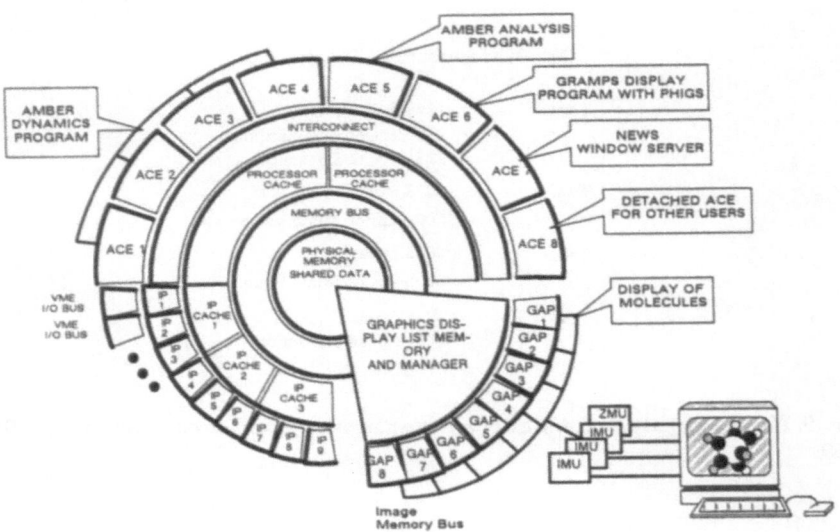

Fig. 12. Throughput: multiple chemistry codes solving the same problem.

achieve both optimal time-to-solution for a single job and high throughput for multiple jobs.

The use of fast parallel dynamics codes will enable researchers to explore more of conformational space, to carry out more modifications using free energy perturbation techniques, and to carry out detailed studies of reaction pathways with combined quantum and molecular mechanics methods.

Table 1 *Computational chemistry load test – simultaneous jobs*

Application/test	# Processors	Time to solution	Under load
Discover			
Test 1	4	30:58	44:23
Test 2	4	34:28	50:56
Test 3	4	31:14	44:51
Test 4	4	31:23	46:36
Mopac			
Test 1	1	34:57	59:24
Test 2	1	37:33	58:04
Test 3	1	26:17	46:07
Ab initio			
Test 1	8	39:00	1:00:35
Test 2	4	24:22	39:23
Total	31	4:50:12	1:00:38

Times are wall-clock times in hours:minutes:seconds.
FX/2828 – September, 1990.

Polypeptide and protein design: Protein engineering

Dedicated to the memory of
C. Levinthal

This section is dedicated to the memory of Professor Cyrus Levinthal, Columbia University, New York, NY, U.S.A. for pioneering studies which led to the birth of computer graphics that has made the visual realization of biological macromolecules possible.

The N-terminal fragment of bovine phosphophoryn, an extracellular mineral matrix protein, shares sequence homology with viral, bacterial and eukaryotic transcriptional and post-translational regulatory proteins

John Spencer Evans* and Sunney I. Chan

Arthur Amos Noyes Laboratory of Chemical Physics, Division of Chemistry and Chemical Engineering, California Institute of Technology, Pasadena, CA 91125, U.S.A.

Abstract

We report here the N-terminal amino acid sequence for the first 23 residues of purified bovine dentine phosphophoryn (BDPP), a polyelectrolyte mineral matrix protein (PMMP), as determined by Edman liquid/gas-phase automated sequencing:

S D P N S X D E D N G D A D A N D S D X N S D...

Uncertainties (denoted as X) exist at positions 6 and 20 which are not resolvable by repeated sequencing runs. These positions may either be occupied by Asp or Ser. The 23-residue terminus possesses short stretches of Asp, Glu or Ser flanked by non-anionic amino acids Gly, Ala, Val, or Asn. This structural motif has been revealed by earlier ^{1}H/^{31}P NMR observations on BDPP in solution (Evans and Chan, 1991a,b). A database search of existing protein sequences (NBRF, EMBL, GENBANK, NEW ENZYME LIST) has uncovered that the 23 amino acid BDPP sequence shares homologies with peptide fragments obtained from (a) Herpes Simplex Virus Type I unique transcriptional activator Vmw 175 protein (52% homology), (b) DNA-directed RNA polymerase, σ subunit (*Salmonella typhimurium*) (43% homology), and (c) the secretory gene product (*sec 7*) of *Saccharomyces cerevisiae* (43% homology). Homology of 39% is also observed between BDPP and cauliflower mosaic virus 55 kDa coat protein. Chou-Fasman and Kyte-Doolittle prediction analyses of the homologous sequences indicate that the structural motifs of these domains are similar. The domains are primarily hydrophilic and may contain stretches of β-sheet, β-turn, or coil structure, but very little α-helix structure. These results are discussed in the light of what is currently known about the evolution of template-mediated biomineral development in organisms.

Introduction

The formation of organism-mediated biominerals involving Ca^{2+}, Mg^{2+}, Fe (III), and other metal ions has presented an intriguing problem with regard to the underlying mechanism. In some systems, notably Mollusca [1,2], vertebrate bone matrix [3,4], and dentine tooth matrix [5,6], a class of polyelectrolyte mineral matrix proteins (PMMP) which possess significant (> 50%) amounts of anionic amino acids such as Asp, Glu, and o-phosphoserine (Pser), have been isolated and characterized. These proteins have

---

* To whom correspondence should be addressed.

been shown to either associate with a calcium mineral phase or mediate its formation from solution [1–6]. It is not known whether these proteins catalyze and/or inhibit the epitaxial growth of calcium mineral phases. It is believed that the anionic domains of the PMMPs act to chelate and coordinate divalent cations in solution and permit additional counter-anions to stabilize the PMMP-cation complex. The interest in these proteins not only centers on their role in mineral formation, but is also directed at their potential to participate in the formation of novel biocomposite materials, including both inorganic and organic polymers.

Of particular interest to the biologist is the evolution of the PMMPs as a class of extracellular proteins. The ability to form a mineral phase may have arisen as a result of evolutionary pressure for organism adaption. It has been speculated that mineral phase formation may have evolved as a metal ion 'reservoir'; this reservoir could then be tapped as the demand increased within the cell for metal ions critical to metabolic, catalytic, and other cellular functions [7,8]. Similarly, the mineral

	Sequence	% Homology	Anionic residues*
BDPP	1 .... 23 SDPNSXD [E] DNGDADAN [D] S [D] XNSD...	n/a	9
HSV-1 Vm175	1516 .... 1538 ...SSSSSDD [E] DEEDADDE D E D PESD...	52	15
Sec 7	99 .... 121 ...DEDEDED [E] DNGDEDDE D V D SSSS...	43	16
Pol σ	193 .... 215 ...DDDEDED [E] EDGDDDAA D D D NSID...	43	17
CaMV	1201 .... 1222 ...EEETSTE [E] DDGSSTSE D S D SESD...	39	12

Fig. 1. N-terminal 23-residue amino acid sequence of BDPP and its comparison with database sequences.

The N-terminal 23-residue amino acid sequence of BDPP was analyzed by the automated Edman gas-liquid phase sequencing technique. BDPP protein (10 µl sample, 1 µg), purified as described (Evans and Chan, 1991a) (> 95% homogeneity), was applied on a polybrene coated glass filter and analyzed on an Applied Biosystems 477A Liquid-Gas Phase Sequencer and PTH analyzer. A total of 35 cycles were performed on a single sample. Cycles were manually analyzed and compared to computer analysis of the cycles. Three runs on parallel samples were performed. Amino acids are denoted by the single-letter code. Data represent the best fit between the three determinations. X denotes uncertainty in that particular cycle; possible amino acids are either Asp or Ser.

The 23-residue N-terminal BDPP sequence was compared to known protein sequences in the GenBank, NBRF, EMBL, and New Enzyme List sequence databanks, using the interactive Genetics Computer Group software programs WORDSEARCH [43] and/or TFASTA [44] on a Digital VAX/VMS 5.7A operating system. The search, which compared two-word databits, included two ambiguities (X) in the BDPP N-terminal sequence at positions 6 and 20, and resulted in the following matches (> 40% homology) to known sequences. Shaded columns outline the consensus sequence of the five proteins. Numbers indicate the residue position in the overall primary amino acid sequence.

*Per 23 amino acids. n/a: not applicable.

phase may have also conveyed a structural improvement in the organism, permitting a cellular scaffolding to form which is thermoresistant as well as resistant to compressive and tensile forces. Such evolutionary mechanical adaptation would permit survival under compressive pressures (e.g., at ocean depths), temperature fluctuations, and allow for the support of multicellular structures capable of locomotion. The evolution of the PMMP cation-binding protein with multiple ligand binding sites is a unique phenomenon. The ability of these PMMPs to mediate mineral phase formation suggests that there may be a common chemical mechanism to PMMP-mineral complexation [9,10].

We have undertaken a study of one such PMMP, phosphophoryn, which is isolated from the dentine matrix of teeth and has been shown to be crucial to the formation of the inorganic calcium phosphate phase of this matrix [5,6,11–13]. The bovine isoform of phosphophoryn (abbreviated as BDPP; 155 kDa) is comprised of 40% Asp and 40% Pser, with the remaining amino acids distributed as Gly, non-phosphorylated Ser, Glu, Ala, Val, Leu, Ile, His, and > 1% of Phe and Tyr [5]. The protein possesses two types of Ca^{2+} binding sites (high and low affinity) per protein molecule [14], and has been shown to precipitate Ca^{2+} and PO_4^{3-} from solution to form amorphous calcium phosphate [12], a precursor of the apatitic calcium phosphate phase found in dentine matrix.

N-terminal Sequence Analysis of Bovine Phosphophoryn

In the course of our work, we have performed an N-terminal sequence analysis of bovine phosphophoryn (Fig. 1), purified as described elsewhere [15], using the automated Edman gas-liquid phase sequencing technique. This method resulted in a 23 amino acid sequence with uncertainties at only two positions: 6 and 20. We were unsuccessful in resolving these ambiguities by repeated sequencing runs: thus the X-position is either Asp or Ser. Since the o-phosphoserine residue is unstable under conditions of acid hydrolysis, the sites of phosphorylation cannot be determined in this analysis. After 23 cycles on the automated Edman sequencer, equimolar amounts of Asp and Ser were released and determination of the actual sequence beyond position 23 was not possible. A preliminary report on the N-terminal sequence of rat incisor phosphoprotein [16] revealed a tetramer of either Asp-Asp-Asp-Asn or Asp-Asp-Pro-Asn. In either instance, Asp is assigned to the first position, whereas the Ser is assigned to this position in the bovine sequence. This discrepancy in the sequence may be due to species divergence in sequence, and the presence of other phosphoproteins in rat dentine [6,15].

The 23-residue terminus possesses short stretches of Asp, Glu or Ser flanked by non-anionic amino acids Gly, Ala, Val, or Asn. In a recent paper [16], we described evidence for similar anionic domains of $(Asp)_n$ and $(Pser-Asp)_n$ flanked by intervening domains containing Gly, Ala, Val, Glu, Ser, and Leu. It is shown that these intervening amino acids respond to changes in the protonation state of the anionic domains [17]. It is interesting that these intervening amino acids also occur in the 23-residue N-terminal sequence (Fig. 1) in exactly the same manner as the rest of the protein. Additionally, the presence of a contiguous Ser-Asp sequence in the

N-terminal domain lends support to our earlier NMR findings of the existence of heteropolymer repeats of (Pser-Asp)$_n$ in BDPP [16]; however, at this time we cannot specify the sites of serine phosphorylation on the 23-mer.

Sequence Homologies between BDPP N-terminal Sequence and other known Protein Sequences

With the BDPP N-terminal sequence data, we have performed a protein sequence databank search of GenBank, EMBL, NBRF Protein, and New Enzyme List, to ascertain if there exist homologies between BDPP N-terminal sequence and other known protein sequences. In particular, we are interested in the existence of stretches of Asp, Glu or Ser flanked by intervening non-ionic amino acids. As shown in Fig. 1, the search yielded 5 matches with varying homology (39–52%) per 23 residues. On the basis of this analysis (Fig. 1), we found that: (1) BDPP showed homology (> 40%) to the anionic/intervening domains found in the following proteins: (a) Herpes Simplex Virus Type I (HSV-1) 175 kDa unique transcriptional activator protein (Vmw 175; also known as ICP-4 protein) [18–20]; (b) DNA-directed RNA polymerase, σ subunit (*Salmonella typhimurium*) [21]; and (c) the secretory gene product 7 (*sec 7*) of the temperature-sensitive mutant of *Saccharomyces cerevisiae* [22] (Fig. 2); and (2) homologies of less than 40% were found between BDPP and Asp-Glu-rich C-terminal region of Cauliflower mosaic virus (CaMV) coat protein (55 kDa precursor of open region IV) [23] (Fig. 1). By substituting Asp or Ser at the 6 and 20 position of the BDPP, the homology could increase to 57% for HSV-1 transcriptional activator (position 6, Asp), to 52% for *sec 7* (position 20, Ser) and 43% for Cauliflower mosaic virus (position 20, Ser). These sequence fragments all possess anionic domains containing Asp or Glu, and non-ionic intervening amino acids such as Ala, Val, Asn, Pro, Gly, Thr, and Ser.

Peptide Structure Predictions

Further analysis of the peptide sequences was performed to determine what other similarities exist at the level of peptide structure prediction (Chou-Fasman [24–26]) including hydrophobicity and hydrophilicity (Kyte and Doolittle [27]). It has been established that three-dimensional structures are more conserved in evolution than primary sequences [28–30]. A two-dimensional squiggle plot [31] detailing helical, β-sheet, turn, and coil regions and hydrophilicity index predictions (Kyte-Doolittle; > 1.3) is presented in Fig. 2 for BDPP N-terminus and the corresponding fragments from HSV-1 Vm175, CaMV coat protein, DNA-directed RNA polymerase σ subunit, and *sec 7* protein. For BDPP, we substituted either Ser or Asp at positions 6 and 20; the presence of either amino acid did not affect the outcome of the Chou-Fasman prediction. In the case of the hydrophilicity plot, the presence of Ser at positions 6 and 20 reduced the hydrophilic moment at those positions by 50% as compared to Asp. The hydrophilicity plots reveal that all of the homologous fragments possess a large percentage of residues (> 70% composition) that would be predicted to be hydrophilic. In addition, for each fragment the Chou-Fasman plots predict the existence of turn

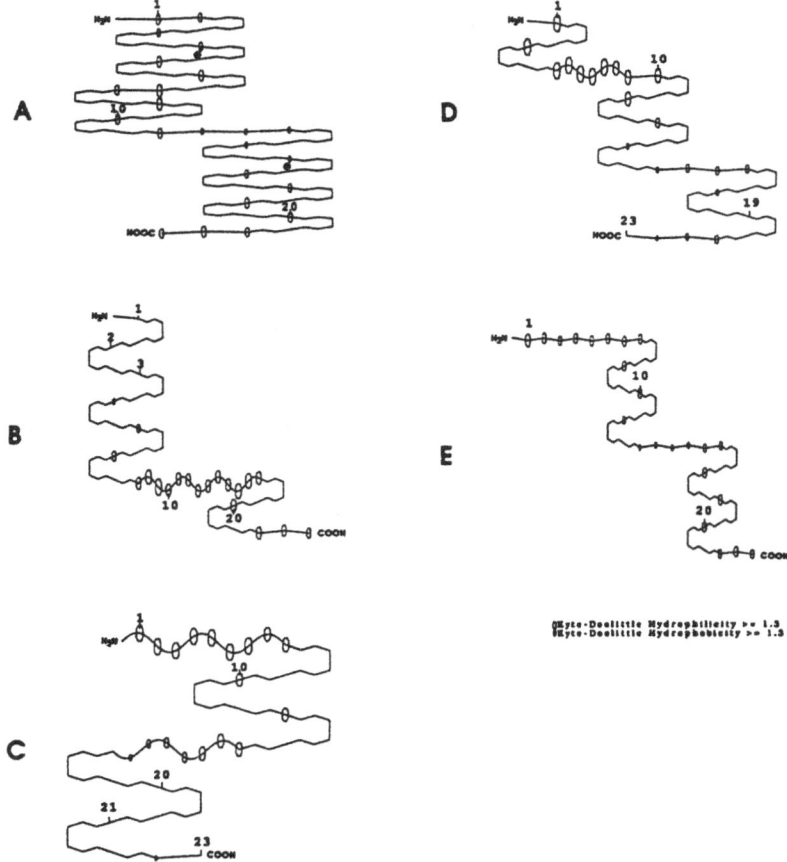

Fig. 2. *Chou-Fasman/Kyte-Doolittle squiggle plots of predicted secondary structures for BDPP and homologous protein fragments.*

Squiggle plots have been constructed for each of the sequence fragments in Fig. 1 using the Genetics Computer Group software programs *PEPTIDESTRUCTURE/PLOTSTRUCTURE [31]*. For comparison, sequences are numbered 1 through 23; the actual sequence positions are given in Fig. 1. Chou-Fasman secondary structures are represented as different wave forms: **helix:** sine wave; **β-sheet:** sharp saw-tooth wave; **turn:** 180° turn; **coil:** dull saw-tooth wave. Kyte-Doolittle hydropathy/hydrophilicity predictions are identified at each residue by either a hexagon (hydrophilic) or a diamond (hydrophobic), with the size of the symbol proportional to the value of hydrophobicity or hydrophilicity. The Kyte-Doolittle threshold for both values has been set to a value greater than or equal to 1.3, as noted at the bottom of the Figure. Potential sites of glycosylation, denoted by a solid black circle, are predicted for sites where the residues have the composition NXT or NXS. When X = D, W, or P residue, then the site is assumed to be a weak glycosylation site; otherwise, it is a strong glycosylation site. In this plot, the unidentified positions 6 and 20 (X) of the BDPP sequence are filled by Ser. Substitution of either Ser or Asp at positions 6 and 20 results in no change in the Chou-Fasman prediction, and only a small decrease in hydrophilicity for the Ser residue.

Legend to figures: A: BDPP 23-residue N-terminal sequence. B: HSV Vmw 175 transcriptional activator peptide fragment. C: Sec 7 secretory protein fragment. D: DNA-directed RNA polymerase, σ subunit fragment. E: Cauliflower Mosaic Virus 55 kDa coat protein fragment.

domains (ranging from 6 to 9 turns per 23 residues), with BDPP possessing the largest number of turn regions (15 turns per 23 amino acids) (Fig. 2). Note that the positioning of the turns varies with the protein sequence. Each protein fragment would be predicted also to have β-sheet and coil (random) structure, with BDPP containing the highest content of both. The presence of β-sheet structure has been shown experimentally to be characteristic of anionic charge domains in PMMP-divalent metal ion or PMMP-crystal complexes [32–34]. Another prediction is that the 23-residue N-terminus of BDPP would contain little if any helix structure, whereas the *sec 7* fragment would contain two helical regions (residue 1-9, 12-18) separated by turn and sheet regions. For the σ subunit and HSV-1 Vm 175 protein fragments, each is predicted to possess a single helical region of varying residue length, flanked by turn regions on either side of the helix (Fig. 2). Thus, although the detailed theoretical arrangement of secondary structure in each fragment is different, there exist similarities in hydrophilicity, β-sheet, coil, and turn content within the 23 amino acid residue fragments.

Structure-Function Considerations

What the above sequence data and peptide structure predictions suggest is that the N-terminal fragment of BDPP shares a number of sequence and domain motif similarities with other proteins which function intracellularly and are not involved in extracellular mineralization. At this time we are not able to determine what other parts of the BDPP sequence beyond the N-terminus may possess sequence homology with these or other proteins. On the other hand, we have come across other interesting similarities that exist between BDPP and these other proteins.

1. *Regulation*. Previous studies have established that several of these proteins are involved in regulatory functions of one kind or another, either at the level of transcription (HSV-1 activator Vm 175 [18–20,35]; DNA-directed RNA polymerase σ subunit [21] or intracellular processing (*sec 7*) [22]. BDPP itself has been postulated to be involved in the regulation of calcium phosphate mineralization. However, based on present information, none of these other proteins are involved directly in Ca^{2+} regulation. We note that none of the database searches has resulted in the detection of any sequence homology between the 23 amino acid BDPP N-terminal fragment and known Ca^{2+} regulatory proteins, such as calcitonin, calmodulin, intestinal calcium-binding protein, or calbindin [36,37], namely, the EF hand or helix-loop-helix class of proteins. Unfortunately, we do not know if this 23-residue peptide domain is involved in divalent cation binding in BDPP, or if the sequences in the other proteins participate in the formation of intra- or intermolecular peptide salt bridges as has been demonstrated in the case of BDPP (Evans and Chan, 1991a).

2. *Phosphorylation*. From the existing body of data available to us, it is evident that most of the proteins in this study are phosphorylated. It is known that *sec 7* does bind Ca^{2+}, and is phosphorylated at a number of Ser residues within its sequence [22], a trait shared by BDPP. Similarly, phosphorylation also occurs in CaMV coat protein (primarily as Pser (84%), PThr (14%) and PTyr (2%)) [38,39] and in the

HSV-1 Vmw 175 transcriptional activator protein which has a serine-rich domain [18,40].

Insights into the Evolution of Polyelectrolyte Mineral Matrix Proteins

Does this sequence homology data shed any insights into the evolution of mineral matrix proteins or PMMPs? For the sake of discussion, let us assume that these regulatory proteins, critical for cellular or viral function, appeared prior to the development of mineralization and the appearance of the PMMP. Two mechanisms by which the extracellular PMMP may have evolved are as follows: (1) It might be that existing intracellular regulatory proteins became activated by extracellular stress signals (see below), and were somehow exported outside the cell, where their sequence allowed them to function fortuitously as regulators of available cation concentrations. (2) Conversely, a 'new' protein might have been generated from mutation or rearrangement of existing genomic fragments endowed with the ability to bind divalent cations to insure the survival of the cell under conditions of low $(M^{2+})_{free}$. In either (1) or (2), the function of the 'new' or exported protein might not be to induce mineralization, but rather to bind divalent cations and regulate the $(M^{2+})_{free}$ outside the cell. The mechanism by which this 'new' protein modulated solution-to-solid transformation might have evolved at a later time.

In earlier reports it was speculated that environmental pressure may have played a role in the evolution of mineralization [7,8]. We would like to offer two possible conditions under which the appearance of the PMMP might have taken place:

1. *The availability of free cations outside the cell.* Changes in the free cation concentration might have acted as a 'shock' stimulus, inducing the transcription of proteins to stabilize the cell.

2. *Temperature change.* An interesting trait which three of these diverse non-BDPP proteins possess is their ability to regulate cellular activity in response to alterations in temperature. Temperature-sensitive mutants exist either at the level of transcription (RNA polymerase σ subunit [21]; HSV-1 Vm 175 transcriptional activator [18–20]), or post-translational secretion (*sec* 7 [22]). It is conceivable that temperature shift might have been one of the signals which induced gene expression of proteins which later developed into PMMPs.

The concept of intracellular calcium mechanisms evolving prior to the appearance of calcareous skeleton development has previously been proposed by Lowenstam and Margulis [8]. In their hypothesis, early organisms, most likely bacteria, evolved a means of exporting Ca^{2+} to protect Mg^{2+}-based enzymes and prevent the precipitation of calcium phosphates intracellularly. The development of intracellular calcium-binding proteins was postulated as one response to this cellular demand for Mg^{2+} homeostasis. However, from the evidence available to us, there appear to be very little if any existing sequence homologies between intracellular Ca^{2+} EF hand proteins, such as calmodulin, and BDPP. We suggest that at the same time or later in evolution to divalent cation export, certain template mineralization molecules, distinct from the EF hand Ca^{2+} proteins, made an appearance; and that this appearance was stimulated by events occurring external to the cell. These molecules included proteins which bind Ca^{2+} and transport or sequester it. This reasoning is consistent with the hypothesis put forth by Lowenstam [41,42], in which mineralization developed from 'pin-point' mineral deposition into defined mineral phases that exhibit contiguous, ordered structures. This progression from random foci to ordered phases most likely paralleled the evolution of the PMMP and other organic templates which could exert control over the precipitation and growth of crystal foci. Admittedly, the exact stimulus for this development of Ca^{2+} export and precipitation regulation has not yet been established. However, our present findings pose an interesting evolutionary question which hopefully will be answered by further research: How did mineralization templates, and mineralization, evolve in response to environmental stress?

Acknowledgements

We thank Professor Heinz Lowenstam for his reading of the manuscript and helpful comments. JSE is a recipient of a National Research Service Award from the NIH (DE-05445-02).

References

1. Weiner, S. (1983) *Biochemistry* **22**, 4139.
2. Addadi, L., Moradian, J., Shay, E., Maroudas, N.G. and Weiner, S. (1987) *Proc. Natl. Acad. Sci. USA* **84**, 2732.
3. Uchiyama, A., Suzuki, M., Lefteriou, B. and Glimcher, M.J. (1986) *Biochemistry* **25**, 7572.
4. Gotoh, Y., Sakamoto, M., Sakamoto, S. and Glimcher, M.J. (1983) *FEBS Lett.* **154**, 116.
5. Stetler-Stevenson, W.G. and Veis, A. (1983) *Biochemistry* **22**, 4326.
6. Tsay, T. and Veis, A. (1985) *Biochemistry* **24**, 6363.
7. Lowenstam, H.A. and Weiner, S. (1989) in *On Biomineralization*, Oxford University Press, New York, NY, pp. 260–271.
8. Lowenstam, H.A. and Margulis, L. (1980) *Biosystems* **12**, 27.
9. Mann, S. (1988) *Nature* **332**, 119.
10. Addadi, L. and Weiner, S. (1985) *Proc. Natl. Acad. Sci. USA* **82**, 4110.
11. Narwot, C.F., Campbell, D.J., Schroeder, J.K. and Valkenburg, M.V. (1976) *Biochemistry* **15**, 3445.
12. Marsh, M.E. (1989) *Biochemistry* **28**, 339.
13. Fujisawa, R., Kuboki, Y. and Sasaki, S. (1987) *Calcif. Tiss. Int.* **41**, 44.
14. Stetler-Stevenson, W.G. and Veis, A. (1987) *Calcif. Tiss. Int.* **40**, 97.
15. Linde, A., Bhown, M. and Butler, W.T. (1980) *J. Biol. Chem.* **255**, 5931.
16. Evans, J.S. and Chan, S.I. (1991a) *Biochemistry*, in press.
17. Evans, J.S. and Chan, S.I. (1991b) *Biochemistry*, in press.
18. McGeoch, D.J., Dolan, A., Donald, S. and Brauer, D.H.K. (1986) *Nucl. Acids Res.* **14**, 1727.
19. O'Hare, P. and Hayward, G.S. (1985) *J. Virology* **53**, 751.
20. Dixon, R.A.F. and Schaffer, P.A. (1980) *J. Virology* **36**, 189.
21. Erickson, B.D., Burton, Z.F., Watanabe, K.K. and Burgess, R.R. (1985) *Gene* **40**, 67.
22. Achsteter, T., Franzusoff, A., Field, C. and Schekman, R. (1988) *J. Biol. Chem.* **263**, 11711.
23. Franck, A., Guilley, H., Jonard, G., Richards, K. and Hirth, L. (1980) *Cell* **21**, 285.
24. Chou, P.Y. and Fasman, G.D. (1974) *Biochemistry* **13**, 211.

25. Chou, P.Y. and Fasman, G.D. (1974) *Biochemistry* **13**, 222.
26. Chou, P.Y. and Fasman, G.D. (1977) *J. Mol. Biol.* **115**, 135.
27. Kyte, J. and Doolittle, R.F. (1982) *J. Mol. Biol.* **157**, 105.
28. Bajaj, M. and Blundell, T. (1984) *Ann. Rev. Biophys. Bioeng.* **13**, 453.
29. Johnson, M.S., Sutcliffe, M.J. and Blundell, T.L. (1990) *J. Mol. Evol.* **30**, 43.
30. Chothia, C. and Lesk, A.M. (1986) *EMBO J.* **5**, 823.
31. Jameson, J.S. and Wolf, E.G. (1988) *CABIOS* **4**, 181.
32. Addadi, L. and Weiner, S. (1985) *Proc. Natl. Acad. Sci. USA* **82**, 4110.
33. Lee, S.L., Veis, A. and Glonek, T. (1977) *Biochemistry* **16**, 2971.
34. Weiner, S. and Traub, W. (1984) *Phil. Trans. R. Soc. London* B **304**, 425.
35. Everett, R.D. (1988) *J. Mol. Biol.* **203**, 739.
36. Levine, B.A. and Dalgarno, D.C. (1983) *Biochem. Biophys. Acta* **726**, 187.
37. Strynadka, N.C.J. and James, M.N.G. (1989) *Annu. Rev. Biochem.* **58**, 951.
38. Martinez-Izquierdo, J. and Hohn, T. (1987) *Proc. Natl. Acad. Sci. USA* **84**, 1824.
39. Hahn, P. and Shepherd, R.J. (1982) *Virology* **116**, 480.
40. Wilcox, K.W., Kohn, A., Skylanskaya, E. and Roizman, B. (1980) *J. Virology* **33**, 167.
41. Lowenstam, H.A. (1978) in *Biogeochemistry of Amino Acids* (Hare, P.E., Hoering, T.C. and King, Jr., K., Eds.) John Wiley and Sons, New York, pp. 3–16.
42. Lowenstam, H.A. (1984) *Palaeontology* **2**, 79.
43. Pearson, W.R. (1990) *Methods in Enzymology* **183**, 63.
44. Doolittle, R.F. (1990) *Methods in Enzymology* **183**, 99.

Computational and database retrieval approaches for determining polypeptide conformation

F.R. Salemme, Z. Wasserman and P.C. Weber

Central Research and Development Dept., Experimental Station, E.I. du Pont de Nemours and Company, Inc., P.O. Box 80228, Wilmington, DE 19880-0228, U.S.A.

Introduction

Computational energy methods are finding wide use for the analysis and prediction of polypeptide and protein structure and properties. An important application involves the use of molecular dynamics simulations to investigate aspects of polypeptide conformational flexibility. Points along a dynamic trajectory may be minimized independently, to give a sampling of alternative low-energy conformations, or alternatively, provide the means for estimating some ensemble property of the structure that is characteristic of the set of dynamic states explored during the simulation. One example of the latter application involves a molecular dynamics simulation of the elastin polypeptide as a function of 'stretch', which was aimed at examining the entropic origins of elastomer entropy.

In more general applications, attempts to predict polypeptide structure using computational energy methods typically lead to a multiplicity of alternative local energy minima. Moreover, many of the interactions that may be important in stabilizing an active polypeptide conformation are furnished by the biological receptor, whose structural details are known only infrequently. In this case, database retrieval approaches offer an attractive alternative to molecular mechanics methods as a means of generating polypeptide conformational models. Determination of binding conformations for major histocompatibility antigens provides an example of this approach.

Computational Studies of Elastin Properties

Elastin is a major bioelastomer characterized by the repetitive occurrence of the pentapeptide sequence Val-Pro-Gly-Val-Gly [1]. Thermoelasticity studies of both natural elastin and synthetic elastin polypentapeptides indicate the elasticity is entropic in origin. Classical theories of polymer elasticity [2] suggest that conformational entropy differences between the stretched and unstretched states of a polymer provide the major component of the elastic restoring force. In the simplest terms, this model suggests that stretched molecules are more structurally organized and conformationally regular than unstretched polymers. However, both natural and synthetic elastin polymers [3] undergo a thermal ordering transition between 20° and 40° Centigrade which correlates with an increase in polymer elastic modulus [4]. This association of increased elastic restoring force and increased polymer order appears contrary to expectations of the simple conformational entropy model.

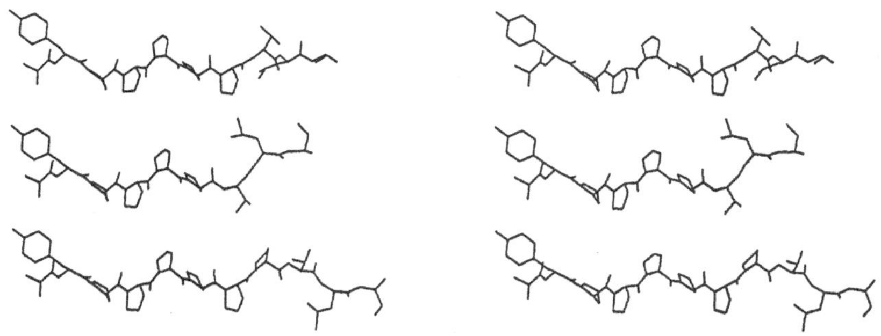

Fig. 1.   Stereo pair longitudinal and axial views of six-pentapeptide segments of the MD-averaged structures for the 'relaxed' and 'stretched' polypentapeptide models (data in Table 1).

Urry proposed an explanation based upon a librational entropy mechanism of elasticity. This mechanism stems from an elastin model of the polymer (Val-Pro-Gly-Val-Gly)$_n$ that is structurally organized as a continuous coil of β-turns interconnected by glycine residues. The fundamental concept is that the elastomeric restoring force originates from reduction of librational entropy in the tripeptides Val-Gly-Val linking successive β-turns as the elastin coil is stretched. An alternative (or complementary) source of elastin entropy changes involves solvent entropy associated with changing hydrophobic surface area of the polymer as it is stretched [5,6]. Stretching can potentially increase hydrophobic side-chain exposure to the aqueous environment, thereby decreasing the entropy of surrounding waters which are assumed to be relatively immobilized in the vicinity of hydrophobic surfaces.

The properties of elastin were investigated by carrying out molecular dynamics simulations of the polymer in various states of extension [7]. The simulations used the AMBER united atom potential function [8]. In each stretched state, a fragment consisting of a spiral polymer of (Val-Pro-Gly-Val-Gly)$_{18}$ was simulated, although statistics were evaluated for only the central 40 residues. The polymers were hydrated with an explicit hydration shell containing from 1200 to 1500 H$_2$O molecules. For each stretched state, an initial conformation was derived by averaging internal torsional coordinates over a 10 ps simulation window, or by molecular modeling methods, followed by energy minimization. Table 1 compares the properties of some of the structures examined, including the original 'unstretched' Urry model [4], together with a similarly 'unstretched' molecular dynamics (MD) averaged model. Although there are several detailed differences in geometry between models, it is notable that the MD averaged model shows nearly exact 3-fold helical

Table 1   *Computed properties of elastin β spiral models*

	Rise per penta-peptide	Penta-peptides per turn	Pitch (Å)	Energy/ residue (kcal/mol)	Entropy/ residue (cal/mol.deg)	Surface area (Å^2)
Urry	3.22	2.63	8.47	−2.4		
Relaxed	2.92	3.00	8.75	−3.1	17.5	2083
Stretched	5.10	2.99	15.24	−2.5	16.4	2501

261

symmetry and also has a somewhat lower energy than the original Urry model with 2.63 pentapeptide units per spiral turn. This seems to illustrate the utility of MD methods to sample low-energy conformations (particularly if structures are slowly cooled during the simulation) and find low-energy states that may be missed by conformational searches. Interestingly, the model retains its average 3-fold helical character when 'stretched' to about 1.5 times its initial length. One potentially interesting consequence of this situation is that it could allow bundles of spirals to stretch without involving intermolecular torsional forces, as is the usual case for coiled-coil biopolymer structural assemblies [9].

Hydrated, unstretched and stretched elastin polymers were each simulated for 100 ps and their internal conformational entropy estimated using the method of Meirovitch et al. [10]. Table 1 illustrates that the stretched structure has a computed decrease in entropy of about 1 cal/mol.deg when the polymer was stretched to 1.5 times its original length. Detailed analysis of backbone angles fluctuations showed that large amplitude torsional motions occur in peptide bonds of residues connecting sequentially adjacent hairpin bends in relaxed elastin, and that many of these reflect correlated crankshaft motions of peptide groups. In contrast, these effects were much reduced in stretched polymer models. In addition to these conformational effects, there are substantial changes in solvent-accessible surface area during the initial stages of elastin stretching (Table 1). Collectively these results suggest that hydrophobic interactions make contributions to elastin entropy at low extensions, but

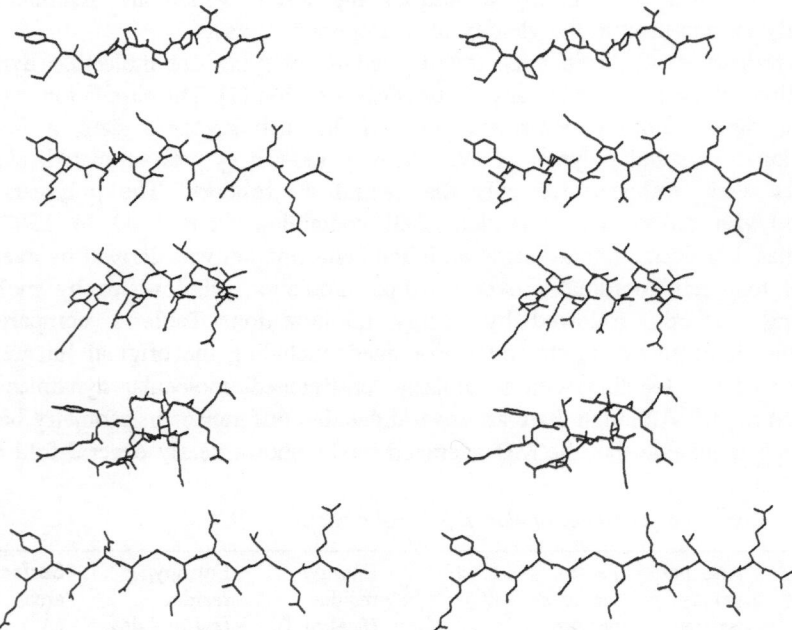

Fig. 2. Top stereogram shows the sequence AYP$_5$TLA, with the central pentaproline sequence organized in the energetically preferred polyproline II conformation, and with the end residue segments AY and TLA oriented in preferred extended conformations. Stereograms below show AYP$_4$TLA and AYP$_6$TLA with the same local conformations for epitopic residues.

that librational mechanisms make larger contributions to the elastic restoring force at longer extensions.

Modeling Major Histocompatibility Complex-Antigen Interactions

Major histocompatibility receptors (MHC) are cell surface molecules found on immune system T-lymphocytes whose function is to present antigenic peptides for recognition by T-cell receptors. The specific interactions formed between the MHC, antigenic peptide, and T-cell receptor result in T-cell restriction, which forms the basis for self-nonself discrimination in higher organisms. Although MHC molecules show a much less extensive pattern of sequence differentiation than immuno-globulins, they are able, nevertheless, to bind a wide range of antigenic peptides. The recent X-ray crystal structure determination of the HLA-A2 MHC molecule shows that the antigen binding site is located in a deep groove formed by a pair of α-helices (one bent) lying on the surface of an antiparallel β-sheet. Amino acid residues that are hypervariable among MHC molecules are predominantly situated at the periphery of the groove, and so provide the basis for MHC-peptide antigen specificity [11].

In order to investigate the limits of antigen binding diversity, Maryanski and Corradin [12] measured the binding properties of a variety of peptides to the mouse MHC H-2Kd. Their data suggested that a common epitopic feature of peptides active in a cytotoxic T-cell assay were the sequence features YX_6TL; e.g. two regions of the peptide separated by an intervening hexapeptide of variable sequence. In order to test this hypothesis, peptides with sequences of the form AYP_nTLA were synthesized and tested in the cytotoxic T-cell assay system. It was found that a peptide with the sequence AYP_5TLA was as effective as the natural peptides in eliciting the cytotoxic biological response. This approach allows definition of the minimal structural determinants that endow HLA peptides with binding specificity, and, in the case of conformationally constrained competitors incorporating oligoproline, makes it possible to draw inferences about the probable conformation of peptides when bound to MHC H2-2Kd.

In contrast to other naturally occurring amino acids, the side chain of proline forms a five-membered ring with the peptide nitrogen, so that backbone torsional rotations are highly restricted. Moreover, owing to interaction between side chains, oligoproline sequences are generally known to form extended, all-trans, 3-fold helical conformations [13]. Thus, the P_5 spacer in AYP_5TLA functions essentially as a rigid connector between the epitopic determinants at the ends of the peptide. In order to determine the backbone and side-chain conformations of the terminal Y,T and L residues of the AYP_5TLA sequence, a library of 48 highly refined protein structures [14] was examined to determine the most frequently occurring conformations for the residues preceding or following proline sequences in globular proteins [15]. The database analysis showed some marked preferences for side-chain orientation for residues adjacent to proline residues in extended polypeptide conformations, particularly for tyrosine (Y) residues. The most probable structures derived from the epitopic regions were then fitted [16] to the central polyproline II helix to define the conformation of the oligoproline antigen mimic. As shown in Fig. 3, these same terminal epitopic conformations are oriented quite differently on oligoproline peptides with different numbers of spacer residues.

The observations that pentaproline is conformationally constrained, and that the competing natural peptide incorporates 6, rather than 5, residues between epitopic residues Tyr...Thr-Leu, allows prediction of a probable conformation of the A24 peptide when bound to MHC H2-2Kd. If it is assumed that the specificity conferring sequences assume similar conformations in both the oligoproline and natural peptides, then the central 6 residues of the natural peptides must differ in conformation from the polyproline helix. Modeling studies show that conformational correspondence of the Y...TL sequences necessitates near correspondence of neighboring residues, so that the difference in the interval backbone conformation is essentially restricted to the central 4 residues of the natural peptide. As illustrated in Fig. 3, this necessitates the introduction of a 3_{10} or similar hairpin bend in the natural peptide in order to achieve near-spatial correspondence with the oligoproline peptide. The probable conformation of the natural peptide when bound to MHC H2-2Kd is consequently one that incorporates a single 3_{10} helical turn, or similar structural kink, in an otherwise nearly extended polypeptide chain.

Although the 'extended-hairpin' motif has not been recognized previously as an important MHC binding conformation, a database search [15] of 48 highly refined

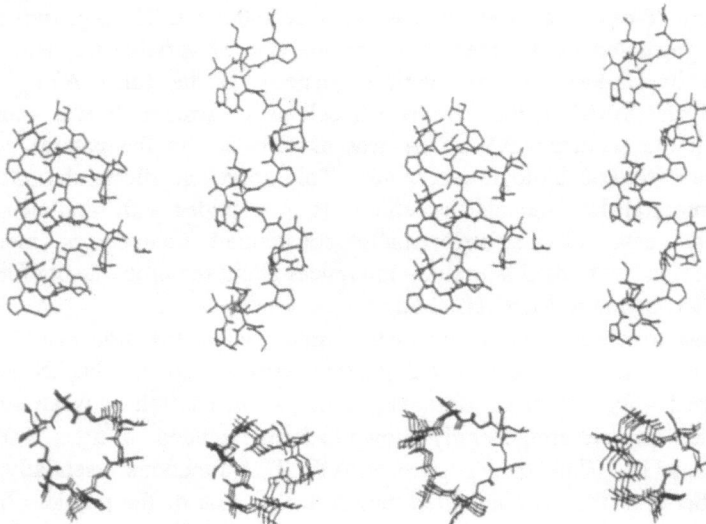

Fig. 3. Top stereogram shows the sequence AYPPPPPTLA, with the preferred conformation suggested from database behavior. The second stereogram shows the A24 peptide sequence [RYLENGKETLQRA], whose binding ability is mimicked, organized in a stable conformation that preserves the distance and approximate orientation of the specificity-conferring side chains in the sequence AYP₅TLA. This conformation consists of an essentially extended polypeptide chain interrupted by a 3_{10} helical turn. The third, fourth, and fifth panels, respectively show the A24 peptide organized as an α-helix, with the conformation experimentally observed in the crystal structure of the homologous region of the HLA-A2 histocompatibility antigen [11] and as an extended polypeptide strand.

protein structures [14] produced 24 hexapeptides of variable sequence, whose backbone atoms fit the extended-hairpin conformation of the natural peptide sequence LENGKE (Fig. 3) with an RMS error of less than 1.0 Å. Although further work is required to test the generality of the present proposal, it seems clear that the extended-hairpin conformation described here could potentially bind deep in the MHC groove, as described for an endogenous peptide in the recent MHC X-ray structure analysis [11].

Summary

Computational tools now provide powerful approaches to studying the structure and properties of polypeptides and proteins. Molecular dynamics methods are particularly useful for investigating the flexibility of structures in the vicinity of local minima, as exemplified by the simulations of elastin. In other cases, where important aspects governing the behavior of the structure may be absent (e.g. the structural details of an MHC antigen binding site), sampling data base fragments of known structures with similar sequence properties can frequently give useful insight into peptide conformational properties that may be difficult to establish using other methods.

References

1. Foster, J.A., Bruenger, E., Gray, W.R. and Sandberg, L.B. (1973) *J. Biol. Chem.* **248**, 2876.
2. Flory, P.J. (1953) *Principles of Polymer Chemistry*, Cornell University Press, Ithaca, NY.
3. Urry, D.W., Haynes, B. and Harris, R.D. (1986) *Biochem. Biophys. Res. Commun.* **141**, 749.
4. Urry, D.W. (1988) *J. Protein Chem.* **7**, 1.
5. Gray, W.R., Sandberg, L.B. and Foster, J.A. (1973) *Nature* **246**, 461.
6. Gosline, J.M. (1978) *Biopolymers* **17**, 677.
7. Wasserman, Z.R. and Salemme, F.R. (1990) *Biopolymers* **29**, 1613.
8. Weiner, S.J., Kollman, P.A., Case, D.A., Singh, U.C., Ghio, C., Alagona, G., Profeta, S. Jr. and Weiner, P. (1984) *J. Am. Chem. Soc.* **106**, 765.
9. Salemme, F.R. (1985) *Ann. New York Acad. Sci.* **439**, 97.
10. Meirovitch, H., Vasquez, M. and Scheraga, H.A. (1987) *Biopolymers* **26**, 651.
11. Bjorkman, P.J., Saper, M.A., Samraoui, B., Bennett, W.S., Strominger, J.L. and Wiley, D.C. (1987) *Nature* (Lond) **329**, 512.
12. Maryanski, J., Verdini, A., Weber, P.C., Salemme, F.R. and Corradin, G. (1990) *Cell* **60**, 63.
13. Sasisekharan, V. (1959) *Acta Crystallogr.* **12**, 897.
14. Brownstone, F.C., Koetzle, T.F., Williams, G.J.B., Meyer, E.F. Jr., Brice, M.D., Rodgers, J.R., Kennard, O., Shimanouchi, T. and Tasumi, M. (1977) *J. Mol. Biol.* **112**, 535.
15. Finzel, B.C., Kimatian, S., Ohlendorf, D.H., Wendoloski, J.J., Levitt, M. and Salemme, F.R. (1989) in *Crystallographic and Modeling Methods in Molecular Design* (Ealick, S. and Bugg, C., Eds.), in press.
16. Kabsch, W.A. (1978) *Acta Crystallogr.*, **A34**, 827.

Structure-activity relationships for 14-membered cyclic dermorphin analogs with two phenylalanines at the third and fourth positions

Murray Goodman* and Toshimasa Yamazaki

Department of Chemistry, 0343, University of California, San Diego, 9500 Gilman Drive, La Jolla, CA 92093-0343, U.S.A.

Introduction

There have been numerous studies of naturally occurring opiate peptides such as enkephalins (Tyr-Gly-Gly-Phe-Leu/Met-OH) and dermorphin (Tyr-D-Ala-Phe-Gly-Tyr-Pro-Ser-NH$_2$). The result of these studies has led to the general agreement of the requirement for the amine and phenolic groups of the Tyr residue at the first position and the aromatic group of the Phe residue at the third or fourth position for biological activity. The nitration at the *para* position of the phenylalanine leads to a pronounced enhancement of activity for the enkephalin analogs while resulting in a decrease of activity for the dermorphin analogs [1]. These results suggest that the aromatic group of the Phe[3] residue in the dermorphin analogs may either interact with a different subsite of the receptor or with a similar subsite in a different orientation from the corresponding group (Phe[4]) in the enkephalin analogs.

Previously we reported the results of the conformational analysis for the 14-membered cyclic enkephalin analogs Tyr-c[D-A$_2$bu-Gly-Phe-(L and D)-Leu] and their partially retro-inverso modified analogs Tyr-c[D-A$_2$bu-Gly-gPhe-(S and R)-mLeu] with a reversed amide bond between residues four and five using ^{1}H NMR spectroscopy and computer simulations [2]. From a comparison of the preferred conformations estimated for these enkephalin analogs with the observed biological activities, we have proposed that relatively extended conformations, in which the two aromatic side chains are oriented in opposite directions from one another with a ~ 14 Å separation, are required for activity at the μ-receptor. On the other hand, folded conformations with a close proximity (< 10 Å) of the two aromatic rings of the Tyr and the aromatic residue at the fourth position assuming nearly parallel side-chain orientation are required for activity at the δ-receptor. Similar folded conformations have been observed for Tyr-c[D-Pen-Gly-Phe-D-Pen]-OH which shows a high activity at only the δ-receptor from both experimental and theoretical investigations [3,4].

Wilkes and Schiller [5] carried out a theoretical conformational analysis of cyclic tetrapeptides, structurally related to the highly μ-receptor-selective dermorphin analog Tyr-c[D-Orn-Phe-Asp]-NH$_2$. According to their results, analogs with high μ-receptor affinity showed a tilted stacking interaction between the Tyr and Phe[3] aromatic rings

---

* To whom correspondence should be addressed.

in low-energy conformations while the same kind of stacking was not possible in low-energy conformers of analogs with poor affinity for the μ-receptor. From these findings, they have concluded that the tilted stacking arrangement of the two aromatic rings may represent a structural requirement for μ-receptor affinity of the cyclic dermorphin analogs. A close proximity of the two aromatic rings has been also observed for Tyr-c[D-Orn-Phe-Asp]-NH$_2$ and Tyr-c[D-Asp-Phe-Orn]-NH$_2$ with high μ-receptor affinity from ^{1}H NMR experiments and molecular dynamics simulations [6].

Since both peptides with the Phe residue at the fourth position (enkephalins) and peptides with the Phe residue at the third position of the peptide sequence (dermorphins) display high opiate activity, it was of interest to prepare opiate peptides containing two Phe residues at positions three and four simultaneously. In order to study structure-activity relationships of opiate peptides with two Phe residues at positions three and four, we synthesized six 14-membered cyclic dermorphin analogs, Tyr-c[D-A$_2$bu-Phe-Phe-(L and D)-Leu], Tyr-c[D-A$_2$bu-Phe-gPhe-(S and R)-mLeu] with a reversed amide bond between residues four and five, and Tyr-c[D-Glu-Phe-gPhe-(L and D)-rLeu] with two reversed amide bonds between residues four and five and between residue five and the side chain of residue two, which are closely related to the enkephalin analogs previously studied incorporating a phenylalanine at the third position in place of the glycine. In this paper, we report an overview of the conformational analysis for the six 14-membered cyclic dermorphin analogs and briefly discuss structure-activity relationships of these molecules.

Materials and Methods

The analogs were prepared in our laboratories by Dr. Odile E. Said-Nejad and Dr. Eduard R. Felder with a combined use of the solid phase technique and peptide synthesis in solution, and purified by reverse phase HPLC. The conformational analyses of the molecules were carried out by ^{1}H NMR and molecular dynamics calculations with nuclear Overhauser effect (NOE) restraints. All of the proton resonances were assigned using two-dimensional homonuclear Hartmann-Hahn and rotating frame nuclear Overhauser (ROESY) experiments. The NOEs were measured in ROESY experiments at mixing times of 75 to 500 ms. Computer simulations were carried out using the DISCOVER program.

Results and Discussion

The in vitro biological activities of the 14-membered cyclic dermorphin analogs are shown in Table 1, measured against the GPI and MVD assays, along with the values for the corresponding enkephalin analogs with a glycine at the third position. These potencies were determined by Dr. Peter W. Schiller and his associates. As compared to [Leu^5]-enkephalin, all the dermorphin analogs are more potent in the GPI assay. With the exception of the two analogs Tyr-c[D-A$_2$bu-Phe-gPhe-R-mLeu] and Tyr-c[D-Glu-Phe-gPhe-D-rLeu] which display reduced activity at the δ-receptor, the

Table 1 *Potencies of 14-membered cyclic dermorphin analogs and their corresponding enkephalin analogs in the GPI and MVD assay*

Compound	GPI[a] IC$_{50}$/nM		MVD[a] IC$_{50}$/nM		MVD/GPI IC$_{50}$-ratio
Tyr-c[D-A$_2$bu-Phe-Phe-Leu]	1.07	± 0.09	4.70 ±	0.92	4.39
Tyr-c[D-A$_2$bu-Phe-Phe-D-Leu]	1.14	± 0.12	13.9 ±	0.9	12.2
Tyr-c[D-A$_2$bu-Phe-gPhe-S-mLeu]	0.518	± 0.099	6.30 ±	1.11	12.2
Tyr-c[D-A$_2$bu-Phe-gPhe-R-mLeu]	17.6	± 2.8	177 ±	31	10.1
Tyr-c[D-Glu-Phe-gPhe-rLeu]	39.0	± 9.3	3.64 ±	0.10	0.09
Tyr-c[D-Glu-Phe-gPhe-D-rLeu]	2.75	± 0.63	49.1 ±	9.9	17.9
Tyr-c[D-A$_2$bu-Gly-Phe-Leu]	14.1	± 2.9	81.4 ±	5.8	5.77
Tyr-c[D-A$_2$bu-Gly-Phe-D-Leu]	66.1	± 8.4	27.1 ±	1.6	0.409
Tyr-c[D-A$_2$bu-Gly-gPhe-S-mLeu]	1.51	± 0.19	7.76 ±	3.17	5.14
Tyr-c[D-A$_2$bu-Gly-gPhe-R-mLeu]	25.5	± 2.0	14.9 ±	5.0	0.584
Tyr-c[D-Glu-Gly-gPhe-D-rLeu]	19.4	± 3.4	313 ±	102	16.1
[Leu^5]enkephalin	246	± 39	11.4 ±	1.1	0.0463

[a]Mean of 3 determinations ± SEM.

remaining analogs show equal or higher potencies in the MVD assay than [Leu^5]-enkephalin.

Comparisons of the biological activities between the 14-membered cyclic dermorphin and the corresponding enkephalin analogs provide some interesting insights. The L-Leu containing dermorphin analog Tyr-c[D-A$_2$bu-Phe-Phe-Leu] is 13–17 times more potent in both the GPI and MVD bioassays than Tyr-c[D-A$_2$bu-Gly-Phe-Leu]. Therefore, both the dermorphin and enkephalin analogs show comparable µ-receptor selectivity. On the other hand, the D-Leu-containing dermorphin analog Tyr-c[D-A$_2$bu-Phe-Phe-D-Leu] is 60 times more potent than Tyr-c[D-A$_2$bu-Gly-Phe-D-Leu] in the GPI assay but only 2 times more potent in the MVD assay, showing higher µ-receptor selectivity while the latter enkephalin analog is non-selective for the µ- and δ-receptors. In the case of the analogs containing a reversed amide bond between residues Phe^4-Leu, Tyr-c[D-A$_2$bu-Phe-gPhe-S-mLeu] displays about the same activity in the MVD assay relative to Tyr-c[D-A$_2$bu-Gly-gPhe-S-mLeu] but is 3 times more potent in the GPI assay, showing higher µ-receptor selectivity than the latter enkephalin analog. Although an enhancement of µ-receptor selectivity is also observed for Tyr-c[D-A$_2$bu-Phe-gPhe-R-mLeu] as compared with Tyr-c[D-A$_2$bu-Gly-gPhe-R-mLeu], this enhancement is due to the greatly reduced activity of the former dermorphin analog at the δ-receptor. The dermorphin analog containing two reversed amide bonds between residues Phe^4-Leu^5 and Leu^5-D-Glu^2 (side chain), Tyr-c[D-Glu-Phe-Phe-D-rLeu], displays 6–7 times higher potencies in both the GPI and MVD assays than the corresponding enkephalin analog, although the activity profiles of these two analogs are quite similar. The analog Tyr-c[D-Glu-Phe-gPhe-rLeu] which is more potent in both the GPI and MVD assays than [Leu^5]-enkephalin shows a higher preference for the δ-receptor over the µ-receptor.

Multiple resonances were observed in the ^{1}H NMR spectra for two of the six dermorphin analogs, resulting from *cis/trans* isomerization about unsubstituted amide

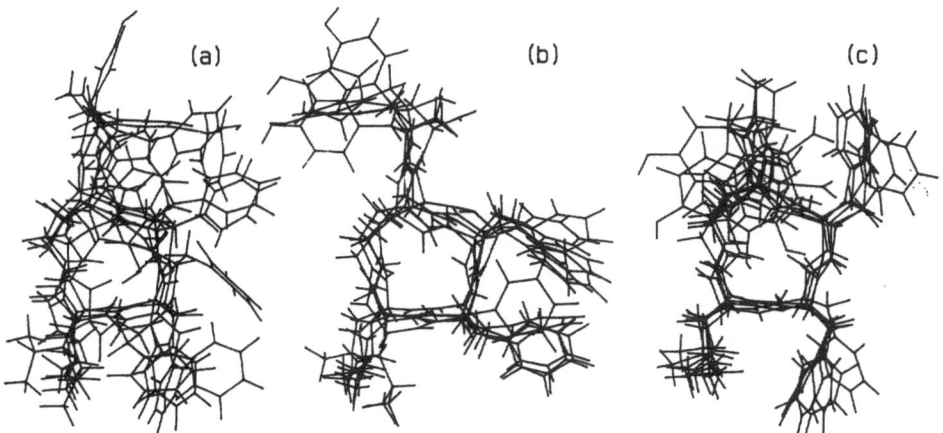

Fig. 1. Superpositions of the structures taken from the molecular dynamics simulations (20 ps): (a) Tyr-c[D-A₂bu-Phe-gPhe-S-mLeu], (b) the major isomer of Tyr-c[D-A₂bu-Phe-gPhe-R-mLeu] with all trans amide bonds (66% at 30°C), and (c) Tyr-c[D-Glu-Phe-gPhe-rLeu].

linkages [7]. The analog Tyr-c[D-A₂bu-Phe-gPhe-R-mLeu] has two additional isomers: one contains a *cis* amide bond between Phe-gPhe (21%) and the other has a *cis* amide bond between R-mLeu-D-A₂bu side chain (13%). The analog Tyr-c[D-Glu-Phe-gPhe-D-rLeu] is composed of only 28% of all *trans* isomer, with 51% of a major isomer containing a *cis* amide bond between D-Glu-Phe and 21% of a minor isomer containing a *cis* amide bond between Phe-gPhe.

The ^{1}H NMR studies show different conformational preferences between the 14-membered cyclic dermorphin analogs and the corresponding enkephalin analogs. The analysis of the temperature coefficients of the amide proton chemical shifts (dδ/dT) and NOEs indicate that Tyr-c[D-A₂bu-Phe-Phe-D-Leu] assumes a type βII turn about the Phe-Phe sequence with an intramolecular hydrogen bond between D-Leu NH and D-A₂bu CO while Tyr-c[D-A₂bu-Gly-Phe-D-Leu] lacks intramolecular hydrogen bonds. The existence of one intramolecular hydrogen bond involving D-A₂bu side chain N$^\gamma$H has been observed for the major isomer of Tyr-c[D-A₂bu-Phe-gPhe-R-mLeu] with all *trans* amide bonds whereas there are no stable intramolecular hydrogen bonds in Tyr-c[D-A₂bu-Gly-gPhe-R-mLeu]. On the other hand, an intramolecular hydrogen bond involving D-A₂bu side chain N$^\gamma$H has been observed for Tyr-c[D-A₂bu-Gly-gPhe-S-mLeu] but no hydrogen bond for Tyr-c[D-A₂bu-Phe-gPhe-S-mLeu]. Although similar intramolecular hydrogen bonding patterns were observed for Tyr-c[D-A₂bu-Phe-Phe-Leu] and Tyr-c[D-A₂bu-Gly-Phe-Leu], the Phe⁴ side chain of the dermorphin analog displays greater flexibility than that of the enkephalin analog. These observations indicate that the replacement of a glycine with a phenylalanine at the third position increases the conformational flexibility of the molecules with the L- and S-residues at the fifth position but reduces flexibility of the molecules with the D- and R-residues at the same position. The small value of dδ/dT observed for D-Glu NH in Tyr-c[D-Glu-Phe-gPhe-rLeu] suggests this amide proton is involved in an intramolecular hydrogen bond.

Superpositions of the structures in molecular dynamics (20 ps) for the Tyr-c[D-A$_2$bu-Phe-gPhe-S-mLeu], Tyr-c[D-A$_2$bu-Phe-gPhe-R-mLeu] (the major isomer with all *trans* amide bonds) and Tyr-c[D-Glu-Phe-gPhe-rLeu] are shown in Fig. 1(a), (b) and (c), respectively. The analog Tyr-c[D-A$_2$bu-Phe-gPhe-S-mLeu] which is superactive at both the μ- and δ-receptors displays large conformational flexibility and adopts both extended and folded conformations required for activity of enkephalin analogs at the μ- and δ-receptors, respectively. The analog Tyr-c[D-A$_2$bu-Phe-gPhe-R-mLeu] which displays activity at only the μ-receptor adopts only extended conformations. In contrast, the highly δ-receptor-selective analog Tyr-c[D-Glu-Phe-gPhe-rLeu] preferentially assumes folded conformations. These results are in complete agreement with our model proposed for the cyclic enkephalin analogs. In addition, it is worthwhile mentioning that a close proximity of the two aromatic rings of Tyr and Phe[3], which seems to be necessary for μ-affinity of dermorphin analogs, is also observed in the folded conformations of Tyr-c[D-A$_2$bu-Phe-gPhe-S-mLeu] and Tyr-c[D-Glu-Phe-gPhe-rLeu]. These results are in agreement with those reported by Schiller et al. [5].

Acknowledgements

The authors gratefully acknowledge the NIH DK 15410, NIH RR03342, NSF BBS86-12359, and a grant of computer resources from the San Diego Supercomputer Center, which helped support this research.

References

1. Schiller, P.W., Nguyen, T.M.-D., DiMaio, J. and Lemieux, C. (1983) *Life Sci.* **33**, 319.
2. Mierke, D.F., Lucietto, P., Schiller, P.W. and Goodman, M. (1987) *Biopolymers* **26**, 1573; Goodman, M. and Mierke, D.F. (1987) in *Protides of the Biological Fluids: Proceedings of Colloquium XXXV* (Peeters, H., Ed.) Pergamon Press, Oxford, p. 457.
3. Hruby, V.J., Kao, L.-F., Petti, B.H. and Karplus, M. (1988) *J. Am. Chem. Soc.* **110**, 3351.
4. Keys, C., Payne, P., Amsterdam, P., Toll, L. and Loew, G. (1988) *Mol. Pharmacol.* **33**, 528.
5. Wilkes, B.C. and Schiller, P.W. (1990) *Biopolymers* **29**, 89.
6. Mierke, D.F., Schiller, P.W. and Goodman, M. (1990) *Biopolymers* **29**, 943.
7. Mierke, D.F., Yamazaki, T., Said-Nejad, O.E., Felder, E.R. and Goodman, M. (1989) *J. Am. Chem. Soc.* **111**, 6847.

Topographical considerations in the design of potent, receptor-selective peptide hormones and neurotransmitters

Victor J. Hruby, W. Kazmierski, T.O. Matsunaga, G.V. Nikiforovich and O. Prakash

Department of Chemistry, University of Arizona, Tucson, AZ 85721, U.S.A.

Introduction

Efforts to develop a rational approach to the design of peptide hormone and neurotransmitter ligands have been hindered because they generally are short, linear polypeptides with many accessible conformations. Our goal has been to develop a general approach utilizing conformational and topographical constraints [1,2], extensive conformational and dynamic investigations, and comprehensive biological assays. We seek to obtain the 'biologically active' conformation(s) (pharmacophore) for interaction of the peptide with specific receptors. Here, we briefly outline two successful applications of this approach which have led to ligands with unique, receptor-specific biological effects.

Results and Discussion

Highly δ opioid receptor-selective cyclic analogs based on enkephalin

Our initial design involved conformational constraints and topographical considerations in which H-Tyr-Gly-Gly-Phe-Met-OH ([Met[5]]enkephalin) was converted via pseudoisosteric cyclization and constraints imposed by the geminal dimethyl effect in a medium-sized ring to a constrained 14-membered ring-containing peptide. This led to the design of [D-Pen[2], D-Pen[5]]enkephalin (H-Tyr-D-Pen-Gly-Phe-D-Pen-OH, DPDPE) [3] which was the most δ opioid receptor selective peptide known.

The conformational properties of DPDPE in aqueous solution have been studied by the use of 2D nuclear magnetic resonance (NMR) spectroscopy (COSY, NOESY, etc.) [4]. The $^3J_{\alpha\beta}$, $^2J_{\alpha\alpha}$ and $^2J_{\beta\beta}$ coupling constants were used to calculate all possible dihedral angles, and distance constraints based upon the observed NOEs provided the basis for molecular mechanics calculations using CHARMM. This led to a conformational model for DPDPE consistent with the NMR data. We have repeated these studies in DMSO-d_6. Though there are chemical-shift differences as a result of the solvent change, basically the chemical-shift parameters remain the same. The changes in coupling constants ($^3J_{H\alpha\text{-}H\beta}$) for the Tyr[1] and Phe[4] residues indicated that in DMSO, these residues are more strongly biased to *trans* and/or *gauche*(−) side-chain rotamers, respectively. In order to better understand the role of side-chain

271

rotamers (topography), and their importance to biological activity, we have performed comprehensive NMR studies for the (S,S)-; (S,R)-; (R,S)- and (R,R)-[β-Me-Phe[4]]DPDPE analogs (II–V), respectively. The best fit chemical shifts and coupling constants of the peptide backbone in these analogs were obtained (Table 1). The large nonequivalence of the chemical shifts of the diastereotopic Gly α-protons, and other NMR parameters indicate that the (S,S)- and (S,R)-β-Me-Phe[4]-DPDPE analogs II and III, and to a lesser extent the (R,S)- and (R,R)-β-Me-Phe[4]-DPDPE analogs IV and V, had backbone conformations similar to DPDPE. The side-chain rotamer populations for Tyr[1] residues were calculated [5]. These results indicate that the *gauche*(+) conformer (+60°) is excluded for the Tyr[1]χ^1 value in these analogs. These NMR observations in conjunction with the biological results demonstrate that the β-Me-Phe[4] residue side chain takes a particular orientation in DPDPE analogs to show high opioid receptor selectivity.

These results have been confirmed by 200 ps molecular dynamics simulations [6]. From this trajectory, 200 configurations were taken at one picosecond intervals and each subjected to energy minimization. A histogram of conformers vs. energy revealed a Gaussian distribution with no discernible preferred family of conformers. However, despite not finding a skewed distribution of low-energy conformers, it was found that the lowest-energy conformers generated had an amphiphilic structure as previously concluded. As with the molecular mechanics studies, the most distinguishing features were the distribution of the side chains of Tyr[1] and Phe[4] and the penicillamine methyls

Table 1   *Proton chemical shifts (ppm) and coupling constants (Hz) of the peptide backbone in DPDPE analogs*

Residue	Peptide	δCHα (ppm)	δNH (ppm)	$J_{NH-C\alpha H}$ (Hz)
D-Pen[2]	I	4.54	8.64	9.1
	II	4.47	8.40	8.6
	III	4.60	8.60	8.0
	IV	4.42	8.70	bs
	V	4.32	8.44	bs
Gly[3]	I	4.38; 3.21	8.58	9.5; 3.1
	II	4.15; 3.10	8.42	9.0; 4.0
	III	4.54; 3.20	8.60	9.0; 4.0
	IV	3.80; 3.65	8.82	6.1; 5.4
	V	3.78; 3.47	8.42	6.1; 5.5
Phe[4]	I	4.30	8.85	8.0
	II	4.20	8.30	8.2
	III	4.38	8.62	7.3
	IV	4.64	7.56	6.3
	V	4.67	7.52	9.4
D-Pen[5]	I	4.33	7.24	8.5
	II	4.18	7.38	7.1
	III	4.25	7.30	8.2
	IV	4.26	7.53	8.2
	V	4.23	7.65	8.9

to form a hydrophobic surface and the backbone moieties to form a hydrophilic surface.

Systematic energy calculation studies also were performed on DPDPE and its four β-MePhe⁴-substituted analogs [7]. The calculations obtained 61, 55, 45, 60 and 72 low-energy structures for DPDPE and (S,S)-; (S,R,)-; (R,S)- and (R,R)-[β-Me-Phe⁴]-DPDPE, respectively. The most distinct conformational features revealed by the calculations are the strong preferences of **g⁻** type rotamers for (S,S)-β-Me-Phe⁴, **t** rotamers for the (S,R)- and (R,S)-β-Me-Phe⁴, and **g⁺** type rotamers for the (R,R)-β-Me-Phe⁴ residues. It is noteworthy also that this systematic conformational search found DPDPE conformers with energies that are significantly lower than those corresponding to structures proposed from NMR data [4,8]. Comparison of peptide backbone structures for each pair of low-energy conformers of DPDPE and its analogs revealed that almost every type of backbone conformer of (S,S)- or (S,R)-[β-Me-Phe⁴]DPDPE is similar to some type of DPDPE backbone, but only a few conformers of (R,S)- or (R,R)-[β-Me-Phe⁴]DPDPE are similar to DPDPE. Recently we have proposed a model for the δ-receptor-bound conformer for DPDPE taking into account the topographical space arrangements of the α-amino group and the $C^ζ$- and $C^γ$-atoms of the Tyr¹ and Phe⁴ side chains, and the $C^α$-atom of the residue in position 2 [7]. It is interesting that the additional comparisons performed in this study with the same atomic centers showed very close similarity of the proposed δ-receptor-bound conformers of DPDPE to low-energy structures of DPDPE with the (S,S)-β-Me-Phe⁴ analog (lowest rms = 0.14 Å, see Fig. 1), reasonable similarity to DPDPE with the (R,R)-β-Me-Phe⁴ analog (lowest rms = 0.85 Å), and poor similarity to DPDPE with (S,R)- (rms ≥ 1.32 Å) and (R,S)-β-Me-Phe⁴ analogs (rms ≥ 3.25 Å). Thus the δ-receptor-bound conformer model proposed [7] is most consistent with low-energy conformers of (S,S)-β-Me-Phe-substituted DPDPE which possesses the space arrangement of Phe aromatic moiety predicted by this model, and is the most δ-selective β-Me-Phe-substituted DPDPE analog.

Design and conformational analysis of somatostatin-derived μ-opioid receptor antagonists

The cyclic tetradecapeptide somatostatin is a regulatory hormone distributed throughout the central nervous system, gastrointestinal tract and pancreas. In addition to its well-established role as a regulatory hormone, somatostatin also has been observed to have weak neurotransmitter-like properties. This pharmacological behavior prompted us to attempt to convert octapeptide analogs to an opioid peptide. This was accomplished via cyclic octapeptide analogs related to somatostatin (1, Table 2), that contain a β-turn template -Phe-D-Trp-Lys-Thr- as a 'scaffold'. Out of all the disulfide bridge combinations: Cys²,Cys⁷; Cys²,Pen⁷; Pen²,Cys⁷; and Pen², Pen⁷, the second one,

H-D-Phe-Cys-Phe-D-Trp-Lys-Thr-Pen-Thr-OH, resulted in high μ/δ and μ/somatostatin ratios in a rat brain receptor-binding assay [9].

Similarly, permutation of Phe and Tyr in positions 1 and 3 gave four possible combinations, with the D-Phe¹,Tyr³ peptide, D-Phe-Cys-Tyr-D-Trp-Lys-Thr-Pen-Thr-OH, exhibiting the desired changes of affinity to all three receptor systems.

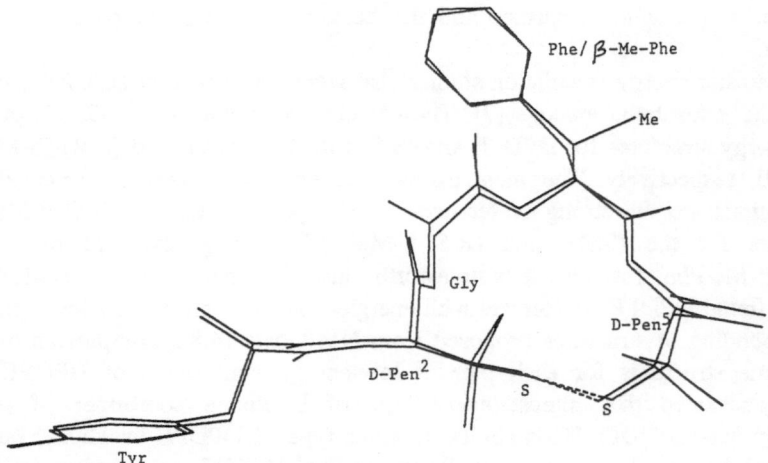

Fig. 1. Comparison of the δ-receptor-bound conformer proposed for DPDPE with one of the low-energy conformers of (S,S)-β-Me-Phe⁴-substituted DPDPE.

Replacement of the C-terminal carboxylate by a carboxamide, resulted in a peptide exhibiting high potency and selectivity towards a μ-opioid receptor [9,10]. Conformational studies using 2D NMR methods [11] strongly supported the maintenance of a preferred β-turn conformation. We then applied topographical considerations for the design of more potent and μ-opioid receptor-selective analogs. This led us to the design of a series of peptides with N-terminal replacement by the conformationally restricted residue D-1,2,3,4-tetrahydroisoquinoline-3-carboxylic acid (Tic) [12,13]. Peptides **1–4** (Table 2) exhibited diverse pharmacological profiles with D-Tic-Cys-Tyr-D-Trp-Lys-Thr-Pen-Thr-NH₂ (**2**) being the most μ-selective and potent opioid antagonist known [11]. On the other hand, peptides **3** and **4** lost significantly in selectivity and potency.

These results prompted us to address several questions: 1. What is the side-chain conformation of the D-Tic residue (only g(−) and g(+) are allowed)?; 2. is Tic altering the backbone conformation of the peptide?; 3. what are the conformational and

Table 2   *Binding potencies and receptor selectivities of substituted analogs of CTP*

Peptide	IC_{50} (nM) [³H]CTOP	IC_{50} [³H]DPDPE
1 D-Phe-Cys-Tyr-D-Trp-Lys-Thr-Pen-Thr-NH₂	3.7 ± 0.8	1,153 ± 116
2 D-Tic-Cys-Tyr-D-Trp-Lys-Thr-Pen-Thr-NH₂	1.2 ± 0.0	9,324 ± 546
3 Gly-D-Tic-Cys-Tyr-D-Trp-Orn-Thr-Pen-Thr-NH₂	278.7 ± 0.5	5,352 ± 503
4 D-Phe-Cys-Tic-D-Trp-Orn-Thr-Pen-Thr-NH₂	1,439.0	> 10,000

topographical differences between peptides **2** to **4** that make them so different pharmacologically?; 4. can one postulate a bioactive conformation on the basis of these ligands?

To answer these questions, we first performed comprehensive 1 and 2D NMR studies on all four peptides. Interproton distances in peptides and proteins can be measured from their NOEs providing isotropic motions and the previous internal distance calibration standards such as Ily (CH^α) and Pro (CH^δ). To overcome difficulties associated with a direct application of this approach, we also used the method of Davis [14]. This involves measurements of longitudinal ($\sigma_{//}$) and transverse ($\sigma_\perp$) cross-relaxation rates, and is based on the observation that $\sigma_{//}$ and $\sigma_\perp$ have different dependencies on τ_c. From these studies, we were able to obtain complete chemical shift assignments including the diastereotopic β-protons [15] as well as 20–40 interproton distances. In the NOE/ROE pattern, we found in each case that there were strong α^i/NH^{i+1}, α^2/α^7 (transannular, across the disulfide bridge) and NH^5/NH^6 cross relaxations, suggesting a β-type turn with a negative helicity of the disulfide bridge [11]. Consequently, we refined the NOE-derived conformations for peptides **1** to **4** with GROMOS utilizing restrained molecular dynamics (RMD) at 800 K (Kazmierski, Ferguson, Hruby and van Gunsteren, unpublished). Resulting structures at 1 ps intervals were 'cooled down' by 5 ps of RMD at 303 K, followed by energy minimization (EM) without constraints. This strategy allowed us to get the molecule to traverse through much of ψ, Φ and ω conformational space (high temperatures), while final energies (303 K, no constraints) were devoid of any artificial NOE potential energy contribution.

Though there were transitions between βII', γ and γ' turns during the dynamics, there was no dramatic alteration in backbone conformation. A peptide bond libration between residue i + 1 (D-Trp) and i + 2 (Lys for **1** and **2**, Orn for **3** and **4**) is responsible for the observed βII' to γ turn transition, noticed for all four peptides simulated here. The βII' turn was the most common conformation. It was seen that while the side chains of Thr^6, Lys^5 and $D-Trp^4$ exhibit significant motions, the side chains of Tyr^3 and $D-Tic^1$ are very restricted (possibly by the environment for Tyr^3, and by cyclization for $D-Tic^1$). The *gauche*(−) side-chain conformation of $D-Tic^1$ in peptide **5** makes this amino acid protrude out from the cyclic peptide, rendering the peptide an extended topography with the three aromatic ($D-Tic^1$, Tyr^3, $D-Trp^4$) rings on one side of the peptide, and the hydrophilic side chains (Thr^6, Lys^5, Thr^8) on the other. Simple attachment of glycine to $D-Tic^1$ in **2** (resulting in peptide **3**) had rather dramatic consequences on the overall shape of the molecule (globular in **3** vs. extended in **2**), while the conformational properties of the backbone are not altered and are described best as either a βII'- or a γ-turn.

Finally, the last peptide simulated in this study, **4**, features an L-Tic in the 3 position instead of Tyr^3. Interestingly, this substitution also leads to a backbone transition from βII' to a γ turn via a peptide-bond plane libration involving the i + 1 and i + 2 residues. Interestingly, the **g**(+) side chain of Tic^3 again results in a folded topography for the peptide with subsequent loss of amphiphilicity.

In summary, all of the peptides **1–4** exhibit similar dynamic properties for their backbone with turns that are in agreement with the NMR data: βII' and some γ. The similar backbone conformations of **1-4** suggest that their significantly different pharmacological properties (Table 2), are related not to their backbones, but rather to the topographical arrangements of their side chains, in particular, those of the aromatic amino acids in positions 1 and 3. On the basis of these results, we suggest that the critical distance and orientation of D-Tic[1] and Tyr[3] side-chain groups represent the bioactive conformational elements for μ-opioid antagonists.

Acknowledgements

This work was supported by grants from the U.S. Public Health Service NS 19972 DA 06284 and DA 04248, and by the National Science Foundation.

References

1. Hruby, V.J. (1982) *Life Sciences* **31**, 189.
2. Hruby, V.J., Al-Obeidi, F. and Kazmierski, W. (1990) *Biochem. J.* **268**, 249.
3. Mosberg, H.I., Hurst, R., Hruby, V.J., Gee, K., Yamamura, H.I., Galligan, J.J. and Burks, T.F. (1983) *Proc. Natl. Acad. Sci. USA* **80**, 5871.
4. Hruby, V.J., Kao, L.F., Pettit, B.M. and Karplus, M. (1988) *J. Am. Chem. Soc.* **110**, 3351.
5. De Leeuw, F.A.A.M. and Altona, C. (1982) *Int. J. Peptide Protein Res.* **20**, 120.
6. Pettitt, B.M., Matsunaga, T., Al-Obeidi, F., Gehrig, C., Hruby, V.J. and Karplus, M. (1990) *Biophys. J.*, submitted.
7. Nikiforovich, G.V., Hruby, V.J., Prakash, O. and Gehrig, C.A. (1990), submitted.
8. Mosberg, H.I., Sobczyk-Kojiro, K., Subramanian, P., Crippen, G.M., Ramalingam, K. and Woodward, R.W. (1990) *J. Am. Chem. Soc.* **112**, 882.
9. Pelton, J.T., Gulya, K., Hruby, V.J., Duckles, S.P. and Yamamura, H.I. (1985) *Proc. Natl. Acad. Sci. USA* **82**, 236.
10. Pelton, J.T., Kazmierski, W., Gulya, K., Yamamura, H.I. and Hruby, V.J. (1986) *J. Med. Chem.* **86**, *29*, 2370.
11. Sugg, E.E., Tourwe, D., Kazmierski, W., Hruby, V.J. and van Binst, G. (1988) *Int. J. Peptide Protein Res.* **31**, 192.
12. Kazmierski, W., Wire, W.S., Lui, G.K., Knapp, R.J., Shook, J.E., Burks, T.F., Yamamura, H.I. and Hruby, V.J. (1988) *J. Med. Chem.*, 2170.
13. Kazmierski, W. and Hruby, V.J. (1988) *Tetrahedron* **41**, 697.
14. Davis, D.G. (1987) *J. Am. Chem. Soc.* **109**, 3471.
15. Kazmierski, W., Yamamura, H.I. and Hruby, V.J. (1991) *J. Am. Chem. Soc.*, in press.

Mimicry and antagonism in biotechnology drug discovery: Recognition peptides in protein scaffolds

Irwin M. Chaiken, Tom J. Graddis, Feng Xian Lu, Michael Brigham-Burke, Sergio Rosé and Daniel J. O'Shannessy

Department of Macromolecular Sciences, SmithKline Beecham, 709 Swedeland Road, King of Prussia, PA 19406, U.S.A.

Introduction

At a time when protein anatomy, folding and recognition are becoming increasingly understood, proteins also are being used increasingly as molecular starting points for drug discovery. Proteins and their component domains can be the therapeutic agents themselves, for example, virus receptor proteins such as CD4 for HIV infection and serum proteases such as tissue plasminogen activator for acute myocardial infarction. Alternatively, the active agents may be inhibitors (or modulators) of proteins, for example, anti-inflammatory agents which neutralize cell adhesion molecules and cytokines.

In both mimicry and antagonism, knowing protein structure can aid in drug design. Proteins can be thought of as composites of conformational frameworks, or scaffolds, in which a limited number of structural elements are incorporated for the 'business end', i.e. surface interaction and consequent function [1]. Thus, if one can identify the critical structural elements on the protein surface which are responsible for interaction and function, these limited regions could be excised and used to create mimics. Alternatively, small molecules which interact with these regions could be used as rationally designed antagonists. The challenges of rational drug design are to determine three-dimensional structures, identify the structural elements required in these structures for either binding or transducing binding into function, and then design recognition molecules which bind to the specific structural surfaces or mimic those surfaces. However, even if one can achieve all of this (certainly not routine), small molecule mimics of proteins and ligands targeted for macromolecular surfaces, when created, are often conformationally flexible and thus not of high enough affinity to be therapeutically viable.

We have been interested in the design of recognition molecules and also in finding ways to restrict their conformations to make them more effective binders. To do this, we have considered the use of host conformational scaffolds into which synthetically designed recognition molecules could be inserted. The basic idea is shown in Fig. 1. The current paper gives a brief report on initial ideas and preliminary experiments aimed to create scaffolded recognition molecules.

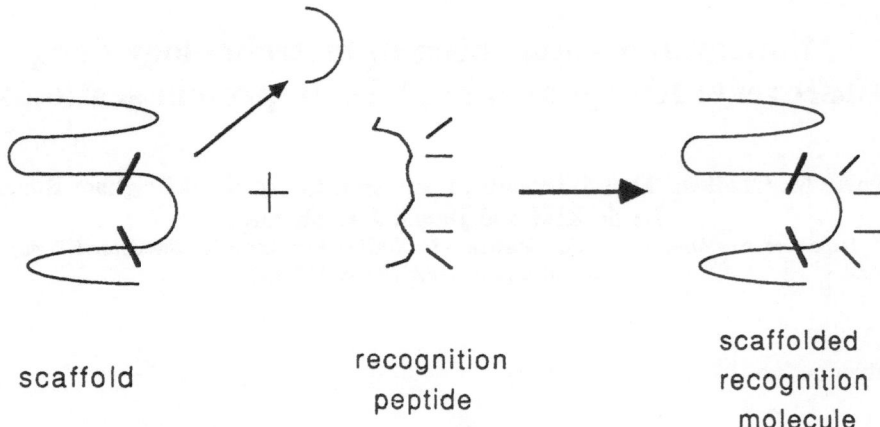

scaffold   recognition peptide   scaffolded recognition molecule

Fig. 1. Schematic diagram depicting the design of recognition molecules through protein scaffolds. The scaffold is a stable folded molecule, in which a contiguous sequence (for example on a loop) can be excised and replaced with a guest sequence without disturbing the scaffold conformation. The guest sequence is a peptide which has an intrinsic propensity (potential or actual) to interact with some other molecular species (for example an enzyme or cell surface receptor). Once inserted into the host scaffold, the guest peptide can assume a more conformationally fixed structure, resulting in enhanced or new recognition properties. The sequence of the guest can be varied and the scaffolded variants screened to select recognition molecules with desirable recognition characteristics (affinity, specificity, etc.).

The scaffold

Any existing protein or subdomain with a well-defined structure could provide the scaffold for a newly designed molecule. Since many proteins have been cloned and their three-dimensional structures determined, scaffolds can be chosen based on the kind of structural element (e.g. α-helix, turn etc.) that is most appropriate as the frame to carry the recognition element. In recent experiments, we have begun to study leucine zipper design and have considered zippers as possible scaffolds. A major reason for this strategy is that zippers, or leucine heptad repeat coiled-coil α-helices, are conformationally stable [2,3] and thus could bring steric order to small peptides inserted into them. Secondly, the sequence pattern of coiled-coil zippers is fairly regular and, we expect, can be simplified and redesigned in order to control their assembly, including affinity and chain orientation. For example, we recently found [4] that a simplified model sequence, shown in Fig. 2, will produce a strong homodimer coiled coil. The major proof for this is from circular dichroism monitoring of α-helix formation (Fig. 2). By inference with other known cases, we assume that the two helices in this simplified coiled coil are parallel in sequence orientation. However, this feature still remains to be demonstrated directly.

Controlled leucine zipper heterodimers likely will be more useful than homodimers as scaffolds for recognition peptides. Here, one zipper of the pair can be synthesized with a recognition element within its sequence. This is shown schematically in Fig. 3. Other zipper scaffolding designs also are possible. Work on heterodimer design currently is in progress.

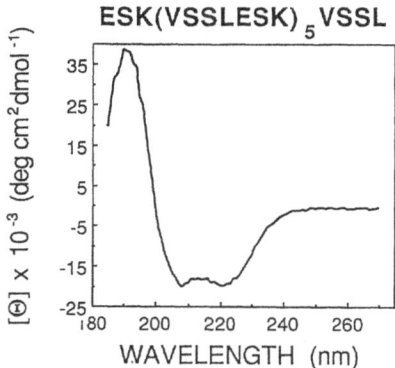

Fig. 2. *Representative far-UV circular dichroism spectrum of 50 μM synthetic leucine zipper homodimer in 10 mM sodium phosphate, pH 7.4. The 42-residue zipper sequence is given at the top in single letter code. The spectral minima at 222 nm and 208 nm suggest significant α-helical content indicative of a coiled-coil structure.*

Recognition peptides

Once a suitable scaffold is designed, we would like to insert recognition peptide sequences into it and use the scaffolding or conformational framework to limit flexibility. The sequence in or adjacent to the recognition peptide could be mutated and sufficiently high-affinity species selected by an affinity capture method. This would allow a search of many local conformational orientations in order to identify those providing optimum binding of the recognition elements. For recognition sequences, we have been interested in the use of antisense (AS) peptides. AS peptides are encoded in the antisense strand of DNA (Fig. 4). These sequences normally are not expressed in cells (phage is an exception). However, in 1969, Mekler [6] predicted that polypeptides encoded in the antisense strand of DNA would recognize 'sense peptides' encoded by the corresponding sense DNA. His prediction was based on indirect theoretical considerations of interaction properties of viral proteins encoded by different strands of nucleic acid and sequence relationships between antibodies and polypeptide antigens. A more recent prediction of antisense peptide recognition

CONTROLLED HETERODIMER

ZIPPER SCAFFOLDS

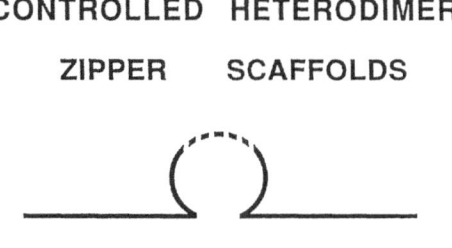

Fig. 3. *Schematic diagram of a leucine zipper scaffold. The zipper heterodimer, represented as parallel lines, forms a coiled-coil structure into which guest sequences may be introduced. A guest peptide sequence, represented as a hatched loop, would be conformationally constrained by the coiled-coil structure of the scaffold.*

de novo Design of Affinity Agents to Native (Sense) Peptides and Proteins

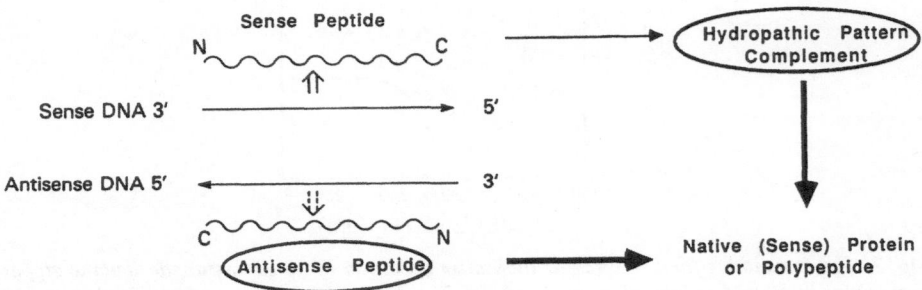

Fig. 4. Scheme showing the relationship of antisense and hydropathic pattern anticomplement peptides to native (sense) peptide sequences and the experimental observation that the former two can bind to the latter. Antisense peptides are those encoded in antisense DNA. Hydropathic pattern anti-complements are peptides synthesized based on the sense peptide sequence directly, with residues in the former being hydrophilic in positions corresponding to residues in the sense peptides which are hydrophobic and vice versa. (Figure taken from ref. 5).

properties was made by Biró [7]. That the sense-antisense peptide interaction actually occurs was observed experimentally by Blalock and his colleagues, initially with the test case of ACTH [8]. This experimental observation in turn has stimulated attempts to examine the mechanistic nature of this interaction, starting with the study of ribonuclease SPeptide by Shai et al. [9].

The experimental results obtained so far with antisense peptides show that these sequences can bind selectively to sense polypeptide sequences in both peptides and proteins (10–21; see for example Table 1). So why bother putting them into scaffolds? The main reasons are (i) the generally observed low affinity, (ii) complex interaction behavior, and (iii) difficulty to detect unique conformations by spectroscopic methods. Dissociation constants observed with AS peptides to AVP are in the range of 10^{-4}–10^{-5}M. Higher affinities (lower K_d values) have been observed in some cases [19], but the lower affinity range is more typical. In addition, AS peptide interactions can show multiple stoichiometry [5,6] and significant degeneracy [5–7]. The overall behavior argues that AS peptides interact with sense peptides so as to form

Table 1 *Comparison of selective affinities of AVP and OT for AVP-AS peptide versus biologically relevant bovine neurophysin II (BNPII) and AVP receptor*

	K_d(M)		AVP receptor
	BNPII	AVP-AS (12-1)	
AVP	2×10^{-5}	5×10^{-5}	4×10^{-10}
OT	2×10^{-5}	5×10^{-4}	720×10^{-10}
AVP/OT affinity ratio	1	10	180

conformationally heterogeneous complexes. Greater conformational uniformity may occur when antisense peptides bind to proteins, which are more constrained into a single conformation.

Restricting the conformation of AS peptides in productive binding modes should improve affinity significantly. Some evidence already exists to support this expectation. The unusually high affinity found for insulin antisense peptide [19] is for a folded protein (insulin) rather than a peptide fragment. Furthermore, AVP bound to receptor binds to the AVP antisense peptide more strongly than does AVP alone [22]. In the two above cases, conformational restriction of the sense partner may effect higher affinity in the AS-sense interaction. A similar effect may be expected if the AS partner is restricted conformationally. Thus, our interest in scaffolds.

By scaffolding antisense peptides in a conformationally constrained protein frame as depicted in Fig. 1, the antisense sequence and flanking sequences can be genetically varied, including by random mutagenesis. This would allow multiple sequence variants to be produced and screened in order to select those with maximal affinity. Data obtained in this way could be extremely helpful to determine just how much sequence degeneracy is allowed for AS peptide recognition. Further, the recently published method of using filamentous phage surface proteins [23,24] to carry recognition peptides could offer a powerful and novel way to scaffold antisense peptides, mutate them (perhaps randomly), and then select for high-affinity peptide sequences by a suitable affinity screen/selection method.

Final comment

Recent results with antisense peptides argue that these peptides have significant selectivity in their binding to sense peptides. However, AS-sense interactions are relatively weak. In addition, AS peptides themselves are generally expected to be flexible and may aggregate to give the multiple stoichiometry sometimes observed in AS-sense complexes. These complications may be ameliorated if AS peptides are incorporated into scaffolds, in which the conformational framework organizes the AS peptides sterically and makes them potentially of higher affinity and selectivity. It remains for future work to experimentally evaluate chimeras of AS peptides in such scaffolds as the leucine zipper.

Acknowledgement

We wish to thank Dr. Ronald Wetzel, of the Department of Macromolecular Sciences, for sharing and discussing his ideas on protein scaffolds that helped shape the scaffold strategies presented in this paper.

References

1. Komoriya, A. and Chaiken, I.M. (1982) *J. Biol. Chem.* **257**, 2599.
2. Cohen, C. and Parry, D.A.D. (1986) *Trends Biochem. Sci.* **11**, 245.
3. O'Shea, F.K., Rutkowski, R., Stafford, W.F. III. and Kim, P.S. (1989) *Science* **245**, 646.
4. Graddis, T.J. and Chaiken, I.M., unpublished results.

5. Chaiken, I.M. (1989) *J. Chromatogr. Biomed. Appl.* **488**, 145.
6. Mekler, L.B. (1969) *Biophysics USSR* (English Translation) **14**, 613.
7. Biró, J. (1981) *Medical Hypotheses* **7**, 981–993.
8. Bost, K.L., Smith, F.M. and Blalock, J.E. (1985) *Proc. Natl. Acad. Sci. USA* **82**, 1372.
9. Shai, Y., Flashner, M. and Chaiken, I.M. (1987) *Biochemistry* **26**, 669.
10. Shai, Y., Brunck, T. and Chaiken, I.M. (1989) *Biochemistry* **28**, 8804.
11. Fassina, G., Zamai, M., Brigham-Burke, M. and Chaiken, I.M. (1989) *Biochemistry* **28**, 8811.
12. Chaiken, I.M. (1988) in *Molecular Mimicry in Health and Disease* (Lernmark, A., Dryberg, T., Terenius, L. and Hökfelt, B., Eds.) Elsevier, Amsterdam, pp. 351–367.
13. Blalock, J.E. and Bost, K.L. (1986) *Biochem. J.* **234**, 679.
14. Mulchahey, J.J., Neill, J.D., Dion, L.D., Bost, K.L. and Blalock, J.E. (1986) *Proc. Natl. Acad. Sci. USA* **83**, 9714.
15. Gorcs, T.J., Gottschall, P.E., Coy, D.H. and Arimura, A. (1986) *Peptides* **7**, 1137.
16. Blalock, J.E. and Bost, K.L. (1988) *Recent Progress in Hormone Research* **99**, 199.
17. Brentoni, R.R., Ribeiro, S.F., Potocnjak, P., Pasqualini, R., Lopes, J.D. and Nakaie, C.R. (1988) *Proc. Natl. Acad. Sci. USA* **85**, 364.
18. Johnson, H.M. and Torres, B.A. (1988) *J. Immunol.* **141**, 2420.
19. Knutson, V.P. (1988) *J. Biol. Chem.* **263**, 14146.
20. Fassina, G., Roller, P.P., Olson, A.D., Thorgeirsson, S.S. and Omichinski, J.G. (1989) *J. Biol. Chem.* **264**, 11252.
21. Ghiso, J., Saball, E., Leoni, J., Rostagano, A . and Frangione, B. (1990) *Proc. Natl. Acad. Sci. USA* **87**, 1288.
22. Lu, F.X., Aiyar, N. and Chaiken, I.M., manuscript in preparation.
23. Scott, J.K. and Smith, G.P. (1990) *Science* **249**, 386.
24. Devlin, J.J., Panganiban, L.C. and Devlin, P.E. (1990) *Science* **249**, 404.

Comparative modeling of proteins in the design of novel cyclic renin inhibitors

Jonathan Greer[a], Charles Hutchins[a], Giorgio Bolis[a,*], Anthony Fung[b] and Hing Sham[b]

[a]Computer Assisted Molecular Design Group and
[b]Cardiovascular Division, Pharmaceutical Products Division, Abbott Laboratories,
Abbott Park, IL 60064-3500, U.S.A.

Introduction

The renin-angiotensin system plays a major role in the control of blood pressure in the body. Angiotensin II, a product of the cleavage of angiotensinogen, first by renin and subsequently by angiotensin-converting enzyme (ACE), is one of the most potent pressor molecules. A variety of ACE inhibitors currently approved and available for the treatment of hypertension provide clear evidence of the efficacy of inhibiting the formation of angiotensin II in the treatment of hypertension. Consequently, significant effort has been devoted to the discovery of inhibitors of renin as potential therapeutic agents which would also prevent the formation of angiotensin II.

Renin is an aspartic proteinase, and the crystal structures of several of these enzymes have been reported [1–4]. Comparative modeling studies [5–11] have been performed by several groups to produce a tentative three-dimensional structure for human renin [12–21] to use in support of the drug design process. We have modeled the structure of human renin using the known structures of three fungal enzymes [1–3] and the mammalian enzyme, porcine pepsin [4]. These modeling studies have been reported [19,20,22].

In this paper, we describe the use of our model structure in the design of novel cyclic compounds which were intended to partially restrict the inhibitors to the conformation when bound in the active site of renin. Cyclic renin inhibitors were first produced by Boger et al. [23,24]. They bridged a peptide inhibitor of renin between the side chains of the P_5 and P_2 residues[**] (Fig. 1) and showed that a 16-membered ring was optimal. In a further modeling study based upon the rhizopuspepsin-pepstatin crystal structure [2], Boger et al. [23,24], linked the P_2 residue side chain to the P'_3 carboxy terminal residue using a disulfide bridge (Fig. 1). Cysteine and homocysteine residues were tried in each case to optimize the potency of the cyclic compounds. Very recently, a macrocyclic series was reported bridging between the side chains of

---

*   Present address: Farmitalia, Milan, Italy
** The nomenclature used to describe the inhibitor residues is that of Schechter and Berger [31]. Residues N-terminal from the scissile bond are labeled P_1, P_2, P_3, ..., P_n with the number increasing as one moves away from the scissile bond. Residues C-terminal are labeled P'_1, P'_2, ..., P'_n, once again increasing as one moves away from the scissile bond.

positions P_2 and P_1' (Fig. 1) [25,26]. The compounds were composed of 13- or 14-membered cyclic lactams.

Using our own model of renin we undertook to design new cyclic compounds to address the following goals:

1. At least maintain and, if possible, enhance inhibitory potency.
2. Maintain the high specificity of the inhibitors for renin. This is essential to avoid unwanted side effects.
3. Increase metabolic stability of the inhibitor. In particular, chymotrypsin was known to cleave between the P_3 and P_2 residues to inactivate the inhibitors.
4. Create novel, patentable compounds.

Results and Discussion

Cyclic compounds connecting P_1 and the main-chain amide of P_2

When the structure of the hexapeptide reduced amide inhibitor was examined in the active site of renin, we found that the P_1 side chain pointed towards the hydrogen of the main-chain amide of His at the P_2 site (Fig. 1). Accordingly, we designed several analogs with different chain lengths connecting the P_1 side chain to this main-chain amide group. Rings of 9, 10, 11, 12, and 14 members were examined graphically and minimized in the active site of the renin model using the energy program VFFPRG of Hagler and co-workers [27]. We decided that compounds with ring sizes between 10 and 14 could fit reasonably well in the active site causing only

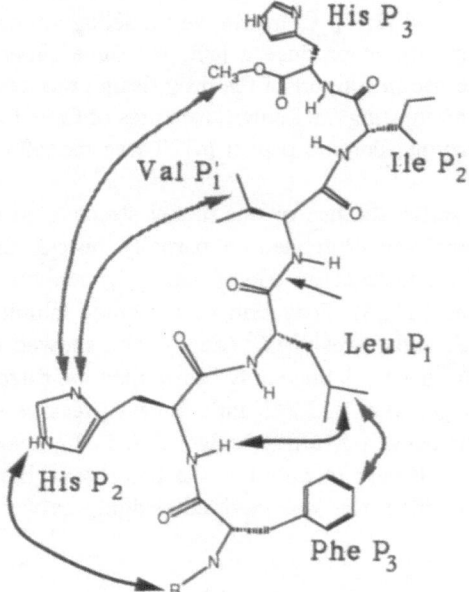

Fig. 1. *Schematic of the renin inhibitor in its conformation in the active site of the enzyme. Notice that it has an extended conformation. The cyclic structures that have been reported in the literature are shown with gray arrows connecting the sites that are bridged. The cyclic series that are described in this paper are shown with solid black arrows.*

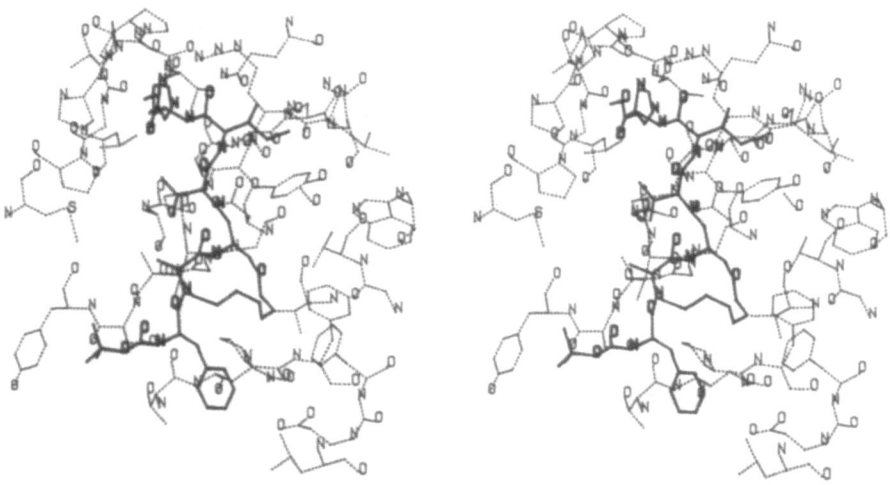

Fig. 2. Structure of the 14-membered ring, 5, bound to the active site of renin. The dashed lines show the various hydrogen bonds that are formed between the inhibitor and the residues on the enzyme.

small changes in the conformation of the flap, including residues 76 through 79* of renin. This flap was known to be very flexible from one aspartic proteinase to the next and probably moves to allow the substrate or inhibitor access to the active site. As the ligand is bound, the flap moves into contact with the inhibitor and forms a variety of hydrogen bonds with it (Fig. 2). These bonds were preserved in these cyclic analogs, with the exception of that between the NH of His P_2 and Thr[77] of the flap.

Accordingly, three analogs were synthesized in this series (Table 1). These include the 10-, 12- and 14-membered ring compounds [20]. The scissile bond was substituted with the reduced amide to produce an inhibitor, which was the standard substitution being employed at the time of this experiment. When the compounds were tested in the renin inhibition assay, the results were rather unexpected. The 10-membered ring compound, 3, the first to be synthesized, was inactive as a renin inhibitor up to 10^{-4} M. The 12-membered ring compound, 4, was tested next and found to be slightly active at 69 μM. The last of the series, the 14-membered ring, 5, was 2.4 μM or within a factor of 2 of the potency of the parent compound (Table 1).

When the compounds were tested for activity against pepsin, no significant inhibition was observed up to 10^{-4} M, so specificity for renin was preserved in this series. The compounds were also tested for susceptibility to digestion by chymotrypsin [20]. Under conditions where the parent had a half-life of 10 minutes (Table 1), the 14-membered cyclic compound, 5, was completely stable to chymotrypsin.

---

* The numbering system used here for renin is that of pepsin, the classical aspartic proteinase studied. This allows facile comparison between different aspartic proteinases without having to convert the residue numbers from one member of the aspartic proteinase family to the next.

Table 1 *Cyclic inhibitors from side chain of P_1 to main-chain NH of P_2*

| | | | Boc-Phe-Ala-Ser^{R}Val-Ile-His-OMe | | |
| | | | $(CH_2)_n$ | | |
n		Ring size	IC$_{50}$ (μM) human renin	% Trans conformation	Chymotrypsin half-life (min)[a]
1	Parent		1.4	100	10
2	NMeAla		1.2	100	–
3	3	10	> 100	0	–
4	5	12	69	20	–
5	7	14	2.4	50	stable

[a] Conditions for the chymotrypsin assay were reported in Sham et al. [20].

Summing the results of this series, compounds were synthesized which were novel, almost equipotent to the parent, highly specific, and stable to metabolic cleavage by chymotrypsin. However, we were puzzled by the result that the 10-membered ring compound was inactive while the 14-membered ring was nearly as potent as the parent. In the modeling on the renin active site, we could see no reason for this striking difference. Consequently, we embarked upon a detailed NMR structure analysis of the compounds of this series to see if we could discern any differences among the three molecules [28]. The NMR spectra showed an important difference between the three analogs. Compound **3**, the 10-membered ring, was found to have a cis-peptide bond between residues P_3 and P_2. Indeed, this peptide bond was found to be completely in the *cis* conformation (Table 1). While it is true that substitution at the amino group of the peptide bond tends to stabilize the *cis* conformation, as is seen with proline residues, nevertheless, it is somewhat surprising in this case since the N-methylated Ala linear analog, **2**, was found to be completely in the *trans* form. The NMR analysis suggested that stabilization of the *cis* form was due to a hydrophobic interaction between the phenyl ring of Phe P_3 and the methyl of the side chain of Ala P_2. This interaction was clearly observable from the ring shift that was present in the NMR spectra for the Ala P_2 methyl in the *cis* conformation.

When the *cis* conformation structure of these cyclic compounds was examined in the active site of the model of renin, it was clear that, unlike the trans, it could not fit. It caused a major overlap of the P_3 site with residues Ala^{218} and Ser^{219} of the enzyme. Consequently, the discovery of the *cis* form explained the inactivity of the 10-membered ring compound. The 12-membered ring analog, **4**, was observed to have about 20% trans, 80% cis. This would explain the small amount of activity that was observed. The 14-membered ring, **5**, on the other hand, was 50% in the *trans* conformation. Therefore, significant activity was observed. Indeed, if we correct for the 50% *cis* form, which is presumably inactive, then the 14-membered ring analog is equipotent with the parent.

This study taught the very important lesson that unproductive conformations of the designed molecule in solution can greatly affect the binding effectiveness of compounds. Consequently, in considering the design process, it is important to carefully assess the energetics of the compounds free in solution as well as in the active site of the enzyme or receptor.

Cyclic compounds connecting the N-terminus of P_3 and the side chain of P_2

After the experience of designing and modeling cyclic compounds between P_1 and the main-chain amide of P_2 we reexamined the renin-inhibitor complex model and wanted to build an inhibitor that bridged the side chains of P_1 and P_3 (Fig. 1). Unfortunately, there was great reluctance to embark on the synthesis of these cyclic compounds because of the crucial importance of the Phe at P_3 and especially the cyclohexylalanine at P_1 to the renin inhibitory potency [29]. In fact, a series of cyclic compounds of this type was prepared by Rich and co-workers as inhibitors of the homologous enzyme, pepsin. Some of these were very potent, i.e. in the nanomolar range, and equipotent to the corresponding linear pepsin inhibitor (D. Rich, personal communication).

As an alternative set of cyclic compounds, we designed a series that connected the N-terminus of position P_3 with the side chain at position P_2. We decided to construct a compound that we thought would be quite rigid, yet fit well in the renin model when examined using energetics. We also designed a parallel series of cyclic molecules that would be somewhat more flexible and thus better able to compensate for the inaccuracies of our renin model. To facilitate the synthesis, we chose molecules that would have an extension on the N-terminus of P_3 which would form a urea linkage with an appropriate residue on the side chain of P_2 [30]. Compounds were

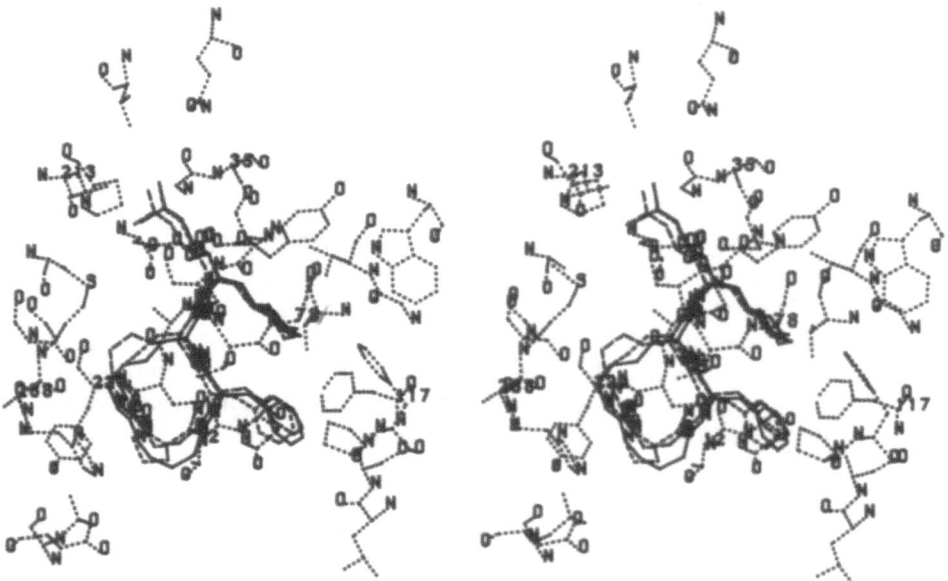

Fig. 3. *The three cyclic inhibitors, 6–8, shown in Table 2, are depicted here bound to the active site of the renin model as minimized to accommodate compound 6. Notice how similar the binding modes are for the three molecules. Nevertheless, their inhibitory potencies are very different. This demonstrates how difficult it is to estimate or calculate inhibitory potency from a model structure (see text).*

287

Table 2 *Cyclic inhibitors from the N-terminus of P_3 to the side chain of P_2*

Structure	IC$_{50}$ (nM)
6	7,000
7	12
8	6.7

designed with 14-, 15- and 16-membered rings. Examples of this series are given in Table 2.

Compound **6** was constructed to maximize the rigidity of the macrocycle yet fit in the active site of the renin molecule (Fig. 3). To rigidify the molecule, a proline-based five-membered ring was included in the macrocycle. When energy minimized in the active site, with the program DISCOVER (Biosym Technologies), this inhibitor was found to fit very well, with very little significant distortion of the active site residues of renin. When the activity was measured, however, it was found to be a disappointing 7,000 nM. Other, more flexible forms, including 15- and 16-membered rings, without the additional constraint of the five-membered ring, were significantly more active with an IC$_{50}$ of 6.7 nM for compound **8** (Table 2).

If one examines the energy-minimized structures of the three molecules of Table 2 in the renin enzyme structure, Fig. 3, one can see that the conformations of the three analogs are very similar. While the structure for the enzyme of only one of the complexes is shown in this figure to avoid confusion, they are also very similar and do not indicate that the expected potencies should be very different. This experiment shows how difficult it is to predict potency based upon a model structure such as that of renin. Perhaps with a more accurate crystal structure, more effective predictions may be achievable.

References

1. James, M.N.G. and Sielecki, A.R. (1983) *J. Mol. Biol.* **163**, 299.
2. Bott, R., Subramanian, E. and Davies, D.R. (1982) *Biochemistry* **21**, 6956.
3. Pearl, L. and Blundell, T. (1984) *FEBS Lett.* **174**, 96.
4. Andreeva, N.S., Zdanov, A.S., Gustchina, A.E. and Fedorov, A.A. (1984) *J. Biol. Chem.* **259**, 11353.
5. Browne, W.J., North, A.C.T., Phillips, D.C., Brew, K., Vanaman, T.C. and Hill, R.L. (1969) *J. Mol. Biol.* **42**, 65.
6. McLachlan, A.D. and Shotton, D.M. (1971) *Nature New Biol. (London)* **229**, 202.
7. Greer, J. (1980) *Proc. Natl. Acad. Sci. USA* **77**, 3393.
8. Greer, J. (1981) *J. Mol. Biol.* **153**, 1027.
9. Greer, J. (1985) *Ann. N.Y. Acad. Sci.* **439**, 44.
10. Feldmann, R.J., Bing, D.H., Potter, M., Mainhart, C., Furie, B., Furie, B.C. and Caporale, L.H. (1985) *Ann. N.Y. Acad. Sci.* **439**, 12.
11. Greer, J. (1990) *Proteins* **7**, 317.
12. Sibanda, B.L., Blundell, T., Hobart, P.M., Fogliano, M., Bindra, J.S., Dominy, B.W. and Chirgwin, J.M. (1984) *FEBS Lett.* **174**, 102.
13. Blundell, T., Sibanda, B.L. and Pearl, L. (1983) *Nature (London)* **304**, 273.
14. Carlson, W., Karplus, M. and Haber, E. (1985) *Hypertension* **7**, 13.
15. Carlson, W., Handschumacher, M., Karplus, M. and Haber, E. (1984) *J. Hypertension* **2**, 281.
16. Carlson, W., Haber, E., Feldmann, R. and Karplus, M. (1984) in *A Model for the Three-dimensional Structure of Renin* (Hruby, V.J., Rich, D.J., Eds.) Pierce Chemical Co., Rockford, IL, pp. 821–824.
17. Akahane, K., Umeyama, H., Nakagawa, S., Moriguchi, I., Hirose, S., Iizuka, K. and Murakami, K. (1985) *Hypertension* **7**, 3.
18. Raddatz, E., Schittenhelm, C. and Barnickel, G. (1985) *Kontakte (Darmstadt)* **3**, 13.
19. Plattner, J.J., Greer, J., Fung, A.K., Stein, H., Kleinert, H.D., Sham, H.L., Smital, J.R. and Perun, T.J. (1986) *Biochem. Biophys. Res. Commun.* **139**, 982.
20. Sham, H.L., Bolis, G., Stein, H.H., Fesik, S.W., Marcotte, P.A., Plattner, J.J., Rempel, C.A. and Greer, J. (1988) *J. Med. Chem.* **31**, 284.
21. Hutchins, C.W. and Greer, J. (1990) *CRC Reviews*, in press.
22. Bolis, G. and Greer, J. (1989) in *Role of Computer-aided Molecular Modeling in the Design of Novel Inhibitors of Renin* (Perun, T.J., Propst, C.L., Eds.) Marcel Dekker, Inc., New York and Basel, pp. 297–326.
23. Boger, J. (1983) in *Renin Inhibitors: Design of Angiotensinogen Transition-state Analogs Containing Statine* (Hruby, V.J., Rich, D.J., Eds.) Pierce Chemical Co., Rockford, IL, pp. 569–578.
24. Boger, J. (1986) in *Renin Inhibitors: Drug Design and Molecular Modelling* (Lambert, R.W., Ed.), pp. 271–292.
25. Rivero, R.A., Greenlee, W.J., Patchett, A.A., Halgren, T.A., Doyle, J.J. and Siegl, P.K.S., *22nd National Medicinal Chemistry Symposium*, 1990.
26. Weber, A.E., Steiner, M.G., Dhanoa, D.S., Fitch, K.J., Doyle, J.J., Lynch, R.J., Halgren, T.A., Siegl, P.K.S., Parsons, W.H., Greenlee, W.J. and Patchett, A.A., *200th National American Chemical Society Meeting*, 1990.
27. Dauber, P., Osguthorpe, D. and Hagler, A.T. (1982) *Biochem. Soc. Trans.* **10**, 312.
28. Fesik, S.W., Bolis, G., Sham, H.L. and Olejniczak, E.T. (1987) *Biochemistry* **26**, 1851.
29. Boger, J., Payne, L.S., Perlow, D.S., Lohr, N.S., Poe, M., Blaine, E.H., Ulm, E.H., Schorn, T.W., LaMont, B.I., Lin, T.-Y., Kawai, M., Rich, D.H. and Veber, D.F. (1985) *J. Med. Chem.* **28**, 1779.
30. Sham, H.L., Rempel, C.A., Stein, H. and Cohen, J. (1990) *J. Chem. Soc., Chem. Commun.*, 666.
31. Schechter, I. and Berger, A. (1967) *Biochem. Biophys. Res. Commun.* **27**, 157.

Helical transitions in peptides containing multiple α,α-dialkyl amino acids

Garland R. Marshall*, Denise D. Beusen, John D. Clark and
Edward E. Hodgkin**

Center for Molecular Design, Washington University, St. Louis, MO 63130, U.S.A.

Introduction

The presence of α,α-dialkyl amino acids, such as α-methylalanine (MeA, aminoisobutyric acid, Aib), in microbial natural products, such as the peptaibol antibiotics, argues strongly for a special role related to function. One aspect is the conformational restrictions imposed by these amino acids [1–6]. Despite the variety of conformations theoretically available to α,α-dialkyl amino acids, the impact of multiple substitutions of these amino acids on the overall conformation of a peptide is dramatic forcing of an α- or 3_{10}-helical conformation in most cases. The crystal structure of alamethicin [7], which contains eight MeA residues out of twenty, is predominantly α-helical, with NMR data [8] supporting a similar solution conformation in methanol. According to Marshall et al. [9], those factors which govern helical preference include the inherent relative stability of the α-helix compared with the 3_{10}-helix, the extra hydrogen bond seen with 3_{10}-helices, and the enhanced electrostatic dipolar interaction of the 3_{10}-helix when packed in a crystalline lattice. The balance of these forces, when combined with the steric requirements of the amino acid side chains, determines the relative stability of the two helical conformations under a given set of experimental conditions.

The primary interest in peptaibol antibiotics is related to their ability to induce voltage-dependent conductance changes in artificial bilayer membranes [10,11]. Transitions between the two helical forms may be responsible, in part, for the sensitivity to voltage seen with pores formed by peptaibol antibiotics, such as alamethicin [10,11]. One motivation to consider the 3_{10}-helix is the activity of emerimicin on bilayers. Its length of 15 residues is insufficient to span the dielectric thickness of the membrane in the α-helical conformation (15×1.5 Å per residue = 22.5 Å), while the 3_{10}-helical conformation (15×2.0 Å per residue = 30 Å) is sufficient. The work of Tosteson et al. [12] showing that a 22-residue segment of the S4 repeat of the sodium channel is capable of forming voltage-gated channels stimulates consideration of the 3_{10}-helix as the active form, as positively charged Lys or Arg occurs at every third residue. A bundle of four amphipathic α-helices has been suggested as a plausible structure for the pore-forming element of voltage-gated channels [13]. Transition between the α- and 3_{10}-helix would change the relative

---

* To whom correspondence should be addressed.
** Present address: British Biotechnology, Ltd., Oxford OX4 5LY, U.K.

orientation of side chains at the channel interface, increasing the opposition of Lys and Arg side chains, and would increase the diameter of the pore by electrostatic repulsion. This helical transition with the associated increase in length per residue of 0.5 Å should also be considered as a possible transduction mechanism for trans-membrane signaling in receptors which have a single transmembrane segment, such as the insulin receptor.

In order to evaluate the feasibility of the 3_{10}-helix having a significant role in biological systems, we have determined the relative stability of model systems containing multiple α,α-dialkyl amino acids in the two helical conformations and determined the transitional energetics through molecular dynamics simulations.

Methods

SYBYL (Tripos Associates Inc., St. Louis, MO) and AMBER [14] molecular modeling software were used. Potentials of mean force were calculated using the method of umbrella sampling [15]. The distance between the α-carbons of residues 1 and 10 was used as the reaction coordinate, R, and was restrained by a harmonic potential. Ensemble averages were corrected in order to correspond to thermodynamic variables using the method of Valleau and Torrie [16].

Results and Discussion

It was necessary to establish a quantitative theoretical basis for understanding the relative stability of α- and 3_{10}-helical structures. In order to characterize possible helical structures, minimizations of oligomers of $(MeA)_n$ and $(Ala)_n$, where n = 1 to 15, were performed [17] in order to separate the intrinsic, or length-independent, stability from end effects. The evaluation of MeA oligomers as a function of length showed an increased stability of approximately 1.94 kcal/mol/residue for the isolated α-helix as compared with the 3_{10}-helix for oligomers of MeA. This compares favorably and refines the estimate of between 0.3 and 3.6 kcal/mol/residue by Prasad and Sasisekharan [4]. The enthalpic difference between the 3_{10}-helix and the α-helix was −2.36 kcal/mol/residue for similar oligomers of alanine, where previous workers [4] had estimated between −3.3 and −3.6 kcal/mol/residue. These calculations are consistent with the small number of observations of 3_{10}-helix in proteins [18] as well as the increased probability of 3_{10}-helix with increased content of MeA residues.

In a 3_{10}-helix, an additional hydrogen bond is present for the same length peptide due to 4-to-1 rather than the 5-to-1 hydrogen bonding of the α-helix. In solution, a length of peptide is reached where the inherent increased stability of the α-helix dominates the energy contribution of the single additional hydrogen bond formed in a 3_{10}-helix. Our calculations [17] in vacuo estimate the difference in end effects between the α- and 3_{10}-helix (primarily the extra hydrogen bond) to be 13 kcal/mol. In short peptides, a 3_{10}-helix conformation is favored in solvents of low dielectric due to one additional hydrogen bond which can be satisfied. As the peptide length increases, the inherent stability of the α-helix compensates for the lost hydrogen bond and the α-helix becomes dominant in solution. Modulation of the 3_{10}- to α-helix transition length by variation of the dielectric constant of the solvent should be experimentally observable. While the exact transition length will depend on the solvent and peptide sequence, we can estimate that most nonpolar solvents would require a length longer than seven residues for the α-helix to be favored.

Evidence from solution NMR studies support helix preference as a function of solvent polarity. Pentapeptides containing repetitive MeA-Ala or MeA-Val sequences showed [19] three intramolecular hydrogen bonds in both $CDCl_3$ and $(CD_3)_2SO$, consistent with 3_{10}-helical structure. Heptapeptides of similar sequence [19] had five hydrogen bonds in $CDCl_3$ (implying a 3_{10}-helix), but only four hydrogen bonds in the more polar $(CD_3)_2SO$, consistent with α-helical conformation. Balaram et al. [20] have reported that two decapeptides, Boc-MeA-Val-MeA-MeA-Val-Val-Val-MeA-Val-MeA-OMe and Boc-MeA-Leu-MeA-MeA-Leu-Leu-Leu-MeA-Leu-MeA-OMe, show the presence of eight intramolecular hydrogen bonds in $CDCl_3$, but only seven hydrogen bonds in $(CD_3)_2SO$. This is consistent with a transformation from 3_{10}-helix to α-helix upon increasing the polarity of the solvent and decreasing the hydrogen-bond strength. Recent NMR studies (Beusen et al., unpublished) of emerimicin 1–9 benzyl ester in which all resonances have been assigned indicate predominantly 3_{10}-helix in $(CD_3)_2SO$, as contrasted with the α-helical structure found in the crystal [9]. It is interesting that the solution data on an emerimicin 2–9 fragment, Z-MeA-MeA-MeA-Val-Gly-Leu-MeA-MeA-OMe, were consistent with the presence of a right-handed α-helix [21] in trifluoroethanol, a hydrogen-bonding solvent, whereas the peptide has a 3_{10}-helical conformation in the crystal.

In the crystal, intermolecular interactions become important and may actually dominate. The predicted enhanced stability of the α-helix in solution once a critical length is reached stands in contrast to the observation of 3_{10}-helical structure in crystals for numerous oligomers of MeA [22] as well as poly-MeA [23] itself. One factor governing the crystal structures in poly-MeA is an improved packing for the 3_{10}-helix as compared with the α-helix arising from the interdigitation of the methyl side chains. Another major aspect is electrostatic, as the 3_{10}-helix has a smaller radius (1.9 versus 2.3 Å) than the α-helix and a similar dipole moment, 35.6 versus 36.8 Debye for the MeA decamer, which would result in a stronger electrostatic interaction during antiparallel helical stacking. The α-helical peptides containing multiple α,α-dialkyl amino acids seen in crystals, therefore, arise when the sequence of residues leads to unfavorable packing of the 3_{10}-helix.

Since the calculations and experimental data both suggest that peptides containing multiple α,α-disubstituted amino acids are capable of assuming both helical forms, it was important to determine more precisely their relative free energies and the transition barrier between the two helices. Geometrical parameters of MeA residues from crystal structures of helical peptides were analyzed to identify a pathway for a transition between the two helical forms. A plot of helical pitch (axial rise per residue) against period (residues per turn) indicated a well-defined reaction coordinate for the transition. The transition coordinate was used with umbrella sampling molecular dynamics simulations of CH_3CO-MeA_{10}-NMe. Potentials of mean force in vacuo describing the internal energy, the Helmholtz free energy and the entropy were generated as a function of helix length, R_0 (Hodgkin, Clark and Marshall, unpublished). The difference in the internal energy, U, between the α-helix minimum (at $R_0 = 14$ Å) and the 3_{10}-helix minimum (when $R_0 = 18$ Å) is 10.4 kcal/mol with a transition state (at $R_0 = 17.5$ Å) 1.5 kcal/mol above the 3_{10}-helix, while the equivalent difference in the Helmholtz free energy, A, is 8.3 kcal/mol, with the transition state 1.3 kcal/mol above the 3_{10}-helix. Thus, the 3_{10}-helix is entropically stabilized by approximately 2.1 kcal/mol. Further simulations in solvent and with varying peptide length are necessary to determine the transition length between α- and 3_{10}-helices with a solvent of a given dielectric.

Conclusions

The choice between α-helix and 3_{10}-helix by peptides containing multiple α,α-dialkyl amino acids depends on length, environment, size and distribution of amino acid side chains. Once a critical length of seven to eight residues is reached in solution, the α-helix is favored, especially in more polar solvents. In the crystal, the electrostatic interactions between the dipoles associated with the aligned amide bonds dominate, leading to antiparallel 3_{10}-helices which are associated head-to-tail. As the side-chain bulk is increased, the reduced radius of the 3_{10}-helix becomes proportionately less of a factor and the inherent stability of the α-helix predominates as well as general packing considerations.

The energy differences between the α-helix and the 3_{10}-helix for decamers of MeA have the same order of magnitude as an additional hydrogen bond. The energetic transition barrier between the two helices is quite low, suggesting that environmental effects, such as solvation, the external electric field and ligand binding, could trigger a conformational transition. While the presence of multiple MeA's decreases the energy difference between the α-helix and the 3_{10}-helix in peptaibol antibiotics and restricts the overall conformations to helical, similar transitions in membrane proteins could be responsible for transduction, as the energy differences between helical types with normal amino acids are well within the range of environmental modulation.

Acknowledgements

This research was supported in part by National Institutes of Health grants GM24483 (GRM) and GM33918 (GRM).

References

1. Marshall, G.R. and Bosshard, H.E. (1972) *Circulation Res.* **30/31** (Suppl. II), 143. An earlier preliminary report was published [Marshall, G.R. (1971) *Intra-Science Chem. Rep.* **5**, 305].
2. Burgess, A.W. and Leach, S.J. (1973) *Biopolymers* **12**, 2599.
3. Pletnev, V.Z., Gromov, E.P. and Popov, E.M. (1973) *Khim. Prir. Soedin.* 224.
4. Prasad, B.V.V. and Sasisekharan, V. (1979) *Macromolecules* **12**, 1107.
5. Paterson, Y., Rumsey, S.M., Benedetti, E., Nemethy, G. and Scheraga, H.A. (1981) *J. Am. Chem. Soc.* **103**, 2947.

6. Smith, G.D., Pletnev, V.Z., Duax, W.L., Balasubramanian, T.M., Bosshard, H.E., Czerwinski, E.W., Kendrick, N.E., Mathews, F.S. and Marshall, G.R. (1981) *J. Am. Chem. Soc.* **103**, 1493.
7. Fox, R.O. Jr. and Richards, F.M. (1982) *Nature* **300**, 325.
8. Esposito, G., Carver, J.A., Boyd, J. and Campbell, I.D. (1987) *Biochemistry* **26**, 1043.
9. Marshall, G.R., Hodgkin, E.E., Langs, D.A., Smith, G.D., Zabrocki, J. and Leplawy, M.T. (1990) *Proc. Natl. Acad. Sci. USA* **87**, 487.
10. Hall, J.E., Vodyanoy, I., Balasubramanian, T.M. and Marshall, G.R. (1984) *Biophys. J.* **45**, 233.
11. Menestrina, G., Voges, K.P., Jung, G. and Boheim, G. (1986) *J. Membrane Biol.* **93**, 111.
12. Tosteson, M.T., Auld, D.S. and Tosteson, D.C. (1989) *Proc. Natl. Acad. Sci. USA* **86**, 707.
13. Montal, M. (1990) *FASEB J.* **4**, 2623.
14. Weiner, S.J., Kollman, P.A., Case, D.A., Singh, U.C., Ghio, C., Alagona, G., Profeta, S. and Weiner, P. (1984) *J. Am. Chem. Soc.* **106**, 765.
15. Beveridge, D.L. and DiCapua, F.M. (1989) in *Computer Simulation of Biomolecular Systems* (van Gunsteren, W.F. and Weiner, P.K., Eds.) ESCOM, Leiden, pp. 1–26.
16. Valleau, J.P. and Torrie, G.M. (1977) in *Statistical Mechanics, Part A: Equilibrium Techniques, Modern Theoretical Chemistry, Vol. 5* (Berne, B.J., Ed.) Chapters 4 and 5, Plenum Press, New York.
17. Hodgkin, E.E., Clark, J.D., Miller, K.R. and Marshall, G.R. (1990) *Biopolymers* **30**, 533.
18. Barlow, D.J. and Thornton, J.M. (1988) *J. Mol. Biol.* **201**, 601.
19. Vijayakumar, E.K.S. and Balaram, P. (1983) *Tetrahedron* **39**, 2725.
20. Balaram, H., Sukumar, M. and Balaram, P. (1986) *Biopolymers* **25**, 2209.
21. Toniolo, C., Bonora, G.M., Benedetti, E., Bavoso, A., Di Blasio, B., Pavone, V. and Pedone, C. (1983) *Biopolymers* **22**, 1335.
22. Bavoso, A., Benedetti, E., Di Blasio, B., Pavone, V., Pedone, C., Toniolo, C. and Bonora, G.M. (1986) *Proc. Natl. Acad. Sci. USA* **83**, 1988.
23. Malcolm, B.R. (1977) *Biopolymers* **16**, 2591; *Biopolymers* (1983) **22**, 319.

Towards the design of structural mimics for proteins using helical peptide modules

K. Uma[a], I.L. Karle[b] and P. Balaram[a],*

[a]*Molecular Biophysics Unit, Indian Institute of Science, Bangalore 560 012, India*
[b]*Laboratory for the Structure of Matter, Naval Research Laboratory, Washington, DC 20375-5000, U.S.A.*

Introduction

The rational design of synthetic peptide mimics for structural motifs in proteins requires the ability to control polypeptide chain stereochemistry [1]. An attractive approach relies on the ability of non-protein amino acids, modified at the C^α carbon atom, to stabilize specific backbone conformations. The conformational properties of α,α-disubstituted amino acids (α,α-dialkylglycines) (Fig. 1) have been most extensively investigated, amongst this class of amino acids [2], with the stereochemical preferences of α-aminoisobutyric acid (Aib), an achiral residue, being the best characterised [3–5]. The impetus for structural studies on Aib peptides resulted initially from the widespread occurrence of this residue in alamethicin and related fungal peptides, which form voltage-sensitive ion channels across lipid bilayer membranes [6–9]. A large number of crystallographic studies have established the overwhelming preference of this residue for conformations characteristic of right- or left-handed helices (3_{10} and α). Figure 2 shows a distribution of experimentally determined backbone ϕ,ψ angles on a Ramachandran map [5]. Clearly, with very few exceptions primarily in small peptides incapable of forming intramolecular hydrogen bonds and in cyclic systems, the ϕ,ψ values cluster around the helical regions ($\phi \sim \pm 60° \pm 20°$, $\psi \sim \pm 30° \pm 20°$). The helical conformations of Aib peptides are also maintained in organic solvents, with NMR and CD studies generally indicating appreciable structural homogeneity [10,11]. The conformational properties of the

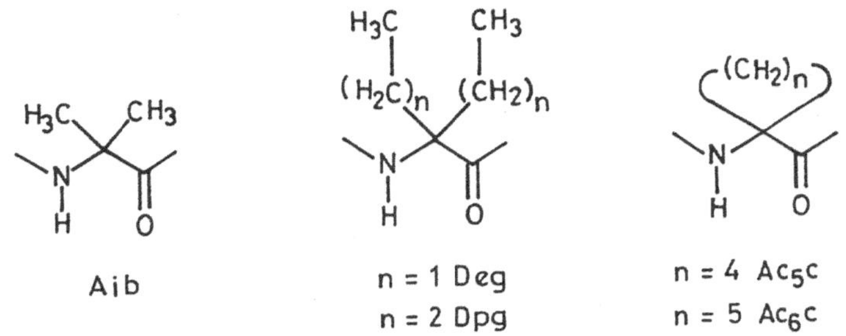

Fig. 1. *Structures of some α,α-dialkylated amino acids.*

* To whom correspondence should be addressed.

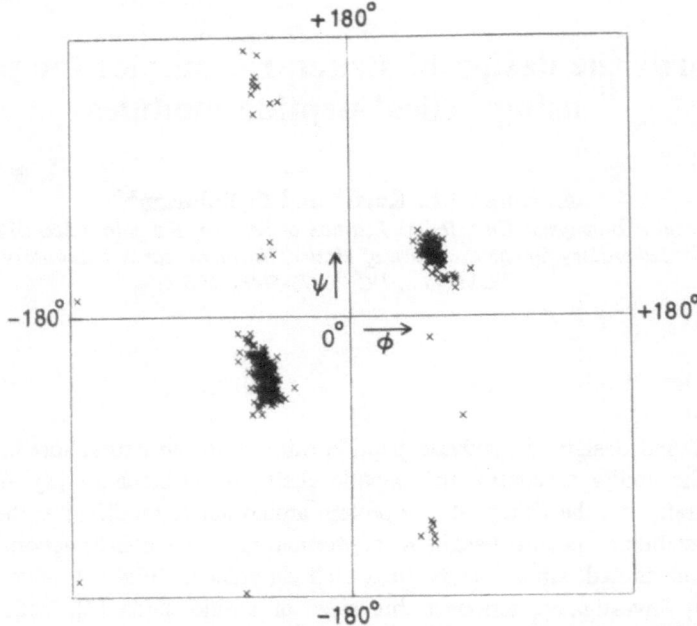

Fig. 2. *Crystallographically observed ϕ,ψ (x) values of 305 Aib residues from 108 independent crystal structures. In the case of achiral peptides crystallizing in the centrosymmetric space group, the sign of the dihedral angles has been chosen arbitrarily.*

dialkylglycines, with longer paraffin or cycloalkane side chains have also been investigated, although fewer peptides have been studied. While the cycloalkane residues support helical conformations [12], the linear alkyl substituents favour fully extended peptide chains [13]. The use of modified amino acids would thus appear to provide a means of influencing local stereochemistry of peptide chains. The use of such residues in constructing rigid oligopeptide units of defined structure is being investigated as part of a general strategy to approach the building of a synthetic protein mimic in a modular ('Meccano set') fashion [5,14,15]. Figure 3 illustrates schematically the design of an α,α-motif using structurally rigid modules and a flexible 'linker' sequence.

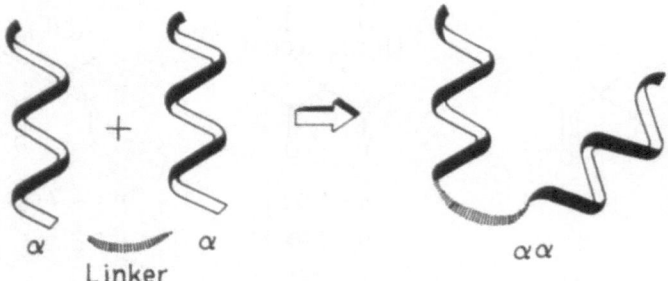

Fig. 3. *Schematic illustration of the 'Meccano set' approach to the construction of an α,α-motif.*

The use of Aib residues in conjunction with nonpolar amino acids permits the generation of long peptides that are soluble in apolar organic solvents like chloroform. In such poorly solvating environments, peptide backbone folding is a function primarily of intramolecular interactions like non-bonding, electrostatic and hydrogen bonding effects. This eliminates consideration of hydrophobic effects on peptide folding, a key element in determining peptide and protein structures in aqueous solution.

Helical Modules

The effects of Aib content positioning and precise sequence on the conformations of oligopeptides have been systematically examined by investigating a large number of peptides ranging in length from 6 to 16 residues, in the crystalline state [16] and in solution [17]. Several features emerge from these studies:

(i)   Even peptides with an Aib content of between 11–15% adopt helical conformations in the solid state. A heptapeptide (Boc-Val-Ala-Leu-Aib-Val-Ala-Leu-OMe) [18] and a nonapeptide (Boc-Val-Ala-Leu-Phe-Aib-Val-Ala-Leu-Phe-OMe, unpublished) are examples of helices containing a single, centrally positioned Aib residue.

(ii)   The precise positioning of Aib residues in peptides containing $\geq$ 20% Aib appears to be without effect and similar helical backbones have been determined for a series of isomeric decapeptides [16].

(iii) Solvent-independent helical conformations are obtained for peptides with $\geq$ 30% Aib, while some fraying of the helix is noted in strongly solvating media like dimethylsulfoxide for sequences with lesser Aib content [17].

(iv)   Large contiguous segments of non-Aib residues are comfortably accommodated into α-helices, with as many as 6 being observed in crystals of the 15-peptide, Boc-Val-Ala-Leu-Aib-Val-Ala-Leu-Val-Ala-Leu-Aib-Val-Ala-Leu-Aib-OMe [15].

(v)   Helical peptide crystal structures reveal novel aspects of packing (parallel, antiparallel and skewed helices) [19–21], solvation [18,22] and subtle conformational heterogeneity [18,23].

Design of an α,α-Motif

Armed with the various helical modules, the next step is to introduce a flexible connecting linker element. As a first step we chose to use the Gly-Pro sequence. The 18-peptide synthesized by solution-phase procedures had the sequence Boc-Aib-Val-Ala-Leu-Aib-Val-Ala-Leu-Gly-Pro-Val-Ala-Leu-Aib-Val-Ala-Leu-Aib-OMe. The central Pro is expected to interrupt the chain of continuous hydrogen bonds and cause a distortion. Gly is a residue frequently found in linker segments (irregular regions) of protein structures (N. Srinivasan and R. Sowdhamini, unpublished). The central stretch of eight non-Aib residues is expected to afford a degree of flexibility at the centre of the molecule. The N-terminal (residues 1–8) and C-terminal (residues 11–18) segments have already been shown to yield helical solid-state conformations in the

297

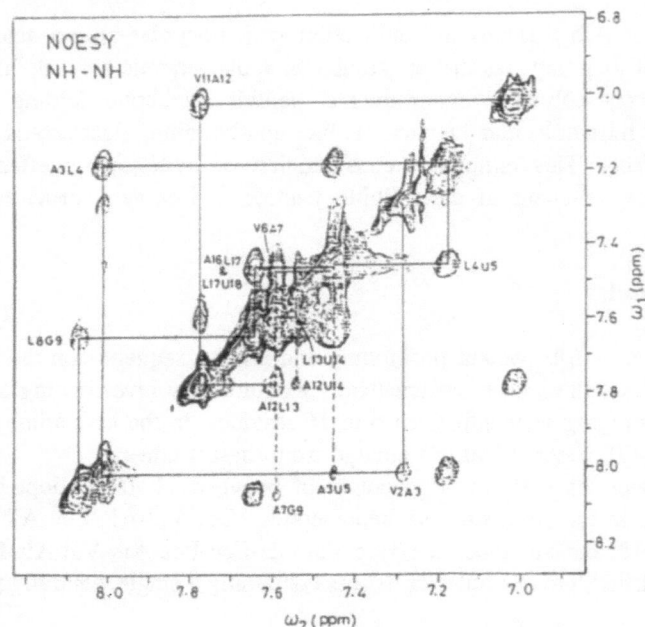

Fig. 4.   Contour plot of the amide region of the 400 MHz NOESY spectrum of the 18-peptide in CDCl₃. Sequential $N_iH \rightarrow N_{i+1}H$ NOEs have been traced.

peptides Boc-Aib-Val-Ala-Leu-Aib-Val-Ala-Leu-Aib-OMe [24] and Boc-Val-Ala-Leu-Aib-Val-Ala-Leu-OMe [18]. The Gly-Pro segment was then expected to function as a helix-breaking linker.

Figure 4 shows part of a 400 MHz NOESY spectrum illustrating NOEs between amide NH resonances. Individual resonance assignments were achieved using a combination of 2D-HOHAHA and NOESY experiments [25]. A clear feature of the NMR results is the observation of a large number of successive $N_iH \rightarrow N_{i+1}H$ NOEs characteristic of a helical conformation. In addition, several $C_i^\alpha H \rightarrow C_{i+3}^\beta H$ NOEs,

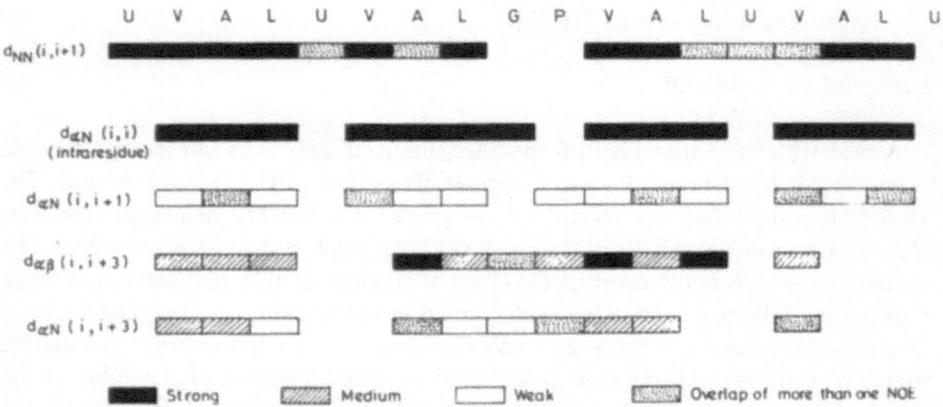

Fig. 5.   Summary of the sequential connectivities obtained from the 2D NOESY data for the 18-peptide. Diagnostic intra- and interresidue NOEs are indicated.

298

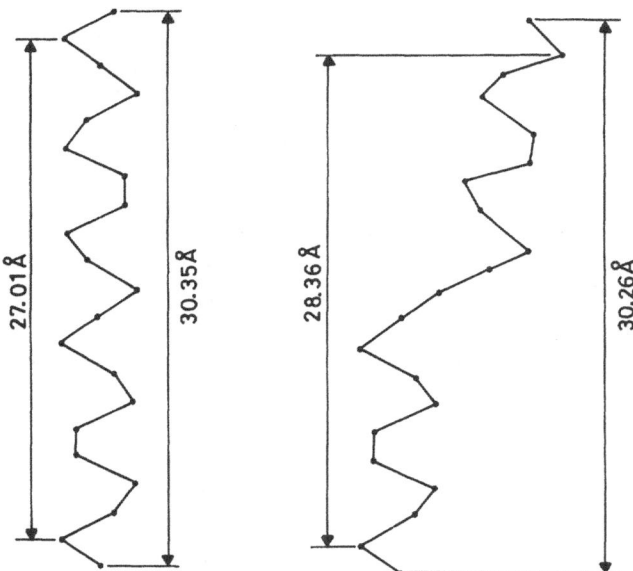

Fig. 6. C^α chain tracings of conformations generated for a 21-residue helix. Left, with all the residues in α_R conformation (ϕ −51.4°, ψ −52.7°). Right, with the central Gly residue in α_L conformation (ϕ 51.5°, ψ 52.7°).

also diagnostic of helices, are observed. Figure 5 schematically summarizes the available NOE data, which are largely consistent with a long, continuous helical conformation.

The solvent accessibility of NH groups in the 18-peptide was probed using temperature dependence of NH chemical shifts in 10% dimethylsulfoxide-chloroform at 270 MHz. 11 of the 17 NH groups are solvent shielded and presumably hydrogen bonded, as seen from their low temperature coefficients ($d\delta/dT \leq 3$ppb/K). Leu [17] NH has a marginally high value of 3.4 ppb/K. The remaining five NH groups are clearly solvent exposed ($d\delta/dT \geq 4.3$ ppb/K). Two of these are the amino terminal Aib[1] and Val[2] NH groups, which are necessarily exposed in a helical conformation. The remaining three are the Leu[13], Aib[14] and Val[15] NH groups. The data clearly suggests that the N-terminal helical segment is stable and well formed, while the C-terminal segment is weaker as indicated by the enhanced solvent exposure of the amide groups.

When the 18-peptide was designed it was hoped that the central Gly[9]-Pro[10] segment would serve to disrupt the continuous chain of hydrogen bonds. If this expectation were realised, then Val[11]NH would be solvent exposed. On the contrary, this resonance has a low temperature coefficient (0 ppb/K), favouring an unbroken helical structure.

The vicinal and geminal coupling constants for the ABX spin system of Gly[9] have been derived from difference decoupling studies and values of J_{AB} = −14.6 ± 0.5 Hz,

J_{AX} = 5.6 ± 0.8 Hz and J_{BX} = 7.6 ± 0.5 Hz have been obtained. Following Barfield et al. [26], these values are consistent with both α_R- and α_L-conformations at Gly^9 i.e. $\phi \sim \pm 50°$, $\psi \sim \pm 50°$. The effect of the two types of conformations at the central residue linking two helical segments was modelled. The C^α carbon tracings of the two structures are shown in Fig. 6. A noteworthy feature is that switching the signs of the ϕ,ψ angles at the central residue has little effect on the overall length of the molecule, but results in a displacement of the axes of the two helical segments.

Single crystals of the 18-peptide were obtained from aqueous methanol in the space group $P4_1$ with a = 9.89 Å, b = 9.89 Å, c = 114.60 Å and Z = 4. The cell dimensions are consistent with a continuous rod-like helical conformation for the molecule.

The results presented above demonstrate that the introduction of a central Gly-Pro segment did not result in breaking the molecule into two distinct helical segments. This is presumably a consequence of the fact that a single Pro residue can be comfortably incorporated into the central portion of helices, with minor distortions [27]. Future attempts at design of α,α-motifs will focus on larger linker segments, the effects of introducing D-residues and other non-protein residues into linkers and the incorporation of additional stabilising interactions between the helical segments.

Acknowledgements

This research was supported by a grant from the Department of Science and Technology, India and in part by National Institutes of Health Grant GM30902 and the Office of Naval Research. We are grateful to Dr. H. Holenweger, Bruker Spectrospin AG, Zurich for recording the NMR spectra. We thank N. Srinivasan for generating Fig. 6 and V.V. Krishnan for help with NMR experiments at 270 MHz.

References

1. Balaram, P. (1984) *Proc. Ind. Acad. Sci. Chem. Sci.* **93**, 703.
2. Toniolo, C. and Benedetti, E. (1988) *ISI Atlas of Science: Biochemistry* **1**, 225.
3. Prasad, B.V.V. and Balaram, P. (1984) *CRC Crit. Rev. Biochem.* **16**, 307.
4. Uma, K. and Balaram, P. (1989) *Ind. J. Chem.* **28B**, 705.
5. Karle, I.L., Flippen-Anderson, J.L., Uma, K. and Balaram, P. (1990) *Current Science* **59**, 575.
6. Nagaraj, R. and Balaram, P. (1981) *Acc. Chem. Res.* **14**, 356.
7. Fox, R.O. Jr. and Richards, F.M. (1982) *Nature* **300**, 325.
8. Mathew, M.K. and Balaram, P. (1983) *Mol. Cell. Biochem.* **50**, 47.
9. Bosch, R., Jung, G., Schmitt, H. and Winter, W. (1985) *Biopolymers* **24**, 961.
10. Vijayakumar, E.K.S. and Balaram, P. (1983) *Biopolymers* **22**, 2133.
11. Balaram, H., Sukumar, M. and Balaram, P. (1986) *Biopolymers* **25**, 2209.
12. Paul, P.K.C., Sukumar, M., Bardi, R., Piazzesi, A.M., Valle, G., Toniolo, C. and Balaram, P. (1986) *J. Am. Chem. Soc.* **108**, 6363.
13. Benedetti, E., Toniolo, C., Hardy, P., Barone, V., Bavaso, A., Di Blasio, B., Grimaldi, P., Lelj, F., Pavone, V., Pedone, C., Bonora, G.M. and Lingham, I. (1984) *J. Am. Chem. Soc.* **106**, 8146.
14. Karle, I.L., Flippen-Anderson, J.L., Uma, K. and Balaram, P. (1989) *Biochemistry* **28**, 6696.

15. Karle, I.L., Flippen-Anderson, J.L., Uma, K., Sukumar, M. and Balaram, P. (1990) *J. Am. Chem. Soc.*, in press.
16. Karle, I.L. and Balaram, P. (1990) *Biochemistry* **29**, 6747.
17. Uma, K. (1990) Ph.D. thesis, Indian Institute of Science, Bangalore.
18. Karle, I.L., Flippen-Anderson, J.L., Uma, K. and Balaram, P. (1990) *Proteins: Structure, Function and Genetics* **7**, 62.
19. Karle, I.L., Flippen-Anderson, J.L., Sukumar, M. and Balaram, P. (1990) *Int. J. Peptide Protein Res.* **35**, 518.
20. Karle, I.L., Flippen-Anderson, J.L., Uma, K. and Balaram, P. (1990) *Biopolymers* **29**, 1835.
21. Karle, I.L., Flippen-Anderson, J.L., Uma, K. and Balaram, P. (1990) *Biopolymers,* in press.
22. Karle, I.L., Flippen-Anderson, J.L., Uma, K. and Balaram, P. (1988) *Proc. Natl. Acad. Sci. USA* **85**, 299.
23. Karle, I.L., Flippen-Anderson, J.L., Uma, K., Balaram, H. and Balaram, P. (1989) *Proc. Natl. Acad. Sci. USA* **86**, 765.
24. Karle, I.L., Flippen-Anderson, J.L., Uma, K. and Balaram, P. (1988) *Int. J. Peptide Protein Res.* **32**, 536.
25. Wüthrich, K. (1986) *NMR of Proteins and Nucleic Acids*, John Wiley and Sons, New York.
26. Barfield, M., Hruby, V.J. and Meraldi, J.P. (1976) *J. Am. Chem. Soc.* **98**, 1308.
27. Karle, I.L., Flippen-Anderson, J.L., Uma, K., Balaram, H. and Balaram, P. (1990) *Biopolymers* **29**, 1433.

The polypeptide 3_{10}-helix

Ettore Benedetti[a], Benedetto Di Blasio[a], Vincenzo Pavone[a], Carlo Pedone[a], Antonello Santini[a], Claudio Toniolo[b], Marco Crisma[b], Fernando Formaggio[b] and Luciana Sartore[b]

[a]*Department of Chemistry, University of Naples, Via Mezzocannone, 4, 80134 Naples, Italy*
[b]*Biopolymer Research Centre C.N.R., Department of Organic Chemistry, University of Padova, Via Marzolo 1, 35131 Padova, Italy*

Introduction

Besides the classical $\alpha(3.6_{13})$-helix, the only other principal helical structure which occurs to any great extent in globular proteins and in the membrane-active, channel-forming peptaibol antibiotics [1] is the 3_{10}-helix [2] with an ideal 3.0 residue repeat (instead of 3.6). For the right-handed 3_{10}-helix the backbone torsion angles are approximately $\varphi = -60°$, $\psi = -30°$, within the same energy region as the α-helix (approximately $\varphi = -55°$, $\psi = -45°$). However, the intramolecular N–H⋯O=C H-bonding schemes are significantly different in the two helices, being of the $i \rightarrow i + 3$ (C_{10}-form or type III (III′) β-bend) in the 3_{10}-helix, while of the $i \rightarrow i + 4$ (C_{13}-form or α-bend), in the α-helix [3].

For a long periodic structure formed by C^{α}-monosubstituted α-amino acid residues the 3_{10}-helix is significantly less stable than the α-helix. Therefore, it is very unlikely that long stretches of the 3_{10}-helix would be observed. In a recent survey of all helices found in 57 of the known protein crystal structures, Barlow and Thornton [4] showed that: (i) 3.4% of the residues is involved in 3_{10}-helices; (ii) the majority of 3_{10}-helices are very short (mean length = 3.3 residues); and (iii) the 3_{10}-helices are generally irregular, in that they have a larger radius and a smaller pitch than expected.

In 1971 Marshall [5] showed by conformational energy calculations that Aib (α-aminoisobutyric acid or $C^{\alpha,\alpha}$-dimethylglycine), the prototype of $C^{\alpha,\alpha}$-disubstituted α-amino acids, is able to promote the onset of 3_{10}-helices, due to steric interactions of the *gem*-methyl groups linked to the α-carbon. In an initial attempt to get information on the effect of main-chain length and Aib content on the relative stabilities of 3_{10}- and α-helices, we present here the results of a structural analysis in the crystal state (by X-ray diffraction) of three $(Aib)_n$ (n = 6,8,10) homo-oligopeptides and three $(Aib$-L-$Ala)_n$ (n = 3–5) sequential oligopeptides.

Results and Discussion

Two X-ray diffraction structures, representative of the six oligopeptides [6–10] studied are shown in Fig. 1. The average parameters characterizing the 3_{10}- and α-helices are listed in Table 1.

The three $(Aib)_n$ homo-peptides, irrespective of their main-chain length, and the

(Aib-L-Ala)$_n$ (n = 3) sequential hexapeptide are completely folded in 3$_{10}$-helices stabilized by the appropriate number of intramolecular N–H···O=C H-bonds. This study has allowed us to characterize for the first time this important peptide secondary structure in great detail (at atomic resolution). Since the observed structures have on average 3.2 residues per turn, they are intermediate between an ideal 3$_{10}$-helix (3.0 residues per turn) and an ideal α-helix (3.65 residues per turn). As a result of this, they have an improved staggering of side chains.

In contrast, in the sequential (Aib-L-Ala)$_n$ octa and decapeptides a large portion of the molecule is α-helical, with an incipient 3$_{10}$-helix either near the N-terminus (in the octapeptide) or near the C-terminus (in the decapeptide). This finding represents the first experimental proof for a 3$_{10}$ → α-helix conformational transition in the crystal state induced only by peptide main-chain lengthening.

The more useful (average) parameters for discriminating between the two types of

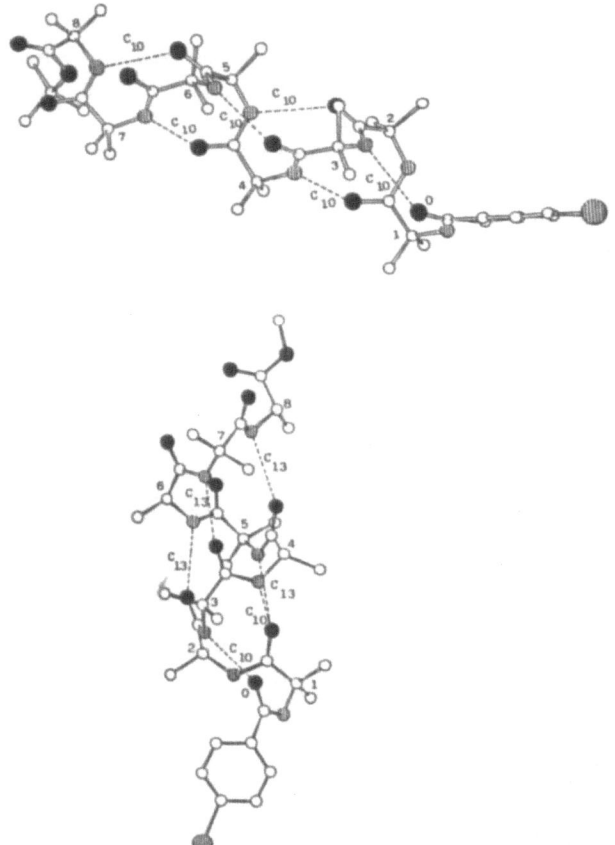

Fig. 1. Top: Molecular structure of pBrBz-(Aib)$_8$-OtBu. In this figure the right-handed molecule is shown. The six intramolecular H-bonds are indicated as dashed lines. Bottom: Molecular structure of pBrBz-(Aib-L-Ala)$_4$-OMe. The six intramolecular H-bonds are indicated by dashed lines. In both molecules the type of intramolecularly H-bonded cyclic structure (C$_{10}$ or C$_{13}$) is indicated.

Table 1 *Average parameters characterizing the helices of the (Aib)$_n$ (n = 6,8,10) and Aib-L-Ala)$_n$ (n = 3,4,5) oligopeptides*

Parameter	-(Aib)$_6$-	-(Aib)$_8$-	-(Aib)$_{10}$-	-(Aib-L-Ala)$_3$-	-(Aib-L-Ala)$_4$-	-(Aib-L-Ala)$_5$-
φ(°)	±56.0	±53.9	±54.1	−62.4	−61.2	−62.6
ψ(°)	±36.2	±28.4	±31.2	−25.2	−36.2	−47.7
$\Delta\omega$(°)	9.2	1.6	3.9	4.0	4.3	4.1
τ(°)	110.9	111.3	111.1	111.3	110.3	109.1
N···O (Å)	3.09	2.99	3.04	3.11	3.12	3.09
N···O=C (°)	127.8	128.9	129.6	124.3	154.1	157.3
Rotation/residue (°)	107.9	116.5	112.1	113.3	103.1	97.6
Axial translation/ residue (Å)	1.98	2.00	1.96	1.99	1.57	1.54
Pitch (Å)	6.61	6.18	6.29	6.33	5.48	5.68
Residues per turn	3.34	3.09	3.21	3.18	3.49	3.69

helical structures seem to be: (i) the ψ torsion angle, −30° (3_{10}-helix) and −42° (α-helix); (ii) the N···O=C angle, 128° (3_{10}-helix) and 156° (α-helix); (iii) the pitch, 6.3 Å (3_{10}-helix) and 5.5 Å (α-helix); and (iv) the number of amino acid residues per turn, 3.2 (3_{10}-helix) and 3.6 (α-helix).

We believe that these results are of great interest to protein biochemists in general and, more specifically, to biophysicists, in view of the observation that the membrane-active, channel-forming, ion-transporting antibiotics of the alamethicin family are 14–20 residues long, Aib-rich peptides (peptaibol antibiotics).

References

1. Benedetti, E., Bavoso, A., Di Blasio, B., Pavone, V., Pedone, C., Toniolo, C. and Bonora, G.M. (1982) *Proc. Natl. Acad. Sci. USA*, **79**, 7951.
2. Donohue, J. (1953) *Proc. Natl. Acad. Sci. USA*, **39**, 470.
3. Toniolo, C. (1980) *C.R.C. Crit. Rev. Biochem.*, **9**, 1.
4. Barlow, D.J. and Thornton, J.M. (1988) *J. Mol. Biol.*, **201**, 601.
5. Marshall, G. (1971) in *Intra-Science Chemistry Reports* (Kharasch, N., Ed.) Gordon and Breach, New York, p. 305.
6. Di Blasio, B., Santini, A., Pavone, V., Pedone, C., Benedetti, E., Moretto, V., Crisma, M. and Toniolo, C. *Struct. Chem.*, in press.
7. Toniolo, C., Bonora, G.M., Bavoso, A., Benedetti, E., Di Blasio, B., Pavone, V. and Pedone, C. (1986) *Macromolecules*, **19**, 472.
8. Toniolo, C., Crisma, M., Bonora, G.M., Benedetti, E., Di Blasio, B., Pavone, V., Pedone, C. and Santini, A. *Biopolymers*, submitted.
9. Pavone, V., Benedetti, E., Di Blasio, B., Pedone, C., Santini, A., Bavoso, A., Toniolo, C., Crisma, M. and Sartore, L. *J. Biomol. Struct. Dyn.*, in press.
10. Benedetti, E., Di Blasio, B., Pavone, V., Pedone, C., Santini, A., Bavoso, A., Toniolo, C., Crisma, M. and Sartore, L. *J. Chem. Soc., Perkin. Trans. II*, in press.

Binding and translocation of Ca^{2+} by calcium channel drugs

Vettai S. Ananthanarayanan

Department of Biochemistry, McMaster University, Hamilton, Canada L8N 3Z5

Introduction

The influx of extracellular Ca^{2+} ions through transmembrane channels in many excitable cells is inhibited by drugs such as verapamil, nifedipine and diltiazem which are extensively used in treating heart diseases. In spite of the similarity in the mechanism of calcium channel blockade by these antagonists [1], problems exist in relating their structure to their function [2] chiefly due to the fact that they represent vastly different classes of organic compounds. Relatively minor chemical modifications can change an antagonist to an agonist [3]. Moreover, many of these compounds also interact with muscarinic, opiate and adreno-receptors [4]. In the present study, we have examined the interaction of different classes of Ca channel antagonists with Ca^{2+} in lipid-like environments with a view to arriving at a common structural basis for their action.

Results and Discussion

The equilibrium binding of Ca^{2+} by the drugs was studied by monitoring spectral changes caused by the addition of $Ca(ClO_4)_2$ to a solution of the drug in acetonitrile (ACN). The binding data were analysed according to the method of Rueben [5] which provides for the presence of both 1:2 and 1:1 Ca^{2+}-drug complexes in equilibrium.

The CD spectrum of diltiazem changes substantially in the presence of Ca^{2+} indicating a significant conformational change that affects the asymmetry of the drug molecule. The binding isotherm at 25°C (Fig. 1, inset) shows the 1:2 Ca^{2+} : drug complex to be the predominant species with a binding constant of 7×10^3 M^{-1}. Ca^{2+} binding by verapamil was followed by difference absorption spectral measurements in the 300–200 nm region and also by NMR. As with diltiazem, the binding curve obtained from difference spectral data shows the 1:2 Ca^{2+}-drug complex to be the major species with a binding constant of 3×10^4 M^{-1} (Fig. 2). ¹H- and ¹³C-NMR data also indicated the 1:2 stoichiometry and, in addition, provided valuable clues about the Ca^{2+} binding sites and the conformational change involving the methylene groups in the middle of the molecule (data not shown). Ca^{2+} binding to flunarizine caused changes in its fluorescence emission spectrum implying the involvement of the piperazine ring. Saturation of binding occurs at a 1:1 Ca^{2+}-drug ratio with a binding affinity of 1×10^4 M^{-1} (Fig. 2). All of the above drugs also bound Mg^{2+} but with less affinity than Ca^{2+}.

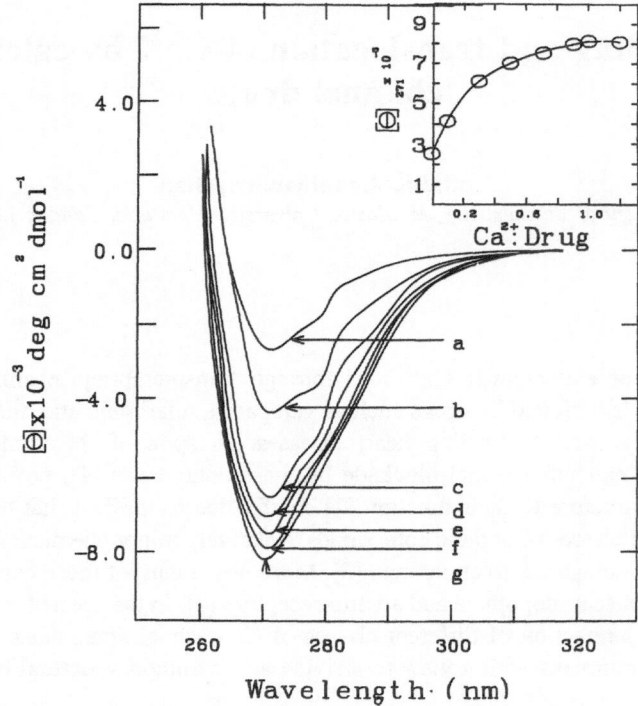

Fig. 1. *CD spectral changes in diltiazem in the presence of Ca^{2+} at $25 \pm 2°C$. The drug in ACN (2 mM) was treated with $Ca(ClO_4)_2$ to yield the indicated mol ratios of Ca^{2+} to drug: a, 0; b, 0.1; c, 0.3; d, 0.5; e, 0.7; f, 0.9 and g, 1.0. Inset: Binding isotherm at the different Ca^{2+}/drug mol ratios. $[\Theta]_{271}$ = molar ellipticity at 271 nm.*

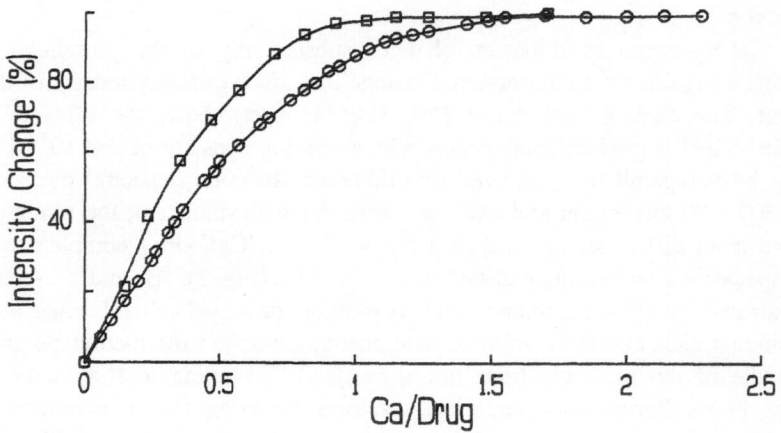

Fig. 2. *Ca^{2+} binding isotherm for: a) verapamil (10^{-4} M) as obtained from difference spectroscopic data (-□-) and, b) for flunarizine (10^{-5} M) as obtained from fluorescence data (-○-). Solvent: ACN, temp: $24 \pm 2°C$.*

In the light of our studies on synthetic peptides [6] and peptide hormones [7], we tested the ability of the calcium channel antagonists to translocate Ca^{2+} across the lipid bilayer. The protocol was essentially similar to that used by Weissmann et al. [8] for demonstrating the Ca^{2+} ionophoretic activity of phosphatidic acid except for trapping Ca^{2+} instead of Arsenazo III inside the vesicle. The data obtained for verapamil, diltiazem and lidoflazine at 18°C, pH 7.2 are shown in Fig. 3. The rates of cation transport were in the order: lidoflazine > verapamil > diltiazem. Transport by flunarizine, which is analogous to lidoflazine but does not contain an amide bond, was not appreciable in this lipid system. We have also verified the transport characteristics of these and other drugs such as nifedipine using Arsenazo III-loaded vesicles (unpublished data). A detailed quantitative analysis of the transport data is now in progress.

The results presented here demonstrate a hitherto unrecognized fact that calcium channel antagonists are themselves Ca^{2+} binders. The specific interaction of these drugs with Ca^{2+} in a nonpolar milieu would suggest a major role for Ca^{2+} in determining the bioactive conformation of these molecules at the site of their interaction with their membrane-bound receptors, namely, the Ca^{2+} channels. It is worth noticing that other drugs such as the adrenergic agents [9] and the antibiotic

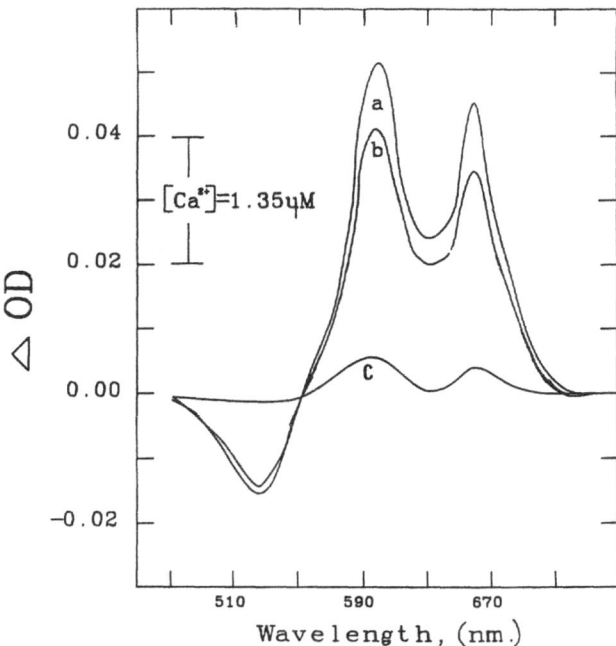

Fig. 3. *Translocation of Ca^{2+} at 18 ± 0.5°C across unilamellar vesicles made of dimyristoyl-phosphatidylcholine (DMPC). Total lipid concentration was 5 mM. External buffer contained 20 μM Arsenazo III, 0.145 M KCl-NaCl and 0.1 M MOPS at pH 7.4. CaCl₂ (5–10 mM) was trapped inside the vesicles. The drug in ACN was added to the sample cuvette of a spectrophotometer. The difference spectra due to Ca^{2+}-Arzenazo III complex were measured at various time intervals. Shown are spectra at 30 minutes obtained for a) lidoflazine; b) verapamil; and c) diltiazem. Lipid-to-drug ratio was 50 for verapamil and diltiazem and 250 for lidoflazine.*

chlortetracycline [10] also bind Ca^{2+} in nonpolar media. Interestingly, the latter drug is also a Ca^{2+} ionophore [11].

We have recently observed that several classes of peptide hormones bind Ca^{2+} stoichiometrically in nonpolar solvents and translocate this ion across synthetic bilayer lipids [7]. Our observation on the calcium channel ligands reported here would indicate that these characteristics are common to many types of extracellular stimulants. The interaction of the stimulant with Ca^{2+}, while insignificant or weak in the aqueous medium, is strengthened in the nonpolar milieu of the lipid bilayer. The Ca^{2+} ionophoretic character of the drugs observed here in an artificial model system may have implications for their in vivo action. As with the other stimulants [7], one can visualize the drug to be capable of carrying the extracellular Ca^{2+} into the lipid where it would be intercepted by the calcium channel protein. We may visualize the formation of the ternary complex, drug : Ca^{2+} : channel, which, in the case of the antagonist drugs, may be a 'dead-end' complex promoting the closed state of the channel. This will be akin to an enzyme-cofactor-inhibitor complex. By analogy, an agonist drug would act like a substrate and lead to the 'product', namely the open state of the Ca^{2+} channel. The similarity between peptide hormones and calcium channel ligands in terms of their Ca^{2+} binding and translocation may provide a clue to the observed cross-reactivity among and between these two types of stimulants [4]. Examination of the interaction of other types of stimulants with Ca^{2+} should yield valuable insights into a possible common link among the endocrine, nervous and other systems.

Acknowledgements

I thank Mr. L. Taylor for his help in the earlier part of this study and Dr. Y. Tian and Mr. S. Pirritano for their help with some of the data presented here. This study was supported in part by the Medical Research Council of Canada.

References

1. Hess, P., Lansman, J.B. and Tsien, R.W. (1984) *Nature* **311**, 538.
2. Triggle, D.J. and Janis, R.A. (1987) *Annu. Rev. Pharmacol. Toxicol.* **27**, 347.
3. Schramm, M., Thomas, G., Towart, R. and Franckowiak, G. (1983) *Nature* **303**, 535.
4. Nayler, W.G. and Dillon, J.S. (1986) *Br. J. Clin. Pharmacol.* **21**, 97S.
5. Reuben, J. (1973) *J. Am. Chem. Soc.* **95**, 3534.
6. Shastri, B.P., Rehse, P.H., Attah-Poku, S.K. and Ananthanarayanan, V.S. (1986) *FEBS Lett.* **200**, 58.
7. Ananthanarayanan, V.S. (1989) in *Advances in Gene Technology, Vol. 9: Molecular Neurobiology and Neuropharmacology* (Rotundo, R.L. et al., Eds.) IRL Press, Oxford, p. 18.
8. Weissmann, G., Anderson, P., Serhan, C., Samuelson, E. and Goodman, E. (1980) *Proc. Natl. Acad. Sci. USA* **77**, 1506.
9. Ananthanarayanan, V.S. and Horne, C. (1990) in *Peptides: Chemistry, Structure and Biology*, Proceedings of the 11th American Peptide Symposium (Rivier, J.E. and Marshall, G.R., Eds.) ESCOM, Leiden, pp. 527–529.
10. Caswell, A.H. and Hutchison, J.D. (1971) *Biochem. Biophys. Res. Commun.* **43**, 625.
11. White, J.R. and Pearce, F.L. (1982) *Biochemistry* **21**, 6309.

Morphology, conformation and stability of Alzheimer β-amyloid peptide fibrils

P.E. Fraser[a], H. Inouye[a], J. Nguyen[a], K. Halverson[b], P.T. Lansbury, Jr.[b], L.K. Duffy[c] and D.A. Kirschner[a]

[a]*Neurology Research, Children's Hospital and Department of Neurology, Harvard Medical School, Boston, MA 02115, U.S.A.*
[b]*Department of Chemistry, Massachusetts Institute of Technology, Cambridge, MA 02139, U.S.A.*
[c]*Chemistry Department, University of Alaska, Fairbanks, AK 99775, U.S.A.*

Introduction

The accumulation of proteinaceous filaments within extracellular and intraneuronal compartments is the pathological hallmark of Alzheimer's disease (AD) [1]. While the protein composition of the intraneuronal paired helical filaments (PHF) appears complex [2], the isolation and analysis of extracellular senile AD plaque fibrils has revealed that a single, small (~ 40-residue) protein [3] - termed the β/A4 protein - is the major amyloid component. Subsequent cloning of the amyloid precursor protein (APP) demonstrated that β/A4 is a fragment of a much larger (695 residues) membrane-spanning glycoprotein [4]. The β/A4 N-terminal corresponds to residue 597 of APP and encompasses the first 28 amino acids of the APP extracellular domain plus 11-14 residues of the predicted transmembrane segment. During normal cell functioning, precursor cleavage occurs at or near position 612 (K-16 of β/A4) [5,6]. Incorrect processing to produce β/A4 - whether a product of faulty or insufficient constitutive enzymes - is thought to be a prime etiological factor in amyloid deposition.

β/A4 assembles into amyloid fibrils that are formed by the interaction of numerous β-sheets aligned in a cross-β conformation [7]. This particular β-arrangement of the protein backbone in AD amyloid may account for its characteristic insolubility [8] and the diagnostic green birefringence observed upon binding Congo red [9]. While β/A4 fragments or synthetic peptide homologues will assemble into structures that are virtually indistinguishable from the naturally-occurring amyloid [10,11], it has not yet been demonstrated which β/A4 residues contribute to the β-sheets or their interactions. To better understand the interactions involved in fibril growth, the factors affecting protein insolubility and ultimately the structure of AD amyloid, we are examining the morphology, conformation and stability of fibrils formed by synthetic peptide homologues of β/A4.

Results and Discussion

Three separate regions of β/A4 were selected for investigation by virtue of their predicted differences in physical-chemical properties. These include sequences that are contained solely within the water-soluble extracellular (residues 6–25) and trans-membrane domains (residues 34–42) and the overlapping sequence between these two regions (residues 22–35).

Morphological features of peptide fibrillar assemblies

Electron microscopy of negative-stained preparations of the peptides revealed that each was capable of assembling into fibrils. The unbranched $\beta(6-25)$ fibrils were of varying length and had widths ranging from 50–80 Å (Fig. 1A). There were few instances of lateral aggregation, but occasional axial staining indicated two or more constituent filaments. Peptide $\beta(22-35)$ assembled into long, rigid-appearing fibrils when compared to $\beta(6-25)$ and displayed helical twisting with a pitch of ~ 1000–1100 Å (Fig. 1B). Fibril width varied from a maximum of ~ 120 Å to a minimum of ~ 50–60 Å. As with $\beta(6-25)$, staining within the fibril suggested that each was composed of at least two individual filaments. Fibrils formed by $\beta(34-42)$ varied greatly in length and appeared as helically twisted ribbons (Fig. 1C) reminiscent of neuritic plaque PHF [12]. The helical twists occurred at intervals of 1200–1400 Å, and the maximum and minimum fibrillar widths were 190–200 Å and 85–95 Å, respectively. Slab-like structures, ~ 90 Å in width, which were apparently single untwisted fibrils, were also observed.

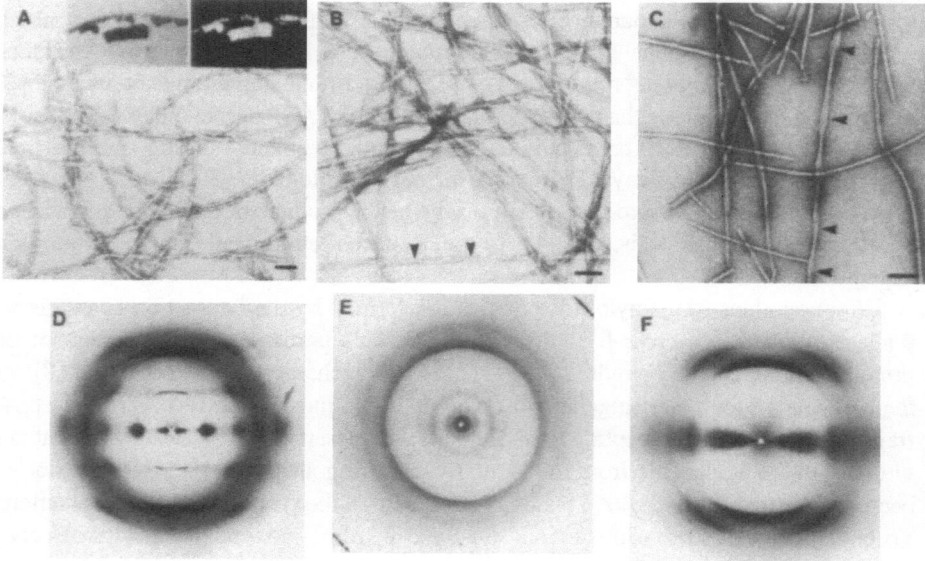

Fig. 1. Negative-stained (2% uranyl acetate) preparations (1–2 mg/ml) of (A) $\beta(6-25)$, (B) $\beta(22-35)$, and (C) $\beta(34-42)$. (A, inset) Congo red stained preparation of $\beta(6-25)$ (left) and viewed through crossed polarizers (right). The arrowheads in panels B and C indicate the helical twisting of the fibrils. Scale bars are 500 Å in (A) and (B), and 1000 Å in (C). X-ray diffraction patterns of (D)$\beta(6-25)$, (E) $\beta(22-35)$, and (F) $\beta(34-42)$ were obtained from 10–15 mg/ml peptide solutions that were air-dried in siliconized glass capillaries. Exposure times varied from 48–72h using nickel-filtered, double-mirror focused CuKα radiation from an Elliott GX-20 rotating anode generator.

Conformation of peptide assemblies

X-ray diffraction has consistently demonstrated a cross-β conformation for native amyloid fibrils from diverse sources [7,13,14]. Cross-β fibrils consist of interacting β-sheets where the hydrogen-bonding distance (4.7 Å) corresponds to the fibril direction and the intersheet (9–11 Å) spacing is oriented orthogonally to the H-bonding direction. In the current study, each of the three peptides examined assembled in the cross-β conformation (Figs. 1D–F).

Prominent wide angle reflections from β(6–25) were characteristic of β-sheets in an orthogonal lattice having dimensions a = 9.4 Å (H-bonding direction), $b \approx$ 6.6 Å (chain direction) and $c \approx$ 10 Å (inter-sheet direction) (see Ref. 10) (Table 1). In β(6–25), the equatorial reflections indexed on a one-dimensional lattice with a ~ 56 Å period. A two-dimensional lattice to give an apparent one-dimensional period can be modelled as either a restricted crystal (i.e., lattice points display a local ordering) [15] or a discrete hollow cylinder (i.e., with lattice points positioned on the circumference) [16]. The observed period corresponds to the distance between the lattice points in the former or to the diameter of the cylinder in the latter. In these models, scattering is largely determined by a zero order Bessel function (J_0). In support of the discrete cylinder model for β(6–25), EM data of fibrils in cross-section indicate a 55–60 Å diameter tubular structure assembled from a pentameric or hexameric arrangement of 25–30 Å diameter globular subunits [17].

In a similar fashion the equatorial reflections of β(34–42) indexed on a one-dimensional lattice with period of ~ 65 Å (Table 1). These reflections also showed a pronounced fanning with ~ 30° angle between the branches in the cross. This indicates that the elongated β-sheets are tilted or twisted by ~ 15° with respect to the fibril axis

Table 1 *Summary of Bragg spacings for X-ray diffraction*

β(6–25)					β(22–35)	β(34–42)		
E	1L	2L	3L	4L	Rings	E	2L	4L
53.5 (vs)	9.4 (m)	4.7 (vs)	3.1 (w)	2.4 (w)	33.4 (m)	~86 (s)	4.7 (vs)	2.4 (m)
28.0 (m)	8.8 (m)	4.6 (m)		2.3 (vw)	19.3 (m)	34.0 (m)	4.1 (w)	
19.6 (w)	7.8 (w)	4.4 (s)			11.7 (m)	22.4 (s)	3.9 (s)	
14.8 (w)	7.1 (vw)	3.9 (vw)			8.1 (w)	16.1 (s)	3.5 (w)	
11.1 (s)	6.0 (vw)	3.7 (m)			4.7 (s)	12.5 (m)		
9.5 (s)	5.0 (m)	3.2 (w)			4.5*(m)	11.0 (m)		
6.6 (vw)	4.5 (vw)	2.8 (vw)			3.8 (w)	9.4 (m)		
5.7 (vw)					3.3 (w)	8.1 (w)		
5.2 (m)						7.3 (w)		
4.3 (m)						6.5 (w)		
3.9 (vw)						5.2 (m)		
3.5 (m)						4.6 (w)		
						4.1 (w)		
						3.2 (w)		
						3.0 (w)		

E, equator; *1L, 2L, 3L, 4L* are first, second, third, and fourth layer lines; *Rings*, circular reflections; *vw, w, m, s, vs* indicate intensities of very weak, weak, moderate, strong and very strong reflections. *, denotes very sharp reflection which may not arise from the peptide β-sheet.

which is consistent with the EM observation of twisted ribbons (Fig. 1C). Similar fanning has been reported for cellulose fibers [18] and collagen [19].

The β(22–35) reflections – while largely circular – showed a distinct accentuation of the hydrogen-bonding (4.7 Å) and intersheet (~ 10 Å) spacings along orthogonal axes which is indicative of cross-β. The 4.7 Å reflection was very sharp, suggesting that the diffracting object was elongated along the hydrogen-bonding direction. Two broad intensity maxima at 11.7 Å and 8.1 Å were seen near the inter-sheet distance of ~ 10 Å spacing and there were small-angle intensity maxima at 33.4 Å and 19.3 Å. Assuming that the object is a tube with a long axis parallel to the hydrogen bonding direction and with a diameter ~ 40 Å, the first four intensity maxima should be at Bragg spacings 33 Å, 18 Å, 12 Å and 9 Å. While the agreement between the observed and calculated spacings is not exact, the model nonetheless represents a satisfactory first approximation.

Congo-red binding of in vitro fibrils

Many amyloid fibrils exhibit an ability to bind the histological stain Congo red (CR), and when subsequently viewed through crossed polarizers display a distinctive apple-green birefringence [9,20]. Solutions of β(6–25), β(22–35) (5 mg/ml in water) and β(34–42) (5 mg/ml in hexafluoroisopropanol) were air-dried on glass slides, fixed with absolute ethanol for 10 min and stained with Congo red (1.5% in 80% ethanol). Excess stain was removed with three ethanol washes. All of the peptide films bound the stain and exhibited the diagnostic green birefringence (Fig. 1A, inset).

CR binding is thought to be cross-β specific, and binding models have been proposed to explain this specificity [21,22]. However, the amyloid fibril-CR interaction at the molecular or atomic level has yet to be elucidated. The specificity of Congo red for amyloid and peptide fibrils (Fig. 1A, inset) coupled with the resolution of fiber diffraction (e.g., β(6–25); Fig. 1D) may provide a vehicle for further structural studies and possibly design of diagnostic/therapeutic aids for use in AD.

Alteration of β(22–35) fibril morphology by pH and detergent

Native amyloid fibrils and PHF are resistant to solubilization by sodium dodecyl-sulfate (SDS) or urea [8]. To determine the factors affecting fibril stability, we examined the morphological changes of β(22–35) assemblies when subjected to chemical perturbations.

The sequence of β(22–35) [E-D-V-G-S-N-K-G-A-I-I-G-L-M] contains three ionizable amino acid side chains, E-22, D-23 and K-28 in addition to the N- and C-termini. The electrostatic charge of these residues was varied by dissolving the peptide in buffered aqueous solutions at pH 2, 6 and 11. Electron microscopy of negative-stained preparations revealed that even at pH 2, when E-22 and D-23 are electrically neutral, β(22–35) formed long narrow fibrils with diameters of ~ 45–60 Å (Fig. 2A). At pH 6, when E-22, D-23 and K-28 are all charged, similar fibrils were observed but with a marked lateral aggregation to create twisted 'rope-like' structures that varied from 150–300 Å in diameter (Fig. 2C). At pH 11, when K-28 is electrically

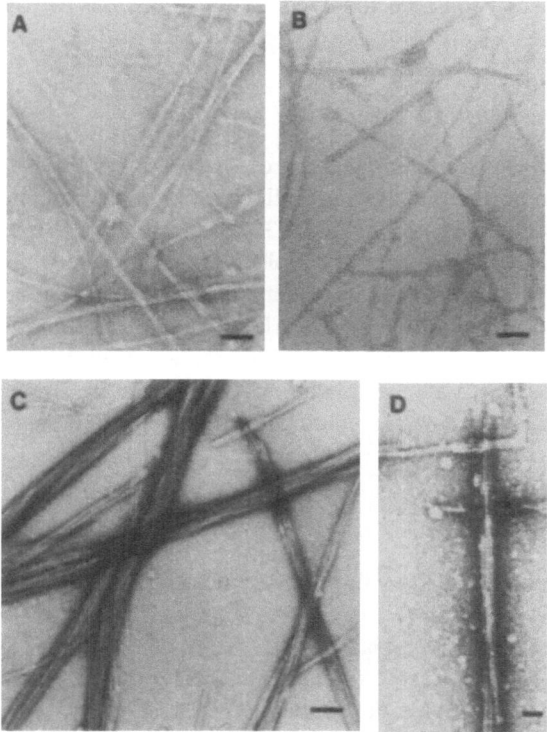

Fig. 2. Negative-stained (2% uranyl acetate) preparations (1–2 mg/ml) of β(22–35) buffered at (A) pH 2 (10 mM glycine), (B) pH 11 (10 mM Tris), (C) pH 6 (10 mM HEPES), and (D) pH 6 and 0.1% SDS. Scale bars are equal to 500 Å.

neutral, the fibrils were shorter and slightly narrower in diameter (35–50 Å) (Fig. 2B) than those at pH 2.

The ability of β(22–35) peptides to maintain their fibrillar morphology even at such pH extremes (2 and 11) suggests that the hydrophobic, putative membrane segment, containing residues 29–42, may constitute a strong force in fibril stability. This notion is supported by solubility studies on β(34–42) [23]. Further, the considerable pH-induced morphological changes indicate that the ionizable amino acid side chains (at pH 6) could be involved in ion pairing that promotes inter-residue interactions and possibly fibril aggregation and twisting.

To test further the contribution of hydrophobic interactions, fibril bundles at pH 6 were treated with sodium dodecylsulfate (final concentration 0.1% SDS). Initially, large ragged fibrils were observed (Fig. 2D) which disappeared completely after continued incubation (several hours at room temperature). Unlike native amyloid fibrils which do not dissolve in SDS, the disruption of the hydrophobic interactions within the β(22–35) fibrils was sufficient to eliminate the lattice of β-sheets. This suggests that additional factors and/or other residues within β/A4 (possibly in the region of 34–42 [23]) must be involved in the insolubility of the in vivo amyloid fibrils.

Conclusions

Peptides from vastly different regions of β/A4 are capable of producing fibrils, albeit with different morphology but all exhibiting an amyloid-defining cross-β conformation. As demonstrated by β(22–35), the stability of these fibrils may be a reflection of hydrophobic interactions coupled with its tendency to form a highly stable β-conformation. Although similar in conformation, the β(22–35) and β(34–42) fibrils do not morphologically resemble native AD amyloid. Therefore, the determinants of amyloid morphology may be contained within the extracellular, N-terminal region of β/A4 which is consistent with the structure of β(6–25) fibrils. Further modelling of the diffraction data should allow for continued refinement of our understanding of the amyloid fibril assembly.

Acknowledgements

The research was supported by the American Health Assistance Foundation (DAK), National Institutes of Health Grant AG-08572 (DAK) from the National Institute of Aging, National Science Foundation grant BNS-8719741 (LKD), the Alzheimer Disease and Related Disorders Association (LKD), the Whitaker Health Sciences Foundation (PTL), the Dreyfus Foundation (PTL), and Merck & Co. (PTL). PEF is a postdoctoral fellow of the Medical Research Council of Canada. Some of the work was carried out in facilities related to the Mental Retardation Research Center of Children's Hospital, and was supported by Core Grant HD-18655 from the National Institutes of Health.

References

1. Selkoe, D.J. (1989) *Annu. Rev. Neurosci.* **12**, 463.
2. Kosik, K. (1989) *J. Gerontol.* **44**, B55.
3. Glenner, G.G. and Wong, C.W. (1984) *Biochem. Biophys. Res. Commun.* **120**, 885.
4. Kang, J., Lemaire, H.-G., Unterbeck, A., Salbaum, J.M., Masters, C.L., Gryeschik, K.-H., Multhaup, G., Beyreuther, K. and Müller-Hill, B. (1987) *Nature* **235**, 733.
5. Sisodia, S.S., Koo, E.H., Beyreuther, K., Unterbeck, A. and Price, D.L. (1990) *Science* **248**, 492.
6. Esch, F.S., Keim, P.S., Beattie, E.C., Blacher, R.W., Culwell, A.R., Oltersdorf, T., McClure, D. and Ward, P.J. (1990) *Science* **248**, 1122.
7. Kirschner, D.A., Abraham, C. and Selkoe, D.J. (1986) *Proc. Natl. Acad. Sci. USA* **83**, 503.
8. Selkoe, D.J., Ihara, Y. and Salazar, F.J. (1982) *Science* **215**, 1243.
9. Glenner, G.G., Eanes, E.D., Bladen, H.A., Linke, R.P. and Termine, J.D. (1974) *J. Histochem. Cytochem.* **22**, 1141.
10. Kirschner, D.A., Inouye, H., Duffy, L.K., Sinclair, A., Lind, M. and Selkoe, D.J. (1987) *Proc. Natl. Acad. Sci. USA* **84**, 6953.
11. Gorevic, P.D., Castano, E., Sarma, K. and Frangione, B. (1987) *Biochem. Biophys. Res. Commun.* **147**, 854.

12. Wisniewski, H.M., Merz, P.A. and Iqbal, K. (1984) *J. Neuropath. Exp. Neurol.* **43**, 643.
13. Termine, J.D., Eanes, E.D., Ein, D. and Glenner, G.G. (1972) *Biopolymers* **11**, 1103.
14. van Andel, A.C.J., Hol, P.R., van der Maas, J.H., Lutz, E.T.G., Krabbendam, H. and Gruys, E. (1986) in *Amyloidosis* (Glenner, G.G., Osserman, E.F., Benditt, E.P., Calkins, E., Cohen, A.S. and Zucker-Franklin, D., Eds.) Plenum, New York, p. 39.
15. Burge, R.E. (1961) *Proc. R. Soc.* **A260**, 558.
16. Waser, J. (1955) *Acta Crystallogr.* **8**, 142.
17. Fraser, P.E., Duffy, L.K., O'Malley, M.B., Nguyen, J., Inouye, H. and Kirschner, D.A. (1990) *J. Neurosci. Res.* (in press).
18. Heyn, A.N.J. (1948) *J. Am. Chem. Soc.* **70**, 3138.
19. Miller, A. and Wray, J.S. (1971) *Nature* **230**, 437.
20. Puchtler, H., Waldrop, F.S. and Meloan, S.N. (1985) *Appl. Pathol.* **3**, 5.
21. Cooper, J.H. (1974) *Lab. Invest.* **31**, 232.
22. Klunk, W.E., Pettegrew, J.W. and Abraham, D.J. (1989) *J. Histochem. Cytochem.* **37**, 1273.
23. Halverson, K., Fraser, P.E., Kirschner, D.A. and Lansbury, P.T. Jr. (1990) *Biochemistry* **29**, 2639.

α-Helices in proteins: Strategies for detecting and analysing coiled-coil motifs

David A.D. Parry[a] and Carolyn Cohen[b]

[a]*Department of Physics and Biophysics, Massey University, Palmerston North, New Zealand*
[b]*Rosenstiel Basic Medical Sciences Research Center, Brandeis University, Waltham, MA 02254, U.S.A.*

Introduction

α-Helices are commonly found in many types of proteins, and these are often packed in bundles related to the coiled-coil structure. In such cases, the helix axes are inclined to one another in a left-handed manner by angles of up to 30°. The rationale for such packing in the α-fibrous class of proteins, as exemplified by tropomyosin, myosin rod and the intermediate filament proteins, lies in the periodic distribution of apolar residues in the amino acid sequence. In particular a heptad substructure has been recognised with the form $(a\text{-}b\text{-}c\text{-}d\text{-}e\text{-}f\text{-}g)_n$ where the a and d positions are filled predominantly by apolar residues [1,2]. The regular distribution of these residues in a right-handed α-helix results in the formation of a left-handed apolar stripe on its surface. Single α-helices with this characteristic, however, are generally unstable in an aqueous environment. Hence, for a stable structure to exist in vivo several α-helices aggregate to shield their apolar residues from water. Closest packing occurs when there is a regular meshing of the apolar side chains from different α-helices. These 'internal' apolar interactions between the α-helices can be maintained over a long range provided that the axes of the helices become supercoiled and wind round one another in a left-handed manner to generate a coiled-coil rope-like structure [1]. This fundamental coiled-coil motif was described in the early 1950s but it is only relatively recently that the same heptad substructure has been recognised as a feature of α-helical bundles in both globular and membrane proteins [3,4]. Consequently the recognition of significant heptad repeats, together with predictive methods for determining secondary structure, provide a simple and rather reliable means of delineating α-helices from sequence data alone. This approach also provides information on probable tertiary arrangements of the α-helices in vivo. Further data on the mode of packing is often available from the recognition of structural periodicities in the linear distribution of specific residue types (e.g. apolar, acidic, basic). In this paper we describe the methods by which such structural data may be obtained from an analysis of the primary structure of a predominantly α-protein.

Results and Discussion

We consider here a selection of the methods by which α-helices may be recognised in a protein sequence and describe as well the packing constraints which are then applicable in the tertiary structure. The problem may be tackled using the following step-by-step approach.

1. Calculate the ratio of charged to apolar residues for any segment of the sequence showing a special or unusual characteristic (e.g. highly charged, apolar, serine-glycine-rich etc.), noting in particular those segments with very high (> 1.0) or very low values (< 0.6). The former are often indicative of a multichain elongated structure (possibly long heptad-containing segments) while the latter generally favour a compact assembly of short structural elements (such as α-helices) in a single chain structure. Preliminary data also suggest that there may be an inverse correlation between the charged/apolar ratio and the number of α-helices that interact with one another. The highest ratios (> 1.0) are found for two-stranded coiled-coils, intermediate ratios (~ 0.8) for three-stranded ropes and the lowest ratios (~ 0.6) for four or more α-helices in a bundle, although variations do occur.

2. Delineate all heptad-containing segments in a sequence using the criterion that at least three out of five consecutive *a* and *d* positions are filled by apolar residues (leucine, valine, isoleucine, methionine, alanine, tyrosine, phenylalanine), paying particular attention to those segments with the highest occupancy of apolar residues in the *a* and *d* sites. Of special significance is the frequency of leucine residues in these key positions. Discontinuities in the phasing of the heptad repeats, which are found in all proteins except tropomyosin, have been interpreted as points at which the axis of the coiled-coil rod may undergo a change in direction – either abruptly or more gradually – possibly to alleviate local packing difficulties. Although the average extent of heptad domains in two- and three-stranded α-fibrous proteins is about 12 and 7 heptads respectively [5], in a globular protein the physical limitations of size imposed by its molecular weight make it unlikely that more than about four heptads will occur consecutively in a single stretch of α-helix. In membrane proteins the highly apolar α-helical transmembrane portions of the sequence are typically about 23 residues long, i.e. a little over three heptads. In such cases it is also valuable to calculate the hydropathy profile [6], since this method provides an independent indication of transmembrane α-helical segments.

3. Apply the Robson [7] and the Chou-Fasman techniques [8] (plus any other appropriate predictive method) to indicate likely stretches of α-helix as well as those pieces of sequence that are unlikely to be α-helical. Of particular importance in the latter case are segments of sequence that are rich in proline, glycine and serine residues, segments that have high predicted flexibilities on the basis, for example, of the Karplus and Schulz data [9], and segments with high β-turn potential.

4. Ascertain the degree of correlation between the predictions in 1, 2 and 3 and, in doing so, attempt to delimit the N- and C-terminal ends of the α-helices. This can be facilitated by reference to the Richardson and Richardson data [10] on the

relative preference values of the various amino acids at particular positions within the α-helix. Furthermore, where the heptad-containing and predicted α-helical regions of an α-helix are long (as is the case in α-fibrous proteins) it is often possible to place limitations on the positions of the ends using a fast Fourier transform (FFT) method. This technique involves calculating fast Fourier transforms corresponding to the linear distributions of apolar, acidic and basic residues in the segment. Some of these residue types, especially the latter two, are generally involved in specifying modes of intermolecular aggregation. Consequently the boundaries of such long regions may be systematically varied to pick out the regions with the highest (scaled) Fourier intensities, thus placing limitations on the probable N- and C-termini. In such analyses particular note is taken of those periods that are common for different residue groupings (see [11] for a summary).

5. Calculate the number and distribution of possible intrachain ionic interactions of the type $i \rightarrow i + 4$ (and also $i \rightarrow i + 3$). Both are favourable in an α-helical segment, especially the former class [12,13]. Such interactions are known to stabilise an α-helix and consequently their presence may induce the adoption of such a conformation. Intrachain ionic interactions are probably of greater importance in a coiled-coil rod than in a compact globular assembly of α-helices, where the apolar interactions would seem to dominate in providing stability. It is also relevant to compare the linear distribution of the intrachain ionic interactions with that of the interchain ionic interactions. A maximum in the latter *always* indicates a parallel in-register arrangement of chains; typical values range from 0.2 to 0.8 ionic interactions per heptad pair [11]. The former may compensate for a low number of interchain ionic interactions and, in doing so, provide stability for constituent α-helices while allowing flexibility of the coiled-coil molecule at that point. It is of interest to note that the M-protein rod, which protrudes from the surface of the streptococcal cell wall, has a decreasing density of interchain ionic interactions with increasing distance from the membrane. This would be compatible with increasing flexibility of the rod.

6. Check that known structural constraints are satisfied. For example, have the residues involved in forming a disulphide bond been characterised? Does the model place these residues in physically realistic spatial positions? (Statistically it has been shown that disulphide bonds occur far more frequently between coil-coil and coil-α residues than between residues that are both in an α-helical conformation [14,15]). Have homologous sequences within the protein been assigned the same or closely related structures? Have sequences in the protein been recognised that are homologous to those in proteins of known tertiary structure? Is the proposed structure compatible with available crystallographic or NMR data? Are there any examples of such motifs as the well-known EF hand? (This Ca^{++} binding site is located between two short pieces of α-helix with a characteristic sequence motif [16]). Have any phosphorylation sites been identified? (Current evidence suggests that these rarely (if ever) occur in α-helices). Has the postulated structure violated any known structural constraints such as the fact that protein chains never produce 'knots' [17]? There are, of

course, many other constraints that could (and should) be built into any modelling process.

7. If a heptad-containing α-helical segment contains a common long-range regularity in the linear distribution of its acidic and its basic residues then a multichain, rodlike, coiled-coil structure with parallel, in-register chains is almost certain. Calculation of intermolecular ionic interactions as a function of relative axial stagger and molecular polarity provides a limited number of possible axial arrangements of the molecules in the filamentous assembly thus formed (see, for example, refs. 18,19). In contrast, the lack of long-range regularities in the sequences of heptad-containing α-helical segments indicates the probable formation of a left-handed bundle of α-helices formed from a single chain. In these cases a short sequence linking consecutive α-helices provides a strong constraint in the establishment of an antiparallel assembly [20]. If the linking segments are long (say > 15 residues) then both parallel and antiparallel assemblies of α-helices become possible in principle (see, for example, the connectivities of 4-α-helical bundles [4]). Dipole considerations, however, strongly favour antiparallel arrangements. Finally we note that the lack of heptad repeats in an α-helical sequence – as in myoglobin – suggests that the α-helices are not packed into bundles.

In the case of highly α-helical proteins or domains we have shown that consideration of some simple principles of protein structure allows the reliable identification of α-helical segments as well as features of their packing. Such an approach not only provides guidelines to evaluate models proposed for proteins but is also useful in the design of new proteins. As more native and synthetic structures are solved and their properties determined, additional principles and constraints will emerge.

References

1. Crick, F.H.C. (1953) *Acta Crystallogr.* **6**, 689.
2. McLachlan, A.D. and Stewart, M. (1975) *J. Mol. Biol.* **98**, 293.
3. Cohen, C. and Parry, D.A.D. (1986) *Trends Biochem. Sci.* **11**, 245.
4. Cohen, C. and Parry, D.A.D. (1990) *Proteins: Structure, Function and Genetics* **7**, 1.
5. Conway, J.F. and Parry, D.A.D. *Int. J. Biol. Macromolecules* (submitted).
6. Kyte, J. and Doolittle, R.F. (1982) *J. Mol. Biol.* **157**, 105.
7. Garnier, J., Osguthorpe, D.J. and Robson, B. (1978) *J. Mol. Biol.* **120**, 97.
8. Chou, P.Y. and Fasman, G.D. (1978) *Adv. Enzymol.* **47**, 45.
9. Karplus, P.A. and Schulz, G.E. (1985) *Naturwissenschaften* **72**, 212.
10. Richardson, J.S. and Richardson, D.C. (1988) *Science* **240**, 1648.
11. Conway, J.F. and Parry, D.A.D. (1990) *Int. J. Biol. Macromolecules* (in press).
12. Marquseee, S. and Baldwin, R.L. (1987) *Proc. Natl. Acad. Sci. USA* **84**, 8898.
13. Perutz, M.F. and Fermi, G. (1988) *Proteins: Structure, Function and Genetics* **4**, 294.
14. Thornton, J.M. (1981) *J. Mol. Biol.* **151**, 261.
15. Fraser, R.D.B., MacRae, T.P., Sparrow, L.G. and Parry, D.A.D. (1988) *Int. J. Biol. Macromolecules* **10**, 106.
16. Kretsinger, R.H. (1987) *Cold Spring Harbor Symp. Quant. Biol.* **52**, 499.
17. Richardson, J.S. and Richardson, D.C. (1989) in *Prediction of Protein Structure and the Principles of Protein Conformation* (Fasman G.D., Ed.) Plenum Press, New York, pp. 1–98.

18. Fraser, R.D.B., MacRae, T.P., Suzuki, E. and Parry, D.A.D. (1985) *Int. J. Biol. Macromolecules* **7**, 258.
19. Fraser, R.D.B., MacRae, T.P., Parry, D.A.D. and Suzuki, E. (1986) *Proc. Natl. Acad. Sci. USA* **83**, 1179.
20. Weber, P.C. and Salemme, F.R. (1980) *Nature* **287**, 82.

Design of crystalline helices of short oligopeptides as a possible model for nucleation of the α-helix

R. Parthasarathy, Sanjeev Chaturvedi and Kuantee Go

Center for Crystallographic Research and Biophysics Department,
Roswell Park Cancer Institute, 666 Elm Street, Buffalo, NY 14263, U.S.A.

Introduction

How proteins start to fold and what clues the primary structure-itself-gives to the folding process has long been a mystery. The relationship of nucleic acid (DNA) sequence to the amino acid sequence, the first part of the genetic code, has been solved long back; the second part of the genetic code, namely how amino acid sequences determine the three-dimensional structure and folding of proteins, is one of the most hotly pursued research topic now (for reviews on protein folding, see refs. 1–3). One of the key elements of the folding process, as originally suggested by Anfinsen [4] and later experimentally observed by Anfinsen and co-workers [5] and by many others [6,7] is the presence of nucleating sites in the polypeptide. These nucleating sites are short fragments of the polypeptide chain that, during the folding process, can rapidly flicker in and out of the conformation that they assume in the final fold; if these fragments have well defined conformations, they seed effectively the folding process. In the initial stages of folding in aqueous solution, the nonpolar side chains are exposed to water and are driven to associate by short-range hydrophobic interactions. This concept provides a basis for experimental studies of the conformations of short oligopeptides, in particular, the α-helices. The α-helices are one of the most abundant and stable structural features of proteins; they are internally hydrogen bonded and are autonomous folding units. In order that the folding of the fragment that serves as a nucleation site in protein is rapid, the fragment should be short, say 10 or 12 residues at most. However, it has been rather widely assumed that short oligopeptides will not have a preferred and well-defined conformation; also, the Zimm-Bragg equation [8] predicts that peptides as long as twenty residues will not show measurable α-helix formation in water regardless of the temperature and amino acid sequence [9]. However, it has been found [10] that a highly immunogenic nonapeptide from flu virus shows structure that is measurable by NMR. Further, α-helices of short peptides of particular sequences have been found to be stable [11,12] in solution. Consequently, it is worthwhile asking: What is the shortest oligomer of natural amino acids that will prefer a reasonably stable α-helical conformation?

The reasons we chose to study α-helices are the following: their abundance in proteins, stability due to intramolecular hydrogen bonding, easy detectibility of their presence in solution using CD or NMR, ability to predict their presence and their importance in biological activities. Since peptides play very vital roles as biological triggers in many biological reactions, any modifications of their biological activities by designing and engineering α-helices with altered sequences are of great therapeutic and pharmacological importance, especially when peptides are being increasingly realized as drugs of the future.

A key feature of the isolated α-helix is that it is not stable in water; consequently, additional interactions are needed to stabilize the helix. These stabilizing interactions may be provided in a number of ways:

1. Interaction with the surrounding fragments, as in globular proteins and in the α-helix bundle, in particular, the 4-α-helical motif that is found in many globular proteins [13].
2. Intramolecular i → i + 4 salt bridges [11,12].
3. Dipolar interactions with the macro dipole by keeping the N- and C-terminal residues acidic and basic, respectively [12].
4. Using amino acid preferences for helices obtained from protein data analyses [14] for N- and C-caps, and other positions along the oligopeptide chain that give rise to favorable initiators and terminators; for example, D and P are favored initiators and G is a favored terminator.
5. Using sequences appropriate for coiled-coils based on the fact that there are heptad repeats in α-fibrous proteins with preferences for various amino acids along the heptad sequence [15,16].
6. Our method is based on stabilizing α-helices by using special hydropathic sequences and by providing intermolecular interactions with the neighboring α-helices and water molecules in a crystal.

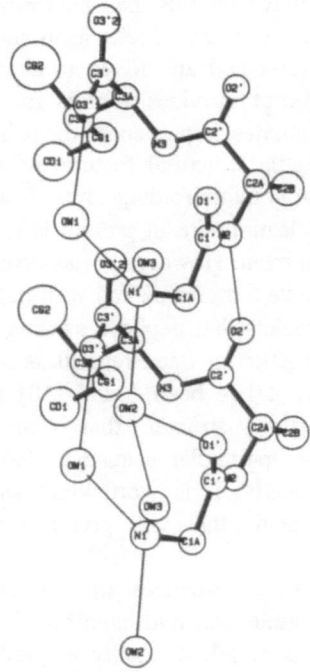

Fig. 1.   Helix formed by GAI and water molecules in the GAI1 crystal.

Results

We discovered two short tripeptide sequences, GAF and GGV that exhibited a helical conformation in the crystalline state [17]. Based on this work of ours on two short peptides, we have been able to predict possible tripeptide sequences that will have helical conformation. We have now designed, synthesized, crystallized (at pH 7.0) and carried out X-ray diffraction studies of three more tripeptides, GAV, GAL and GAI, that show stable helical conformation in the crystalline state [18]. These tripeptides GAF, GGV, GAV, GAL and GAI all are, in the crystal structure, zwitterions and each tripeptide molecule assumes a conformation similar to a reverse turn stabilized by water molecules. What is remarkable is that in the crystal the neighboring molecules together with water molecules form extended helices similar to a right-handed α-helix. The folded conformation similar to an α-helix is a characteristic property of these sequence-related tripeptides. Whether this property is reflected when they are a part of a longer polypeptide remains to be tested by synthesizing longer oligomers (see, however, ref. 19). The helices are all amphipathic. One side of the helix is hydrophobic and involves the side chains; the other side involves the bound water molecules and hydrogen bonding, and is hydrophilic. The helices of these tripeptides are stabilized by three important factors: i) a selected sequence with increasing hydrophobicity, ii) interactions from adjacent molecules in the crystal and iii) interaction with specific water molecules.

GAI forms two types of crystals, GAI1 and GAI2; GAI1 is isomorphous with GAV and forms a helix (Fig. 1); GAI2 is not helical and imitates a cyclic peptide and traps a water molecule (Fig. 2). The side-chain conformation of GAI2 is different from that for GAI1 and indicates that the restriction of side-chain rotamer conformation may be important in determining the helix-forming tendencies.

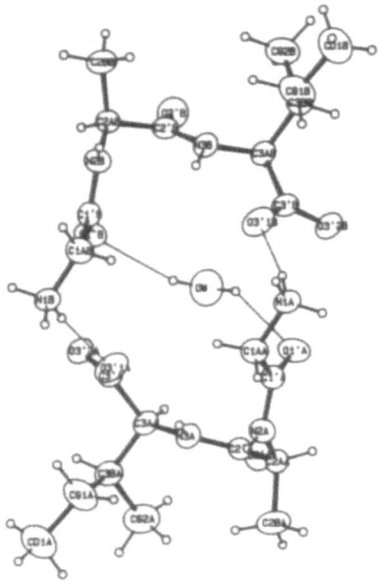

Fig. 2.   GAI is not helical in the crystal GAI2. Two molecules of GAI trap a water molecule.

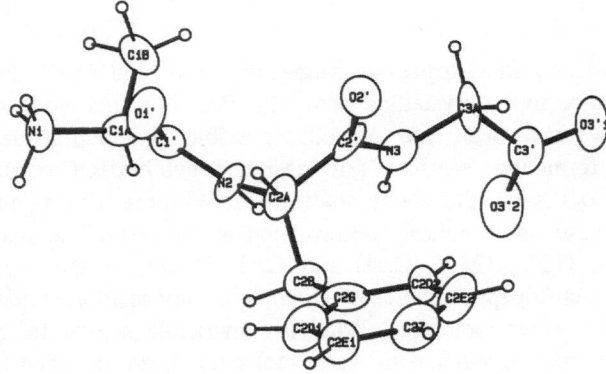

Fig. 3.   AFG is not helical in the crystalline state unlike GAF.

Most of the designs of α-helical peptides are based on the helical preferences of amino acid monomer residues; however, our design of crystalline helices shows that helix forming tendencies of a particular amino acid depends on the sequence context in which it occurs. For example, GAF yields a crystalline helix but AFG containing the same amino acids with a different order in the sequence yields a crystal which does not contain the helical conformation (Fig. 3).

References

1. Anfinsen, C.B. and Scheraga, H.A. (1975) *Adv. Protein Chem.* **29**, 205.
2. Baldwin, R.L. (1986) *Trends Biochem. Sci.* **11**, 6.
3. King, J. (1989) *Chem. Eng. News*, April 10, 32.
4. Anfinsen, C.B. (1973) *Science* **181**, 223.
5. Sachs, D.H., Schechter, A.N., Eastlake, A. and Anfinsen, C.B. (1972) *Proc. Natl. Acad. Sci. USA* **69**, 3790.
6. Creighton, T.E. (1978) *Prog. Biophys. Mol. Biol.* **33**, 231.
7. Kim, P.S. and Baldwin, R.L. (1982) *Annu. Rev. Biochem.* **51**, 459.
8. Zimm, B.H. and Bragg, J.K. (1959) *J. Chem. Phys.* **31**, 526.
9. Bierzynski, A., Kim, P.S. and Baldwin, R.L. (1982) *Proc. Natl. Acad. Sci. USA* **9**, 2470.
10. Dyson, J., Cross, K.J., Houghton, R.A., Wilson, I.A., Wright, P.E. and Lerner, R.A. (1985) *Nature* **318**, 480.
11. Marqusee, S. and Baldwin, R.L. (1987) *Proc. Natl. Acad. Sci. USA* **84**, 8898.
12. Shoemaker, K.R., Kim, P.S., York, E.J., Stewart, J.M. and Baldwin, R.L. (1987) *Nature* **326**, 563.
13. DeGrado, W.F., Wasserman, Z.R. and Lear, J.D. (1989) *Science* **243**, 622.
14. Richardson, J.S. and Richardson, D.C. (1988) *Science* **240**, 1648.
15. Cohen, C. and Parry, D.A.D. (1986) *Trends Biochem. Sci.* **11**, 245.
16. Hodges, R.S., Semchuk, P.D., Taneja, A.H., Kay, C.M., Parker, J.M.R. and Mant, C.T. (1988) *Peptide Res.* **1**, 19.
17. Ramasubbu, N. and Parthasarathy, R. (1989) *Biopolymers* **28**, 1259.
18. Parthasarathy, R., Chaturvedi, S. and Go, K. (1990) *Proc. Natl. Acad. Sci. USA* **87**, 871.
19. Scatturin, A., Tamburro, A.M., Rocchi, R. and Coletta, M. (1966) *Gazz. Chim. Ital.* **96**, 1383.

A novel crepe ribbon representation for protein structures

N. Pattabiraman* and Keith B. Ward

Code 6030, Laboratory for the Structure of Matter, Naval Research Laboratory, Washington, DC 20375–5000, U.S.A.

Introduction

Learning the principles governing tertiary folds in protein structures is very important for understanding the relationships between structure and function in these molecules. In the Brookhaven Protein Data Bank [1] there are coordinates for more than 500 protein structures obtained from single crystal X-ray diffraction studies. To represent various secondary structural fragments in protein structures, Richardson [2] used ribbon spirals around a cylinder for α-helices and flat twisted ribbons with arrows at the C-terminus for β-helices. Variations of the ribbon representations were generated [4,5] by cubic B-spline curves for a series of 'guide points' that were spaced equally along the desired ribbon width depending upon the secondary structure of the peptides. Lesk and Hardman [3] used cylinders instead of ribbons for α-helices. In their representation, the cylinders were generated using an idealized α-helix whereas ribbons were constructed using spline curves. In order to generate the ribbon representation, the secondary structure classification for each residue must be assigned. This information is often presented in the literature report describing the crystal structure. Such descriptions give qualitative information about the inter-relationships between various secondary structures, but most often do not quantify the nature of the local as well as global folding in a protein structure. However, a number of methods to describe various secondary structures have been proposed in the literature [6–8]. In addition, approximate helical axes have been obtained by least-squares fitting of short probe helices to backbone atoms in successive residues [9–11]. Using the method developed in describing nucleic acid structure, a P-curve [12] for protein structure was developed using successive peptide planes. This method was based on the peptide plane linking residues instead of the residue itself. By smoothing the local coordinate system for the whole protein, the authors described the P-curve and did not quantify the relationships between various secondary structures. In this report, we present a novel representation of protein structures called a *crepe ribbon* in which each piece of the ribbon is constructed using the coordinates of the backbone atoms (N, C^α and C) of each individual residue. In our representation, the preassigned secondary structure information is not used to generate the crepe ribbon. Instead, from this new representation it is possible to make this assignment in an objective manner and to quantify the inter-relation-

* Permanent address: GEO-CENTERS Inc., 10903 Indian Head Highway, Fort Washington, MD 20744, U.S.A..

ships between various fragments of protein structure and thereby compare and contrast foldings in protein structures.

Using the internal parameters of bond lengths, bond angles and torsion angles of the monomeric unit of a peptide or nucleotide, it is possible to calculate helical parameters based upon a local helical coordinate system [13]. In our method we fragmented the protein structure of N residues into N-2 fragments because phi and psi torsion angles are undefined for the first and the last residues respectively. For the ith fragment, the atoms used in the calculation were N_i-C_i^{α}-C_{i+1}-N_{i+1}. From the internal parameters for each unit and a helix generating program [14], we determined the local helix origin, O_i, of a local orthogonal coordinate system in which the atom N_i is on the x-axis, the z-axis is along the direction of the helix axis and the y-axis is orthogonal to the x and z axes in a right-handed sense. The perpendicular distance between the atom C^{α} and the local z-axis is defined as the 'local' radius for the ith unit. For extended conformations of a peptide backbone, the local radius (0.57 Å) is smaller than the local radius (2.28 Å) for the α-helix conformation. For an ideal α-helix the origins of the local coordinate systems for the residues lie on a straight line. The end points of ten vectors are generated by rotating the perpendicular vector from C^{α} to the local z-axis by $\theta/10°$ about the z-axis and translating by h/10 Å along the z-axis, where θ and h are the local helical twist and unit height for a residue. The ray tracing program, Raster3d [16], was then used to generate the crepe ribbon model by dividing the quadrangles formed by adjacent vectors into triangles. The crepe ribbon model may be color-coded according to the local radius or any other geometrical, physical, or chemical properties of each amino acid residue. A black and white version of the crepe ribbon was generated by assigning a gray scale value between 0 (white) and 1 (dark) for each triangle depending upon the value of the z-coordinates of one of the verticles of the triangle.

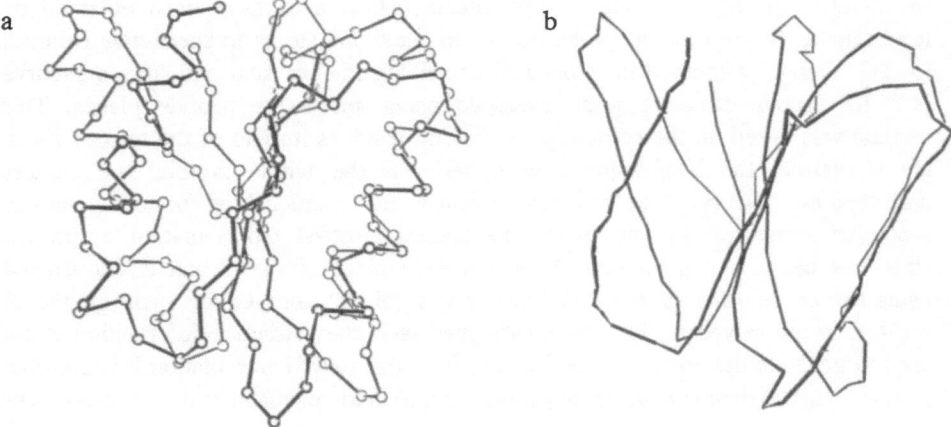

a   b

Fig. 1.   (a) Backbone Ca atom tracing of flavodoxin. (b) Locus of the local origins obtained for each residue of flavodoxin.

Results and Discussion

Fig. 1a shows a backbone C^α atom tracing of flavodoxin. The coordinates were taken from the structure deposited (3FXN) in the Brookhaven data bank [1,15]. The darker atoms and bonds are in the foreground. Various secondary structures such as α-helix (left and right side of the figure) and β-sheet (in the middle) can be clearly seen with this backbone C^α atom representation. In Fig. 1b, the locus of the local origins obtained by our method for each residue are shown. Since the positions of the local origins are derived from the coordinates of backbone atoms of each residue, certain regions of this line joining the local origins are approximately straight because they represent residues in nearly ideal α-helices and β-sheets in the structure. Fig. 1b clearly shows the path and the direction by which the protein folds.

We then calculated the virtual bond angle between three successive local origins and identified the sets of origins whose virtual bond angle is greater than or equal to 165°. Table 1 lists the segments satisfying this criterion, which we have chosen to assign regions of α-helix and β-sheet. The residues that are reported by the authors to be part of an α-helix and β-sheet in the crystal structure are also listed for comparison. The residues assigned as β-sheet from this work agree with that of the reported residues in the crystal structure except for a few bends in the strands. In case of α-helices, even though the assigned residues fall within the regions of residues originally reported with the crystal structure, due to local distortions, the residues at the N- and C-termini of the helices deviate from the ideal α-helix. In particular, three residues in the N-terminus do not exhibit the parameters of an ideal α-helix. Upon examination, the torsion angles for these three residues were found to

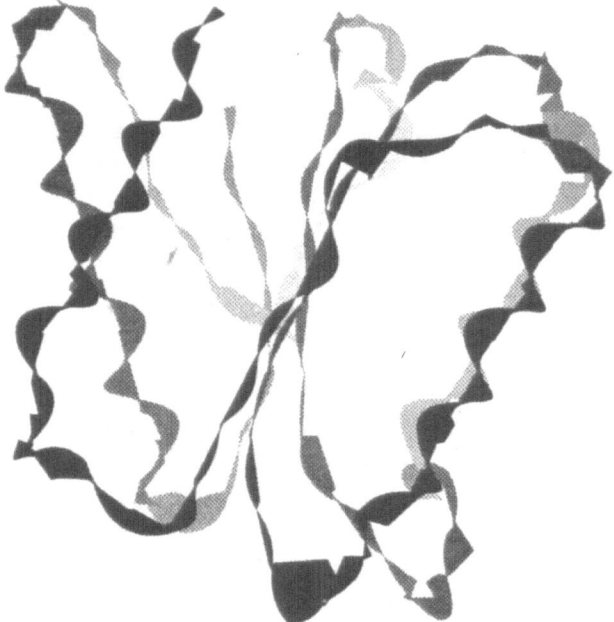

Fig. 2. Crepe ribbon representation of flavodoxin protein molecule.

327

Table 1 *Comparison of secondary structures reported with the crystal structure data of flavodoxin (oxidized form), 3FXN, and assigned from this work*

Original assignment reported with the structure	From this work
Residues involved in α-helix	
10 to 27	14 to 16 and 18 to 23
65 to 73	68 to 71 and 72 to 74
93 to 104	97 to 99 and 100 to 104
124 to 136	121 to 123 and 129 to 132
Residues involved in β-sheet	
1 to 5	2 to 5
29 to 35	28 to 33 and 33 to 35
49 to 53	48 to 53
80 to 88	80 to 87
109 to 119	108 to 110, 113 to 115 and 115 to 118

deviate significantly from the values of phi (303°) and psi (313°) angles for an ideal α-helix.

Figure 2 shows a black and white version of the crepe ribbon representation for flavodoxin (3FXN). The gray scale of the ribbon changes from the front to the back of the molecule. The darker part of the picture is in the foreground. From Fig. 2 it appears that a long crepe ribbon has been twisted and bent to form this representation. The various helical (with large ribbon thickness) and β-strand (with smaller ribbon thickness) regions of the molecule are apparent.

We believe there will be a number of uses for this crepe ribbon representation. First of all, this representation will be used to depict the folding of the polypeptide chain. Secondly, by fitting a least-squares line to the coordinates of the local origins for the segments listed in Table 1, it is possible to calculate parameters describing the interrelationships between various segments by the procedure described by Richards and Kundrot [11]. Using the coordinates of the local origins one can calculate the origin-origin distance plot similar to the C^{α}-C^{α} distance plots [17]. The origin-origin distance plots can be used to compare and contrast the folding of the same protein with different conformations or from different sources. The details of the results of our calculations and examples of how this novel representation can be useful will be published elsewhere.

Acknowledgements

This work was supported in part by the United States Army Medical Research & Development Command. Figures 1a and 1b were generated using SCHAKAL88B [18].

References

1. Berstein, F.C., Koetzle, T.F., Williams, G.J.B., Meyers, E.F., Brice, M.D., Rodgers, J.R., Kennard, O., Shimanouchi, T. and Tasumi, M. (1977) *J. Mol. Biol.* **112**, 535.

2. Richardson, J.S. (1982) *Adv. Protein Chemistry* **34**, 167; Richardson, J.S. (1985) *Methods in Enzymology* **115**, 341.
3. Lesk, A.M. and Hardman, K.D. (1982) *Science* **216**, 539.
4. Carson, M. and Bugg, C.E. (1986) *J. Mol. Graph.* **4**, 121; Carson, M. (1987) *J. Mol. Graph.* **5**, 103.
5. Priestle, J.P. (1988) *J. Appl. Cryst.* **21**, 572.
6. Levitt, M. and Greer, J. (1977) *J. Mol. Biol.* **114**, 181.
7. Lifson, S. and Sander, C. (1980) *J. Mol. Biol.* **139**, 627.
8. Rose, G.D., Gierasch, L.M. and Smith, J.A. (1988) *Adv. Protein Chem.* **37**, 1.
9. Louie, A.H. and Somarjai, R.L. (1983) *J. Mol. Biol.* **168**, 143.
10. Barlow, D.J. and Thornton, J.M. (1988) *J. Mol. Biol.* **201**, 601.
11. Richards, F.M. and Kundrot, C.E. (1988) *Proteins: Structure, Function and Genetics* **3**, 71.
12. Sklenar, H., Etchebest, C. and Lavery, R. (1989) *Proteins: Structure, Function and Genetics* **6**, 40.
13. Sugeta, H. and Miyazawa, T. (1967) *Biopolymers*, **5**, 673.
14. Pattabiraman, N. (1979) Ph.D. Thesis, Indian Institute of Science, Bangalore.
15. Smith, W.W., Burnett, R.M., Darling, G.D. and Ludwig, M.L. (1977) *J. Mol. Biol.* **117**, 195.
16. Bacon, D. and Anderson, W.F. (1988) *J. Mol. Graph.* **6**, 219.
17. Phillips, D.C. (1970) *Biochem. Soc. Symp.* **313**, 11.
18. Kneller, E. (1988) SCHAKAL88B, a program for the graphics representation of molecular and crystallographic models, Kristallographisches Institut der Universität, Freiburg, FRG.

Molecular modeling of the interactions between *Escherichia coli* DNA polymerase I and substrates

Janardan Yadav[a], Prem Narayan Yadav[b], Edward Arnold[c],
Swamy Laxminarayan[a] and Mukund J. Modak[b]

[a]*Division of Academic Computing Services, Department of Information Services and Technology and*
[b]*Department of Biochemistry and Molecular Biology, University of Medicine and Dentistry of New Jersey, 185 South Orange Ave, Newark, NJ 07103, U.S.A.*
[c]*Center for Advanced Biotechnology and Medicine and Department of Chemistry, Rutgers University, 679 Hoes Lane, Piscataway, NJ 08854, U.S.A.*

Abstract

Molecular modeling studies have been carried out with the Klenow fragment of *E. coli* DNA polymerase I (pol I) to understand the binding mechanism of deoxynucleoside triphosphate (dNTP) in the active site of enzyme in the presence of DNA template-primer. A model has been constructed to assess the steric interactions of the DNA template-primer and dNTP substrates with Klenow fragment. Starting with the C^{α} coordinates of the Klenow fragment, the polypeptide backbone was completed and side chains were added using the SYBYL molecular modeling package. This model satisfies many biochemical and genetic data reported for the binding properties of DNA template-primer and deoxynucleoside triphosphate substrates to DNA polymerase I.

Introduction

DNA polymerase I (pol I) of *Escherichia coli* provides a simple model for the study of the enzymatic reactions involved in DNA replication [1]. The Klenow fragment is derived from *E. coli* DNA polymerase I [2]. It represents two thirds of the entire polymerase I structure and retains full polymerase activity [3]. The molecule is folded into polymerase and 3'-5' exonuclease activity domains according to the X-ray structural data [3,4]. Recent biochemical experiments by chemical modification [5-10], photoaffinity labeling [11] and site-directed mutagenesis [12] suggest that Lys^{635}, Lys^{758}, Arg^{682}, His^{881}, Tyr^{766}, Arg^{841}, Asn^{845}, Arg^{668}, Gln^{849} and Asp^{882} residues play important roles in the polymerase reaction. Generally, Lys^{635}, Arg^{690}, Gln^{849} and Arg^{841} appear to play roles in the binding of DNA template-primer whereas Arg^{682}, His^{881}, Tyr^{766}, Lys^{758} and Asn^{845} are proposed as residues required in the binding of substrate nucleotides and/or in the polymerization reaction.

The availability of the above data and that of C^{α} coordinates of Klenow fragment of pol I [2] encouraged us to initiate molecular modeling studies to provide insight into the structural, electrostatic and hydrophobic complementarities between some of the amino acid residues and both template-primer and deoxynucleoside triphosphate (dNTP) substrates. We present here our current model that has been obtained utilizing the biochemical information on the binding of dNTP and DNA template-primer substrates to the Klenow fragment of pol I with the help of SYBYL molecular modeling software.

Molecular Modeling and Computational Details

The geometries of the deoxynucleosides were taken from the consensus structure of B-DNA (A.R. Srinivasan, personal communication). The internal geometry for pyrophosphate ester was taken from Saenger [13]. The pyrophosphate group was added to the deoxynucleoside monophosphate to obtain the deoxynucleoside triphosphate (dNTP). The C^α coordinates of the Klenow fragment of pol I, taken from the Brookhaven Protein Data Bank (entry PDB1DPI, 2), served as starting point in developing a complete atomic model of the enzyme. The complete coordinates for the backbone atoms were generated and the side chains were added. The structure of the Klenow fragment was minimized by MAXIMIN2 using Tripos force field parameters [14] keeping the positions of C^α atoms fixed to remove unfavourable contacts between non-bonded atoms in the distal side chains of the backbone. We believe that much of the structure of the Klenow fragment derived in this manner is quite realistic, since similar treatment of C^α coordinates of bovine pancreatic trypsin inhibitor (BPTI) with the use of SYBYL software [14] yielded a structure which coincided with the position of the backbone and C^β in the reported crystal structure (PDB4PTI, 15) of BPTI. Nonetheless, some distal side-chain atom positions are less precise.

Strategy for making the Complex of Pol I, dNTP and DNA Template-primer

Our strategy of orienting dNTP substrate into the active site cleft was based on the results of chemical modification studies of this enzyme. It was shown that His[881] is the site for cross-linking of the base moiety of dNTP [7] and Arg[682] was reported to be responsible for the triphosphate moiety binding by use of affinity labeling of pol I by fluorosulphonylbenzoyladenosine [5]. Therefore, we oriented the triphosphate ester chain and base groups of the dTTP towards Arg[682] and His[881] of the pol I surface respectively in such a way as to minimize possible steric hindrance between atoms of the two molecules as well as to optimize the potential chemical contacts. Lys[758] and Tyr[766] were also reported to participate in dNTP binding [9,11]. The position of Lys[758] does not seem suitable to make contact with any position of dNTP and was excluded from further considerations. The Tyr[766] was oriented to make a hydrogen bond with the base of dNTP. The complex so obtained with Arg[682], Tyr[766] and His[881] residues and dTTP as a representative of dNTP is depicted in Fig. 1.

In the polymerization reaction enzyme binds to the dNTP substrates in the presence of the DNA template-primer. Therefore, we constructed a $(dA)_{25}$-$(dT)_{14}$ template-primer model duplex of B-DNA [16] using the technique provided in SYBYL. Binding between DNA template-primer and enzyme is quite complex since multiple contact points are involved. Therefore, we decided to use the guidance of the dNTP-enzyme complex to position the 3'-OH terminus of primer nucleotide. The model template-primer was then positioned into the cavity of pol I in a manner that permits

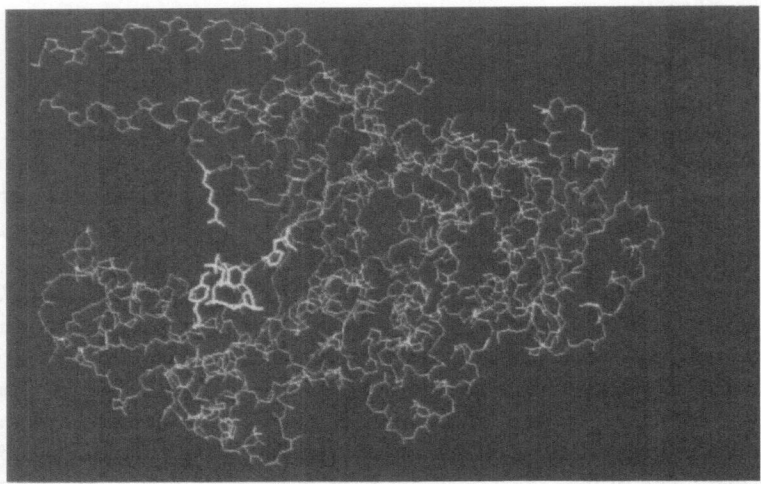

Fig. 1. Model of the Klenow fragment of E. coli *DNA polymerase I complexed with deoxythymidine triphosphate. The critical amino acids Arg[682], Tyr[766] and His[881] are highlighted. These residues of the pol I surface, are guiding points for the positioning of the dTTP in the active site cleft. The side chains of the remaining residues are not shown.*

interaction/vicinity of some of the residues reported to be involved in the binding of the template-primer to enzyme (e.g. Arg[841], Gln[849], Arg[690]) and causes minimum steric hindrance with the surrounding. Figure 2 shows the position of some of the amino acid residues in the backbone of the enzyme which are reported to be involved in the interaction of enzyme with DNA template-primer and which were considered in orienting the template-primer in the cavity. In order to mimic the biological requirement that the base of dNTP substrate be hydrogen-bonded to the appropriate template base (Watson-Crick base pair), we prepared a model that encompasses both the dNTP and template-primer. In this model $(dA)_{25}$-$(dT)_{14}$ and dTTP served as template-primer and substrate nucleotide respectively.

The 3'-OH of the primer end was brought near to the α-phosphate of the dTTP. Furthermore, the oxygen of the cleavable pyrophosphate bound to $P(\alpha)$ of the dTTP was oriented linearly opposite to the oxygen of 3'-OH of the primer terminus based on the expected stereochemistry of the reaction. Orientation of the triphosphate moiety according to this scheme is also suggested by Beal et al. in an RNA polymerase system [17]. The distance between 3'-OH and $P(\alpha)$ was adjusted to a short contact distance in order to poise the groups for the ensuing phosphodiester bond formation. Since we treated the nucleic acid as a rigid body, the rest of the template-primer model was automatically positioned into the cleft. The single-stranded template overhang in the direction of the 3'-5' exonuclease portion of pol I and 7 base pairs of the B-DNA duplex were accommodated by the cleft of the polymerase binding site. This is in agreement with the earlier modeling findings of

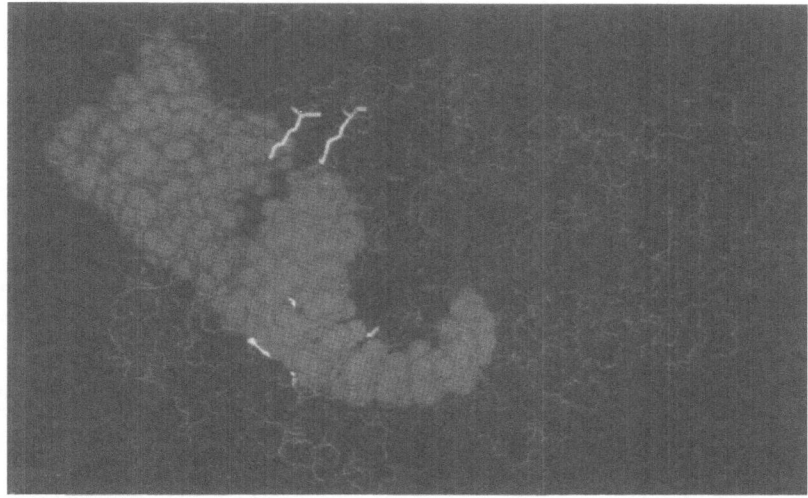

Fig. 2.   Complex of the Klenow fragment with DNA template-primer (CPK model). The template strand is $(dA)_{25}$ and primer strand is $(dT)_{14}$. Side chains of selected amino acids which might be involved in nucleic acid recognition are highlighted.

Joyce and Steitz [3]. For further modeling, the template-primer and dTTP were treated together as a single rigid body in the cavity. The position and the orientations of the template-primer-dTTP were further systematically adjusted within the pol I to optimize the contacts between the two molecular surfaces which seems to satisfy the requirements of the prepolymerization complex. In this context, it should be pointed out that the polymerization reaction is dependent on the presence of divalent metal ion. For simplicity, we have not yet included metal ion in this model. However, there is sufficient space to accommodate either an Mg^{2+} or Mn^{2+} ion to interact with the β- and/or γ-phosphate groups. Work in this direction is in progress.

Results and Discussion

The critical amino acid residues of the pol I involved in the recognition/polymerization of the DNA template-primer and dNTP substrates, are highlighted in Fig. 3 which represents the current model of dTTP-template-primer-pol I complex. This model serves as a starting point for future studies designed to: (i) unravel the other amino acid residues which may make contacts with DNA template-primer and dNTP substrates; (ii) investigate the interaction for each contact; (iii) elucidate rules for molecular recognition to propose an interaction model for this complex; and (iv) predict additional residues of pol I which may participate in the catalytic reaction. In the procedure that we have used in this work, however, the location and conformation of the amino acid side chains in pol I might deviate from the actual structure. The

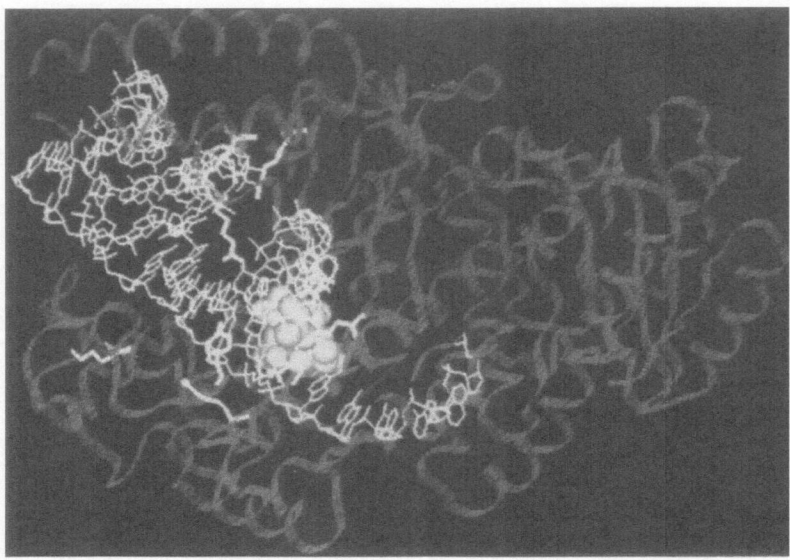

Fig. 3.   Model of deoxythymidine triphosphate-template-primer-DNA polymerase I complex based on the available biochemical and genetic data for the binding properties of DNA template-primer and deoxythymidine triphosphate with DNA polymerase I. The dTTP is shown in the CPK representation.

SYBYL modeling package [14] puts the amino acid side chains in their most preferred orientations based on the structural survey of orientation in many proteins [18,19; Dr. David Mosenkis, Tripos Associates, private communication]. Therefore, an extensive conformational energy search is currently being carried out on the amino acids in the vicinity of the substrates using SYBYL [14], DISCOVER [20], AMBER [21] and CHARMm [22] force fields in an effort to find realistic orientations of these residues in the pol I structure. It has been suggested that application of several force field parameters in conformational studies allows the location of more low-energy minima which might be missed if only one force field is used [23]. It is our hope that we may be able to identify most of the low-energy minima of these side chains in the present energy-search calculations and that one of them will correspond to the actual bioactive structure which is required in the recognition of the substrates. Nonetheless, modeling studies like the present one, can be helpful even if the X-ray and/or NMR coordinates are available for a molecule because the computational methods provide energy maps for the different conformers contrary to the single minimum in X-ray and NMR methods.

For example, it might be expected that considerable conformational change of the Klenow fragment might occur upon binding of template-primer-dNTP substrates, and such calculations may help to identify some of the potential changes. An important assumption used in constructing the present model was that the DNA is uniformly in the B-form in the complex. Possible deviations of the DNA structure from the assumed, including alternative conformational forms (e.g. A-DNA), helix curvature and the presence of kinks, also need to be considered. Further refinement of the present model in the light of these points and more recent biochemical, genetic and structural data on pol I is in progress.

Acknowledgements

The authors thank Dr. Roger Williams, Professor T.P. Singh and Dr. A.R. Srinivasan for helpful discussions and critical comments. We gratefully acknowledge the willing and helpful cooperation, interest and support of Dr. Leslie Michelson, Director of Academic Computing Center, University of Medicine and Dentistry of New Jersey. E.A. thanks the Alfred P. Sloan Foundation for a research fellowship. The research was supported in parts by NIH grants (AI-27690 to E.A. and NIGMS-36307 to M.J.M.).

References

1. Kornberg, A. (1980) *DNA Replication*, W.H. Freeman, San Francisco.
2. Ollis, D., Brick, P., Hamlin, R., Xuong, N.G. and Steitz, T.A. (1985) *Nature* 313, 762.
3. Joyce, C.M. and Steitz, T.A. (1987) *TIBS* 12, 288.
4. Freemount, P.S., Ollis, D.L., Steitz, T.A. and Joyce, C.M. (1986) *Proteins: Structure, Function and Genetics* 1, 66.
5. Pandey, V.N. and Modak, M.J. (1988) *J. Biol. Chem.* 263, 6068.
6. Pandey, V.N., Kaushik, N.A., Pradhan, D.S. and Modak, M.J. (1989) *J. Biol. Chem.* 265, 3679.
7. Pandey, V.N., Williams, K.R., Stone, K.L. and Modak, M.J. (1987) *Biochemistry* 26, 7744.
8. Mohan, P.M., Basu, A., Basu, S., Abraham, K.I. and Modak, M.J. (1988) *Biochemistry* 27, 266.
9. Basu, A. and Modak, M.J. (1987) *J. Biol. Chem.* 262, 9601.
10. Basu, A., Basu, S. and Modak, M.J. (1988) *Biochemistry* 27, 6710.
11. Rush, J. and Konigsberg, W.H. (1990) *J. Biol. Chem.* 265, 4821.
12. Polesky, A.H., Steitz, T.A., Grindly, N.D.F. and Joyce, C.M. (1990) *J. Biol. Chem.* 265, 14579.
13. Saenger, W. (1983) *Principles of Nucleic Acid Structure* (Cantor, C.R., Ed.) Springer-Verlag, New York, p. 86.
14. SYBYL Molecular Modeling Software, Tripos Associates, Inc., St. Louis, MO.
15. Marquart, M., Walter, J., Deisenhofer, J., Bode, W. and Huber, R. (1983) *Acta Crystallogr., Sect. B* 39, 480.
16. Arnott, S. and Hukins, D.W.L. (1972) *Biochem. Biophys. Res. Commun.* 47, 1504.
17. Beal, R.B., Pillai, R.P., Chuknyisky, P.P., Levy, A., Tarien, E. and Eichhorn, G.L. (1990) *Biochemistry* 29, 5994.
18. Ponder, J.W. and Richards, F.M. (1987) *J. Mol. Biol.* 193, 775.
19. Claessens, M., Cutsem, E.V., Lasters, I. and Wodak, S. (1989) *Protein Engineering* 2, 335.
20. DISCOVER Molecular Simulation Program, Biosym Technologies, San Diego, CA.
21. Weiner, S.J., Kollman, P.A. and Nguyen, D.T. (1986) *J. Comp. Chem* 7, 230–252.
22. CHARMm Molecular Mechanics and Molecular Dynamics Package, Polygen Corporation, Waltham, MA.
23. Yadav, J.S., Barnickel, G., Labischinski, H. and Bradaczek, H. (1982) *J. Theoret. Biol.* 95, 167.

Molecular modeling of peptides on microcomputers

S. Scott Zimmerman

Department of Chemistry, Brigham Young University, Provo, UT 84602, U.S.A.

Introduction

Until recently most molecular modeling of peptides was carried out on mini-computers, mainframe computers, or supercomputers. Using these large computers has several drawbacks: (a) the expense of buying and maintaining hardware; (b) the expense of computational time; (c) the slow turn-around time between submitting a batch job and receiving the results; (d) the lack of a user-friendly interface; and (e) the lack of easily accessible graphic display. Some researchers and several software companies have tried to remedy these drawbacks by developing molecular modeling software that runs on microcomputers. None of the software, however, includes the ECEPP/2 system developed by Scheraga and co-workers [1-3]. ECEPP/2 is one of the best and most popular sets of empirical functions, parameters, and geometries for molecular modeling of peptides [4-5].

For these reasons, we have recently developed the computer program PepCAD that runs on Apple Macintosh and on IBM-compatible microcomputers [6]. The aim of PepCAD is to bring molecular modeling of small peptides within the grasp of anyone who has a microcomputer, including researchers, teachers, and students.

Results and Discussion

The first version of PepCAD that we developed [6] runs on any Macintosh with at least 512K RAM and two disk drives, although we recommend a Macintosh II with a hard disk drive. More recently, we developed a version of PepCAD that runs on IBM-compatible MS-DOS computers.

PepCAD was written using the C programming language. The Macintosh version was programmed with LightspeedC (Think Technologies Inc., 135 South Road, Bedford, MA 01730) and the IBM version was programmed with Turbo C (Borland International, 1800 Green Hills Road, Scotts Valley, CA).

PepCAD uses the ECEPP/2 functions, parameters, and geometries [1-3] to compute the total conformational energies and to carry out energy minimization. Unlike ECEPP/2, PepCAD also allows the user to interactively display peptide conformations on the computer graphics screen, change dihedral angles, rotate the peptide about the X, Y, and Z axes, and reduce or enlarge the peptide on the screen. PepCAD also supports a non-graphical batch mode for carrying out minimization from large numbers of starting conformations.

PepCAD has the following general advantages: (a) low initial outlay for the purchase of hardware; (b) low computational cost; (c) fast turn-around time for most

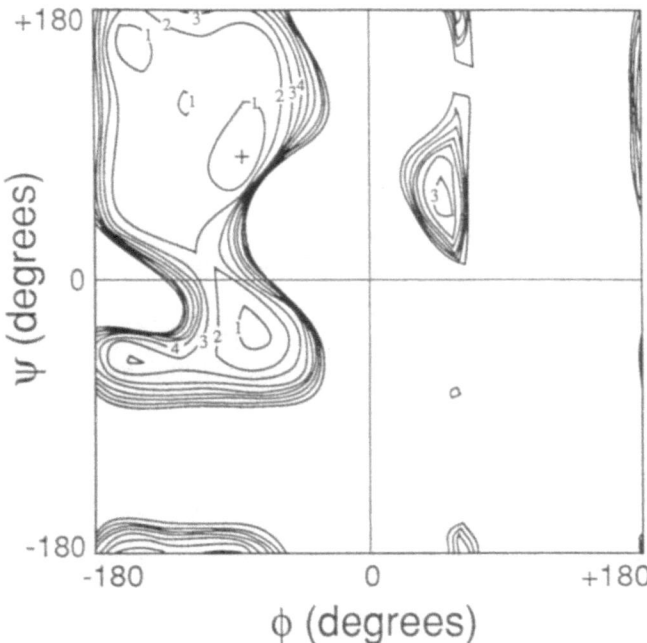

Fig. 1. The conformational energy contour map of the alanine single residue (N-acetyl-N'-methylalanineamide). The + marks the global minimum. Lines of constant energy are drawn from 1 to 9 kcal/mol above the global minimum.

operations; (d) user-friendly interface; (e) easy access to graphic display of peptides, and (f) many capabilities and features that go beyond the original ECEPP/2.

New features of PepCAD that have been added since the original publication of the program [6] include the following: (a) the ability to create conformational energy contour maps (IBM version only) [7]; (b) the ability to carry out statistical mechanical analyses of low-energy conformations to determine statistical weights, bend probabilities, average end-to-end distances, etc. (see ref. 8); and (c) the ability to combine low-energy minima of smaller peptides to make starting conformations of larger peptides using the 'build-up method' of Scheraga and co-workers (see ref. 9). Figure 1 shows a conformational energy contour map of *N*-acetyl-*N'*-methylala-nineamide as determined by PepCAD.

The major disadvantage of PepCAD is slow computation time. Users can partially overcome this drawback by using modern personal computers with fast micro-processors and fast math coprocessors.

We have recently used PepCAD to perform molecular modeling of the chemotactic tripeptide formyl-Met-Leu-Phe and three analogs [10] and of the pentapeptide Ac-Arg-Asn-Cys-Tyr-Asn from α-purothionin [11].

337

Acknowledgements

This research was funded by the Brigham Young University Development Fund and the Brigham Young University Department of Chemistry.

References

1. Momany, F.A., McGuire, R.F., Burgess, A.W. and Scheraga, H.A. (1975) *J. Phys. Chem.* **79**, 2361.
2. Némethy, G., Pottle, M.S. and Scheraga, H.A. (1983) *J. Phys. Chem.* **87**, 1883.
3. Sippl, M.J., Némethy, G. and Scheraga, H.A. (1984) *J. Phys. Chem.* **88**, 6231
4. Roterman, I.K., Gibson, K.D. and Scheraga, H.A. (1989) *J. Biomol. Struct. Dyn.* **7**, 391.
5. Roterman, I.K., Lambert, M. Gibson, K.D. and Scheraga, H.A. (1989) *J. Biomol. Struct. Dyn.* **7**, 421.
6. Feller, D.C., Delmoe, E.F. and Zimmerman, S.S. (1989) *Computers Chem.* **13**, 337.
7. Davidson, R.B. and Zimmerman, S.S. (1990) *J. Chem. Inf. Comp. Sci.* **30**, 174.
8. Zimmerman, S.S., Shipman, L.L. and Scheraga, H.A. (1977) *J. Phys. Chem.* **81**, 614.
9. Scheraga, H.A. (1989) *Chimica Scripta* **29A**, 3.
10. Feller, D.C. and Zimmerman, S.S. (1989) *Int. J. Peptide Protein Res.* **34**, 229.
11. Feller, D.C., Zimmerman, S.S. and Vernon, L.P. (1989) *Int. J. Peptide Protein Res.* **34**, 487.

Theory and simulation of protein folding: A personal view of problems and prospects

Rajmund L. Somorjai

Institute for Biological Sciences, National Research Council of Canada, M-54, Rm. 1113, Montreal Rd., Ottawa, Ont., Canada K1A 0R6

Introduction

It is generally recognized that understanding protein folding is one of the most challenging problems of molecular biology. It is equally clear that, despite enormous and continuing effort, progress is disappointingly slow. Ultimately, we would like to start with the linear amino acid sequence and from it predict the 3-dimensional native structure. At the moment, this goal seems elusive at best.

There are unavoidable experimental limitations that confront most protein folding studies. Notable among these is that the monitoring of the folding process is usually indirect, and that the folding-unfolding transition has a predominantly all-or-none nature. The former indicates that the experiments are macroscopic, the latter is the reason why it is notoriously difficult to identify, trap and study folding intermediates. Thus the theoretician has the additional burden of relating his microscopic simulation results to macroscopic (and as far as he is concerned) incomplete experiments.

Any viable theory of folding will have to reconcile two essential and apparently contradictory experimental facts. These are that folding/unfolding is a reversible equilibrium process, and that folding/unfolding is fast. The former led to the Thermodynamic Hypothesis [1], which asserts that under native external conditions the combined protein + environment system is stable and in its global free energy minimum. The latter implies that the folding process is somehow guided or controlled, such that the complete random searching of the protein's vast conformation space is effectively circumvented. We may postulate that this control is encoded in the linear sequence; I suggest that the control appears as a specific set of physical constraints [2]. Thus the goal of the theoretician (and the experimentalist for that matter!) is to unravel and understand the nature and physical manifestation of these constraints. To predict successfully the native, 3-dimensional structure of a protein from its amino acid sequence, we have to identify these constraints, and use them explicitly in our simulations.

There are definite clues as to how the protein manages to avoid exhaustive search of its conformation space. The most important of these is that protein structures are *organized*. Furthermore, this structural organization is modular or pseudohierarchical. Given the structural hierarchy (well-observable in the X-ray conformations), it is both tempting and reasonable to postulate that protein folding itself is a hierarchical (or, more accurately, multistage) process. In addition, the observed hierarchical subunits must be implicated in the kinetics/dynamics of folding. It is now generally accepted that protein folding can be regarded as a multistage process of self-assembly, with some type of feedback mechanism to rearrange and stabilize the subunits involved. Many arguments support this notion [3]. A much more detailed discussion can be found in refs. 2 and 3.

Discussion

Given this somewhat cursory experimental/theoretical background, let us see how the *simulation* of the folding process fared in the hands of the theoreticians.

The first surprising fact to note is that despite the plausibility and appeal of the qualitative ideas sketched above, there has been no serious attempt to implement them and simulate the protein folding process. There are many reasons for this failure, some conceptual, some technical. The conceptual difficulty stems from the definition of what protein folding is, as adopted by many theoreticians. According to this definition, the solution to the protein folding problem is equivalent to finding an algorithm which predicts the native, 3-dimensional structure from the linear amino acid sequence.

One extreme form of this algorithmic approach would require a set of *rules*, which when applied to the linear sequence, would give the unique native structure (This approach is the obvious generalization of many of the secondary structure prediction algorithms, which combine rules that were deduced from statistical information). The technical problem with this approach is that the 3–400 currently known 3-dimensional structures do not provide enough raw data from which useful rules could be extracted. Although potentially important, I shall not comment further on this, at the moment, premature line of inquiry.

Less obviously algorithmic are all approaches that are based on some version of *minimization*. Despite their origin in the Thermodynamic Hypothesis, they are algorithmic because the folding *process* is largely or completely ignored. The assumption is that if we are given the primary sequence, and appropriate microscopic atom-atom potentials (bonded and nonbonded, two-body and possibly many-body), then by finding the *global* minimum of a classical many-body, multi-variable potential constructed from these, one in fact recovers the native 3-dimensional structure of the protein. More complete treatments recognize the need for including somehow, explicitly or implicitly, interactions with the solvent, and the importance of finding the global *free* energy of the system.

Even if this approach were correct in principle, there are a number of formidable technical problems to overcome. Let us confront them one by one.

1. Accuracy and validity of the empirical potentials

This is a legitimate concern, since the empirical potentials were generally fitted to small-molecule crystal and thermodynamic data, and even if they were then further calibrated to X-ray data, we don't know how good they are far away from the experimentally determined structure. Thus there is no assurance that the global minimum for this potential corresponds to the native conformation.

2. *Large number of variables*

An N-residue protein has 10–12 N atoms, i.e. 30–36 N Cartesian coordinates, even in the absence of solvent. A small protein of 100 residues thus has 3000–3600 variables. If only rotations about covalent bonds are considered, the number of variables reduces to about 5 N, still considerable. We have to keep in mind that to calculate the energy of any conformation, in general as many as $M^2/2$ pair-interactions have to be computed, where M is the number of atoms.

3. *Large number of local minima*

This has become known as the multiple-minimum problem [4]. To demonstrate how formidable this can be, take a simplified representation of a protein molecule by assuming that the only degrees of freedom are rotations about covalent bonds. Assume further that the rotational barriers about each bond are at most threefold. Then for an N-residue protein estimates of the number of local minima range from $O(10^N)$ to $O(3^{5N})$, enormous even for moderate N. Out of this astronomical number, we would have to find the lowest, global minimum, and with certainty. As a *brute force* minimization problem, this is clearly beyond the capabilities of the largest extant and near-future supercomputers. Yet a protein molecule, a dedicated, microminiaturized, highly parallel analog computer, does it routinely in seconds to minutes! Before we see how one might emulate the folding protein molecule, let us consider the *conceptual* problems.

The problem can be stated concisely: does the global free energy of a protein in its proper physiological environment correspond to its native structure? If the answer to this question is unequivocally yes, then simulation attempts to find the *un-constrained* global free energy, arduous as they may be, are worthwhile. If, on the other hand, the native structure is the outcome of the folding *process*, i.e. it is intimately tied to its folding history, then we have to know more. We have to know what were the controls/constraints during the folding process that helped the protein to avoid dead-end pathways and to correct/edit folding errors. I believe that the second, less palatable answer is closer to the truth: the native structure is a *constrained* global free energy, and we do not know yet what the constraints are and how they act, both in space and time. The essence of Wetlaufer's arguments [5] is the same.

Viewed in the above light it is easy to see why the conventional minimization-oriented simulation methods are likely to fail: downhill progress along the potential hypersurface (whether monotonic or not) is *not* and *cannot be* equivalent to the folding process, as is implicitly claimed. In this context it should be emphasized that to a great extent the multiple-minimum problem is an algorithmic artifact. One encounters the plethora of local minima and their intervening barriers during simulation because the folding process was equated with traversing an *invariant* multidimensional landscape while descending from high- to low-energy regions of conformation space. In reality, this landscape *changes* as folding conditions change. In terms of simulation, this simply means that different folding environments correspond to and require different parametrizations of the potential hypersurface. This implies in turn that the location, extent and even number of the local minima vary with changing external conditions.

Given these caveats and arguments, let us see in more detail what are the achievements, if any, of minimization-based methods. It should be emphasized from the outset that *static* (purely local) minimizers would mimic a physical process at absolute zero temperature, or one that was very rapidly quenched to it. Protein folding takes place at finite temperatures and pressures. At room temperature barriers of the order of kT (0.6 kcal/mol) are transparent to conformational motion. However, most static minimizers will be trapped by any of these shallow minima, usually the one nearest to the starting point. Hence their usefulness is confined to quickly approaching the lowest accessible point of the potential hypersurface when quenching rapidly from a higher temperature during a Monte Carlo (MC) or Molecular Dynamics (MD) simulation. The latter are examples of 'dynamical' or nonlocal, *exploratory* minimizers, based on statistical mechanical concepts. Both operate at finite temperatures (and pressures, etc.). Temperature is either an externally imposed parameter (MC) or is computed from the kinetic energy (MD). Both are characterised by the fact that at finite temperatures they sample continually the available configuration space: they are neither local, nor terminating.

Both methods can sample much larger regions of configuration space, because at finite T neither produces strict, monotonic descent towards a local minimum: barriers of the order kT are transparent to them. Of course, barrier heights >> kT are impenetrable: both methods are classical. To surmount high barriers, higher T is needed. This is at the heart of *simulated annealing*, a relatively recent development [6]. The continuous variable version is a succession of statistical equilibrations at gradually decreasing temperatures. The method is a particular member of a class of stochastic global minimizers; they are guaranteed to find the global minimum, provided that an appropriate cooling schedule is observed [7]. The schedule is critical: too fast and one gets trapped in a local minimum; too slow, and the search becomes exorbitantly expensive. In practice, nobody can afford the theoretically optimal rate of cooling.

A promising variant of MC has been used with some success [8]. It assumes implicitly that proteins are characterized by a hierarchy of minima. Thus the coarser, or smoothed hypersurface that is defined by the local minima *alone* also has a minimum which is near the global minimum. The method uses the speed of static minimizers to find local minima. Each new minimum found is a possible state in a coarse-grained but otherwise conventional MC procedure.

This brings me to a class of methods that I consider most promising. Their operating philosophy is: if the potential $V(x)$ to be minimized is too difficult (e.g. multi-extremal), replace it with a sequence of parametrized, smoothed functions $V(x;\alpha_i)$, where the α_i are the smoothing parameters, and $V(x;\alpha_i \rightarrow 0) = V(x)$. In the Russian literature [9] the $V(x;\alpha_i)$ are called *smoothed functionals* because they are defined as multidimensional integrals of $V(x)$ with respect to a smoothing kernel $h(y;\alpha_i)$, with $h(y;\alpha_i \rightarrow 0) = \delta(y)$, Dirac's delta function. A concise discussion of various properties of smoothed functionals is given in Rubinstein [10]. The general idea is to choose an α large enough so that most local minima are smoothed out, find the minimum, decrease α, and repeat the process until $\alpha = 0$. The method was rediscovered recently and independently by Piela and coworkers [11], motivated by a diffusion equation method, corresponding to h(θ) being a multidimensional Gaussian kernel with respect to the dihedral angles {θ}. It has not yet been applied to large biopolymers.

Recently I have developed a quantum-mechanics-inspired global minimization method [12], (called the QQ method), that can be reinterpreted as a generalization of the smoothed functional methods. One may call it an active, adaptive kernel, multiparameter, parallel smoother, to distinguish it from the smoothed functional approach, which is a passive, fixed kernel, single parameter, sequential smoother. A complete suite of programs is in the last stages of construction and debugging, and is being applied to a simplified model (Crippen's pseudo-atom representation) of a 36-residue protein.

In principle, it is possible to simulate in its entirety a fully microscopic version of the protein-folding process (either by Monte Carlo or by Molecular Dynamics calculation). In practice (and in the foreseeable future!) it is a pipe-dream. The simulation would have to start with the denatured protein + solvent + denaturant + environment (temperature, pressure etc.) system. External conditions (e.g. denaturant concentration, pH, etc.) would then be gradually changed, and after each change the *new system* reequilibrated. This process of change and equilibration would continue until native conditions were recovered.

Would the protein be in its native state at the conclusion of this massive simulation? One would hope so. However, since a full quantum-mechanical treatment is and will be out of the question, the need for a more complete representation (and hence understanding) of the various types of empirical interactions we are forced to use, becomes particularly acute. The knowledge of their dependence on macroscopic, environmental parameters seems particularly important.

Since a proper microscopic simulation of the folding process is computationally prohibitive, judicious simplifications are of paramount importance. We have to strip away all unnecessary details and focus on the essential features. In particular, this involves the search for, and imposition of, those constraints that are absolutely necessary for the proper control of the folding process. An example of relevant simplification is Crippen's construction of residue-residue potentials, to be used in a representation of a protein as a point-and-link unbranched polymer [13]. Each residue is replaced by a pseudo atom and adjacent residues are connected by virtual C_α-C_α bonds.

The best illustration of the power of judicious simplification and selection of constraints is the minimalist modelling of folding by Skolnick and co-workers [14]. They represent a protein, similarly to Crippen, as a connected set of beads, but they confine these to a tetrahedral lattice. Excluded volume effects are included, but bond distances and bond angles are fixed, while torsion angle space is discretized. As a consequence, there are only three possible rotational states for the bonds, planar *trans* (*t*) or *gauche* plus (g^+) or *gauche* minus (g^-). Appropriate combinations of these states can mimic both β-strands (t-t-...) and right-handed α-helices (g^--g^--...). There are only two kinds of beads, hydrophobic or hydrophilic. Additional constraints include the specification of the allowed interactions. For the local, interactions they assume that ε_g is the intrinsic energy of the *gauche* state relative to the *trans* state ($\varepsilon_g > 0$ for β-strands). Thus there is marginal preference built in for secondary structure. Long-range interactions are confined to nonbonded nearest neighbors, for which the nature of the interaction energy depends on the identity of the two beads: attractive (ε_h) if both are hydrophobic, repulsive (ε_w) if at least one of them is hydrophilic. Finally, a cooperativity parameter (ε_c) mimics the effects of hydrogen bonding and local peptide dipole interactions. This permits nonbonded, second-nearest neighbor coupling when a pair of adjacent beads are associated with *trans* conformations. Putative bend regions are weakly hydrophobic. By construction, there is a lowest free-energy conformation.

Given the above types of interactions, a dynamic Monte Carlo technique is used to solve a master equation. The local rearrangements of the model are confined to 3-bond and 4-bond kink motions and 2-bond end flips, with the relative probabilities of the various motions fixed ab initio. Despite (and because of) this drastically reduced and simplified model of protein folding, they succeeded in unfolding/refolding low-resolution versions of both a six-strand Greek-key β-barrel and a four-member α-helical bundle as a function of temperature. In addition, interesting qualitative conclusions could be reached about folding pathways, the assembly process and the free-energy reaction coordinate.

I have purposely spent some time on the details of Skolnick's simulation approach to emphasize that important qualitative information can be gleaned from very simple models, provided that the simplifications are to the point and provided that physically relevant constraints are applied. Note the nature of the interactions retained and the constraints on the types of motions employed. Both were constructed utilizing derivative information (hydrophobicity, secondary structure preference, cooperativity, etc.) that was assembled from accumulated knowledge of folding experiments and theories. It is this *structured* aspect of simulation that is missing from studies based on minimization approaches.

What are then the prospects? I believe that our best hope is a multilateral attack, using, on the one hand, qualitative, rather crude but fast model approaches such as Skolnick's, and on the other hand, some version of smoothed functional or QQ method or their combination, (which seem to mimic the changing potential hypersurface of a folding protein) for a hopefully more quantitative understanding. We are to face a lot of problems, but the potential rewards are well worth the effort.

References

1. Anfinsen, C.B. (1973) *Science* **181**, 223.
2. Somorjai, R.L. (1990) in *Protein Engineering: Approaches to the Manipulation of Protein Folding* (Narang, S.A., Ed.) Butterworths, Boston, MA, pp. 1–19.
3. Somorjai, R.L. and Narang, S.A., *J. Theor. Biol.* (submitted).
4. Gibson, K.D. and Scheraga, H.A. (1988) in *Structure and Expression, Vol. 1, From Proteins to Ribosomes* (Sarma, M.H. and Sarma, R.H., Eds.) Adenine Press, Schenectady, NY, pp. 67–94.
5. Wetlaufer, D.B. (1973) *Proc. Natl. Acad. Sci. USA* **70**, 697.
6. (a) Kirkpatrick, S., Gelatt, C.D. Jr. and Vecchi, M.P. (1983) *Science* **220**, 671.
   (b) Vanderbilt, D. and Louie, S.G. (1984) *J. Comput. Phys.* **56**, 259.
   (c) van Laarhoven, P.J.M. and Aarts, E.H.L. (1987) *Simulated Annealing: Theory and Applications*, Reidel Publ. Co., Dordrecht.

(d) Corana, A., Marchesi, M., Martini, C. and Ridella, S. (1987) *ACM Trans. Math. Software* **13**, 262.

7. Gidas, B. (1985) *J. Stat. Phys.* **39**, 73.
8. Li, Z. and Scheraga, H.A. (1987) *Proc. Natl. Acad. Sci. USA* **84**, 6611.
9. (a) Zakharov, V.V. (1970) *Engineering Cybernetics* **4**, 637.
   (b) Antonov, G.E. and Katkovnik, V.Ya. (1971) *Automation and Remote Control (USSR)* **4**, 990.
   (c) Katkovnik, V.Ya and Kulchitsky, O.J. (1972) *Automation and Remote Control (USSR)* **8**, 1321.
10. Rubinstein, R.Y. (1981) *Simulation and the Monte Carlo Method*, John Wiley & Sons, New York, pp. 263–272.
11. Piela, L., Kostrowicki, J. and Scheraga, H.A. (1989) *J. Phys. Chem.* **93**, 3339; Kostrowicki, J. and Piela, L. JOTA (submitted).
12. Somorjai, R.L. *J. Phys. Chem.* (May, 1990).
13. Crippen, G.M. and Snow, M.E. (1990) *Biopolymers* **29**, 1479.
14. Skolnick, J. and Kolinski, A. (1990) *J. Mol. Biol.* **212**, 787, 819 and references therein.

Compact protein conformations

Robert L. Jernigan[a] and David G. Covell[b]

[a]*Laboratory of Mathematical Biology, Bldg. 10, Room 4B-56, Division of Cancer Biology and Diagnosis, National Cancer Institute, National Institutes of Health, Bethesda, MD 20892, U.S.A.*
[b]*Advanced Scientific Computing Laboratory, Bldg. 430, NCI-FCRDC, Program Resources, Inc., Frederick, MD 21702, U.S.A.*

Introduction

Calculations to precisely place large numbers of atoms at their optimal positions are not generally feasible. The development of higher order approaches to molecular structure is essential in order to achieve a more complete understanding of the complexities of biological macromolecules themselves, as well as their interactions with other molecules and their assembly into biological structures.

If a protein is confined to the space defined by the size and shape of the known native conformation, what are the characteristics of its conformations? We have found, by this enforced constraint, that it is possible to enumerate all conformations at the level of one point per residue on a lattice. These conformations are highly variable in many of their characteristics; however, one striking feature is the high probability of the ends of the chain being adjacent. Also, it has been remarked by others that in globular proteins the ends are often close together [1]; so this feature is real and appears to be a result of the compactness constraint. Is it possible to evaluate critically these lattice conformations, when only one point per residue is given? We have previously derived effective residue-residue interaction potentials. These were obtained from the numbers of proximate non-bonded residues in crystal structures, on the assumption that these placements, on average, must reflect their effective energies of interaction. By applying these energies to all compact conformations as enumerated for five small globular proteins, the native conformation is always found among the best few per cent.

Method

Generation of folded conformations of proteins
Regular lattices are generally useful for dividing space and treating protein conformations [2,3]. They furthermore provide a set of appropriate points upon which one can generate conformations. A major problem in treating any biological macromolecule is how to reduce the number of conformations that must be considered. Here for proteins, we consider only the possible forms in a box that has the size and overall shape of the known structure, with little detail for each of the residues [4-6]. One of our principal goals has been to develop a method that permits the investigation of overall chain-folding patterns. This procedure is useful for

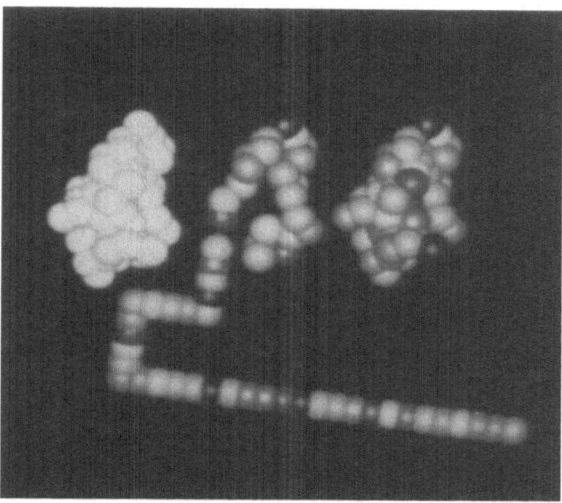

Fig. 1. *Conformation generation. On the left the light spheres are the lattice points upon which the chain of specified sequence is to be placed. In the middle is shown a partially folded chain where the various shades represent various residue types. On the right is shown one completely folded conformation. In the present method all possible ways of placing the chain are generated. Coordinates shown are for trypsin inhibitor.*

considering more complete sets of overall conformations than is possible with all-atom approaches. We are considering proteins at the coarse-grained level of one point per residue.

Choice of lattice points

A variety of lattice types have been evaluated [6]; the best fits are obtained with a face-centered cubic lattice. Other types of lattices considered were simple cubic lattices and body-centered cubic lattices. The lattice points are defined initially by fitting α-carbon positions; a simple cubic lattice gives fits to these atom positions in the range of 1.4 to 1.6 Å RMS deviations, and a face-centered cubic lattice yields RMS deviations in the range of 0.9 to 1.1 Å. There are two reasons for the different qualities of fit with the three lattices: 1) as you move from simple cubic to body-centered cubic to face-centered cubic lattices there are more neighboring positions from which to choose, and 2) the observed virtual bond and torsion angles connecting sequential α-carbon atoms, have preferred values [7] that are fit best with the face-centered cubic lattice. The observed virtual bond angles are centered near 90 and 120 degrees; the face-centered cubic lattice offers, among its choices, good values at 90 and 120 degrees. Observed torsion angles are centered near 45 and 210 degrees; and the face-centered cubic lattice affords good values at 45 and 215 degrees.

Generation and assessment of compact, folded protein conformations

For small proteins, it is then possible to enumerate all conformations rather than resort to less certain procedures such as Monte Carlo. Here (see Fig. 1) we use known structures to define sizes and shapes; however, experimental methods could yield this information about unknown structures. Specifically, the procedure is to enumerate all conformations upon a lattice in a space restricted to the individual protein's known compact conformational space. Then, all possible chain excursions are generated upon these fixed lattice points. The present method achieves large reductions in the number of potential chain conformations in two ways: 1) by restricting the space to the exact size of the protein and 2) by not permitting multiple occupancy. For example, consider a chain of 8 bonds in 2 dimensions on a simple cubic lattice, with three choices at each added bond. If the space were unrestricted, with no volume elements, then there would be 3^7 forms. However, if multiple occupancy is forbidden and chains are built on a 3 × 3 square, then only the following 3 conformations can be achieved, if mirror images, sequence reversals and rotations are not considered:

```
G—F—E         I   F—E         G—H—I
|   |         |   |   |        |
H—I  D        H—G  D           F—E—D
     |                |            |
A—B—C         A—B—C            A—B—C

  a               b               c
```

This is an extremely significant reduction. If all orientations, sequence directions and mirror images are considered, then there are 40 conformers. Consider the same nine points, but designate them physically in some way such as 1 at the bottom left, 2 at the bottom middle, etc. An adjacency [8] matrix, C, can be defined with a unit entry if sites are nearest neighbors on this physical lattice, and zero otherwise. Examination of the value of the element c_{ij} in the product C^k indicates the number of ways in k steps, to pass from lattice point i to j, without volume exclusion. In this way it is possible to calculate approximately the number of ways of going from lattice point i to j. These values can be summed for a constant number of steps for all adjacent lattice sites in order to calculate an approximate distribution function for close contacts, or ring closure. If the element-by-element product is indicated by the symbol ':' so that in R = A : B, we have $r_{ij} = a_{ij} b_{ij}$, then this result for N lattice points is given by $D = C : C^{N-1}$. The distribution is calculated by summing these individual elements, $W_0 = \Sigma_i, \Sigma_j d_{ij}$. With this method we obtain the results shown in Fig. 2. The results show a remarkable independence of the box's shape but a strong dependence on the total number of points in the box. The numbers may be regarded as an upper bound because of the neglect of excluded volume. Approximations can be devised by expanding the matrix C to prevent short-range volume exclusions, such as all rings of 4 bonds. Such approximations can be tested against completely enumerated chains.

Further large reductions in the numbers of conformations can be effected if maximum numbers of core hydrophobic interactions are preferred. In the case above, conformation 'a' has three non-bonded contacts with the central residues, that could

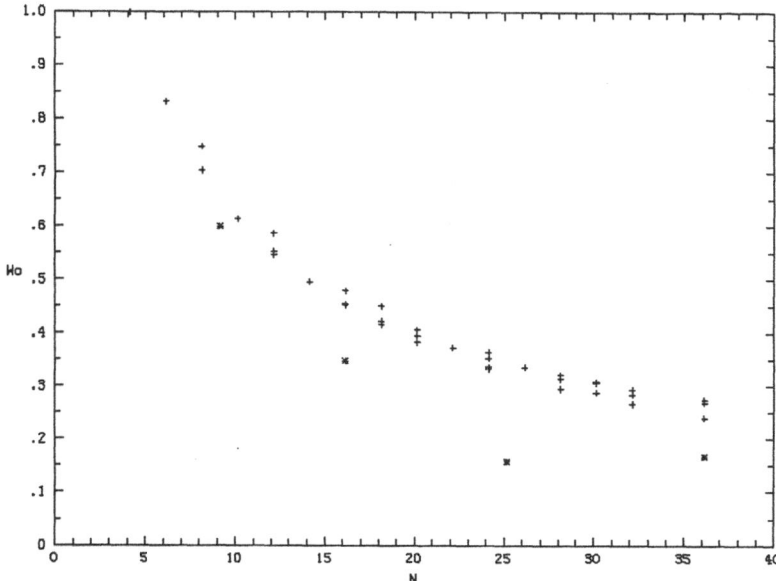

Fig. 2. Probabilities of contacts between chain ends for chains of length N calculated from adjacency matrices are given by the + symbols. This provides an upper bound to the probability because there is no excluded volume consideration in the calculation. Both two- and three-dimensional square, rectangular and cubic boxes are included. Results from the exact enumeration of chains for the 3 × 3, 4 × 4, 5 × 5, and 6 × 6 square boxes are indicated by the ∗ symbols.

be assumed to be hydrophobic. In this way, the 3 × 3 conformations are reduced to one best case. For a 6 × 6 square in two dimensions there would be 35 chemical bonds. An unrestricted lattice would yield 3^{34} or 1.6×10^{16} conformations. Constraining it to a square with volume exclusion reduces the number of conformations to 57337, or 458696, if all orientations, sequence directions and mirror images are included. However, if we maximize the number of non-bonded core (hydrophobic) interactions then there would be only 5 best conformations.

Evaluation of conformations

Conformations are evaluated in terms of residue-specific non-bonded contact energies that favor, principally, hydrophobic interactions. These have been derived from a set of known structures by counting the frequencies of physically close non-bonded pairs and their solvent exposures. We have estimated the contact energies between residues in proteins directly from the observed numbers of residue-residue contacts in crystal structures. Our model is lattice-like with sites occupied by either residues or 'effective solvent molecules' that are composed of a group of actual solvent molecules of total size equal to that of an average residue. In total, 42 globular proteins were used for the numerical evaluations [9]. The interaction values can be averaged and compared directly with the Dayhoff substitution matrix, and a strong correlation is observed (Miyazawa and Jernigan, unpublished). We count the numbers n_{ij} of all pairs of contacts between different types of residues i and j. The total contact energy of any conformation is taken here to be simply the sum of the energies of all contact pairs, $E = \Sigma \Sigma e_{ij} n_{ij}$, where e_{ij} is the energy of contact between the ith and j^{th} types of residues, above the energies of interaction of i with solvent and j with solvent, and n_{ij} is the number of occurrences of these non-bonded pairs in the conformer.

Results

For the five small proteins in Table 1, we find 10^2 to 10^4 possible compact conformations on the face-centered cubic lattice. These conformations include a wide variety of conformations. Subsequent evaluations of these conformations with the residue-residue contact energies find the native structures within the best 2% of all conformers generated. The present method is simple and general and can be used to determine a small group of most favorable overall arrangements for the folding of specific amino acid sequences within a restricted space. This method can be used to obtain a relatively small number of good folded conformers that subsequently can be refined and evaluated with more detailed methods. It is notable that the probabilities for contacts between the amino- and carboxyl-chain termini are unusually high in the samples generated as shown in ref. 6.

Discussion

The present method permits one to obtain a much broader range of folded conformations than with other methods. The non-bonded residue-residue contact energies can be used to discriminate well among conformations, and the exact native conformation is always found within the best 2% of all conformations. We are using

Table 1   *Location of native conformation in samples[a]*

Protein	# Conformations	Percentile ranking of native conformation
Avian pancreatic polypeptide 1PPT	832	1.8
Crambin 1CRN	15,408	0.3
Rubredoxin 1RXN	2,258	1.4
Ferredoxin 1FDX	1,952	0.2
Philippines Sea snake neurotoxin 1NXB	3,000[b]	0.7

[a] Details of calculations given in ref. 6.
[b] Incomplete, only one starting position.

a similar lattice approach to investigate binding between a protein and peptide [10], as well as three-dimensional RNA folding (Covell, Lustig and Jernigan, unpublished).

One new way to approach protein folding would be to combine the distribution functions for contact pairs at different sequence separations with all possible contact energy pairs, to locate nucleation sites. In this case we would favor hydrophobic pairs phe-phe, phe-ile, phe-leu, etc. together with those having high probabilities for close approach in a compact space. Not all interactions would be equally favored because of the constraints introduced by the sequence, packing and chain stiffness. A combinatorial procedure could be useful to learn what new contacts are implied by others and which groups of contacts are spatially incompatible.

We would like to generate conformations of proteins of unknown structure on the basis of experimental information. In particular, a limited number of X-ray reflections or data from electron microscopy could give overall shape information that could be used directly with this method. Other experimental information about specific long-range pairs such as that provided by two-dimensional NOE NMR could be used to accelerate the generation of suitable conformations.

Intrinsically, chain molecules can assume enormous numbers of conformations. One plausible point of view about folded proteins is that there may be only a few compact conformers consistent with the known features of protein structure, such as their hydrophobic interactions.

Acknowledgement

These calculations utilized the facilities of the Advanced Scientific Computing Laboratory's Cray X-MP. Research was sponsored, at least in part, by the National Cancer Institute, DHHS, under contract NO1-CO-74102 with Program Resources, Incorporated. The contents of this publication do not necessarily reflect the views or policies of the DHHS, nor does mention of trade names, commercial products, or organizations imply endorsement by the U.S. Government.

References

1. Thornton, J.M. and Sibanda, B.L. (1983) *J. Mol. Biol.* **167**, 443.
2. Dill, K.A. (1985) *Biochemistry* **24**, 1501.
3. Chan, H.S. and Dill, K.A. (1989) *Macromolecules* **22**, 4559.
4. Jernigan, R.L., Sarai, A., Mazur, J. and Covell, D.G. (1988) *Int. Conf. Supercomput.* **1**, 197.
5. Covell, D.G. and Jernigan, R.L. (1989) *Int. Conf. Supercomput.* **2**, 357.
6. Covell, D.G. and Jernigan, R.L. (1990) *Biochemistry* **29**, 3287.
7. Levitt, M. (1976) *J. Mol. Biol.* **104**, 59.
8. Christofides, N. (1975) *Graph Theory, An Algorithmic Approach*, Academic Press, New York.
9. Miyazawa, S. and Jernigan, R.L. (1985) *Macromolecules* **18**, 534.
10. Jernigan, R.L., Margalit, H. and Covell, D.G. (1991) in *Theoretical Biochemistry and Molecular Biology, A Computational Approach*, Adenine Press, Schenectady, NY.

Protein folding controlled by chemically shifting the temperatures of inverse temperature transitions

Dan W. Urry

Laboratory of Molecular Biophysics, School of Medicine, University of Alabama at Birmingham, P.O. Box 300, University Station, Birmingham, AL 35294, U.S.A.

Introduction

The molecular system

There occur repeating peptide sequences in many proteins. Such repeating sequences provide an opportunity uniquely to gain insights into the relationships between structure and function. When a particular sequence is synthetically repeated many times, it may be referred to as a protein-based polymer (The first symposium on protein-based polymers occurred as part of the April 23–29, 1990 National American Chemical Society meeting and was organized by D. Kaplan, M. Marron, and D. Tirrell who developed the term). In a protein-based polymer the repeating sequence may be as small as a few residues or it may be hundreds of residues as in the repeating domains of transmembrane pumps, channels and multi-subunit enzymes and other proteins. A particularly interesting set of repeating sequences occur in the mammalian elastic protein [1–5]. The most striking repeating sequence seen in bovine and porcine tropoelastins [3,4] can be written $(Val^1-Pro^2-Gly^3-Val^4-Gly^5)_n$ or simply poly(VPGVG).

The molecular conformation of poly(VPGVG) is clearly one in which there is a repeating Type II Pro^2-Gly^3 β-turn with the 10-atom hydrogen-bonded ring occurring between the Val^1 C-O and the Val^4 NH as shown in Fig. 1a [6]. These repeating β-turns on raising the temperature of aqueous solutions are considered to wrap up into a helical structure referred to as a β-spiral in which the dominant helical interturn interactions are hydrophobic and involve, most prominently, the $Val^1\gamma CH_3$ moieties interacting with the $Pro^2\beta CH_2$ moieties [7]. It is proposed that it is dominantly the $Val^1\gamma CH_3$ moieties of repeat i interacting with the $Pro^2\beta CH_2$ moieties of repeat $i + 3$ as shown in Fig. 1. Efforts to check this detail of the proposed β-spiral structure are underway.

The inverse temperature transition

An inverse temperature transition is said to occur when, on raising the temperature of a molecular system (usually in an aqueous solvent) the molecular system, commonly an oligomer or polymer, becomes more ordered. The repeat sequences of elastin provide the most classic demonstration of an inverse temperature transition. The second most striking repeating sequence, particularly apparent in the human, porcine and bovine tropoelastins [3–5], is one in which an Ala residue is introduced between residues one and two of the above repeating pentamer, i.e. $(Ala^1-Pro^2-Gly^3-Val^4-Gly^5-$

Val[6])$_n$. When the cyclododecapeptide of this repeating sequence is prepared, i.e., cyclo(APGVGV)$_2$, it is readily seen to be soluble (molecularly dispersed) in water at low temperature, but on raising the temperature, it crystallizes [8]. Thus this cyclic analog, which has exactly the same amphiphilic balance of hydrophobic (apolar) and polar (peptide) moieties as in the linear high polymer, goes from a state of complete intermolecular disorder to the state which symbolizes intermolecular order, the crystalline state. A molecular description of this interesting example is underway [9]. Also the cyclopentadecapeptide, cyclo(VPGVG)$_3$, aggregates on raising the temperature; the crystal structure of this molecular system has been determined [10], and it is described as the cyclic conformational correlate of the linear high polymer [11], the structure of which is described in Fig. 1. Furthermore, it has also been demonstrated that the polypentapeptide itself self-assembles to form anisotropic fibers comprised of fibrils and of yet smaller filaments [6]. It is now quite clear that these repeating sequences exhibit inverse temperature transitions; they increase order on increasing temperature in an aqueous solvent.

The explanation for this inverse phenomenon has been with us for over half a century [12–16]. Water surrounding hydrophobic groups at lower temperature is more ordered than bulk water. These waters of hydrophobic hydration are often referred to as clathrate or clathrate-like water in analogy to the pentagonal dodecahedral cages of water that have been observed in the crystal structures of alkane gas hydrates [17,18]. In this perspective the more-ordered water surrounding the hydrophobic (apolar side chains such as those of the Val and Pro) residues becomes less-ordered bulk water on raising the temperature as these side chains associate during the ordering of the polypeptide, intramolecularly and intermolecularly. With the current amphiphilic

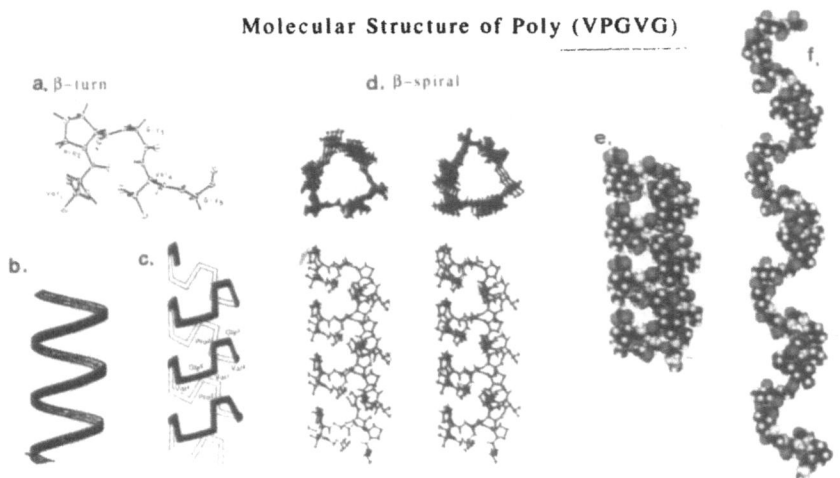

Molecular Structure of Poly (VPGVG)

Fig. 1. Molecular structure of poly(VPGVG). a. β-turn with Pro[2] and Gly[3] at the corners and with the hydrogen bond between the Val[1]C-O and the Val[4]NH. b. Ribbon helical representation. c. Helical representation with addition of β-turn spacers. d. Detailed structure showing approximately 3 pentamers/turn. Stereo pair axis view (above) and side view (below). e. Space-filling model of folded state showing interturn hydrophobic contacts. f. Space-filling model of extended state.

polypeptides it is easy to show (see below) that increasing the hydrophobicity, as in adding a CH_2 moiety, lowers the temperature of the transition whereas decreasing the hydrophobicity as for example in removing a CH_2 moiety increases the temperature of the transition.

Thermomechanical transduction

The molecular process, described as an increase in intramolecular order resulting in the folding of poly(VPGVG), can be thought of as a thermally driven mechanical process of folding as occurs in going to Fig. 1e from Fig. 1f. If a weight could be attached to the bottom of the extended structure in Fig. 1f and the temperature raised, the molecule would fold up as in Fig. 1e and lift the weight. This would be thermomechanical transduction. Such a molecular perspective can be tested macroscopically by forming poly(VPGVG) into a viscoelastic state of about 60% water, 40% peptide by weight and cross-linking with γ-irradiation to form an elastomeric band [6]. As shown in Fig. 2A, a weight can be attached and the temperature can be raised resulting in the thermally driven contraction and lifting of the weight [19]. Thus the molecular perspective developed on the basis of physical, primarily spectroscopic, studies [6] on poly(VPGVG) of thermally driven folding is demonstrated in this most simple mechanical assay of the lifting of a weight.

A generalized hydrophobicity scale and chemomechanical transduction

It was noted above that when the hydrophobicity of the polymer is changed, the temperature range over which the thermally driven folding (contraction) occurs also changes in a predictable way. There have been developed many hydrophobicity scales based on indirect properties such as solubilities in organic solvents [20], partitioning into lipid bilayer membranes [21], changes in surface tension [22], the probability of being buried or on the surface of a globular protein [23,24], etc., but none have been developed directly on the basis of changing the heat and temperature of an inverse

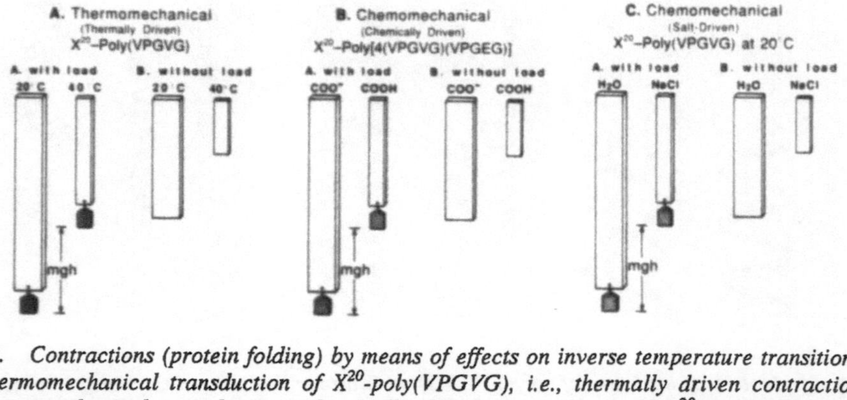

TRANSDUCTION

Fig. 2. *Contractions (protein folding) by means of effects on inverse temperature transitions. A. Thermomechanical transduction of X^{20}-poly(VPGVG), i.e., thermally driven contraction. B. Chemomechanical transduction (chemically driven contraction) of X^{20}-poly[4(VPGVG), (VPGEG)] using the increase in proton concentration to drive contraction by conversion of polypeptide COO^- moieties to COOH. C. Chemomechanical transduction of X^{20}-poly(VPGVG) by means of a salt-driven contraction in which increases in concentration of NaCl lower the temperature of the inverse temperature transition.*

temperature transition. Such work is in progress in this laboratory. Out of what we have seen thus far comes a generalized sense of hydrophobicity which allowed us to propose and to demonstrate that on a generalized scale the COOH moiety is more hydrophobic than the COO⁻ moiety. If this is true, then the thermally driven contraction, for a polymer such as poly[4(VPGVG),(VPGEG)] where E = Glu which contains the COOH functional side chain, will occur at a lower temperature for COOH and a higher temperature for COO⁻. And, if this is correct, it should be possible to remain at a fixed intermediate temperature, to lower the pH to form COOH and have the contracted state and to raise the pH to form COO⁻ and have the extended state [25]. This concept of chemically changing the temperature of an inverse temperature transition is a recently proposed mechanism for mechanochemical coupling or chemomechanical transduction [25]. It is depicted in Fig. 2B and, after having been proposed [25], was experimentally demonstrated [26]. This may be referred to as polymer-based chemomechanical transduction.

Solvent-based chemomechanical transduction

Instead of adding or subtracting a CH_2 moiety or chemically converting a functional side chain of a polypeptide to a more or less hydrophobic state, it is possible to alter the composition of the solvent and change the temperature and heat of the inverse temperature transition. Thus adding sodium chloride to the aqueous solution lowers the temperature (and increases the heat) of the inverse temperature transition (C-H. Luan, T. Parker, K.U. Prasad and D.W. Urry, in preparation). And again this has been shown to be a means of achieving chemomechanical transduction [27] as depicted in Fig. 2C.

Accordingly, poly(VPGVG) and its analogs provide simple assay systems with which to develop an understanding of the driving forces that cause protein-based polymers, and proteins themselves, to achieve function in response to a change in the concentration of a chemical.

Results

Dependence of the temperature of inverse temperature transition on the balance of CH_2 and peptide moieties

As schematically shown in Fig. 3A, the contraction of 20 Mrad γ-irradiation cross-linked poly(VPGVG), indicated as X^{20}-poly(VPGVG), occurs on raising the temperature from 20° to 40°C with the transition being centered near 30°C. If a single CH_2 moiety is added per pentamer as in X^{20}-poly(IPGVG) where I = Ile, the thermally driven contraction now occurs between 0° and 20°C with the transition being centered near 10°C. Thus, adding a CH_2 moiety lowers the temperature by 20°C. The removal of two CH_2 moieties as in poly(VPGAG) results in the transition occurring between 60° and 80°C (data not shown), i.e., the removal of two CH_2 moieties raises the transition temperature by 40°C. Again, the shift is about 20°C per CH_2. Also, it is schematically shown that the removal of a hydrophobic Val residue per repeat as in X^{20}-poly(VPGG) causes the temperature of the transition to be raised some 20°C. It would seem that the apolar isopropyl side chain of a Val residue and the polar peptide backbone of a single residue balance to be equivalent to one CH_2 moiety.

Dependence of the temperature of the inverse temperature transition on the degree of ionization of the COOH moiety

In phosphate-buffered saline (0.15 N NaCl and 0.01 M phosphate), the transition temperature for the inverse temperature transition of cross-linked poly-[4(VPGVG),(VPGEG)], as shown by following contraction, varies remarkably with pH. In general, the conversion of four COOH moieties per 100 residues to COO⁻ raises the temperature of the transition by 50°C from near 20° to near 70°C [28]. In Fig. 3B, the thermally driven contraction is seen to be centered near 20°C at pH 2.1 and near 50°C at pH 4.5. The simple conversion of a COOH to COO⁻ is more than 10 times more effective on a molar basis in shifting the temperature of an inverse temperature transition than the removal of a CH_2 moiety.

Dependence of the heat of the inverse temperature transition on the degree of ionization of the COOH moiety

In Fig. 3B was demonstrated the change in the temperature of the inverse temperature of poly[4(VPGVG), (VPGEG)] with the change in ionization of the carboxyl moiety. In Fig. 4A is shown the change in the heat required to drive the inverse temperature transition as COOH is converted to COO⁻ [29]. At low pH, where essentially all of the carboxylic side chains are protonated, the endothermic heat of the transition is approximately 1 cal/mol. As this is an inverse temperature transition, the endothermic heat is interpreted to be primarily the energy required to convert the more-ordered waters of hydrophobic hydration to bulk water. When the pH is raised, the heat required to drive the transition is less; at pH 4.5, where there would be about 2 carboxylate anions (COO⁻) per 100 residues, the heat required to drive the transition has been reduced to 0.27 cal/mol. The heat now required to destructure the water

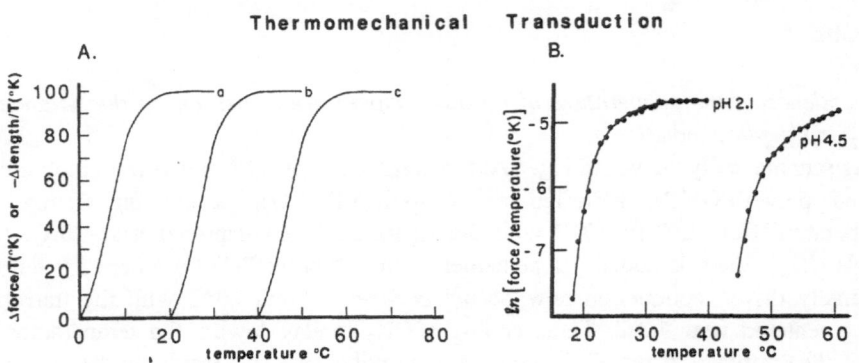

Thermomechanical Transduction

Fig. 3. Dependence of temperature of inverse temperature transition followed by thermally driven contraction. A. Schematic representations of effect of changing the amphiphilic balance of apolar hydrophobic side chains and polar backbone peptide moieties on the temperature of an inverse temperature transition. Curve a is for X^{20}-poly(IPGVG) which has one more CH_2 moiety than X^{20}-poly(VPGVG) in curve b. Curve c is for X^{20}-poly(VPGG) which is the removal of the hydrophobic valyl residue. B. Effect of changing pH on thermally driven contraction of X^{20}-poly[4(VPGVG),(VPGEG)].

surrounding the hydrophobic groups has been reduced to almost one-fourth due to the presence of two COO⁻ moieties per 100 residues. Our interpretation is that the chemical work required to convert two COOH moieties to COO⁻ moieties per 100 residues results in the destructuring of the waters of hydrophobic hydration such that now the complete destructuring can be completed with approximately one-fourth of the previously required heat. Thus, what had at pH 2 been achieved thermally has now largely been achieved chemically.

Cooperativity of the inverse temperature transition

When the experiment on X^{20}-poly[4(VPGVG),(VPGEG)] is an acid-base titration with the elastomer held between a pair of clamps at a constant applied force, the result is most interesting as shown in Fig. 4B [30]. When there is a zero force maintained throughout the titration (that is, the sample is not stretched), the pK_a is very near 4.0 and the curve is steeper than for a standard titration curve (solid line). When the elastomeric band is stretched to a specific applied force and then that force is held constant through the titration, the pK_a increases to just over 4.8, and the titration curve becomes even steeper. This demonstrates a remarkable cooperativity in which three-fourths of the titration is completed with only a 0.5 change in pH. Interestingly, the change in chemical potential, $\Delta\mu$, which is a change in Gibb's free energy per mol, e.g., $\Delta G/\Delta n$ can be calculated with 50% ionization as the reference point, i.e., $\Delta\mu = -2.3RT\Delta pK_a \approx 2RT$. Similarly, when a Monod-type analysis of the change in cooperativity indicated by the change in steepness of the titration curves is attempted, the change in Gibb's free energy per mol of ligand (H⁺), i.e., $\Delta G/\Delta n$, is again about 2RT (S.Q. Peng and D.W. Urry, unpublished). Thus, what is being observed in this protein-based polymer is a cooperativity that has previously been observed in multi-subunit enzymes and other functional proteins.

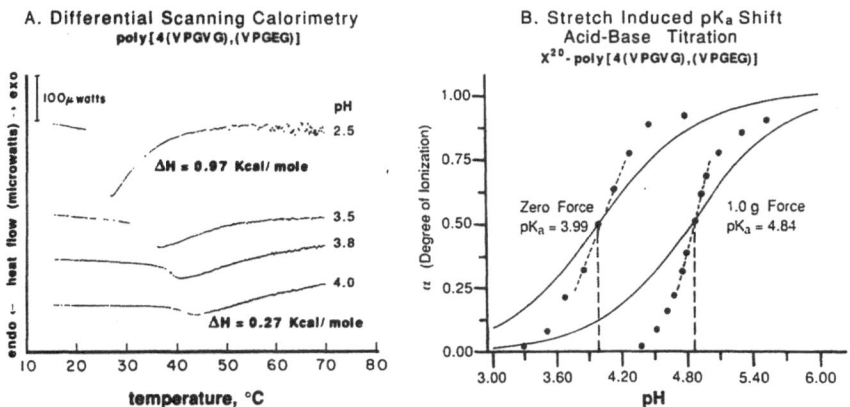

Fig. 4. A. Differential scanning calorimetry of poly[4(VPGVG),(VPGEG)] at low pH (2.5) where almost all carboxyl side chains are protonated and at a raised pH (4.0) where about half of the COOH moieties have been converted to COO⁻ moieties. Note part B for a relevant titration curve. See text for discussion. B. Stretch-induced pK_a shift demonstrated by means of an acid-base titration for X^{20}-poly[4(VPGVG),(VPGEG)]. The pK_a is essentially 4.0 when there is no force applied to the elastic matrix whereas the pK_a is raised on application of a stretching force. See text for discussion. Reproduced with permission from refs. 29 and 30.

Discussion

The working interpretation of the data in Fig. 4A and B is that there exists a hydration-mediated apolar-polar interaction free energy. It results from a competition for hydration free energy between the hydrophobic (apolar) and polar (in this case COO⁻) moieties, and it occurs in structure-limited and water-limited matrices such as in globular proteins, and in these elastic matrices. In short, the structure of water surrounding the COO⁻ moiety which gives rise to its free energy of hydration is different from the structure of water surrounding the Val and Pro side chains which also results in a free energy of hydration. When there is adequate water and distance between these apolar and polar moieties, each receives its full free energy of hydration. But when they become sufficiently proximal, the interceding water molecules cannot simultaneously provide the free energy of hydration for both moieties. In Fig. 4A in the process of achieving adequate waters of solvation, the COO⁻ moieties have destructured much of the water surrounding the hydrophobic moieties. In Fig. 4B, on stretching of the hydrophobically folded β-spirals, the hydrophobic side chains become exposed and compete for the limited amount of water of hydration making water less available to COO⁻ moieties. This means that a lower proton concentration is required before ionization of the carboxyl can occur. But once

Table 1 *Comparison of chemomechanical transduction efficiencies*

$$\eta = \frac{w}{\Delta\mu \cdot \Delta n}$$

w = work = $f\Delta l$, $\Delta\mu$ = change in chemical potential
Δn = change in moles of polymer species.
$\Delta\mu \cdot \Delta n$ = the chemical energy

$\eta_{c,c}$ = efficiency of Charge–Charge repulsion mechanism
(as exemplified by polymethacrylic acid)

w: $\Delta l \approx 1/2$; $f \approx 1000$ x dry weight

Δn: $> 40(COO^- \rightarrow COOH)$ per 200 backbone atoms

$\Delta\mu$:   ($\Delta\alpha$ of $\approx 0.6 \rightarrow \Delta pH \approx 2.5$) $\cong 3.5$ kcal/mole

$\eta_{a,p}$ = efficiency of Apolar–Polar repulsion free energy mechanism
(as exemplified by X^{20} – poly|4(VPGVG),(VPGEG)|)

w: $\Delta l \approx 1/2$; $f \approx 1000$ x dry weight

Δn: $4(COO^- \rightarrow COOH)$ per 300 backbone atoms

$\Delta\mu$:   ($\Delta\alpha$ of $\approx 0.8 \rightarrow \Delta pH \approx 0.5$) $\cong 0.7$ kcal/mole

RESULT: $$\frac{\eta_{a,p}}{\eta_{c,c}} > 10$$

a few COOH moieties convert to COO⁻, they begin to cooperatively destructure the waters of hydrophobic hydration and the remaining carboxyls ionize more readily. Thus the competition for hydration is observed. Importantly, even though there is more water present after stretching, it is dominantly waters of hydrophobic hydration and consequently the ionization is less favored.

Comparison of chemomechanical transduction efficiencies

The previously described polymer-based mechanochemical systems utilized polymethacrylic acid, $[-CH_3CCOOH-CH_2-]_n$ [31]. When high polymers of polymethacrylic acid are cross-linked to form a matrix at low pH, it is contracted and, on raising the pH, it swells or expands due to charge-charge repulsion. This matrix can contract to nearly one-half of its extended length on lowering the pH, and it can pick up weights that are a thousand times its dry weight [31,32]. This achievable work is very nearly the same for X^{20}-poly[4(VPGVG),(VPGEG)], as indicated in Table 1. There are very significant differences, however, in the chemical energy required to achieve the mechanical work. For the polymethacrylic acid system some 50 to 60 carboxylate anions per two hundred backbone atoms are required to achieve full extension by charge-charge repulsion, and these must be lowered to 0 to 10 carboxylate anions per two hundred backbone atoms to achieve full contraction, i.e., some 40 to 50 carboxylate anions per 200 backbone atoms must be protonated to achieve the above noted work. For X^{20}-poly[4(VPGVG),(VPGEG)], on the other hand, only four carboxylate anions must be protonated per 300 backbone atoms to achieve full contraction. Thus the Δn required is an order of magnitude less for the protein-based polymer. Also the required change in chemical potential, $\Delta\mu$, to achieve the contraction is much less due to the cooperativity of the transition as seen in Fig. 4B. To achieve the required change in ionization for polymethacrylic acid requires a ΔpH of about 2.5 [33], whereas a ΔpH of 0.5 would be sufficient in the protein-based polymer (30). Thus the efficiency for converting chemical work into mechanical work is more than ten times greater for the new mechanism demonstrated by X^{20}-poly[4(VPGVG),(VPGEG)].

Since each chemically induced conformational change that results in protein function can be considered to be chemomechanical transduction, and since living organisms derive the required chemical metabolic energy from the food intake, it seems reasonable to expect that protein function would have evolved to utilize the most efficient available mechanism whenever it is suitable to achieve function. Accordingly, we expect the above-described mechanism to be significant in protein function, particularly where that function is the result of chemically induced changes in tertiary and quaternary structure, and where those changes are dominated by the energetics due to changes in exposure of hydrophobic groups to the aqueous milieu.

Acknowledgements

This work was supported in part by National Institutes of Health Grant HL29578 and The Office of the Naval Research Contract No. N00014-89-J-1970.

References

1. Smith, D., Weissman, N. and Smith, D. (1968) *Biochem. Biophys. Res. Commun.* **31**, 309.
2. Sandberg, L.B., Weissman, N. and Smith, D.W. (1969) *Biochemistry* **8**, 2940.
3. Sandberg, L., Leslie, J., Leach, C., Torres, V., Smith, A. and Smith, D. (1985) *Pathol. Biol.* **33**, 266.
4. Yeh, H., Orstein-Goldstein, N., Indik, Z., Sheppard, P., Anderson, N., Rosenbloom, J., Cicila, G., Yoon, K. and Rosenbloom, J. (1987) *J. Collagen Rel. Res.* **7**, 235.
5. Indik, Z., Yeh, H., Ornstein-Goldstein, N., Sheppard, P., Anderson, N., Rosenbloom, J., Peltonen, L. and Rosenbloom, J. (1987) *Proc. Natl. Acad. Sci. USA* **84**, 5680.
6. Urry, D.W. (1988) *J. Protein Chem.* **7**, 1.
7. Chang, D.K., Venkatachalam, C.M., Prasad, K.U. and Urry, D.W. (1989) *J. Biomol. Struct. Dyn.* **6**, 851.
8. Urry, D.W., Long, M.M. and Sugano, H. (1978) *J. Biol. Chem.* **253**, 6301.
9. Luan, C-H., Krishna, N.R. and Urry, D.W. (1990) *Int. J. Quantum Biol. Symp.*, in press.
10. Cook, W.J., Einspahr, H.M., Trapane, T.L., Urry, D.W. and Bugg, C.E. (1980) *J. Am. Chem. Soc.* **102**, 5502.
11. Urry, D.W., Trapane, T.L., Sugano, H. and Prasad, K.U. (1981) *J. Am. Chem. Soc.* **103**, 2080.
12. Edsall, J.H. (1935) *J. Am. Chem. Soc.* **57**, 1506.
13. Frank, H.S. and Evans, M.W. (1945) *J. Chem. Phys.* **13**, 507.
14. Kaufman, W. (1959) *Adv. Protein Chem.* **14**, 1.
15. Tanford, C. (1980) in *The Hydrophobic Effect: Formation of Micelle and Biological Membranes,* John Wiley & Sons, New York.
16. Edsall, J.T. and McKenzie, H.A. (1983) *Adv. Biophys.* **16**, 53.
17. Stackelberg, M.V. and Muller, H.R. (1951) *Naturwissenschaften* **38**, 456.
18. Swaminathan, S., Harrison, S.W. and Beveridge, D.L. (1978) *J. Am. Chem. Soc.* **100**, 5705.
19. Urry, D.W. (1990) *Mat'l. Res. Soc.* **174**, 243.
20. Nozaki, Y. and Tanford, C. (1971) *J. Biol. Chem.* **246**, 2211.
21. Engelman, D.M., Steitz, T.A. and Goldman, A. (1986) *Annu. Rev. Biophys. Biophys. Chem.* **15**, 330.
22. Bull, H.B. and Breese, K. (1974) *Arch. Biochem. Biophys.* **161**, 665.
23. Wertz, D.H. and Scheraga, H.A. (1978) *Macromolecules* **11**, 9.
24. Chothia, C. (1976) *J. Mol. Biol.* **105**, 1.
25. Urry, D.W. (1988) *J. Protein Chem.* **7**, 81.
26. Urry, D.W., Haynes, B., Zhang, H., Harris, R.D. and Prasad, K.U. (1988) *Proc. Natl. Acad. Sci. USA* **85**, 3407.
27. Urry, D.W., Harris, R.D. and Prasad, K.U. (1988) *J. Am. Chem. Soc.* **110**, 3303.
28. Urry, D.W. (1988) *Int. J. Quantum Chem.: Quantum Biol. Symp.* **15**, 235.
29. Urry, D.W., Luan, C-H., Harris, R.D. and Prasad, K.U. (1990) *Polymer Preprints. Am. Chem. Soc. Div. Polym. Chem., Inc.* **31**, No. 1, 188.
30. Urry, D.W., Peng, S.Q., Hayes, L., Jaggard, J. and Harris, R.D. (1990) *Biopolymers,* **29**, in press.
31. Katchalsky, A., Lifson, S., Michaeli, I. and Zwick, M. (1960) in *Size and Shape of Contractile Polymers: Conversion of Chemical and Mechanical Energy* (Wasserman, A., Ed.) Pergamon Press, New York, pp. 1–40.
32. Kuhn, W., Hargitay, B., Katchalsky, A. and Eisenberg, H. (1950) *Nature* **165**, 514.
33. Katchalsky, A. (1951) *J. Polymer Science* **7**, 393.

Hydrophobic characteristics of folded proteins

P.K. Ponnuswamy

Department of Physics, Bharathidasan University, Tiruchirapalli 620 024, India

The Surrounding Hydrophobicity Scale

In the native folded state of a protein molecule, each amino acid residue acquires a characteristic environment. This environment is nothing but the interlocked side chains of the residues appearing within a volume of an optimal size. Any quantity that measures the character of this volume should then, to a great extent, reflect the character of the residue at the centroid of the volume. Our investigations showed that a radius of 8 Å is the most suitable value for the volume and hydrophobicity is the most suitable property to characterize the contents of the volume [1,2]. An empirical parameter, H_p, describing the hydrophobic character of the surrounding volume for a residue in a protein is defined as the sum of Tanford-Jones hydrophobic indices of those residues that come within the volume and an average parameter, $< H_p >$, to this residue is computed from the H_p values noted for that residue in a set of 21 selected protein crystals. The $< H_p >$ values of the twenty residues [2,3] and the Tanford-Jones indices [4,5] of the corresponding amino acids (both in kcal) are: Ala (12.28, 0.87), Asp (10.97, 0.66), Cys (14.93, 1.52), Glu (11.19, 0.67), Phe (13.43, 2.87), Gly (12.01, 0.10), His (12.84, 0.87), Ile (14.77, 3.15), Lys (10.80, 1.64), Leu (14.10, 2.17), Met (14.33, 1.67), Asn (11.00, 0.09), Pro (11.19, 2.77), Gln (11.28, 0.00), Arg (11.49, 0.85), Ser (11.26, 0.07), Thr (11.65, 0.07), Val (15.07, 1.87), Trp (12.95, 3.77), Tyr (13.29, 2.67). The $< H_p >$ values (the first entries in the brackets) forming the so-called 'surrounding hydrophobicity scale', in a way, quantitatively measure the effective hydrophobic bonding the respective residues can make in the protein environment. This scale stands out to be one of the very best compared to 37 other scales [6].

Influence of Protein Environment on the Residue Character

A comparison of the Tanford-Jones scale and the surrounding hydrophobicity scale clearly indicates the extent to which the hydrophobic character of each residue is modified in the protein environment with respect to its character in the organic solvent environment. In Fig. 1 we plot side by side the side-chain contributions of 19 amino acids/residues as noted in the two scales, keeping the glycine/glycyl values as references. The first observation is that whereas in almost all the amino acids, the side chains enhance the hydrophobic bonding character in comparison to that of glycine (barring the cases of Asn, Gln, Ser and Thr for which the values do not differ significantly from that of Gly) in the Tanford-Jones scale, it is not so in the modified scale; all the polar residues and the residue proline become less hydrophobic than the glycyl residue in the $< H_p >$ scale; this means that when an amino acid forms part of a polypeptide chain, the behavior of the neutral backbone is highly influenced by the nature of its side chain, and invariably, it loses its hydrophobic character when its side chain is of polar nature. Thus, the side chain does not just add up to the hydrophobic factor of the backbone, as one would presume. Secondly, the four nonpolar residues Ala, Cys, Met and Val acquire relatively much stronger hydrophobic character, whereas the three aromatic residues Phe, Trp and Tyr acquire relatively much weaker hydrophobic character in the protein than in the organic environment. Surprisingly, His and Leu behave identically in both environments.

Hydrophobic Levels of Secondary Structures

In order to study the hydrophobic environment of residues in the different secondary structural segments in protein molecules, average surrounding hydrophobicity values for the 20 residues were computed by considering the α-helical, β-strand, and β-turn segments in the 21 protein molecules [3]. These parameters, termed as $< H_\alpha >$, $< H_\beta >$ and $< H_t >$, are plotted in Fig. 2, keeping the respective $< H_p >$ values as reference levels. A study of these plots brings out the following facts: (i) the residues Cys, His, Val, Trp and Tyr lose, whereas the residues Ala, Glu, Phe, Gly, Lys, Asn, Pro and Thr strengthen their hydrophobic bonding character when they are accommodated in an α-helix; the residues Asp, Ile, Leu, Gln, Arg and Ser are not influenced in any way in their hydrophobic environment while occurring in an α-helix; (ii) excepting Ile, all the 19 residues strengthen their hydrophobic character when they are accommodated in a β-strand; (iii) excepting Glu, Met and Gln which enrich slightly and Phe, Asn and Ser which remain unaffected, all other residues weaken their hydrophobic environment when they appear in β-turns. These relative hydrophobic characteristics of the residues of the secondary structural elements clearly indicate the general environment of these elements: the helix will have an alternating hydrophobic character, the strand will have a more or less uniform strong hydrophobic character, and the turn will have a weak hydrophobic character. These characteristic/relative hydrophobic levels could be used as a guiding factor in resolving the ambiguities while deciding a particular secondary structure over the other in prediction algorithms [7].

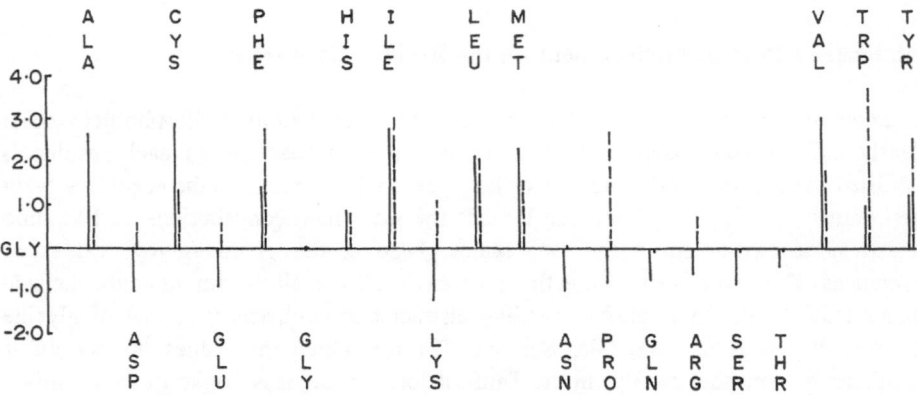

Fig. 1.   *Surrounding hydrophobicity (——) and Tanford-Jones hydrophobicity (– – –) scales with glycine as reference.*

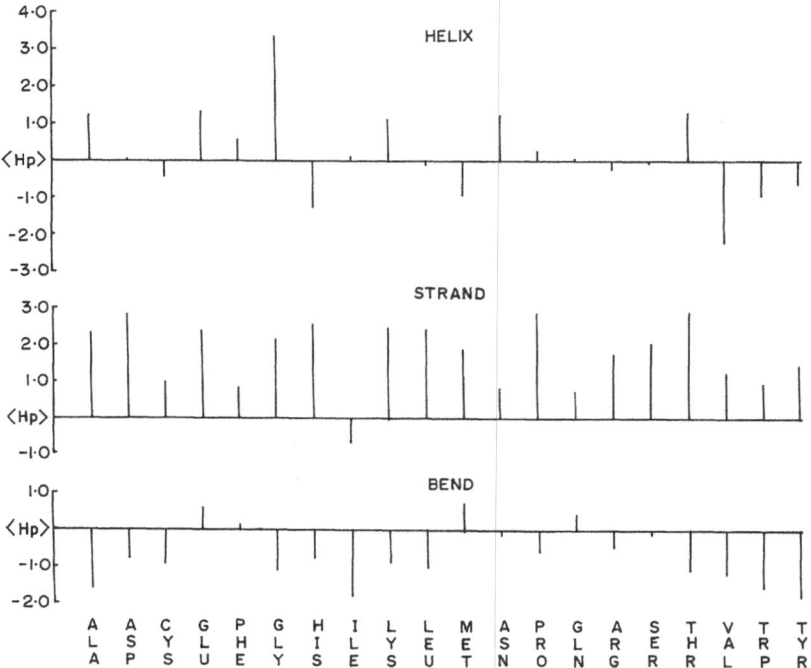

Fig. 2. *Hydrophobic character of amino acid residues while in α-helical, β-strand, and β-turn segments in proteins.*

Helices and Strands in Definite Lengths

A further study was carried out by grouping the secondary structural elements into segments of definite lengths [8]. The α-helical segments were considered in two ways, longitudinally [9] and transversely, and the β-strands in two ways, residue-position-wise and surface-wise. Data of 276 α-helical and 446 β-strand segments as observed in 55 and 48 proteins, respectively, were considered. In the longitudinal arrangement of the helix, the residues were considered into four units along the direction of the helical axis, and in the transverse arrangement, tri-residue units were taken along the polypeptide chain. In Figs. 3a-b, these arrangements are indicated. The averages of the $< H_p >$ values for each of the four units in the helical parts, and for each of the tri-residue units in the strand parts were computed. Only helices of lengths varying from 4 to 20 residues, and strands of lengths varying from 2 to 8 were considered for the present study.

Longitudinal and Transverse Character of Helices

The four units of the α-helix form a parallelopiped as seen in Fig. 3a. The opposing faces of the helix were defined as follows: the unit which has the highest overall average hydrophobicity was taken as the principal part of one side of the helix, the very opposite unit was taken as that of the opposite side, and the other two units (individually and jointly) in combination with each of the above two opposing units. We treated the helix in this fashion so as to examine its amphiphilic character more critically. The hydrophobic strength of the four longitudinal units as measured with the $< H_p >$ scale is plotted against the unit numbers in Fig. 4a. This plot was obtained by taking into consideration all the helices irrespective of their lengths, and it indicates the variation of $< H_p >$ in any α-helical part of the peptide chain: the mean hydrophobicity of unit 2 starting with the second residue is a minimum, and the one which begins with the fourth residue has the highest value, both of them being on opposing sides; it is also noted that units 4 and 3, when considered together, can form a characteristic hydrophobic side of the helix.

The variations of $< H_p >$ in the transverse units of α-helices of varied lengths were computed and plots drawn by considering together the helices of equal length. Typical plots for two segment lengths are shown in Fig. 4b. From these plots, it has been found that in all cases, $< H_p >$ generally increases in going from the N-terminus to the C-terminus of the helix; although the overall hydrophobic character falls at the C-termini of most of the segment groups, it does not fall below the values at the

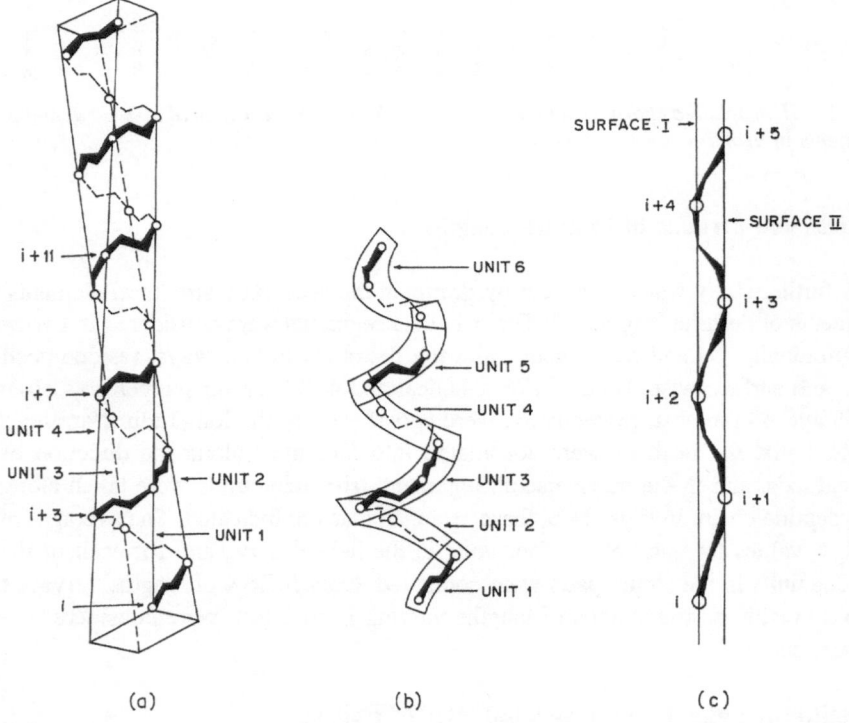

(a)   (b)   (c)

Fig. 3. Different faces of an α-helix and a β-strand. a) Longitudinal arrangement of an α-helix: from the N-terminus residues i, i + 4, i + 8, ..., i + 1, i + 5, i + 9,..., i + 2, i + 6, i + 10, ... and i + 3, i + 7, i + 11, ..., respectively, form Units 1, 2, 3 and 4; these units align in four rows as shown and form the shape of a parallelopiped with a square cross-section; b) transverse arrangement of an α-helix; and c) the two faces of a β-strand.

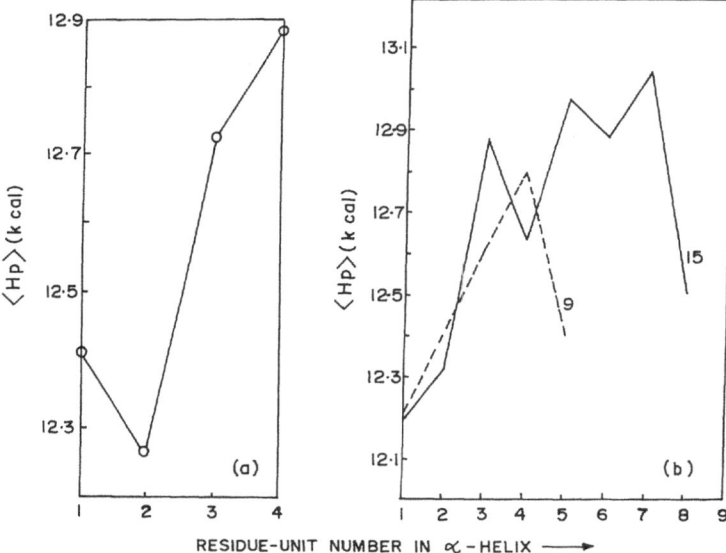

Fig. 4. *Variation of surrounding hydrophobicity in α-helices: a) in the longitudinal units; and b) in the transverse unit. The typical plots shown in b are drawn for helices of equal lengths; the lengths of the segments are given in the plots.*

corresponding N-termini; a maximum occurs asymmetrically nearer to the C-terminal part for many segment groups.

Characteristics of β-Strands

The variations of the average $< H_p >$ for the different residue positions along the β-strands of different lengths are shown in Fig. 5a (only three typical plots are shown). From these plots we noted the following points: (i) in the case of strand groups with not more than 5 residues, the $< H_p >$ tends to be invariably maximum at the second residue position; (ii) for the 7- and 8-residue strand groups a trough is noted approximately at the middle; (iii) exceptionally, the 6-residue strand has the maximum at the center. In general, these plots revealed the unambiguous result that the hydrophobicity of the N-terminal region is dominating over that of the C-terminal region of the β-strands, a fact just opposite to that noted for the α-helices. The plot with a thick line in Fig. 5a corresponds to the case where all strands are put together. This plot also preserves the $< H_p >$ maximum at the second residue position. These results indicate a general fact that the major parts of the 2- to 5- and 7-residue β-strands are highly hydrophobic, whereas the 6- and 8-residue strands are relatively less hydrophobic. Another fact is that the β-strands mostly terminate with less hydrophobic or polar residues.

The hydrophobic fields of the two opposing faces (see Fig. 3c) of the β-strands are shown in Fig. 5b. For the majority of the strand groups the $< H_p >$ of surface II dominates over that of surface I. In certain cases, the differences between the

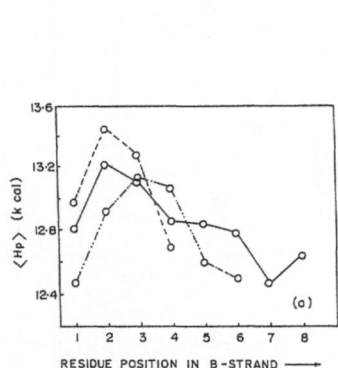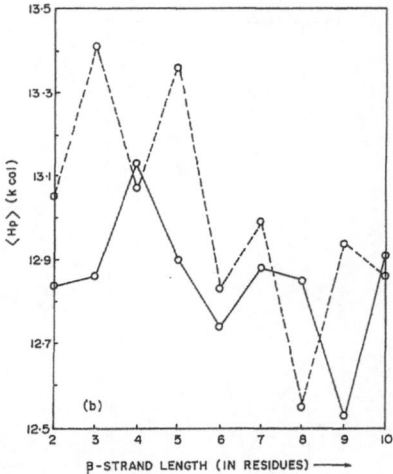

RESIDUE POSITION IN β-STRAND ——→   β-STRAND LENGTH (IN RESIDUES) ——→

Fig. 5. Variation of surrounding hydrophobicity in β-strands: a) the plots corresponding to strands of residue lengths 4 (– – –) and 6 (– ·· –) are shown, the thick line plot corresponds to the average picture of all strands taken irrespective of their lengths; b) surrounding hydrophobicity vs strand length profiles for the two surfaces of β-strands (—— surface I; – – – surface II).

averaged hydrophobicities of the two faces are very small. The fact that even surface I (which is relatively less hydrophobic) of the strands with residues up to 8 is not highly polar indicates the interior seeking property of the β-structure. Moreover, the $< H_p >$ of surface II tends to decrease gradually along two separate paths as the chain length increases. In general, most of the small β-strands tend to have a rich hydrophobic environment. A statistical study shows that about 24% of strands have their surface I, about 38% of strands have their surface II, and about 27% of strands have both their surfaces hydrophobicity values that exceed the overall average. Only 11% of strands have both their surfaces a hydrophobicity value which is below the net average.

References

1. Manavalan, P. and Ponnuswamy, P.K. (1977) *Arch. Biochem. Biophys.* **184**, 476.
2. Manavalan, P. and Ponnuswamy, P.K. (1978) *Nature* **275**, 673.
3. Ponnuswamy, P.K., Prabhakaran, M. and Manavalan, P. (1980) *Biochim. Biophys. Acta* **623**, 301.
4. Tanford, C. (1962) *J. Am. Chem. Soc.* **84**, 4240.
5. Jones, D.D. (1975) *J. Theor. Biol.* **50**, 167.
6. Cornettee, J.L., Cease, K.B., Margalit, H., Spouge, J.L., Berzofsky, J.A. and Delisi, C. (1987) *J. Mol. Biol.* **195**, 659.
7. Cid, H., Bunster, M., Arriagada, E. and Campos, M. (1982) *FEBS Lett.* **150**, 274.
8. Muthuswamy, R. and Ponnuswamy, P.K. (1990) *Int. J. Peptide Protein Res.*, in press.
9. Efimov, A.V. (1979) *J. Mol. Biol.* **134**, 23.

The localization of solvent in protein crystals

Benno P. Schoenborn and Xiaodong Cheng

Center for Structural Biology, Department of Biology, Brookhaven National Laboratory, Upton, NY 11973, U.S.A.

Abstract

In protein crystallography, it has been customary to ignore the contribution of bulk solvent by omitting the low-order diffraction data in refinement procedures. These data contain, however, important information concerning the gross features of the unit cell contents, particularly the solvent structure. The contribution of the solvent to the low-order structure factor terms has been evaluated by dividing the solvent volume into shells extending outward from the protein surface. Two hydration layers in myoglobin crystals have thus been characterized, which allows a better evaluation of the protein's surface structure, improved placement of bound water, ion molecules and a better overall fit to the observed data as well as an improved solvent density Fourier map.

Introduction

An attempt will be made in this paper to correlate the general information on hydration of proteins with crystallographic solvent analysis. Kellenberger [1] suggested, on the basis of electron microscopic studies, that water-soluble proteins are surrounded by a tightly bound layer of water molecules. Small angle neutron scattering showed that the observed radius of gyration of a protein is larger than that predicted for the protein alone [2], suggesting that water is sufficiently associated with protein to 'tumble' with it in solution. However, it has been difficult to demonstrate the existence of water layers in protein crystals. Protein crystallographic analyses on various proteins and particularly some myoglobin derivatives only 'see' water molecules that are well localized in three dimensions, but the agreement between water sites found in different experiments is generally not very good (Takano [3,4], Phillips [5,6], Hanson [7], Kossiakoff [8]). Bound water molecules represent only a small fraction of the total solvent [9] and they are often arranged in clusters, but do not seem to form a layer surrounding the protein. The extent to which a presumed water layer is unexchangeable was investigated by Lehmann et al. using deuterated ethanol-water [10] and dimethyl sulfoxide-water [11] solutions. Nonpolar ethanol and dimethyl sulfoxide were shown to bind to hydrophobic sites in preference of water molecules. Numerous non-crystallographic techniques have been used to estimate the hydration of proteins in an attempt to elucidate or improve solvent structure. For details refer to recent reviews of water structure analysis [8,12-16].

Structure Analysis

In protein crystallographic analysis, bound water molecules are treated often as though they 'belong' to the protein. The rest of the bulk solvent is ignored, and to minimize the effect of this omission the low-order diffraction terms are not used in refinement procedures. An analysis of the low-order diffraction data can, however, provide information on the general distribution of solvent and protein within the crystal. To use the low-order crystallographic data in structural refinement, a solvent evaluation procedure was developed that uses the average scattering of the solvent [17]. As demonstrated below, this procedure can be used to evaluate the scattering density of the solvent as a function of distance expanding outward from the protein's surface [16]. Analysis shows that the water layer surrounding the protein has little mobility, as demonstrated by a low liquidity factor. This procedure further allows the calculation of phases for the low-order diffraction data that enhances the localization of bound water so that all data can be included in the refinement calculations.

The measured structure factor (F_o) is described by the two crystal components of solvent $(\vec{F}_s)$ and protein $(\vec{F}_p)$: $F_o = |\vec{F}_s + \vec{F}_p|$. When the initial crystallographic analysis has reached the stage where a good description of the protein's structure has been obtained, the solvent component of the structure factors is obtained from the difference between the observed and calculated structure factors. Since the solvent has a relatively high liquidity factor, it contributes only to the structure factors with low $\sin \theta$ values and only low-order diffraction data need to be considered. In this analysis, the solvent component $(\vec{F}_s)$ is divided into n structure factor terms $(\vec{F}_{sn})$ corresponding to n shells: $\vec{F}_s = \Sigma_n \vec{F}_{sn}$.

To calculate the structure factors for the n shells, the unit cell of the crystal is divided into a three-dimensional grid. Grid points belong to the protein if a particular point falls within the van der Waals spheres of the atoms of the protein. Grid points external to the protein belong to a given shell, depending on the distance from the surface of the protein. The grid points belonging to a given shell form the coordinate loci used for the structure factor calculations for that shell [16]. The structure factors $(\vec{F}_{sn})$ for each shell and for each Miller index (hkl) are given by:

$F_{sn} = \rho_{sn}\exp(-B_{sn}\sin^2\theta / \lambda^2) \Sigma_{xyz}e^{-2\pi i(hx + ky + lz)_n}$ with xyz being the grid solvent coordinates for the nth shell. Each shell has two variable parameters, the average scattering density of the solvent (ρ_{sn}) and the liquidity factor (B_{sn}) that describes its mobility. This liquidity factor is based on the root-mean-square displacement around a solvent grid point. It now remains to find the best values for ρ_{sn} and B_{sn} for n shells that minimize ΔF: $\Delta F(\rho_{sn}, B_{sn}) = |(\vec{F}_o - \vec{F}_p) - \vec{F}_s(\rho_{sn}, B_{sn})|$.

The best values for ρ_{sn} and B_{sn} are found by an iterative numerical procedure that changes these values in small increments and tests for the lowest ΔF [16]. This minimization is well conditioned, since the number of structure factors that contribute to this minimization is relatively large (~ 2000) compared to the few shells (~ 20).

This procedure is particularly powerful for neutron protein crystallographic analysis since the average neutron scattering density of the solvent can be large and easily adjusted by mixing heavy and light water to obtain a given scattering density [17].

Neutron diffraction data for myoglobin crystals with different solvent scattering densities were collected and used to analyse the solvent liquidity and scattering density as described above. The early crystallographic analysis provided the basic atomic positions of the protein that were used in this analysis [7]. Structure factors ($\vec{F}_{sn}$) for about twenty shells were calculated, with initial scattering densities corresponding to the composition of the solvent that was used in crystallization [16]. The coordinates of the solvent grid were determined, using the known atomic positions of the protein. The best observed scattering densities and liquidity factors of the shells for three data sets (one carbonmonoxymyoglobin and two metmyoglobins) were then determined. The results for myoglobin are plotted in Fig. 1, with the best ρ_{sn} and B_{sn} as a function of distance from the surface of the protein.

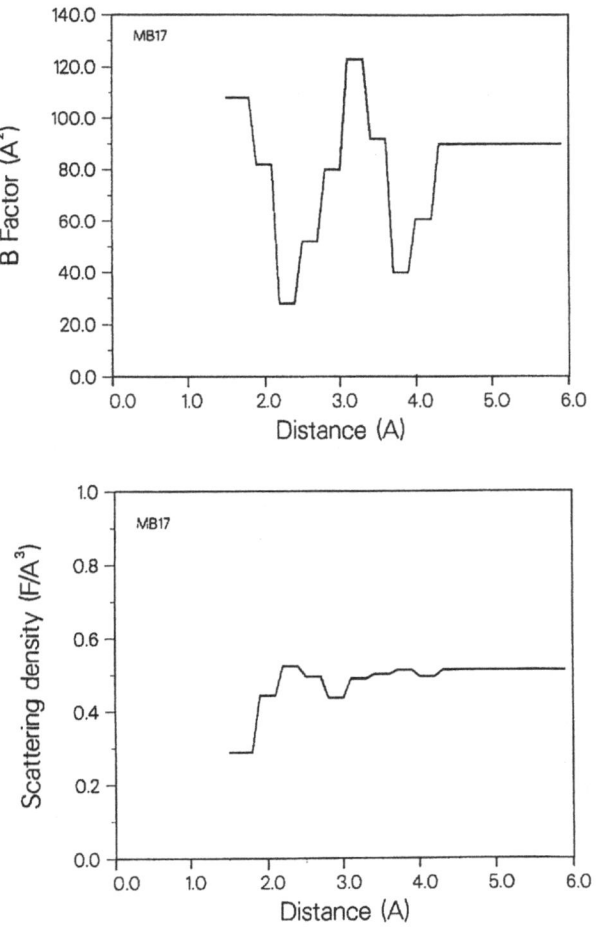

Fig. 1. Graphic presentation of the best scattering densities and liquidity factors for carbonmonoxymyoglobin with about 90% D_2O solvent composition.

Conclusions

These data show that two layers of solvent molecules exist that have low mobility as given by the low liquidity factors. The first layer is located at a distance of 2.3 Å from the protein, which is the expected van der Waals distance of water molecules bound to atoms on the protein's surface. The scattering densities are highest at this distance, suggesting that the bulk mass density of the innermost layer is higher than those further out from the protein. The uniform scattering density for each shell, however, includes the scattering density of any ion located within that shell. The other layer is at about 3.9 Å, which is consistent with the non-hydrogen-bonded water molecules in contact with the protein's surface atoms [16]; this distance is also consistent with a 2nd shell bound to the inner water shell.

With the known best parameters for the solvent shells, modified phases for the observed structure factors can be calculated which improve the Fourier map features on the surface of the protein that depict bound water and ion molecules. The phases (Φ_t) for F_o's are obtained from the calculated structure factors ($\vec{F}_t$) for protein and solvent shells using the expanded structure factor form: $\vec{F}_t = \vec{F}_p + \vec{F}_s$. For the low-order data the contributions from the protein and the solvent are of the same order of magnitude, and large phase shifts are observed. F_t, which included both contributions from the protein and solvent parts, is much closer than F_p to the observed total structure factor (F_o) [16].

The Fourier maps calculated use the observed structure factors (F_o) and the calculated phases (Φ_t) based on all known features of the particular protein. The small Fourier section in Fig. 2 shows the effect of such phasing on the information obtainable, and shows that the improved phasing model greatly facilitates the localization and orientation of water and ion molecules. The solvent-phased Fourier

a   b

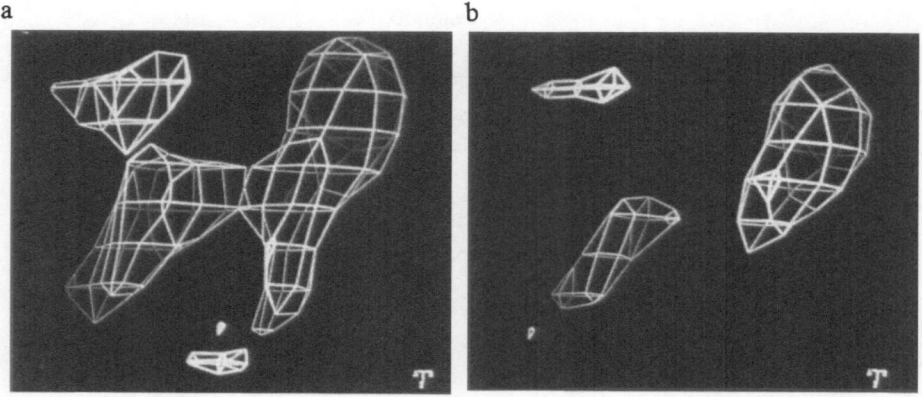

Fig. 2. A detail Fourier map of the surface of carbonmonoxymyoglobin, showing a feature on the left bottom that corresponds to a COO⁻ group of one of the two propionic acids of the heme. a. The Fourier was calculated with ($2F_o - F_p$) coefficients with protein phases (Φ_p) and corresponds to the classical case.
b. The total phases (Φ_t) including solvent and protein contributions were used with the same coefficients as in (a). The contours depicted above were interpreted to contain two water molecules (D_2O) and an ammonium ion (ND_4^+).

map shows features that are completely absent in the conventional map, and generally yields a greatly improved topography on the surface of the protein without changing the clarity of the internal features of the protein.

These new calculated structure factors ($\vec{F}_s$) that represent the solvent shell model can also be used to modify the observed structure factors (F_o) so that they exclude the contribution of the solvent and represent only the protein. These modified terms (F_{om}) are then used in a restrained least-square refinement to refine the protein that allows the inclusion of all structure factors: $F_{om} = |\vec{F}_o - \vec{F}_s|$.

The inclusion of the low-order data in refinement is advantageous, since these data are generally the strongest with the best counting statistics and they condition the least-square refinement well. The inclusion of the solvent model greatly improved the overall R-value to a uniform 11% and led to changes in the surface configuration of the protein particularly the placement of bound water and ions. A detailed analysis of the hydrogen bonding within the protein, protein to water and water to water associations was performed and will be described elsewhere.

Acknowledgements

This research was done under the auspices of the Office of Health and Environmental Research and calculations were performed under the supercomputing program of the US Department of Energy. We thank A.M. Saxena for critically reading the manuscript. The MBCO data was collected in collaboration with A.C. Nunes and J.C. Norvell. The original myoglobin coordinates were provided by J.C. Kendrew and H.C. Watson.

References

1. Kellenberger, E. (1978) *Trends Biochem. Sci.* **3**, N135.
2. Ibel, K. and Stuhrmann, H.B. (1975) *J. Mol. Biol.* **93**, 255.
3. Takano, T. (1977) *J. Mol. Biol.* **110**, 537.
4. Takano, T. (1977) *J. Mol. Biol.* **110**, 569.
5. Phillips, S.E.V. (1980) *J. Mol. Biol.* **142**, 531.
6. Phillips, S.E.V. (1984) in *Neutron in Biology, Basic Life Sciences*, Vol. 27 (Schoenborn, B.P., Ed.) Plenum, New York, p. 305.
7. Hanson, J.C. and Schoenborn, B.P. (1981) *J. Mol. Biol.* **153**, 117.
8. Kossiakoff, A.A. (1985) *Annu. Rev. Biochem.* **54**, 1195.
9. Savage, H. (1986) in *Water Science Reviews*, Vol. 2 (F. Franks, Ed.) p. 1.
10. Lehmann, M.S., Mason, S.A. and McIntyre, G.J. (1985) *Biochemistry* **24**, 5862.
11. Lehmann, M.S. and Stansfield, R.F.D. (1989) *Biochemistry* **28**, 7028.
12. Kuntz, I.D. and Kauzmann, W. (1974) *Adv. Protein Chem.* **28**, 239.
13. Cooke, R. and Kuntz, I.D. (1974) *Annu. Rev. Biophys. Bioeng.* **3**, 95.
14. Finney, J.L. (1979) in *Water: A Comprehensive Treatise*, Vol. 6 (F. Franks, Ed.) Plenum, New York, p. 47.
15. Edsall, J.T. and McKenzie, A.A. (1983) *Adv. Biophys.* **16**, 53.
16. Cheng, X. and Schoenborn, B.P. (1990) *Acta Crystallogr.* **B46**, 195.
17. Schoenborn, B.P. (1988) *J. Mol. Biol.* **201**, 741.

The development of molecular mechanics parameters for carbohydrates

K. Tasaki and J.W. Brady

Department of Food Science, Stocking Hall, Cornell University, Ithaca, NY 14853, U.S.A.

Introduction

Analytic, semi-empirical energy functions of the type used in molecular mechanics calculations [1–3] are only a simplistic caricature of the complex way in which the quantum mechanical energy of a biological system varies with nuclear positions. Nevertheless, the intuitive and easily conceptualized nature of such descriptions of biopolymers, along with their computational simplicity, make empirical energy functions extremely useful. Numerous molecular mechanics simulations have demonstrated that empirical potential functions can reproduce many facets of physical behavior, even for systems as complex as proteins in aqueous solution [2]. However, the results of these simulations depend strongly on the various adjustable parameters which appear in the empirical energy functions. Therefore, it is clearly important to use the most accurate values possible for these parameters, and considerable effort is spent on their improvement. Several extensively tested and widely used force fields have been developed for the study of proteins and nucleic acids [4–6].

Although carbohydrates are more abundant than all other biopolymers combined, they have not received the same exhaustive study as have proteins. In addition to their well-known structural and energy storage roles in such molecules as chitin, cellulose, glycogen and starch, carbohydrates have been found to be very important in molecular recognition as the oligosaccharide components of glycoproteins. Thus, understanding and modeling a wide variety of biological systems will require the accurate treatment of carbohydrate molecules in molecular mechanics calculations. For this reason, producing improved parameter sets for this class of molecules is of importance.

Carbohydrates present a difficult parameterization challenge in molecular mechanics studies for a variety of reasons. Since they contain combinations of functional groups, such as vicinal hydroxyls, not ordinarily encountered in proteins, a simple transfer of protein parameters to these molecules may not always be appropriate. Among the most important characteristics of carbohydrates is their hydrogen-bonding behavior. Because of their large number of strongly hydrogen-bonding hydroxyl groups, the conformations of carbohydrate molecules are often dominated by the requirements of these hydrogen bonds [7]. In particular, the atomic partial charges and non-bonding radii and well depths selected for hydrogen-bonding atoms will significantly affect the thermodynamics of carbohydrate solvation. Unfortunately, appropriate environmentally-averaged atomic partial charges are particularly difficult to assign in a function which does not allow for the dynamic polarization of atoms in response to environment.

Perhaps the most interesting of the special structural features of carbohydrates is the fact that in cyclic sugars, the so-called 'anomeric' carbon atom is simultaneously bonded to two electronegative oxygen atoms, one being the ring oxygen and the other a hydroxyl group (see Fig. 1a). As a result of this special bonding situation about the asymmetric anomeric carbon atom, sugars exhibit a variety of unusual behavior, collectively called anomeric effects [8], which include a greater stability for axial substituents at the anomeric carbon than would be expected on steric grounds. Anomeric effects were first invoked by Edward to explain the greater stability of α-anomers of methyl pyranosides toward acid hydrolysis [9], and subsequently Lemieux [10] observed that pyranosyl fluorides preferred the α configuration, which is also true of most other electronegative substituents on the anomeric carbon. Anomeric effects, which have been found in a variety of quantum mechanical calculations [11,12], are believed to result from a back-donation of lone pair electronic density from the oxygen atoms into C-O σ* orbitals [8]. Such orbital interactions are difficult to represent adequately within the framework of many molecular mechanics energy functions.

Molecular mechanics conformational energy calculations for oligosaccharides have a long history [13–15], but these calculations have had only limited success, possibly due in part to the severe, and often inappropriate, approximations employed, such as the neglect of hydrogen bonding or monomer flexibility. For full molecular dynamics simulations of glycoproteins in aqueous environments, it will be necessary to use functions which not only reproduce carbohydrate behavior, but are also compatible with the protein and solvent functions selected. This paper considers examples drawn from some of the recent molecular dynamics simulations of carbohydrates [16–22] to illustrate the sensitivity of calculated properties to small changes in energy-function

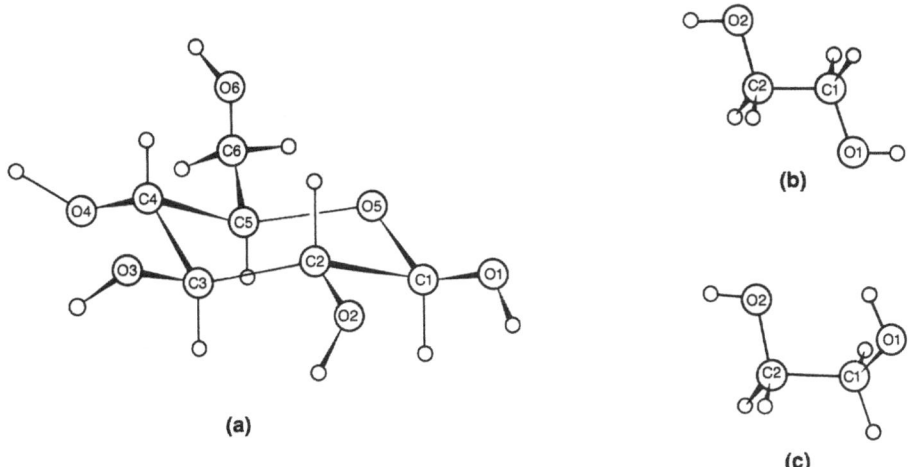

(a)

(b)

(c)

Fig. 1. (a) β-D-glucopyranose; (b) ethylene glycol in the trans *(tTt) conformation; (c) ethylene glycol in the* gauche *(tGg') conformation.*

parameterization, and suggests how these simulations are being used to devise improved parameter sets.

Anomeric ratio in D-glucopyranose

The anomeric carbon atoms of furanoid and pyranoid sugars have two possible configurations, which are designated α and β in the systematic nomenclature. The cyclization reaction of the inherently flexible and reactive linear aldehyde or ketone forms of carbohydrates to form ring structures is acid or base catalyzed in aqueous solution, and is easily reversible. When sugars are dissolved in water, an equilibrium becomes established between the two possible pyranoid forms, as well as between the two possible furanoid structures and the linear open chain. The equilibrium proportions of these various species (called tautomers) are determined by their relative free energies, with the linear form usually present only in trace amounts. For the most stable $^4C_1$ conformation of D-glucopyranose, the α and β configurations correspond to axial or equatorial positions for the C1 hydroxyl group, respectively (see Fig. 1a).

The relative concentrations of the various sugar tautomers are thought to be largely determined by the anomeric effect, which implies a preference for axial forms. However, for D-glucose, D-galactose, and D-xylose in aqueous solution at room temperature, the β anomer actually predominates, with the ratio in D-glucose being 64% β:36% α [23]. Since this ratio is known to be solvent-dependent [24,25], and since the observed preference is often not that which is expected to be favored by the anomeric effect, there must be a substantial solvent contribution to the anomeric equilibrium. Because the various anomeric equilibria are well known and an important feature of carbohydrate behavior, it would be desirable to develop sugar potential energy functions which are able to reproduce this behavior.

In principle, the free energy difference between the two pyranoid anomers of D-glucose in aqueous solution, and hence their relative concentrations, can be determined from free energy simulations [2,3] for a given potential energy function. Such a calculation (Ha, S.N., Gao, J., Tidor, B., Brady, J.W. and Karplus, M., unpublished results) has been attempted for a preliminary energy parameterization for glucose [26] using a CHARMm-type potential energy function [4]. A non-physical pathway between these two isomers was selected which had the effect of gradually 'mutating' α-D-glucopyranose into the β-anomer through the use of a coupling parameter λ which ranged from 0 to 1 in an energy relation of the type

$$V_\lambda = \lambda V_\beta + (1 - \lambda) V_\alpha \quad (0 \leq \lambda \leq 1)$$

An energy equation of this type which is linear in λ has the advantage that it allows the resolution of the total free energy difference into its component terms. Simulations of this transformation in aqueous TIP3P [27] solution found that with the selected potential energy parameters there was indeed a substantial intramolecular energy term favoring the α-anomer, mimicking the anomeric effect, apparently arising primarily from an unfavorable parallel arrangement of the C5-O5 and C1-O1 bond dipoles in the β-anomer (see Fig. 1a). However, this intramolecular term was counterbalanced by a solvent-solute contribution of nearly equal magnitude favoring the β-form. This approximate cancellation correctly predicts a very small energy difference between the two forms, although the α-anomer was calculated to be the most stable form by a small amount. While the error in these types of calculations is possibly large, with the potential energy parameters initially chosen the stabilization of the α-anomer appears to be too large relative to the solvent interaction term. This result suggests that decreasing the unfavorable electrostatic interaction in the β-anomer by changing the oxygen partial charges might result in the correct equilibrium ratio. Such an alteration will at the same time, however, adjust the relative hydrogen bond strengths, thus also changing the solute-solvent contributions. Further simulations to test the dependence of this anomeric equilibrium on oxygen partial charge are underway. It is quite possible that the oxygen partial charges are configuration dependent; selecting the appropriate values will require that they be scaled relative to the water-water and water-solute interactions for the chosen water model in order to correctly balance the anomeric stabilization term with the solvent contribution. It is hoped that through a series of similar molecular dynamics calculations, a combination of parameters can be found which gives the correct anomeric distribution, and which might then be taken as an improved molecular mechanics treatment of the anomeric effect.

Exocyclic rotameric distribution in glucose

Because the electronic distributions in hydroxyl groups must depend to some extent upon their degree of hydrogen bonding, it is not clear that the appropriate partial charges for their atoms can be deduced from ab initio studies of isolated or only partially hydrogen-bonded molecules. For this reason, previous energy function parameterizations have used charges initially suggested by quantum mechanical calculations but subsequently adjusted if necessary to give reasonable interaction energies in solution simulations [28]. As might be expected, partial charges developed in this manner will depend upon the energy function used to describe the solvent water molecules. As a result, mixing such charges with an inappropriate water model can give rise to incorrect behavior. A good example of this problem is provided by the rotameric distribution of the exocyclic hydroxymethyl group in D-glucopyranose. This group in principle has three relatively low-energy conformations, designated as tg, gg, and gt, specifying the position of the O6 atom relative to the O5 and C4 atoms, respectively. NMR experiments reveal that the tg conformation does not occur to any significant extent in aqueous solution, and surveys of the known crystal structures of small carbohydrates find only gt, and occasionally gg, conformations. However, when the structure of a single, isolated α-D-glucopyranose molecule is studied by molecular mechanics simulations [22,26], this exocyclic group exhibits a pronounced preference for the tg conformation. This apparent artifact of the calculations arises because the tg orientation allows the formation of an internal hydrogen bond between the C6 and C4 hydroxyl groups, partially satisfying the bonding requirements of these groups, which normally make hydrogen bonds to crystal neighbors or solvent water molecules.

It might be expected that this primary alcohol group would assume its experimentally observed gt orientation in molecular dynamics or Monte Carlo simulations which explicitly include water molecules, because of the possibility of hydrogen bonds to the solvent. Several molecular dynamics simulations of glucose molecules in solution have now been reported [19,22, and Ha, S.N., Gao, J., Tidor, B., Brady, J.W. and Karplus, M., unpublished results]. The first of these studies [19] employed a preliminary CHARMm-type potential energy function [26] which includes atomic partial charges loosely based upon those found by Jorgensen to be appropriate for solutions. When this potential function was used to simulate α-D-glucopyranose in SPC [29] solution, this exocyclic group surprisingly underwent a spontaneous transition to the intramolecularly hydrogen-bonded geometry, where it remained, indicating that this internal bond out-competed similar hydrogen bonds to the solvent. However, using this same sugar potential in simulations which employed the TIP3P function [27] for the water molecules, the tg conformation was totally destabilized to the extent that in numerous simulations begun with the tg conformation, the molecule invariably underwent spontaneous transitions to the gt, or rarely gg, orientations during the equilibration portion of the calculation, with no return to the tg conformation (Ha, S.N., Gao, J., Tidor, B., Brady, J.W., and Karplus, M., unpublished results). The TIP3P water model, while similar to the SPC function, is the solvent model for which the Jorgensen charges were actually derived. The observed difference in the orientational distribution behavior apparently is the result of mixing charges and water models, giving incommensurate relative hydrogen bond strengths for water-water, water-solute, and solute-solute interactions. This problem does not indicate any inherent difficulty with the SPC water model *per se*; similar simulations of β-D-glucopyranoside using a GROMOS-like sugar energy function, presumably more consistent with the SPC model, found mostly gt orientations in solution, with very little tg probability [22].

Recent molecular dynamics simulations of the carbohydrate analogue ethylene glycol (Figs. 1b and 1c) have further demonstrated the dependence of rotameric distributions upon atomic partial charges and the water model used, indicating the importance of the competition between solvent-solvent, solvent-solute, and solute-solute hydrogen bonds. A series of such MD simulations have been undertaken for various atomic partial charges, van der Waals radii, and water models in a systematic search for the combination which produces the most physical behavior [30]. For example, extensive simulations with one particular choice of atomic partial charges for the glycol atoms found the solute to exist only in the *trans* conformation in SPC solution (Fig. 1b), but to have nearly equal populations of *trans* and *gauche* (Fig. 1c) conformations in TIP3P solution.

Conclusions

It may prove to be particularly difficult to develop suitable potential energy function parameters to describe the unique functional groups of carbohydrate molecules. Problems include a generally poor experimental data base, thermodynamic properties which depend upon extensive hydrogen bonding, and unique bonding arrangements which produce effects which do not easily fit into simple molecular mechanics potential energy functions. Recent molecular mechanics studies have demonstrated the importance of including molecular flexibility in parameterization calculations. The dominant role of hydrogen bonding is also clear, both in crystalline and solution environments. Separating the complex hydrogen bonding of even a simple sugar, which could amount to many kilocalories per mol, from the much smaller anomeric, exoanomeric, and *gauche* effects will require careful simulations of the thermodynamic properties of a number of small analogue systems such as ethylene glycol solutions.

Acknowledgements

The authors gratefully acknowledge the assistance of A.D. French and F. Momany, and the helpful discussions of our research collaborators, M. Karplus, S. Ha, B. Tidor, and J. Gao.

References

1. Burkert, U. and Allinger, N.L. (1982) *Molecular Mechanics*, ACS Monograph Series No. 177, American Chemical Society, Washington, DC.
2. Brooks, C.L., Karplus, M. and Pettitt, B.M. (1988) *Proteins: A Theoretical Perspective of Dynamics, Structure, and Thermodynamics, Advances in Chem. Phys., Vol. LXXI,* Wiley-Interscience, New York.
3. McCammon, J.A. and Harvey, S.C. (1987) *Dynamics of Proteins and Nucleic Acids,* Cambridge University Press, Cambridge.
4. Brooks, B.R., Bruccoleri, R.E., Olafson, B.D., States, D.J., Swaminathan, S. and Karplus, M. (1983) *J. Comput. Chem.* **4**, 187.
5. van Gunsteren, W.F., Berendsen, H.J.C., Hermans, J., Hol, W.G.J. and Postma, J.P.M. (1983) *Proc. Natl. Acad. Sci. USA* **80**, 4315.
6. Weiner, S.J., Kollman, P.A., Nguyen, D.T. and Case, D.A. (1986) *J. Comp. Chem.* **7**, 230.
7. Jeffrey, G.A. (1990), in *Computer Modeling of Carbohydrate Molecules,* ACS Symposium Series No. 430 (French, A.D. and Brady, J.W., Eds.) American Chemical Society, Washington, DC.
8. Tvaroska, I. and Bleha, T. (1989) *Adv. in Carbohydrate Chem. Biochem.* **47**, 45.
9. Edward, J.T. (1955) *Chem. Ind.,* 1102.
10. Lemieux, R.U. and Chu, P. (1958) Abstracts, 133rd Nat. Meeting of the Am. Chem. Soc., San Francisco. 31N.
11. Jeffrey, G.A., Pople, J.A., Binkley, J.S. and Vishveshwara, S. (1978) *J. Am. Chem. Soc.* **100**, 373.
12. Wiberg, K.B. and Murcko, M.A. (1989) *J. Am. Chem. Soc.* **111**, 4821.
13. Rao, V.S.R., Sundararajan, P.R., Ramakrishnan, C. and Ramachandran, G.N. (1967) in *Conformation in Biopolymers, Vol. 2* (Ramachandran, G.N., Ed.) Academic Press, London, pp. 721–737.
14. Brant, D.A. (1972) *Annu. Rev. Biophys. Bioeng.* **1**, 369.
15. Rees, D.A. and Smith, P.J.C. (1975) *J. Chem. Soc., Perkin Trans.* **2**, 830.
16. Brady, J.W. (1986) *J. Am. Chem. Soc.* **108**, 8153.
17. Post, C.B., Brooks, B.R., Karplus, M., Dobson, C.M., Artymiuk, P.J., Cheetham, J.C. and Phillips, D.C. (1986) *J. Mol. Biol.* **190**, 455.
18. Ha, S.N., Madsen, L.J. and Brady, J.W. (1988) *Biopolymers* **27**, 1927.
19. Brady, J.W. (1989) *J. Am. Chem. Soc.* **111**, 5155.
20. Koehler, J.E.H., Saenger, W. and van Gunsteren, W.F. (1987) *Eur. Biophys. J.* **15**, 197; (1987) *Eur. Biophys. J.* **15**, 211; (1988) *J. Biomol. Struct. Dyn.* **6**, 181; (1988) *J. Mol. Biol.* **203**, 241.

21. Yan, Z.-Y. and Bush, C.A. (1990) *Biopolymers* **29**, 799.
22. Kroon-Batenburg, L.M.J. and Kroon, J. (1990) *Biopolymers* **29**, 1243.
23. Shallenberger, R.S. (1982) *Advanced Sugar Chemistry*, AVI Publishing Co., Westport, CT.
24. Franks, F. (1987) *Pure Appl. Chem.* **59**, 1189.
25. Praly, J.P. and Lemieux, R.U. (1987) *Can. J. Chem.* **65**, 213.
26. Ha, S.N., Giammona, A., Field, M. and Brady, J.W. (1988) *Carbohydr. Res.* **180**, 207.
27. Jorgensen, W.L. (1981) *J. Am. Chem. Soc.* **103**, 335.
28. Alagona, G. and Tani, A. (1988) *J. Mol. Struct.: Theochem.* **166**, 375.
29. Berendsen, H.J.C., Postma, J.P.M., van Gunsteren, W.F. and Hermans, J. (1981) in *Intermolecular Forces* (Pullman, B., Ed.) Reidel, Dordrecht, pp. 331–342.
30. (a) Schwartz, M. (1977) *Spectrochim. Acta* **33A**, 1025.
    (b) Maleknia, S., Friedman, B.R., Abedi, N. and Schwartz, M. (1980) *Spectrosc. Lett.* **13**, 777.

On the use of conformationally dependent geometry trends from ab initio studies to determine empirical parameters for the CHARMm molecular mechanics force field

Frank A. Momany[a], V. Joseph Klimkowski[b] and Lothar Schäfer[c]
[a]Polygen Corp., 200 Fifth Ave., Waltham, MA 02254, U.S.A.
[b]Eli Lilly Corp., Indianapolis, IN 46032, U.S.A.
[c]Chemistry Dept., University of Arkansas, Fayetteville, AR 72701, U.S.A.

Introduction

Quantum chemical geometry determinations of dipeptides carried out over the past decade have shown [1–3] that important structure-conformation relations can be found in systems of this kind. That is, as one moves from one point in ϕ, ψ space to another, where ϕ and ψ denote the torsions about N-C$^\alpha$ and C$^\alpha$-C', respectively, bond distances and angles may vary significantly. Thus, conformational geometry maps are as important as energy contour maps to characterize dipeptide conformational behavior.

Comparisons with X-ray structures of small peptides, as far as they are possible, seem to indicate that structural trends similar to those found in calculated geometries [1–3] also exist in real systems. Thus, they should be included in methods of empirical peptide modeling (EPM).

Many schemes currently used in EPM, such as AMBER [4], INSIGHT [5], or CHARMm [6], optimize geometry as they optimize energy. However, since different procedures frequently use different parameters and even different expressions for potential energy, they often relax the same conformation of the same molecule to significantly different local geometries. Thus, in current EPM, geometries are often used which may not be realistic.

If one wishes to improve the performance of EPM in this regard, one faces the difficulty that the experimental data base available for parameter refinements is sparse and often inconclusive. Therefore, we have recently proposed to include local geometry trends from ab initio dipeptide studies in the data base used for empirical force field parameter refinements. Specifically, the parameters used in the CHARMm [6,7] program were adjusted [8] to approximate the most important 4-21G trends in the bond angles of glycine and alanine dipeptides. This project is a continuation of that general scheme in which we now have included bond angle and conformational trends from more than one hundred 4-21G structures of various types of molecules in the refinement of the QUANTA/CHARMm parameter set [7].

Results and Discussion

To test the accuracy of the local geometries that are obtained with the new parameters, the structures of several small peptides for which highly resolved X-ray crystal structures are known, were studied by a variety of computational techniques. Specifically, such systems were selected for refinement, whose conformations in the solid state are close to the conformational energy minima obtained by our calculations. As an example we present in Tables 1 and 2, results of structure refinements performed on cyclo-(Gly-Pro-Gly-D-Ala-Pro) (GPG-D-AP).

The X-ray structure of GPG-D-AP [9] shows the molecule with a familiar Type II bend encompassing Pro^2-Gly^3 and a γ-bend at Pro^5. All peptide bonds are in the *trans* conformation and are nearly planar. Two hydrogen bond interactions occur across the backbone ring, as N_4H---O_1 and N_1H---O_4. The molecule has relatively small thermal ellipsoids associated with its backbone atoms. The packing in the crystal lattice shows molecules connected by hydrogen bonds between O_3 of one molecule and N_3H of a molecule related by a twofold screw parallel to the c axis. The N_3---O_3 distance is 2.90 Å and no other strong polar interactions are available for crystal stabilization.

In the empirical conformational analysis the molecule was first treated as an isolated system in the vacuum state (at 0 K), and energy minimization of the solid state

Table 1 *Experimental and calculated dihedral angles (ϕ, ψ, ω) for the molecule cyclic-[Gly-Pro-Gly-D-Ala-Pro]*

Residue name	Dihedral angle exptl	Dihedral angle CHARMm vacuum	Δ	Δ Ref. 10 PARM19 vacuum	Δ Solvated E min	Δ Solvated dynamic average	Δ Crystal E min
Gly	83	80	3	7	4	8	15
	−134	−107	−27	−44	−16	−16	−17
	174	170	4	1	2	3	6
Pro	−52	−47	−5	16	−5	−5	−1
	126	96	30	57	36	29	1
	−179	173	16	2	3	5	5
Gly	74	76	−2	−12	−5	−3	−19
	12	19	−7	−70	−24	−24	−22
	177	179	2	10	3	1	7
D-Ala	134	145	−11	38	−3	−8	−9
	−69	−72	3	27	8	5	14
	178	172	6	6	6	7	14
Pro	−86	−76	10	14	9	10	3
	70	61	9	13	4	1	8
	−160	−164	4	13	3	5	1
Average dihedral deviation			9	22	9	9	9
RMS			0.31	–	0.27	0.26	0.30

Δ = Experimental – calculated dihedral angles.
RMS = root-mean-square deviation between backbone heavy atoms of the X-ray and calculated atomic coordinates.

conformation was carried out. All gradients on the atoms were reduced to < 1E-4 (kcal/mol/Å). The dihedral angles which resulted from this calculation are presented in Table 1 and compared with the X-ray structure. The average deviation in dihedral angles is 9°. In a similar calculation [10] made with earlier CHARMm (PARM19) parameters the average deviation was 22°.

Selected bond angles for the optimized structure are compared with experimental values in Table 2. In presenting these results the physical differences between the calculated structure (vacuum calculation at 0 K) and the thermal average crystal structure at room temperature [9] must be kept in mind. Calculations are currently underway that will make it possible to compare the crystal structure with molecular-dynamics-averaged structures carried out with the crystal environment at the temperature of the diffraction experiment.

In order to determine whether the X-ray structure corresponded to the global energy minimum calculated with the new CHARMm force field, a series of conformational searches was carried out using a search method described as 'conformational flipping of peptide groups'. In this procedure a set of several peptide groups is flipped by 180° in a random fashion, or in a concerted manner, changing the ψ angle of the ith residue and the ϕ angle of the $(i + 1)$st residue in an anticorrelated motion, and the resulting structures are energy minimized. A second series of

Table 2 *Comparison of experimental and calculated bond angles for the molecule cyclo-(Gly-Pro-Gly-D-Ala-Pro)*

Angles	Gly[1]	Pro[2]	Gly[3]	D-Ala[4]	Pro[5]
C'-N-C$^\alpha$	119.2	121.5	119.5	120.3	121.4
	120.5	120.5	120.0	121.9	121.8
	-1.3	1.0	-0.5	-1.6	-0.4
N-C$^\alpha$-C'	107.9	110.5	115.6	108.9	107.2
	104.9	111.6	116.5	110.3	110.1
	3.0	-1.1	-0.9	-1.4	-2.9
C$^\alpha$-C'-N	119.0	113.6	118.7	116.9	113.5
	119.5	115.7	116.9	117.6	114.7
	-0.5	-2.1	1.8	-0.7	-1.2
C$^\alpha$-C'-O	118.8	122.5	118.3	121.2	123.3
	118.5	121.8	120.4	120.4	122.2
	0.3	0.7	-2.1	0.8	1.1
C'-C$^\alpha$-C$^\beta$		111.8		112.8	112.2
		110.9		110.5	110.5
		0.9		2.3	1.7
N-C$^\alpha$-C$^\beta$		104.3			103.1
		103.4			103.8
		0.9			-0.7
C$^\delta$-N-C$^\alpha$		111.8			111.0
		111.7			111.1
		0.1			-0.1

Values in vertical order are experimental, calculated, and difference.

conformations was studied by changing the *trans* peptides at the Pro residues to *cis* peptides, and reoptimizing the resulting structures. All of the reported investigations led to the discovery of a number of additional low-energy structures of GPG-D-AP, but none of these conformations was lower in energy than the conformer identified in Tables 1 and 2.

Next we examined the effect of solvation on the conformation of GPG-D-AP. An 8 Å layer of water (TIP3P) [11] was placed around the structure and the total complex (GPG-D-AP + water) was energy minimized. The dihedral angles after this operation differed from the vacuum results, but the conformation still resided in the same global energy minimum. The average deviation after solvation was 9° (see Table 1) and the root-mean-square (RMS) deviation of calculated atomic coordinates (backbone atoms without hydrogens) from the X-ray structure was 0.27 Å. In order to observe temperature effects on conformation, the solvated complex was run through a dynamics heating, equilibration, and simulation series of 0.9 ps. No constraints were placed on the water molecules. The average coordinates of the simulation section at 300 K were used to calculate a set of average dihedral angles and these were compared to the X-ray structure. The average deviation was again 9° (see Table 1) relative to the X-ray structure, but individual dihedral angles differed from previous results.

One final series of studies included the packing of the molecule in its crystal environment with 103 symmetry-related molecules. All lattice axis lengths and internal coordinates were energy-minimized (lattice angles were retained at 90° for the $P2_12_12_1$ symmetry) with a dielectric constant of 1.0 and no shielding factors. Electrostatic cutoffs were 20 Å and the final gradient for the lattice minimization was 0.005 kcal/mol/Å. The changes in lattice axes after minimization were −0.05, 0.19, and 0.44 Å, for the a-, b-, and c-axes respectively. The average deviation of calculated and observed dihedral angles of GPG-D-AP after lattice minimization was again 9° (see Table 1) and the molecule retained the basic global structure with minor deviations from experimental values.

The studies reported here indicate that the X-ray crystallographic structure is the global minimum-energy structure calculated with the new CHARMm force field. Environmental effects by solvation or crystal packing forces, or dynamics simulation at 300 K, appear to play only minor roles in determining the conformational ground state of this particular cyclic peptide. Further, an increase in temperature of the solvated state in the dynamics simulation did not result in new preferred conformations.

Crystal packing studies at the experimental conditions of the X-ray structure are being carried out on a series of molecules to further test the new force field parameters. It is clear that more extensive tests will have to be performed to establish the full utility of the modified parameters. Nevertheless, the results obtained so far in this and similar cases make it possible to conclude that force fields which include ab initio geometry trends in their derivation perform better than those whose derivation does not include such information.

The saving of computation time when simple functionals like CHARMm are used, rather than more complex expressions of potential energy, is an advantage in analysis of complex biomolecular systems such as proteins and nucleic acids. Further tests with cyclic peptides whose calculated global energy minima differ from the solid state conformations are in progress. Such systems are expected to be a valuable source of information in the basic factors underlying the conformational properties of polypeptides.

References

1. (a) Schäfer, L., Van Alsenoy, C. and Scarsdale, J.N. (1982) *J. Chem. Phys.* **76**, 1439; (b) Klimkowski, V.J., Schäfer, L., Momany, F.A. and Van Alsenoy, C. (1985) *J. Mol. Struct.* **124**, 143; (c) Scarsdale, J.N., Van Alsenoy, C., Klimkowski, V.J., Schäfer, L. and Momany, F.A. (1983) *J. Am. Chem. Soc.* **105**, 3438; (d) Schäfer, L., Klimkowski, V.J., Momany, F.A., Chuman, H. and Van Alsenoy, C. (1984) *Biopolymers* **23**, 2335.
2. Siam, K., Klimkowski, V.J., Van Alsenoy, C., Ewbank, J.D. and Schäfer, L. (1987) *J. Mol. Struct.* **152**, 261.
3. Siam, K., Kulp, S.Q., Ewbank, J.D., Schäfer, L. and Van Alsenoy, C. (1989) *J. Mol. Struct.* **184**, 143.
4. Weiner, P.K. and Kollman, P.A. (1981) *J. Comp. Chem.* **2**, 287.
5. Dauber-Osguthorpe, P., Roberts, V.A., Osguthorpe, D.J., Wolff, J., Genest, M. and Hagler, A.T. (1988) *Proteins: Structure, Function, and Genetics* **4**, 31.
6. Brooks, B.R., Bruccoleri, R.E., Olafson, B.D., States, D.J., Swaminathan, S. and Karplus, M. (1983) *J. Comp. Chem.* **4**, 187.
7. Parameter Handbook, Quanta Release 3.0, June 1990, Polygen Corp. Waltham, MA 02254.
8. Momany, F.A., Klimkowski, V.J. and Schäfer, L. (1990) *J. Comp. Chem.* **11**, 654.
9. Karle, I.L. (1978) *J. Am. Chem. Soc.* **100**, 1286.
10. Lynn, T.E. and Kushick, J.N. (1984) *Int. J. Peptide Protein Res.* **23**, 601.
11. Chandrasekhar, J. and Jorgensen, W.L. (1985) *J. Am. Chem. Soc.* **107**, 2974.

Incomplete equilibration: A source of error in free energy calculations

Herman J.C. Berendsen

Laboratory of Physical Chemistry, University of Groningen and BIOSON Research Institute, Nijenborgh 16, 9747 AG Groningen, The Netherlands

1. Introduction

A serious limitation of molecular dynamics or Monte Carlo simulations of biological macromolecules is the limited extent of statistical sampling one can obtain in practice with these methods. Proteins in solution are fluctuating structures, even when they have a well-defined conformation. A single conformation consists of a set of closely related *substates*, each substate being defined as an average structure about which the molecular coordinates vibrate. Substates may differ from each other by a different set of dihedral angles of side chains and a different hydration structure or network of hydrogen bonds. The lifetime of such substates may range from picoseconds to nanoseconds. In the course of time, say a nanosecond to microsecond, a protein will traverse several substates and still remain in a single conformation. The definition of *conformation* must be specified in each case, but it comprises at least a set of backbone dihedral angle ranges. Some macromolecules, like most proteins, have a defineable stable conformation with a lifetime of days; others, like many oligopeptides, have no defineable states with lifetimes exceeding milliseconds. In the latter case the distinction between conformation and substate is not relevant.

Molecular simulations of large molecules, be it Molecular Dynamics (MD) or Monte Carlo (MC), necessarily sample only a limited set of substates, and a limited set of configurations within each substate. Even if the sampled configurations are representative for the complete ensemble and thus reasonable ensemble averages are obtained, the *extent* of accessible phase space is not known unless further assumptions are made. It is precisely the extent of accessible phase space that determines the entropy of a state, and hence its free energy. These properties are given by integrals over phase space rather than by ensemble averages and will suffer in a systematic way from incomplete sampling. By integrating over a reversible path from a known state to the unknown state, one can avoid the computation of phase space integrals and obtain free energy differences from ensemble averages only. Thus the incomplete sampling problem seems to have been avoided. However, it is unlikely that a principal problem can be avoided by choosing a different computational approach and it seems worthwhile to investigate this problem and look for compensation of errors and for possible hidden assumptions in the methods used.

In the following we shall first consider the determination of entropy and free energy from phase space integration and the systematic and random errors arising from incomplete sampling (Section 2). We shall find that a *configurational relaxation time* exists that plays a crucial role in the accuracy of free energy determinations. Then we shall investigate the thermodynamic integration technique along a reversible path and the possible systematic effects of insufficient sampling on the resulting free energy difference (Section 3). Finally, possible remedies will be discussed to optimize the accuracy of free energy determinations within a given computational effort (Section 4).

2. Phase Space Integration

Assume that a simulation has provided us with an accurate description of the probability density $P(r^N)$ of the system in Cartesian coordinate space $r^N = r_1, r_2, ...,$ r_N (N is the number of particles). We note that this is only possible in *non-diffusive* systems which are characterized by stable average positions of each (distinguishable) particle. Then the following relations hold:

$$Z = (2\pi kT)^{3N/2} h^{-3N} \prod_{i=1}^{N} m_i \int dr^N \exp[- V(r^N)/kT] \tag{1}$$

$$A = -kT \ln Z \tag{2}$$

$$G = A + pV \tag{3}$$

$$P(r^N) = \frac{\exp [-V(r^N)/kT]}{\int dr^N \exp[-V(r^N)/kT]} \tag{4}$$

$$U = \frac{3}{2} NkT + \int dr^N V(r^N) P(r^N) \tag{5}$$

$$S = \frac{3}{2} Nk [1 + \ln (2\pi kT/h^2)] + k \sum_{i=1}^{N} \ln m_i - k \int dr^N P(r^N) \ln P(r^N) \tag{6}$$

where Z is the canonical partition function, A the Helmholtz free energy, G the Gibbs free energy, U the internal energy, and S the entropy. Of Eq. (6), the first two terms arise from the distribution in impulse space and the last term is the spatial contribution or *configurational entropy*. The 'difficult' term is the configurational entropy since it is an integral over accessible configurational space.

In practice, the distribution in configurational space can best be expressed in internal coordinates such as bond lengths, bond angles and dihedral angles, q_i, i = 1, 2, ..., $M \leq 3N - 6$. In this way the coordinates of overall translation and rotation of the molecule, as well as any internal variables that are treated as constraints, can be easily separated. The entropy now becomes:

$$S = \frac{1}{2}Mk [1 + \ln (2\pi kT/h^2)] + \frac{1}{2}k \int dq^M P(q^M)\ln\{ \det G(q^M)\} +$$

$$-k \int dq^M P(q^M) \ln P(q^M) \tag{7}$$

where G is a metric tensor containing the mass:

$$G_{ij} = \sum_{l=1}^{3N} m_l \frac{\partial x_l}{\partial q_i} \frac{\partial x_l}{\partial q_j} \tag{8}$$

385

Although the metric tensor complicates the calculations, its influence is often small or cancels if differences between substates are considered. The important term, again, is the configurational entropy

$$S_c = -k \int dq^M P(q^M) \ln P(q^M) \tag{9}$$

The problem in evaluating S_c is that it involves a multidimensional distribution function $P(q^M)$ which is based on a limited number of sample points. The number of really independent samples in an MD run of 100 ps is only of the order of 100, while we may be concerned with a 1000-dimensional space. A very sparse coverage of space indeed! Therefore further assumptions about $P(q^M)$ must be made.

One can make an upper estimate of the entropy by assuming a distribution function that maximizes the entropy while satisfying data obtained from the simulation. This *maximum entropy* assumption has a good basis in information theory: since the entropy is a measure of the *uncertainty* an observer has about the distribution function, the distribution with maximum uncertainty, but consistent with experimental data, is the least biased distribution one can choose. Any other distribution would require more knowledge than is available and would therefore be prejudiced. The simulation allows us to assemble data on the *mean-square structural fluctuation matrix*

$$Q_{ij} = \overline{(q_i - \overline{q}_i)(q_j - \overline{q}_j)} \tag{10}$$

where bars indicate time averages over the MD run. Given $\mathbf{Q}$, the entropy-maximizing distribution function is a *multivariate Gaussian* with $\mathbf{Q}$ as its covariance matrix. It can now easily be shown [1] that

$$S_c = \frac{1}{2} kM (1 + \ln 2\pi) + \frac{1}{2} k \ln (\det \mathbf{Q}) \tag{11}$$

Thus, only the determinant of $\mathbf{Q}$ is required to produce this estimate of the entropy. We note that $\mathbf{Q}$ is an ensemble average rather than an integral over configuration space.

The question we now wish to address is the systematic error in S_c due to incomplete sampling of configurational space which produces incomplete knowledge of the elements of the fluctuation matrix $\mathbf{Q}$. In order to do this, we shall consider the simple one-dimensional case of a coordinate x, which fluctuates in a normal distribution $p(x)$ around a mean μ, with variance σ^2. Its normalized autocorrelation function is $\rho(\tau)$. A one-dimensional harmonic oscillator in a stochastic bath would behave like this. The question we address is: Given that we observe x(t') over a time span $0 < t' < t$ (where t is much larger than the correlation time of x) and we evaluate the mean-square fluctuation

$$\overline{(\Delta x)^2} = t^{-1} \int_0^t \{x(t') - \overline{x}\}^2 dt' \tag{12}$$

where

$$\overline{x} = t^{-1} \int_0^t x(t')dt' \tag{13}$$

what then is the *expectation value* of this mean-square fluctuation over the distribution $p(x)$:

$$E\{\overline{(\Delta x)^2}\}$$

and how much does this expectation value deviate from the variance σ^2?

The evaluation is straightforward. We first note that

$$E\{x(t)\} = \mu \tag{14}$$

$$E\{x^2(t)\} = \mu^2 + \sigma^2 \tag{15}$$

$$E\{x(t)x(t+\tau)\} = \mu^2 + \sigma^2\rho(\tau) \tag{16}$$

with $\rho(-\tau) = \rho(\tau)$.

Now we evaluate

$$E\{\overline{(\Delta x)^2}\} = E\{x^2\} - E\{(\overline{x})^2\} = \mu^2 + \sigma^2 - E\{(\overline{x})^2\} \tag{17}$$

$$E\{(\overline{x})^2\} = t^{-2} \int_0^t dt' \int_0^t dt'' \, E\{x(t')x(t'')\}$$

$$= t^{-2} \int_0^t dt' \int_0^t dt'' \, [\mu^2 + \sigma^2\rho(t'' - t')]$$

$$= \mu^2 + \sigma^2 t^{-1} \int_{-\infty}^{\infty} \rho(\tau) \, d\tau \tag{18}$$

The last step is an approximation and assumes that t is much larger than the time after which $\rho(\tau)$ has decayed to zero. Defining the *correlation time* τ_c of x as

$$\tau_c = \int_0^{\infty} \rho(\tau) \, d\tau \tag{19}$$

Eqs. (16), (17), and (18) combine to

$$E\{\overline{(\Delta x)^2}\} = \sigma^2 \left(1 - \frac{2\tau_c}{t}\right) \tag{20}$$

This result indicates that σ^2 will be systematically underestimated from a mean-square fluctuation. Such underestimation is well known in statistics and is the basis of the common practice to estimate the variance of a distribution as $n/(n-1)$ times the mean-square deviation observed from n independent samples. A statistical analysis of the expected error in the mean of a correlated stochastic variable similarly involves the correlation time [2].

Returning to Eq. (11), we see that the entropy is determined by $\frac{1}{2}$ k ln det **Q**. The determinant of **Q** is equal to the product of eigenvalues of the matrix **Q**. The eigenvalues are obtained by diagonalizing the matrix **Q**, which means a transformation from the original coordinates q to a new set of coordinates q' that fluctuate independently of each other. Such coordinates that have a diagonal fluctuation matrix are called *quasi-harmonic normal coordinates*, defining quasi-harmonic normal modes. For the expectation value of det **Q** we can write

$$E\{\det Q\} = \prod_{i=1}^{M} E\{\overline{(\Delta q_i')^2}\} = \prod_{i=1}^{M} \sigma_i^2 (1 - \frac{2\tau_i}{t}) \tag{21}$$

where σ_i^2 is the variance of q_i' and τ_i its correlation time. Thus we underestimate the entropy by

$$\frac{1}{2}k \sum_{i=1}^{M} \ln(1 - \frac{2\tau_i}{t}) \approx -\frac{k}{t} \sum_{i=1}^{M} \tau_i = -\frac{kM}{t}<\tau_c>$$

or

$$S = S_{sim} + \frac{kM}{t}<\tau_c> \tag{22}$$

where S_{sim} is S determined from Eq. (10). Here $<\tau_c>$ is an average correlation time of the quasi-harmonic modes, M is the number of internal degrees of freedom and t is the observation time.

The question can be asked if the systematic error in the mean-square fluctuation (Eq. 20) is relevant compared to the random error in this quantity. R.W. Pastor [this volume] gives for the random error in the variance of a Gaussian variable the quantity $\sigma^2\sqrt{6\tau_c'/t}$, where τ_c' is the correlation time of $x^2(t)$. Under the assumption of pairwise decomposition of Gaussian variables and exponential decay of its correlation function we can derive that $\tau_c' = \frac{1}{2}\tau_c$, so that

$$random\ error\ in\ \overline{(\Delta x)^2} = \sigma^2\sqrt{3\tau_c/t} \tag{23}$$

while, according to Eq. (20),

$$systematic\ error\ in\ \overline{(\Delta x)^2} = -\sigma^2 2\tau_c/t \tag{24}$$

Since $t \gg \tau_c$, it seems that the random error exceeds the systematic error and we shouldn't worry about the latter. This is not true, however, because for a large number of independent degrees of freedom the systematic errors add linearly while the random errors add quadratically. Thus, for M degrees of freedom,

$$random\ error\ in\ \det Q = \sigma^2\sqrt{3\tau_c M/t} \tag{25}$$

while

$$systematic\ error\ in\ \det Q = -\sigma^2 2M\tau_c/t \tag{26}$$

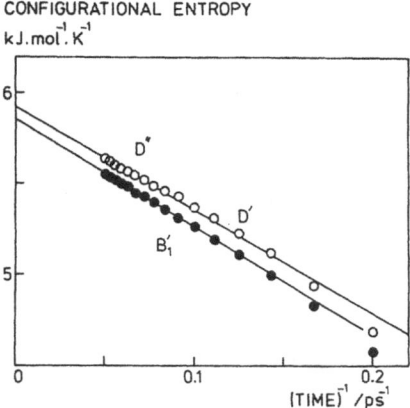

CONFIGURATIONAL ENTROPY
$kJ.mol^{-1}.K^{-1}$

Fig. 1. Configurational entropy of two somatostatin conformers, showing the systematic error proportional to the inverse simulation time (from ref. 3).

Now, for several hundred degrees of freedom, the systematic error easily exceeds the random error.

The systematic error of Eq. (22) can be made visible by plotting S_{sim} versus t^{-1}. Such a plot is given in Fig. 1, from a simulation of various substates of a cyclic dodecapeptide (an analogue of somatostatin), published in 1984 [3]. Two substates, generated by quenching from a high-temperature MD run, were each simulated for 20 ps, during which side-chain motions occurred, but the backbone conformation remained stable. The t^{-1} error is clearly shown. The number of internal degrees of freedom was 250, which yields according to Eq. (22) for the structural relaxation time τ_c a value of 2.8 ps. This large value is due to the slow side-chain dihedral angle changes.

The systematic error discussed thus far is a result of limited statistical sampling from a homogeneous distribution. But what happens if the simulation is trapped in one of many possible substates? In that case the sampling is biased to a particular region of configurational space and hence is non-homogeneous. The system is *non-ergodic* over the time span of the simulation; it does not access other probable regions. It is clear that such a simulation does not contain any information on the possible states that have not been accessed. The proper treatment would be to generate as many substates as possible (preferably *all*), evaluate the free energy G_i of each and determine the free energy of the conformation that comprises all substates from

$$G = \Sigma_i w_i G_i + kT \Sigma_i w_i \ln w_i \tag{27}$$

where

$$w_i = \exp(-G_i/kT) / \Sigma_j \exp(-G_j/kT) \tag{28}$$

389

is the statistical weight of the ith substate. The first term in (27) is the weighted average of the free energy of all substates; the second term is the contribution from a mixing entropy of substates.

If we have only knowledge over a limited time t in which a limited number n(t) of substates are accessed, we need further assumptions in order to extrapolate the error. The best we can hope for is that similar errors occur for similar cases we compare, and hence cancel if we consider differences.

3. Thermodynamic Integration

Thermodynamic integration involves the determination of free energy differences between two states by simulating a *reversible path* from one (reference) state to the other (unknown) state [4,5]. The path is characterized by a *coupling parameter* λ that changes from 0 (reference state) to 1 (unknown state). Reconstruction of the free energy difference relies on knowledge of the derivative of the free energy with respect to λ at intermediate states, either from a number of equilibrium simulations at several intermediate states or from a continuous simulation in which λ is slowly changed from 0 to 1 (*slow growth* method [6]). For an ensemble at constant temperature and volume, when the Hamiltonian $\mathcal{H}$ of the system is changed by modifying the potential energy V, this derivative is given by

$$\frac{\partial A}{\partial \lambda} = <\frac{\partial \mathcal{H}}{\partial \lambda}>_\lambda = <\frac{\partial V}{\partial \lambda}>_\lambda \tag{29}$$

For an ensemble at constant temperature and pressure, A is replaced by G.

We see that the free energy derivative only requires an ensemble average and not an integral over phase space. Thus the sampling problem seems less severe. Nevertheless, if the path involves an entropy change and many degrees of freedom are involved, a systematic error is expected, just as in the case of phase space integration. A simple example may illustrate this: Consider a harmonic oscillator, in which the force constant is changed from f_1 to f_2. This involves an entropy change. Then we see that

$$<\frac{\partial V}{\partial x}> = \frac{1}{2}(f_2 - f_1)<x^2> \tag{30}$$

which involves an average fluctuation just as we encountered in the case of the phase space integral in the quasi-harmonic oscillator approximation. Hence similar systematic errors are expected.

Figure 2a illustrates what errors may occur in a slow-growth computation of a free energy difference. During slow growth a certain volume of configuration space, indicated by the tube connecting state 1 with state 2, will be accessed and sampled. In state 1 the actually accessed volume will correspond to a *substate* of the accessible state 1. A similar partial access will occur over the whole path, including the end point in state 2. The free energy difference will be correct only if the same systematic error is made on both sides, so that the errors cancel. For the slow-growth method an additional requirement is that the system remains in equilibrium along the whole path.

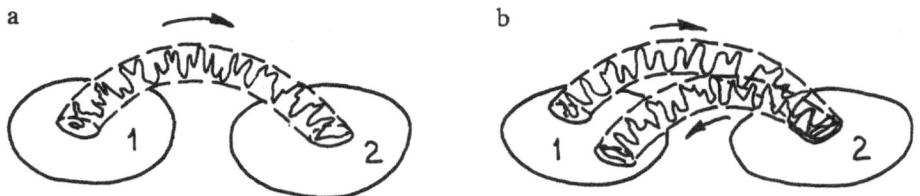

Fig. 2.   (a) Slow-growth path from accessible state 1 to accessible state 2. Only a substate on both sides is actually sampled during the simulation. (b) Irreversible hysteresis occurs when the original substate is not recovered by backtracking from state 2 to state 1.

So two types of errors may occur:

a.   There is a time lag between the Hamiltonian at $\lambda(t)$ and the system configuration, because the system has a structural relaxation time causing a delay in following $\lambda(t)$. This shows in the correlation time of the variable $<\partial V/\partial\lambda>$ and will lead to a systematic error in the free energy difference. This delay will cause a hysteresis. Its effects have been analysed by Wood [7] and can be corrected for as long as the MD run is (much) longer than the relaxation time.

b.   The accessed substate is not sufficiently representative for the full state. This may lead to the system getting *trapped in substates*, which is quite commonly observed in biological macromolecules. Specifically, the system may get trapped in a particular hydrogen-bonded network including water molecules; hydration structures equilibrate quite slowly. One may observe a hysteresis when the path is reversed, as shown in Fig. 2b, because the system does not return to the original substate. In fact, this error is a more severe case of the delay error, mentioned above, now with a relaxation time that exceeds the total MD run time. No possibility exists to correct for this error from the existing run.

4. How to Avoid Sampling Errors in Thermodynamic Integration

As shown above, sampling errors are related to the structural relaxation time of the system. In thermodynamic integration methods, one can choose the pathway between initial and final state at will, even if it involves physically unrealistic intermediate states. While we earlier assumed that a monotonous and preferably linear change of free energy with time in a slow-growth experiment would be an optimal choice [8], it is now apparent that a *path with short relaxation times* is to be preferred. This point has been worked out recently by Mark et al. [9], who proved that intermediate states with reduced energy barriers provide a much more efficient path for free energy determination. The system studied was a simple model liquid consisting of 64 butane-like molecules that were changed from a *trans* preference to a *gauche* preference by modifying the dihedral angle interaction function. Linear interpolation produced a path along which the *trans-gauche* barrier was high everywhere. Slow growth over a linear path, even over a total time of 500 ps, produced an enormous hysteresis (2.93 kJ/mol forward, 0.51 kJ/mol backward, yielding 1.7 ± 1.2 kJ/mol). A modified path involving a reduction of the *trans-gauche* barrier to 25% of its original value, reduced the hysteresis considerably, yielding

1.62 ± 0.21 kJ/mol in a total run of 150 ps. Although the free energy change over the path is now far from linear (it first rises to 8 kJ/mol when the barrier is reduced and then falls to 1.7), the results are much more accurate. The improvement is entirely due to the reduction of the structural relaxation time which is apparent from the increased rate of dihedral transitions.

The lesson to learn from this success is that a path should be constructed from initial to final state that minimizes the structural relaxation times along the path. When hydrated biological macromolecules are involved it may be advantageous to reduce the strength of hydrogen bonding with and between water molecules along the path to avoid trapping in specific hydrogen-bonding networks. To our knowledge, such a strategy has not yet been applied in practice.

Acknowledgements

I wish to thank several collaborators who have contributed to the free energy and entropy development in our laboratory over the years: Alfredo DiNola and Olle Edholm on phase space integration, Harm Zwinderman on substrate binding to proteins, Johan Postma and Tjerk Straatsma on thermodynamic integration methods in aqueous systems, Alan Mark on the fast-relaxing paths, and Wilfred van Gunsteren on everything between ideas and results.

References

1. Karplus, M. and Kushic, J.N. (1981) *Macromolecules* 14, 325.
2. Straatsma, T.P., Berendsen, H.J.C. and Stam, A.J. (1986) *Mol. Phys.* 57, 89.
3. DiNola, A., Berendsen, H.J.C. and Edholm, O. (1984) *Macromolecules* 17, 2044.
4. Berendsen, H.J.C., Postma, J.P.M. and van Gunsteren, W.F. (1985) in *Molecular Dynamics and Protein Structure* (Hermans, J., Ed.) Polycrystal Book Service, Western Springs IL, p. 43.
5. Torrie, G.M. and Valleau, J.P. (1975) *J. Chem. Phys.* 63, 2334.
6. Postma, J.P.M. (1985) Ph. D. Thesis, University of Groningen.
7. Wood, R.H. (1990) submitted.
8. Straatsma, T.P., Berendsen, H.J.C. and Postma, J.P.M. (1986) *J. Chem. Phys.* 85, 6720.
9. Mark, A.E., van Gunsteren, W.F. and Berendsen, H.J.C. (1990) *J. Phys. Chem.*, in press.

Free energy calculations on protein stability: The Thr^{157} → Ala^{157} mutation of T4 Lysozyme

Liem X. Dang and Peter A. Kollman

*Department of Pharmaceutical Chemistry, University of California,
San Francisco, CA 94143, U.S.A.*

Abstract

We present application of free energy perturbation approaches to protein stability. First, we present calculations on a Thr^{157} → Ala^{157} mutant and we are able to reproduce the greater stability of the native sequence. Second, we have repeated the Thr^{157} → Val^{157} mutation reported earlier, this time with a much smaller set of moveable residues of the protein. The results are consistent with those previously reported.

Introduction

T4 Lysozyme has become the paradigm for the study of the dependence of protein stability on protein sequence and three-dimensional structure [1]. Both the X-ray structures and the thermodynamics of denaturation of the protein and many mutants are available [2]. Thus the question can be posed: can theoretical molecular dynamics/free energy perturbation methods simulate the relative free energies of protein stability and the differences in X-ray structures of native and mutant enzymes? These free energy methods have been shown to be very useful in studies of solvation free energies [3,4], effect on drug structure of protein-ligand binding [5] and effect of site-specific mutations on enzyme ligand binding and catalysis [6,7].

A number of applications of these methods to protein stability have been presented [3,8,9]. One of the inherent difficulties in making such a comparison is the fact that no structural model is available for the denatured enzyme. Thus a free energy calculation on protein stability must make an unproven assumption about the denatured protein structure. The nature of the calculation can be summarized by the following model:

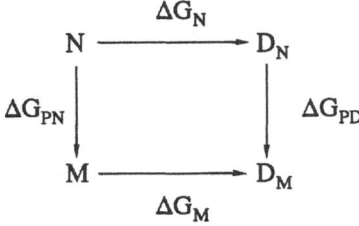

The relative free energy of denaturation of native (N) and mutant (M) protein is $\Delta\Delta G = \Delta G_N - \Delta G_M$. Theoretical studies can be carried out in which one mutates the native protein into the mutant (ΔG_{PN}) and the denatured native protein D_N, into the denatured mutant, $D_M(\Delta G_{PD})$. In the case of the protein simulations, one can use the X-ray structure to carry out the calculation. For the denatured form, no structure is available. We have used a tripeptide model of sequence ANB which is a part of the native sequence and mutated this into sequence AMB in solution. In the specific application reported here, we used residues 156–158 (Gly-Thr-Trp) of T4 Lysozyme with blocked N- and C-terminal groups, and mutated Thr $\rightarrow$ Ala in solution to determine an approximate value of ΔG_{PD}. Then, the value of ΔG_{PN} was determined in two different simulations that mutated $Thr^{157} \rightarrow Ala^{157}$ and $Ala^{157} \rightarrow Thr^{157}$ in the protein. We have also repeated the calculation of ΔG_{PN} for the $Thr^{157} \rightarrow Val^{157}$ mutation [9] using a much smaller number of moveable residues, or 'belly'.

Both of these calculations were inspired by comments made by colleagues. Arieh Warshel of U. of Southern Cal., at the Protein Engineering Meeting in Kobe, suggested the $Thr^{157} \rightarrow Ala^{157}$ mutation might pose much more of a challenge than the Thr $\rightarrow$ Val mutation reported earlier. So we took up this challenge and carried out such a calculation. Secondly, at a meeting on Protein Structure and Drug Design in Alabama, Brian Matthews, of the U. of Oregon, noted his unease at the lack of structural reversibility in our simulations [4]. In our mutation of $Thr^{157} \rightarrow Val^{157} \rightarrow Thr^{157}$, the Thr^{157} in the final structure had not reformed all the native hydrogen bonds and had moved further from the native structure than ideal. Although we plan further restrained dynamics simulation on this system, with constrained distances to force reproduction of the Thr^{157} and Val^{157} structures, we have attempted to address the issues in a more limited way here. In particular, we wished to know whether, when we carried out the $Thr^{157} \rightarrow Val^{157} \rightarrow Thr^{157}$ mutation with a much smaller protein moveable zone, we would determine similar ΔG_{PN} values as before.

Methods

The native protein simulations on the $Thr^{157} \rightarrow Ala^{157}$ mutation involved the placement of a cap of 266 water molecules with a radius of 15 Å from the Thr^{157} C^β atom. The system was minimized and then equilibrated for 6 ps with a nonbonded cut-off of 8 Å, at 300 K, with a nonbonded pair list update every 20 timesteps. During these simulations, only those residues and water molecules within 15 Å of atom C^β of residue 157 were allowed to move with the use of the 'belly' option within AMBER [10] 3.0 of Singh et al. The denatured protein simulations were modelled with the use of residues 156–158 of T4 Lysozyme with a $COCH_3$ group prior to residue 156 and an $NHCH_3$ group following residue 158. This tetrapeptide was placed in a box consisting of 736 water molecules. The system was minimized and then equilibrated for 6 ps with periodic boundary conditions at a constant pressure of 1 atm and a constant temperature of 300 K. Finally, the free energy simulations were carried out on these systems using the slow-growth procedure and the free energies for both the forward ($\lambda = 1 \rightarrow 0$) and reverse ($\lambda = 0 \rightarrow 1$) directions were obtained. In all calculations, a total of 40 ps was used in each direction with a timestep of 1 fs. SHAKE [11] was used to constrain all bond lengths to their equilibrium values. The perturbation group [4] was taken to be the whole amino acid residue 157, but only interactions between this residue and other residues were included in the evaluation of $\Delta\Delta G$. Thus, we are assuming that the intra-group energies are similar in the native and denatured proteins. The mutations of $Thr^{157} \rightarrow Val^{157}$ were handled similarly, except only residues 155–159 of the T4 Lysozyme and waters are moveable.

In the calculation on the Thr $\rightarrow$ Val mutation, we started with the X-ray structure and minimized the energy only for residues 94 and 151–164 as well as all waters. We then equilibrated the system with 5 ps of dynamics, using the same 'belly'. Finally, we reduced the mobile zone to residues 155–159 of the protein and all the water molecules.

Results

The results of the free energy calculations are shown in Table 1. As one can see, mutating Thr $\rightarrow$ Ala in water involves a free energy change of 5.81 ± 0.20 kcal/mol, mutating Thr $\rightarrow$ Ala in the protein involves a free energy change of 7.80 ± 0.12 kcal/mol. These quoted standard deviations are based on averaging the free energies for forward and backward simulations and do not refer to standard deviations within a given simulation.

The calculated $\Delta\Delta G$ of 1.99 ± 1.16 kcal/mol is in reasonable agreement with experiment (1.4 kcal/mol). What is the cause of the greater stability of the Thr^{157} protein, compared to Ala^{157}? It is clear from the structure of the native protein that Thr^{157} is involved in a network of hydrogen bonds, as noted in Table 2 and Fig. 1a (in Ref. 9): the Thr HO^{γ} donates a proton to the carboxylate oxygen of Asp^{159} and OG accepts hydrogen bonds from Asp^{159} NH and Thr^{155} HO^{γ}. Despite the length and poor angle of the $^{157}HO^{\gamma}$ ... $^{159}CO_2$-hydrogen bonds discussed by Alber et al., [2] and given that a charged residue is involved, its existence is not unreasonable. Is this hydrogen-bonding network the key to the greater stability of the Thr^{157} protein? When one considers the electrostatic and van der Waals components to ΔG_{PN} and ΔG_{PD}, one finds specifically $\overline{\Delta G}_{PN}$ (elec) = 8.6 kcal/mol and $\overline{\Delta G}_{PN}$ (vdW) = −0.80 kcal/mol, whereas $\overline{\Delta G}_{PD}$ (elec) = 4.9 kcal/mol and $\overline{\Delta G}_{PD}$(vdW) = 0.9 kcal/mol. Thus, the electrostatic and van der Waals components are of opposite sign, and interestingly, both are significantly different from those we have determined for

Table 1 *Calculated free energies (kcal/mol)*

Mutation	ΔG_{PN}	ΔG_{PD}	$\Delta\Delta G$
Thr $\rightarrow$ Ala[a]	7.80 ± 0.12	5.81 ± 0.20	1.99 ± 0.16
Experiment			1.40
Thr $\rightarrow$ Val[b]	6.34 ± 0.57	4.70 ± 0.14	1.63 ± 0.35
Experiment			1.60

[a] Mutation carried out with 15 Å 'belly'.
[b] Mutation carried out including only residues 155–159 and all water molecules in the 'belly'.

Table 2 *Key distances in the simulation of the Thr^{157} → Val mutation in the smaller mobile zone (Å).*

Atom pair	X-ray (Thr^{157})[a]	After Thr → Val mutation[b]	After reverse mutation[c]
157OG1–159OD1	3.4	6.2	6.1
155OG1–157OG1 (CG1)	2.8	3.5	3.0
159N–157OG1 (CG1)	3.4	3.9	3.8
155OG1–151O	2.9	2.9	2.7
155N–151O	2.9	2.8	2.9
157N–155OG1	3.1	3.5	3.3

[a] Distance in the X-ray structure, refs. 1 and 2.
[b] Distance after mutation of Thr^{157} → Val with small mobile zone (residues 155–159).
[c] Distance after mutation Thr^{157} → Val → Thr with small mobile zone.

the Thr → Val mutation. We plan to further analyze the reasons for those components in subsequent work.

We repeated the Thr^{157} → Val mutation with the small mobile zone, finding (Table 1) a $\Delta\Delta G = 1.63 \pm 0.35$, in excellent agreement with our earlier simulation (1.91 ± 1.1) and experiment (1.6). In addition, as found before, the repulsive van der Waals component in the protein was the key component for differential stability. In Table 2, we report key distances from the X-ray structure, the structure after forward mutation of Thr^{157} → Val and the structure after reverse mutation of Val^{157} → Thr, starting with the structure after forward mutation. As one can see, the three distances involving residue 157 increase upon the Thr → Val mutation, and then decrease back toward their value in the X-ray structure upon Val → Thr mutation. Thus, although the structural reversibility is not perfect, the trends are correct. The exception is the aspartic-acid side chain of residue 159, which, upon equilibration, moves out to hydrogen bond more effectively with water molecules. Even the X-ray structure does not suggest a strong Thr^{157} OH-Asp^{159} CO$_2$-interaction, so it is quite reasonable that, in solution, in contrast to the crystal this group becomes more solvent-exposed.

Discussion and Conclusions

The results presented here suggest that, in some cases, one can use the thermodynamic cycle/perturbation method to analyse the effect of protein stability of site-specific mutation. However, a number of caveats should be emphasized. First, in contrast to the application to some cases of protein-ligand interactions, there is considerably more uncertainty in what structure to use to evaluate ΔG_{PD}, where one transforms the 'denatured protein' into its site-specific mutant. We have made a simple attempt at this by using a tetrapeptide model for this region of the denatured protein. Obviously, in a 40 ps simulation one can not span all the relevant conformation of the very floppy tetrapeptide. Furthermore, even if one could, it is not clear how relevant this sample would be to the actual denatured protein. Thus, all one can say at this point is that our result is reasonable and gives numbers that are in good agreement with experiment. One should note that our procedure may well give an 'upper bound' for the amount of solvent exposure in the denatured protein, so that the calculation using this procedure are likely to overestimate $\Delta\Delta G$. Our calculations on Thr $\rightarrow$ Ala does overestimate it, but the error bars in the calculation make this overestimation not definitive. Other cases must be studied to see if the procedure consistently overestimates $\Delta\Delta G$.

A second uncertainty in the calculations is the structural hysteresis in the calculation of the native protein mutation ΔG_{PN}. There is a larger uncertainty in this calculated number, and it is an average of four different mutations. Again, in 40 ps, one expects to limit sampling of the configuration of the system. Nonetheless, Thr[157] is located on the outside of the protein and it is not unreasonable to speculate that in solution, as opposed to the crystal, the hydrogen-bonding structure might be somewhat different. Again, further analysis of this is required to address these issues, but to get a definitive answer may require orders of magnitude more simulation than carried out here.

A third caveat concerns molecular mechanical force field and simulation protocol, involving the use of a simple molecular mechanical model and keep the part of the protein further than 15 Å from the mutation frozen. This 'belly approach' appears at least reasonable and has seemed to work effectively in many different systems. Again, the molecular mechanical parameters are clearly far from perfect, but the use of the same parameters in ΔG_{PN} and ΔG_{PD} may allow for significant cancellation of errors.

In summary, we have carried out free energy simulations on the Thr[157] $\rightarrow$ Ala[157] and Thr[157] $\rightarrow$ Val[157] mutation in T4 Lysozyme. The calculations are quite successful in reproducing the experimental $\Delta\Delta G$ of protein stability, and this success suggests, at least in this case, a tripeptide model is adequate to represent the denatured protein.

Acknowledgement

Most of the calculations were carried out at the San Diego Supercomputer Center through supercomputer support provided to P.A.K. by the NSF (DMB-87-14775 and CHE-85-10066). Research Support from the NIH (GM-29072) to P.A.K. is acknowledged as is a grant from Burroughs Wellcome Co. to P.A.K.

References

1. Alber, T. and Matthews, B.W. (1987) *Methods in Enzymology* **154**, 511.
2. Alber, T., Dao-pin, S., Wilson, K., Wozniak, J.A., Cook, S.P. and Matthews, B.W. (1987) *Nature* **330**, 41; Grutter, M.G., Gray, T.M., Weaver, L.H., Alber, T., Wilson, K. and Matthews, B.W. (1987) *J. Mol. Biol.* **197**, 315.
3. Bash, P.A., Singh, U.C., Langridge, R. and Kollman, P.A. (1987) *Science* **236**, 564.
4. Singh, U.C., Brown, F.K., Bash, P.A. and Kollman, P.A. (1987) *J. Am. Chem. Soc.* **109**, 1607.
5. Bash, P.A., Singh, U.C., Brown, F.K., Langridge, R. and Kollman, P.A. (1987) *Science* **235**, 574.
6. Hwang, J.-K. and Warshel, A. (1987) *Biochemistry* **26**, 2669.
7. Rao, S.N., Singh, U.C., Bash, P.A. and Kollman, P.A. (1987) *Nature* **328**, 551.
8. Wong, C.F. and McCammon, J.A. (1986) in *Structure, Dynamics and Function of Biomolecules* (Ehrenberg, A. and Rigler, R., Eds.), Springer-Verlag, Berlin.

9. Dang, L.X., Merz, K. and Kollman, P. (1989) *J.Am. Chem. Soc.* **111**, 8505.
10. Singh, U.C., Wiener, P.K., Caldwell, J.W. and Kollman, P.A. (1986) AMBER(UCSF), version 3.0, Department of Pharmaceutical Chemistry, University of California, San Francisco.
11. Ryckaert, J.P., Ciccotti, G. and Berendsen, H.J.C. (1977) *J. Comp. Phys.* **23**, 326.

Indexes

Author index

Subject index